Sewerage and Sewage Treatment

Other books by L. B. Escritt

SEWERAGE ENGINEERING, Contractors Record Lndn, 1940

REGIONAL PLANNING, G. Allen & Unwin Ltd., Lndn, 1943

SURFACE DRAINAGE, John Crowther Ltd., 1944

SEWERAGE DESIGN AND SPECIFICATION, Contractors Record, Lndn, 1946

THE MUNICIPAL ENGINEER, G. Allen & Unwin Ltd. Lndn, 1949

SEWAGE TREATMENT: DESIGN AND SPECIFICATION, Contractors Record, Lndn, 1950

A CODE FOR SEWERAGE PRACTICE, Contractors Record, Lndn, 1950

SURFACE-WATER SEWERAGE, Contractors Record, 1951

BUILDING SANITATION, Macdonald & Evans Ltd. Lndn, 1953

SEWERAGE AND SEWAGE DISPOSAL, CR Books Ltd. Lndn, 1st Ed. 1956, 2nd Ed. 1962, 3rd Ed. 1965

THE WORK OF THE PUBLIC HEALTH ENGINEER, Macdonald & Evans Ltd. Lndn 2nd Edn. 1949, 3rd Edn, 1959

(WITH DR. S. F. RICH)

PUMPING-STATION EQUIPMENT AND DESIGN, CR Books Ltd. Lndn, 1962

DESIGN OF SURFACE-WATER SEWERS, CR Books Ltd, Lndn, 1964

SEWERS AND SEWAGE WORKS WITH METRIC CALCULATIONS AND FORMULAE, G. Allen & Unwin Ltd. Lndn, 1971

ESCRITTS' TABLES OF METRIC HYDRAULIC FLOW, G. Allen & Unwin Ltd. Lndn, 1970

PUBLIC HEALTH ENGINEERING PRACTICE (Vols I & II) MacDonald & Evans Ltd. Lndn., 4th Edn. 1972

Note: Contractors Record and CR Books are now published under the imprint of Applied Science Publishers.

Sewerage and Sewage Treatment

International Practice

Leonard B. Escritt

Edited and Revised by

William D. Haworth

A Wiley-Interscience Publication

JOHN WILEY AND SONS LIMITED

Chichester · New York · Brisbane · Toronto · Singapore

Library of Congress Cataloging in Publication Data:

Escritt, L. B. (Leonard Bushby)
  Sewerage and sewage treatment.
  Wiley–Interscience publication.
  Includes index.
  1. Sewerage. 2. Sewage—Purification. 3. Sewage disposal. I. Haworth, William D. II. Title.
TD645.E7224   1983   628'.2   83–1300
ISBN 0 471 10339 X

British Library Cataloguing in Publication Data:

Escritt, Leonard B.
  Sewerage and sewage treatment.
  1. Great Britain—Sewage—Purification
  I. Title   II. Haworth, William D.
  628.3'0941   TD745
ISBN 0 471 10339 X

Filmset by Composition House Limited, Salisbury, Wiltshire, England
Printed by Page Brothers (Norwich) Limited.

To R. W. Horner

Acknowledgements

Permission to use copyright material was kindly granted by the following:

Institution of Public Health Engineers (Figures 13–19, 21, 38–40, 65, 82, 105–8, abstracts on pp. 143 (pt.), 144, 239, 273B (pt.) 273C–D, 346, 350–3)

Institution of Civil Engineers (Table 36, abstracts on pp. 309, 367 (pt.) 368, 369 (pt.) 373–4)

Institute of Water Pollution Control (abstracts on pp. 317, 322A–D)

W. G. Gerald Snook, Consulting Engineer (Figures 41–43, abstracts on pp. 200 (pt.) 202 (pt.) 204 (pt.) 205–209)

Hydrographer to the Navy (Finance Branch, Hydrographic Dept., Min. of Defence) abstracts on pp. 1962A and B (pt.)

Dr D. Mara, *Sewage Treatments in Hot Climates*, John Wiley and Sons Ltd., (Figures 99, 100. Abstracted matter p. 443 pt. 446–7)

Van Nostrand Reinhold Co., *Handbook of Water Resources and Pollution Control* (1976), edited by Gehm and Bregman © 1976 by Van Nostrand Reinhold and Company. Reprinted by permission of the publisher. (14, I. J. Kugelman *Status of Advanced Water Treatment*, Figures 81, 83–4, Table 55, abstracts on pp. 209A, 348A, 349, 357–60 – 15, *Sludge Handling and Disposal*, by Gehm, Figures 94, 96, Tables 60, 61, abstracts pp. 385, 407–18)

Ann Arbor Science Publishers Inc., National Conferences for Individual On Site Wastewater Systems, Ann Arbor, Michigan (3rd Conference 1976, abstracts on pp. 459 and 460 – 5th Conference 1978, abstracts on pp. 459, 460, 461A, B.

Contents

Preface

The posthumous manuscript which forms by far the larger part of this volume was passed on to me by my friend R. W. Horner, a former colleague of the author, and to whom the book is dedicated, I at once agreed with him that having regard to its title, and the high reputation of the author, there must be a great deal in it of interest and value to practising, or embryo, engineers working in the public health field on both sides of the Atlantic, and by inference those in other parts of the world as well! My close association with it since that time has not altered my opinion that although legislation both here and in the US has had a considerable impact on those working in the public health field, the basic problems facing engineers are still the same, and the solutions to them evolved by Mr Escritt—based on his keen intellectual application and wide practical experience—are worthy of the greatest respect, and offer solutions which further research may modify to a minor degree, but will not alter in any significant respect. Only where new materials, new methods, or differing 'requirements' have developed since 1973, when the original work was completed, have I attempted to adjust or add to the original text. In all my endeavours I have had before me, the author's concept, as expressed in his own brief Preface, as follows:

Over the last century and a half, sewerage and sewage disposal have evolved from virtually nothing to a highly-developed science in several parts of the English-speaking world. The practices of the different countries have influenced one another, but they are not identical. They must, of course, depend on climate, population density and other local factors. But what has come about as the result of accumulated experience, research by dedicated individuals, and developments by commercial firms, eventually tends to become codified. Tradition dies hard, and not everything that is generally accepted can be justified by reason. If, therefore, the collected knowledge of engineers on both sides of the Atlantic can be brought together, compared and valued, some benefit must be derived. That is the intention of this book.

Over the years since the later 1930s engineers in the UK have looked forward to the appearance of each successive 'Escritt' as, apart from the ageing *Reports of the Royal Commission on Sewage Disposal*, and technical papers and articles, little information in book form was available on our particular subject, this, of course did not apply in the US, where a constant stream of valuable books on 'applied science' has always been available, although differences in units, and a more liberal use of water—leading to lower average strength of sewage—tended to limit their appeal to engineers over here. Although most of the 'Escritts' are now out of print, copies of them can still be found, and are frequently referred to, in most engineering offices in this country.

It so happens that the setting-up of the new Water Authorities, and reorganization of local government in England and Wales, caused an upheaval in the consulting field, which has had to turn its main interest overseas, where the problems of the developing countries have provided an alternative to local authority work and where, incidentally, engineers are often in contact with their counterparts from America and Europe. This all occurred just after the completion of Mr Escritt's work, but the comparison of ideas and practice put forward in this book could be useful in promoting a better understanding amongst engineers of different nationalities wherever they are working.

Having gained his early training and experience in the consulting field, Mr Escritt served the LCC, later the Greater London Council (GLC), at the time when big improvements were being engineered at the major treatment works at Beckton and Crossness. This—coupled with widely varied experience on works for large housing estates, schools and institutions—provided him with the widest possible field for the study and solution of public health engineering problems. His work was certainly his vocation, and he spent his evenings compiling his books. He valued practical experience above all else, and although his earlier works show little departure from 'standard practice', he became increasingly irritated by official pronouncements on engineering by those he considered lacking in this respect. He expressed his dissatisfaction, and at the same time analysed and pointed out, through the medium of papers and discussion, any apparent fallacy in such pronouncements. In particular, he resisted all attempts to improve on the work of Roseveare and Bilham in the calculation of storm water run-off, and was an advocate of the 'rational method', as developed in the US. Likewise he was sceptical of some aspects of 'Marston' theory—not without reason, as guidance on more accurate determination of loads on buried pipes is still awaited. The wide acceptance of 'Marston' on both sides of the Atlantic has led me to temper some of his original observations on this subject! Also the vogue for long sea outfalls and the valuable work on jet discharge and dilution have largely developed in recent years and I have felt bound to add significantly to the chapter on sea outfalls.

At the same time as the reorganization of local government ended the reign of the 'Ministry', as arbiters of good practice in this country, the 1972 amendments to the Clean Water Act in the US led to the setting-up of the Environmental Protection Agency which has assumed a very similar roll to the Ministry, though it offers specific guidance based on large-scale experimental work and laboratory studies to all those applying for the generous grants available for upgrading existing works in conformity with the act. I have drawn on their pronouncements in making additions to the chapters on sewage treatment, and on the recent authoritative work edited by Gehm and Bregman to bring up-to-date the comparisons between British and American practice.

Chapters 16 and 18 have been considerably extended to include the latest developments. The latter now bears the title 'Disposal of Sewage from Rural Areas', rather than 'Isolated Habitations', and has been used to describe the approaches being made to sewage treatment in developing countries, as well as embracing the particular problems in the US, where rising costs are prohibiting the extension of main drainage to many growing suburbs and pleasure resorts. Escritt never intended to do more than give a general idea of the 'Control and Treatment of Industrial Wastes' (Chapter 19), so I have added little to this apart from widely extending the list of references which readers may use to pursue the subject in more detail.

In deference to British practice all units, formulae and tables have been metricated. I have been helped in this task by reference to Mr Escritt's *Sewers and Sewage Works, with Metric Calculations and Formulae* which appeared in 1971. Though covering roughly the same ground, this book is much briefer than the present text and is confined entirely to British practice. Although not so fully committed to metric units, I believe official policy in the US is tending in this direction, so they should be reasonably well understood. A list of metric equivalents appears as an appendix, and to this I have added some useful conversion factors, in accordance with the SI code. Metrication introduced a major problem in regard to Table 3. Mr Escritt originally included tables based on the Scobey Formula for 'concrete pipe sewers', but in metricating them I have used his own formula, which is a slight modification of Scobey, which gives fair agreement with several other widely used formulae; and is applicable to "used" pipes of steel, iron, concrete, stoneware, or asbestos cement. These cover a reasonable range of pipe sizes, but for wider application reference can be made to the original tables, based on Scobey's Formula in Imperial units, which now appear as an appendix for the benefit of 'non-metric' engineers or to *Escritt's Tables of Metric Hydraulic Flow*, published by George Allen & Unwin Ltd, London, or *Sewer and Water-Main Design Tables British and Metric*, published by Maclaren & Sons Ltd, London, which again revert to the Scobey Formula, unadjusted.

My thanks are due to the late Mrs V. P. Escritt, who generously supported my endeavours; to Alan Reed of W. S. Atkins and Partners, who arranged for the flow tables mentioned above to be prepared by United Computing Systems Ltd. of Epsom; also to Mr I. B. Muirhead, former Secretary of the Institution of Public Health Engineers; the librarians at the Institution of Civil Engineers and the Royal Society of Health; to Peter Worthington of DOE for help in access to their library, and last but not least to Miss D. R. Nicholson, until recently Secretary to my old Employers, Messrs Lemon and Blizard, without whose efforts my amendments would never have been completed!

CHAPTER 1
Principles of Drainage and Sewerage

Once the word sewer referred to a watercourse usually provided for the drainage of rain water and, in England, there were Commissioners of Sewers for the control and maintenance of such sewers. The discharge of foul matter to ditches was prohibited but later it became permitted and then obligatory. In course of time the ditches were covered and converted to underground culverts or pipes. and thus the first sewerage systems originated.

In America the word sewer now means an artificial conduit to carry off rain water and/or waste matter. In England and Wales the definition is more specific, being defined in the Public Health Act, 1936 as follows

'sewer' does not include a drain as defined in this section but, save as aforesaid, includes all sewers and drains used for the drainage of buildings and yards appurtenant to buildings.

A drain is defined

'drain' means a drain used for the drainage of one building or of any buildings or yards appurtenant to buildings within the same curtilage.

These legal definitions are accepted as technical also.

Sewage is the contents of sewers, or liquid, whether or not it is clean or foul, that has been in a sewer. In England and Wales there are private sewers belonging to landowners who are responsible for them, and public sewers vested in and maintained by the local authority, until 1972, when public sewers became vested in the new Water Authorities, but for the most part, still maintained by the local authorities under 'Agency Agreements'. Owners and occupiers of premises have the right to discharge their domestic drainage to the public sewers. Sewerage is the term for the science and practice of the design and construction of sewers.

Artificial drainage of all kinds is the assistance of natural drainage. It is the improvement of watercourses that carry away rainwater or water combined with filth in such a manner as to make them unobjectionable and not deleterious to health. For this reason modern sewers, although they are usually constructed underground, are designed, from the point of view of hydraulics, as if they were open channels. Although they are usually round pipes or culverts, the crown of the pipe or the arch of the culvert may be considered as being merely there to prevent the earth from falling in and, except in comparatively rare conditions of surcharge, they are not designed with the intention of confining the flow of sewage, this work being done by the invert and sides.

Sewerage systems

All sewers which take both surface run-off and waste matter are now known internationally as combined sewers. Once all sewers were combined, but later practice was, where practical or desirable, to lay separate sewers for surface water, and other separate sewers—known as sanitary sewers in America or soil sewers in Britain—for the reception of domestic and industrial waste and from which surface water was excluded. This is known as the separate system. In England and Wales there is also what is known as the partially separate system in which both sanitary and surface-water sewers are provided but a proportion of surface-water may be discharged to the sanitary sewers at the discretion of the local authority.

The advantages claimed for the combined system of sewerage are simplicity and reduced lengths of both drains and sewers with, in many instances, reduced capital costs. It is understood that at one time this system was officially favoured for these reasons, and schemes for separate sewerage approved with reluctance. At the present time British official opinion appears to be reversed and separate sewerage is favoured. In Holland, however, separate sewers are considered unduly extravagant and are seldom installed. In Canada most modern systems of sewers are separate. The advantages claimed for separate sewers are:

(1) shorter lengths of the large-diameter sewers are required for surface-water disposal, for where the sewers are separate, surface water can be discharged to the nearest watercourse, whereas in a combined scheme the sewers must be of large diameter all the way to the treatment works unless storm overflows, now considered undesirable, are provided;[1]

(2) separate sewers for soil sewage need to be made no larger than necessary to take the peak rates of soil flow, and can therefore be made self-cleansing, unlike combined sewers which, being constructed to large diameters so as to accommodate storm water and at gradients which produce self-cleansing velocities at times of storm only, tend to become silted and foul in dry weather;

(3) the separate disposal of storm water does away with the problem of storm-water treatment at the disposal works.

If not carefully examined, these arguments appear to be ample justification for the adoption of separate systems in practically all instances. When, however, specific proposals are so examined and both separate and combined systems considered; and particularly when, as in the instance of a local authority housing estate, the costs of the house drains as well as the sewers in the roads are estimated, it is frequently found that the separate system is unduly expensive. Not only does cost militate against the adoption of a separate system, but the advantages of the system often prove to be less real than is generally believed.

First, whereas admittedly, shorter lengths of large-diameter sewers are required when storm water can be discharged to the nearest watercourse, the total reduction is not so very great and seldom, if ever, effects a reduction of cost that anywhere near counterbalances the additional cost required for providing two sets of sewers and drains. Also, where there is a separate system, the different sewers and drains have to cross one another in various places. This usually means that the soil sewers have to be laid throughout at greater depths than would otherwise have been necessary, adding to cost and increasing the difficulty of arriving at the sewage-treatment works at a conveniently high level.

As to the argument that combined sewers must be foul during dry weather, this is unsound. It is true to say that only an improperly designed combined system (and many are improperly designed) becomes foul.

Where early designers tended to err was where they laid combined sewers at gradients required to produce self-cleansing velocities in times of storm only, whereas in a first-class combined system the gradients should be such that self-cleansing velocities are produced by the daily peak rates of flow of sanitary sewage during dry weather. This first-class design may mean slightly steeper gradients than would be required for the sanitary sewers of a separate system, and the trunk sewers would perhaps be at lower levels than required for separate soil sewers. However, a satisfactory compromise can be secured by laying combined sewers at the gradients that would be required for the separate sanitary sewers, thereby obtaining reasonably satisfactory velocities during dry-weather flow, and certainly self-cleansing conditions at all other times, without making the sewers unduly deep.

The position is different where, whatever system is adopted, sanitary sewage would have to be pumped. Then the impracticability of pumping large quantities of storm water may make a separate system desirable.

The disposal of storm water at sewage-treatment works is seldom a problem and should not involve costly work.

The conclusion is that, wherever sewerage is proposed, the merits and demerits of the alternative systems should be considered without prejudice and, in cases of doubt, comparative estimates prepared. Broadly, wherever existing sewage is to be extended, the existing system should be adhered to unless there is good reason for doing otherwise. Generally the combined system, apart from usually being the least expensive, is particularly applicable to industrial zones and large towns in countries such as America, where dry-weather flows are high, or Great Britain, where storm intensities are not high. In those parts of the world where storm intensities are extremely high, reasons for adopting the separate system are that dry-weather flows would be too small in proportion to storm flows for the sewers to keep clean; also the type of surface-water sewers could well be in the form of channels covered by slabs and not buried culverts, and these would not be suitable for combined sewers. In areas where surface water soaks away owing to the nature of the ground and there are no nearby streams or ditches, soakaway systems are indicated. In such a case neither combined sewerage nor normal separate sewerage would be satisfactory, for neither would divert large flows of surface water to streams, possibly distant and not capable of taking the flow.

Building drainage

Drainage and sewerage commence at the developed site, where the drains connect from the indoor sanitation of the buildings or from the surface-water gullies, known in America as street inlets, constructed of brickwork or concrete. The English road gully may be of cast iron, pre-cast concrete or vitrified clay-ware, having a water seal and normally set in concrete. Of late years plastics gulleys have also found favour with road authorities in various parts of the world, due to their lighter weight and ease of handling (see Figure 1). There are also many designs of gully used for the drainage of paved areas, for taking sullage from waste and ventilating pipes of buildings, for receiving rain water from vertical rain-water pipes or for separating oil or petrol from drainage of paved areas near garages. The word gully in common use in this sense in Great Britain does not appear in dictionaries except in obsolete hyphenated words such as gully-hole or with different meanings.

The drainage of buildings, while closely related to sewerage, has developed on separate lines and, as a result, is usually executed to standards of design and workmanship differing from, and probably not acceptable for, main sewers. The differences most noticeable to sewerage engineers, when they see building drainage schemes for the first time, are the use of old-fashioned rules for gradients of pipes, the arrangement of junctions and bends in lines of pipes without manholes, and the sub-standard quality of pipe-laying, jointing and testing. These matters, particularly the design, can be improved when an experienced civil engineer undertakes drainage works.

It is of the utmost importance that sanitary drains should be laid to the highest standards of workmanship so as to avoid stoppages caused by large solids being held up by irregularities of the invert. English practice has been backward in that, in the approval under the former Model Byelaws, local authorities clung to an archaic rule-of-thumb relating to the gradients of drains in the belief that steep gradients would make up for poor workmanship.[2] In fact, increased gradient is not as helpful as is generally believed, for wherever there is little flow it causes decreased depth and tends to induce the stranding of large solids. The current Building Regulations, in the latest version of 1976,[3] go far to remedy the situation although for 'small flows' (i.e. individual properties), the 'rule-of-thumb' figure for 100 mm down-pipes still persists!

Excessive gradients add to the cost, not only of drains but also of sewerage, by increasing depth of the sewers to which the drains connect and for this reason it is desirable that practice relating to the gradients of drains should be

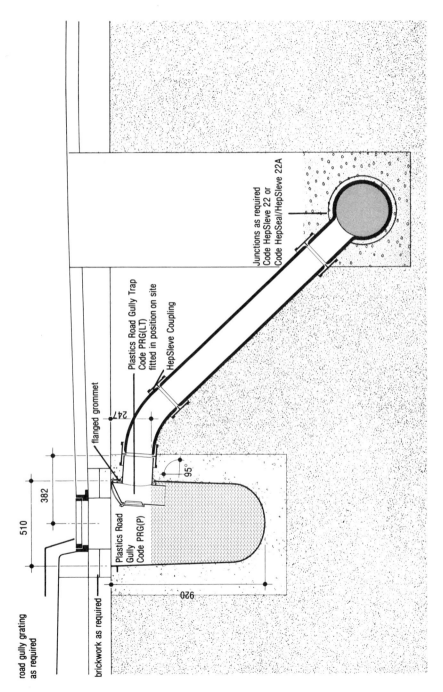

Figure 1　Plastics road gulley (size: 150 mm) (Courtesy Hepworth Plastics Limited)

improved. The recommendation is that drains be laid to the gradients known to be satisfactory for sewers, as given in Table 1. This is more or less in line with the minimum requirements recommended in the American Standard Plumbing Code, published by the American Society of Mechanical Engineers, 1949, which are to the effect that drains larger than 75 mm diameter shall fall at a gradient of not less than 1 in 96, but that where conditions do not permit so great a fall a lesser slope may be permitted, provided that the velocity computed by Manning's formula, with a value of $n = 0.013$, will not be less than 0.6 metre per second. This means that the gradients 1 in 119 for 100 mm pipes and 1 in 204 for 150 mm pipes would be permitted.

There are some differences between American and British drainage practice. Some of the United States do not permit sanitary drains of less than 150 mm diameter. In England and Wales a sanitary drain or private sewer must not have a diameter of less than 100 mm and they are seldom larger than this except where

Table 1 Minimum gradients for circular drains and sewers

Diameter (mm)	Length divided by fall	Discharge (m^3/min)
100	104*	0.36
150	174*	0.83
200	252	1.51
225	291	1.91
300	410	3.31
375	545	5.24
450	685	7.59
525	813	10.00
600	965	13.20
675	1113	16.60
750	1284	20.90
825	1439	25.20
900	1613	30.22
975	1758	34.70
1050	1970	41.69
1200	2272	52.48
1350	2621	66.07
1500	3023	83.20
1650	3389	100.00
1800	3799	120.25

* As far as can be ascertained the optimum gradient for a carefully laid 100 mm drain that is draining no more than one water closet is 1 in 78, or the optimum gradient for a 150 mm drain in the same circumstance, 1 in 69. This is because these gradients produce a velocity of 0.61 m per second on a closet flush of 0.09 m^3/min and yet at a depth of flow sufficient to transport solids.

an increase of diameter is required by the flow. Thus, the vast majority of British drain pipes sold are 100 mm diameter.

Where sewerage is on the separate system, drains must also be on the separate system, except where part of the surface water is permitted by the local authority to be discharged to the sanitary sewers (partially separate system). Where sewers are on the combined system, drains may also be on the combined system. This, however, is not essential, for although sanitary sewage must be discharged to the sanitary sewers, surface water may be discharged to ditches, natural watercourses or soakaways, and in some instances separate surface-water drains are laid and connected to combined drains, via an interceptor or gully.

Capacities of sanitary drains

The peak flows in 100 mm sanitary drains are usually the equivalent of many times dry-weather flow: for example, every drain receiving the flow from one water-closet or bath may be considered to receive a peak daily flow of not less than 0.075 m^3/min. The peak daily flow likely to occur can be estimated as follows:

Firstly, the full demand of water of all the fittings discharging to the drain is calculated in accordance with the scale:

$\frac{1}{2}$-in. bib tap	0.0113 m^3/min
$\frac{3}{4}$-in. bib tap	0.0255 m^3/min
1-in. bib tap	0.0453 m^3/min
Water closet	0.007 m^3/min*
Urinal	Nil

Above the figure of 1.6 m^3/min the simultaneous demand can be taken as being one-fifth of the full demand. (Where showers are provided, the flows from these are taken in full and added to the simultaneous-demand flows for other fittings.)

The surface-water drains on, say, a factory site should be designed to the maximum rate of run-off due to the storm having a duration of 14 min (USA) or about 11 min (Great Britain) appropriate to sewer design for the locality, except where any risk of flooding inside buildings and damage to stock could be involved. Then, three times the above figure should be allowed; for example, 100 mm/h have been allowed in Great Britain for the design of roof gutters, rain-water down pipes and drains inside buildings, in cases where the drains serving paved areas have been designed on 25 mm/h of rainfall, or a little more.

The American Standard Plumbing Code recommended in 1949 that the design of storm drains should be on the basis of 100 mm of rain per hour on roofs, but if in any political subdivision the maximum rainfall were more or less than 100 mm/h, the design figures must be adjusted accordingly. No guidance was given as to what was meant by maximum rainfall or to what parts of the United States the standard could be applied without adjustment.

---

* The discharge of a British water closet is seldom less than 0.091 m^3/min. The above figure is based on the filling rate of the flushing tank.

Requirements of sanitary drains

Every inlet to a sanitary drain, other than an inlet provided for the purpose of ventilation, must be properly trapped. Inlets inside buildings are trapped by the water seal, which is part of the fixture; and in the case of sanitary fixtures in Great Britain there must be no other trap between the fixture and the drainage system. In the United States some cities require a house trap between the indoor sanitation and the drains. This is not permitted in Great Britain, although in the older systems in London and other cities they still exist and cause trouble by frequent blocking.

The wastes from waste fixtures may (in the one-pipe system) discharge to the soil pipe of a water closet or slop sink. In Britain byelaws usually require that if (in the two-pipe system) the wastes from waste fixtures discharged other than to the soil pipe of a water closet or slop sink, they must, at the point where they are brought to ground level, discharge above the trap, but preferably below the grating, of a trapped gully with grating.

The connections from ground-floor water closets, slop sinks and urinals, and from gullies to which waste pipes discharge, should be connected direct to manholes (not junction pipes) by straight lines of drain never exceeding 6 m and preferably not exceeding 3 m in length, and should be brought into the manhole with stoneware channel bends set in the benchings and arranged to discharge over the main channel obliquely in the direction of flow. There should be no bends other than those immediately beneath the S-traps of water closets or slop sinks, or at the bottoms of soil pipes. It is desirable that all connections from fixtures in buildings or direct from the bottom of soil pipes should be of cast iron or other suitable material, not stoneware or concrete, as far as the nearest manhole outside the building. If unplasticized PVC is used in such locations short lengths of pipe with flexible joints should be introduced where the pipes emerge from the building, or join the manhole.

All sanitary drains other than urinal wastes should not be less than 100 mm internal diameter, or 150 mm where so required by the sanitary authority. Urinal wastes should be of chemical lead, polyethylene or of other acid-resisting material throughout, in order that they may be cleansed with hydrochloric acid, and of 60 mm internal diameter where this is permitted.

In no case should it be permitted for any soil or waste connection or ventilating pipe to join a drain at a junction pipe; in every case such connection should be direct to an inspection chamber. In no case should it be permitted for a soil or waste connection to connect to a sewer other than via a manhole on the drainage system. Every drainage system should have one such final manhole in which (where interceptors are provided) the interceptor should be installed and to which a fresh-air inlet connection should be made. Where this manhole cannot be within 6 m of all soil pipes, waste gullies or ground-floor fitments, additional manholes should be provided. Neither lampholes nor rodding eyes should be used on sanitary drains as substitutes for manholes. No dwelling house should have more than one sanitary and more than one surface-water connection to a sewer.

Drains may join other drains or sewers other than at manholes, provided that, where a drain joins a drain or a sewer at a junction pipe, there must be a manhole on the lateral drain within 12 m of the junction. Where a drain joins another drain or a sewer at a junction pipe, a 45° junction (not a T-pipe) should be used and a bend inserted between the drain and the arm of the junction.

As far as practicable, all drains should be laid in straight lines from manhole to manhole but, in contrast to sewerage practice, latitude may be permitted in that bends not sharper than 45° may be made just outside manholes (preferably on the downstream side) because where several lateral connections discharge to a manhole a straight main channel invert is desirable, and this prevents a bend in the manhole. Apart from this latitude, bends or changes of gradient should not be permitted other than in manholes and, in particular, there should be no bends or junctions on drains under buildings, other than at manholes.

Drains under buildings should be constructed of cast iron (jointed with lead) or other unbreakable material, supported throughout on concrete. Local authorities have permitted clay pipes or concrete pipes surrounded by concrete, but a concrete surround is by no means a guarantee that leakage of liquid or foul air will not take place. Access to drains inside buildings should be in the form of bolted hatch boxes, absolutely air-tight and water-tight, set in inspection chambers. Where a drain passes under a building, there should be a manhole immediately at each end, outside the building.

Intercepting traps are desirable in the last manhole on the drainage system before connection to the sewer in those areas only

(1) where intercepting traps have been provided in the past and the sewers concerned are not ventilated via the ventilating pipes of the drainage system;
(2) where drains of other than cast iron or other unbreakable material exist under buildings.

Where interceptors are not provided, sanitary drains are ventilated by the flow of air between the outlet to the sewer and the ventilating pipe, which should connect to the topmost manhole at the head of every drain. Where interceptors are provided a ventilating pipe should connect to the lowest manhole on the system. This should preferably be a pipe carried up the wall of a building above eaves level, but where this is not practicable a fresh-air inlet may be used. If more than one drain discharges to the lowest manhole in such a manner that there is a free flow of air between the ventilating pipes at the tops of the drains via the drains and the lowest manhole, a fresh-air inlet or other ventilator at the lowest manhole is not necessary.

Requirements of surface-water drains

Technically, it is satisfactory for a rain-water pipe to connect to a surface-water drain via an access shoe or untrapped gully, so as to ventilate the surface-water drain. This may conflict with local opinion. Surface-water drains should be trapped from either waste drains or sanitary drains by discharging above the trap of a gully. Where surface-water drains discharge to watercourses they should be protected against the access of rodents by wire grids or tidal flaps.

Connections may be made to surface-water drains at junctions instead of at manholes; also the heads of surface-water drains may be brought up as rodding eyes. It is, however, desirable for surface-water drains to be provided with a sufficient number of suitably placed manholes, rodding eyes, and access shoes to permit rodding—particularly of those drains which discharge to sanitary drains.

Rain-water pipes connecting directly to sanitary or combined drains must discharge over the traps of gullies.

Minimum diameters of 75 mm and 100 mm have been adopted for surface-water drains, according to circumstances.

Construction of drains

Drains are constructed of concrete pipes, clay pipes, asbestos-cement pipes and cast-iron pipes but in Britain cast-iron pipe of a different specification is used for drains as compared with sewers. The Code of Practice on Building Drainage (CP301: 1971)[4] also included pitch-fibre pipes and U-PVC pipes. Although satisfactory at normal depth, these materials may distort under heavy external loading, and should not be used for unusually deep or shallow connections. Since CP301 recognizes Marston Theory for design of drains, in common with Authorities in the US and elsewhere, pipes with flexible joints are now often laid with granular bed and haunch, the well-known 'Class B' bedding, following the same technique as with main sewers. Thus, special attention should be given to pipe connections at chambers. If rigid joints have been used a flexible joint surrounded with pea gravel should be provided within 0.6 m of each wall face. Even with flexibly jointed pipes this technique should be followed, and two joints within a short distance of a structure are even better if any degree of differential settlement is anticipated.

No concrete or clay pipe with mortar joints should be laid on earth fill, granular fill or hardcore, but should be supported throughout the length of the drain and of the barrel of each pipe on the undisturbed trench bottom or, where this is unsuitable, on concrete bedding. Cast-iron drains, where not laid on natural ground, should be supported on concrete and should be so arranged that they do not bear the weight of any superimposed structures.

Inspection chambers

Shallow manholes or inspection chambers on drains are usually made to such proportions as permit the proper arrangement of connections and access for rodding, etc., from ground level. Where manholes on drains are so deep that they have to be entered for maintenance purposes, they should be constructed to the standards required for sewer manholes. Except where manholes come under roads or drives, light (nominally) air-tight covers may be used.

Where drops are required in drain manholes it is usual to provide cast-iron vertical drops constructed of cast-iron soil pipe erected inside the manhole, and having accesses to permit both horizontal and vertical rodding.

Subsoil drains

Subsoil drains used to consist of clayware field drain pipes (generally the most suitable material) or other suitable butt-jointed pipes. Various forms of porous concrete pipe are now frequently used as also are plastics pipes, in long lengths, often laid by the mole-plough method. Where they connect to a surface-water, sanitary or combined drain, a trap should be provided. In place of manholes, catch pits with open bottoms below drain invert level are provided on subsoil drains. Where subsoil drains discharge to watercourses they should be protected against the access of rodents by wire grids or tidal flaps, which also prevent direct ingress of flow from the watercourse.

Sewerage

All small sewers should be laid in straight lines and at constant gradients and diameters between manholes, as otherwise they may be difficult to clean and also hard to locate should they have to be repaired. In Great Britain it is considered that there may be bends on plan (not changes of gradient) on sewers that are larger than 750 mm diameter. All changes of direction, gradient and size, and all steps down in the invert on the line of sewer or between the main and any incoming sewer, are made at manholes. Where there are bends on large sewers protection against scour should be made on the outside of the bend if the velocity is high or wherever it may be easy to provide such protection, e.g. at manholes.

Drains may connect directly to the sewer, being brought in above the crown so that there shall be no surcharge of the drains when the sewer is running just full, and angled in the direction of flow; but no sewer should connect to another sewer except at a manhole.

The practice as regards distances between manholes appears to be almost the same in the United States as in Great Britain. In the United States it is considered that manholes should not be more than 90 m or, at the most, 150 m apart and, preferably, more closely spaced because long distances between manholes involve greater danger to men, either of being caught by flood water or because of the longer distances that casualties may have to be transported. In Great Britain the Ministry of Health laid down many years ago the requirement that manholes on public sewers should not be more than 110 m apart and those on privately owned sewers not more than 90 m apart. Later, approval was given to manholes on large sewers being at up to 180 m apart, but the writer considers that this is an undesirable practice. With smaller-diameter pipes the spacing of manholes is related to convenience in clearance by rodding.

Sewer calculations are made on the assumption that the hydraulic gradient is the gradient of the crown of the sewer. Where there is a change of diameter on a sewer the different sizes should meet at the same crown level with a step down in the outgoing invert. Where a lateral sewer enters a main sewer the crown of the lateral should not be lower than that of the main sewer, otherwise the hydraulic gradient during flowing-full conditions will be above the gradient of the crown of the lateral sewer. In cases where adequate fall is not available this rule is sometimes broken, and sewers of different sizes connected with their inverts at the same level so as to obtain a somewhat better velocity when the sewers are flowing only partially full. In such cases the hydraulic gradient at full condition must be taken from the crown of the smaller sewer at its lower end to the crown of the larger sewer at point of discharge, and not assumed to be the actual gradient of the invert.

Sewers should not be too small. In the United States some authorities permit 150 mm diameter; but most frequently the adopted minimum is 200 mm. In England and Wales it is permitted for a privately owned sewer to be as small as 100 mm diameter but local authorities now generally adopt 150 mm as the minimum diameter for public sewers. Formerly several consulting engineers adopted the practice of making 175 mm the minimum diameter and this practice would appear to be continuing because about 30% of the manufacturers of clay pipes to the new standard for clay pipe sewers include 175 mm diameter pipes in their lists, whereas only about 10% of those lists include 200 mm diameter pipes. More on the construction of pipe sewers will be given in Chapter 6.

There is no maximum size for sewers. The author has been in ancient sewers up to 3.5 m diameter and he has included in a report a proposal for 6.5 m sewers. There are 8.8 m diameter sewers at St Louis, USA.

Sewers must be laid to gradients sufficient to ensure 'self-cleansing' conditions If the velocity is less than 200 mm/s virtually all solids may be expected to settle and stoppages result. At a velocity of 0.3 m/s heavy detritus will remain settled but most organic material will be carried away by the flow. This velocity is not good enough for a sewer and would soon lead to trouble. If the flow in a carefully constructed sewer when partly full is not less than 0.6 m/s, that sewer can normally be expected to keep clean. By reference to Table 3 (p. 59) it can be calculated that if a sewer flows at 0.6 m/s when nearly half full, it should discharge at the velocity of 0.76 m/s when flowing full. Thus, practice in Great Britain is generally to design sewers to flow full at not less than 0.76 m/s and this is usually satisfactory. On the other hand, because of the great difference between flowing-full velocities in combined sewers compared with the average or dry-weather velocities, it is necessary in such cases, with the aid of Table 3, to find the gradient necessary to maintain self-cleansing velocities at average daily flow, and design accordingly. Table 1 gives recommended minimum gradients for most drains and sewers other than combined sewers. Several such tables have been published and no one can say which is the best, but the engineer, having accepted a scale of velocities, should make it a practice not to stray therefrom.

Rising mains from pumping stations will usually keep clean if the minimum velocity is 0.6 m/s. At a velocity of 0.3 m/s they are liable not only to become heavily silted but also to collect a deep deposit of greasy material in the crown. This combination of silt and grease reduces the capacity of the main severely. This also applies to inverted siphons and, for this reason, the velocities in rising mains and inverted siphons generally should not be permitted to drop below 0.76 m/s. There are also economic velocities for rising mains which will be discussed in Chapter 5.

It has been thought that scour of sewers causing damage would increase at high velocity, but research has proved that in the straight pipe increase of velocity lifts heavy particles from the invert and therefore scour does not increase. There is therefore no need to restrict the gradient of, or velocity in, any small pipe sewer; but it is desirable to use erosion-resisting materials such as vitrified brickwork on the outsides of bends and to limit velocities in rising mains and pumping-station pipework in which, of course, bends are unavoidable.

There is scouring of the invert of very old sewers in course of time, particularly if unsuitable materials were used. For example, the Northern Outfall Sewers of London, England, the inverts of which were constructed of a class of brick not now considered suitable, are known to have scoured to a depth of about 12 mm and have had to be repaired. Cast iron resists scour, but nevertheless sliding stones have been known to produce grooves in a cast-iron invert to a depth of at least 20 mm.

There is a reason to consider maximum velocity in large sewers which can be entered by men. The velocity when such a sewer is flowing partly full should never be so great as to endanger the men who work there.

Location of sewers

Sewers are laid with at least 1.2 m of cover under roads and 0.9 m elsewhere. They are not laid at greater depth without good reason, because of the cost of excavation, and the engineer has to attempt to keep his sewers with cover as near to these depths as possible, without introducing too many manholes. There are, however, frequent occasions when cost is reduced by increasing gradient—and consequently excavation—and reducing pipe size.

The reason for the minimum cover for pipes is threefold. First, the cover gives protection to the pipes against spot loads due to traffic etc., and accidental interference due to work on the surface such as road repair or deep cultivation of fields. Secondly, if sewers are too near the surface they are not sufficiently deep for house connections or road gullies to connect into them conveniently. Thirdly, although sewage, being warm, is not so likely to freeze as is town water, sewers —like water pipes—should have sufficient cover to protect them from frost.

All sewers should be laid sufficiently deep to pick up all drain connections and where necessary to pass under one another. In determining the depth of a sewer to serve the drains of a property, allowance should be made for a fall of the drain of not less than, say, 1 in 78 (or other gradient according to the practice of the local authority) from 0.6 m below ground level at the most distant gully etc., to 0.3 m above the crown of the sewer, thus allowing for the proper arrangement of the connecting junction pipe and bend, or channel-junction at manholes. In built-up areas this may well involve minimum depths to invert of at least 1.5 m. For this and other practical reasons, depths up to 3 m will probably be unavoidable. Anything above this depth should be avoided, if possible, as construction costs are likely to escalate rapidly above this depth. Even where laying connecting links across fields it behoves the engineer to consider carefully whether building development which will, of course, have a tendency to follow the provision of sewers, is likely to take place and whether sewer depths are sufficient to enable house connections to be made without difficulty. In this regard new development is often easier to handle than existing, as houses may be grouped and connected to the main sewer by branch or 'rider sewers', at some future date.

In the past, sewers have been most commonly laid in roads and as far as possible under the centre of the carriageway. This practice still applies to residential roads, and when the carriageways are narrow and initially not of very expensive materials the practice still has many advantages. Where wide roads of expensive materials are concerned it may be advisable to lay two sewers—one in each verge or footpath on either side of the road—thus avoiding heavy reinstatement or maintenance costs, or difficulties when making house or gully connections. When this is done the sewer at one side of the road should be considered as the main sewer, while the sewer on the other side should be constructed to the minimum size and connected under the road to the main sewer before the number of properties or gullies draining to it requires an increase of size.

Although town sewers are normally laid in carriageways or verges there are occasions when easements through private properties secure marked economies. In such circumstances British practice is for the engineer to show on his drawings the sewers crossing private properties and in due course, before the work is carried out, the property-owners concerned receive way-leave notices from the sewerage authority. While deviation from line of the carriageway occasionally proves advantageous or even necessary (for example, sewers often have to be laid across fields from a developed area to the sewage-disposal works), the practice of laying the majority of sewers in front or back gardens tends to lead to administrative and maintenance difficulties, and should generally be avoided. Such methods may be applied when all the properties on an estate belong to one owner, particularly if that owner is the local authority.

In circumstances where large sewers are compelled to cross private properties it may be advisable to arrange for the purchase of land occupied by them and the reservation of 'service lanes', i.e. strips of land available for repairs and maintenance and similar purposes. These, again, should not be resorted to too extensively, for they are only occasionally used and add considerably to the difficulty of estate development. The widths of the 'service lanes' should be such that the trenches will not cause danger to the foundations of adjacent buildings. It should be appreciated that the construction of sewers in private land requires an easement or working width, of about 10 m if the sewers are shallow, or up to 15 m if they are deep.

In the United States the locations of all underground services such as sewers, water mains, gas mains and cables are as recommended in *Location of Underground Utilities*, a manual of practice published by the American Society of Civil Engineers.[5] In Great Britain similar recommendations are made in *The Location of Underground Services*, a joint publication of the Institution of Civil Engineers and the Institution of Municipal Engineers.

The engineer's work (preparation of schemes)

Engineers concerned in the preparation of schemes of sewerage and sewage treatment may be direct employees of governments, local authorities, or more recently in England and Wales, of one of the new water authorities. They may also be with a firm of consulting engineers called in by any one of the above as likely to have wider or more specialised knowledge to bring to bear on their particular problems. In this case, they will first require to supplement their special knowledge by acquiring a thorough understanding of local conditions by making preliminary investigations on the lines described, under the following headings.

The first thing to be known is the problem; i.e. just what advice is required by the client. This forms the terms of reference of the consulting engineer who, having been called in to advise on a particular aspect of a town's sewerage, may then fully investigate everything relating thereto and finally give an opinion in the form of a report. The local authority, or other client, may then decide to act on the advice of the engineer and call upon him to design works in accordance with the proposals he has set forth.

The first stage of an investigation that precedes a report consists of a visit or visits to the site, examination of the terrain (maybe in the company of the local engineer) and the collection of existing data in the form of maps, drawings and statistics. This part of the work should always be done by the consulting engineer or by a senior and experienced member of his staff, who knows just what particulars are likely to be of use. The investigator, once on the site, might find that considerations other than those mentioned by the authority were worth taking into account. The following list is given as a reminder of the main items of information that are usually required, in a developed country such as Great Britain.

Population

Particulars of total numbers; tendency to decrease or increase; seasonal variation (e.g. summer influx at watering places); housing schemes; densities of building; town-planning proposals and restrictions (any plans available); types of houses in different areas; areas that will not be developed.

Water supply

Quantity consumed per head of population; type of supply and extent; trade consumption; name, address, and telephone number of principal officer of the water authority; source of supply; positions of large mains liable to be affected by the scheme.

Electric supply

Name, address and telephone number of the local electricity authority; voltage and phase of current; positions of mains and amperage available; regulations, kVA (kilovolt ampere) charges and unit charges.

Roads

Name of highway authority; particulars of public and private roads; highway authority's charges for restoration; particulars of surfaces and foundations—for surface-water sewer schemes, particulars of widths and lengths of roads in different areas.

Industries

Any particulars available of trade wastes, types of industry, quantities, times of discharge and temperatures.

Geology

Any available information on boreholes, trench records, etc.; notes of observations on natural exposures and cuttings; information on level of subsoil water; information on landslides, mining rights, mining subsidence, and other risks.

Rivers and streams

Particulars of rivers and streams liable to be affected by drainage scheme; rough estimate of quantity flowing in them; information regarding any water supply taken from them; name, address, and telephone number of river authority; standard of sewage effluent required by river authority.

Sites for works

Notes on suitable sites for pumping stations, outfall works, etc., with particulars of subsoil, subsoil water level, flood level, normal water level in stream for outfall.

Existing works

Positions, levels and sizes of existing sewers, whether combined, separate or partially separate; dry-weather flow; storm flow, infiltration; trade waste; strength of sewage. Details of existing pumping stations and machinery and sewage-treatment works; particulars of storm overflows (if any); their number, position and rate at which they commence to discharge; particulars of existing sea outfalls; normal tide levels; maximum recorded tides. Information regarding earth—and chemical—closet emptying services, cesspool drainage, soakaways and surface-water drains of roads and their positions of discharge to streams. Methods of cesspool emptying and sewer flushing. Anything at all unusual in local practice.

Materials

Particulars of local stone for building; supplies of engineering bricks; aggregate for concrete; media for trickling filters, and prices; prices of crude oil, coal, and gas.

Names, addresses, and telephone numbers

These, in full, of all officers, authorities, landowners, and parties liable to be interested in the scheme, should be obtained.

In Great Britain for a 'bird's eye view' of a district, the 1:25,000 contoured maps are very good, and will give a quick impression of the lie of the land and any irregularities which may help or hinder the routing of gravity sewers. The 1:10,000 scale maps are useful for large areas, but as soon as any degree of detail is required, reference must be made to 1:2500 which is large enough to enable sewer routes to be fairly accurately determined.

In less-developed countries enquiries to be made may be a lot less detailed, and the scheme may well be concerned entirely with new housing development. In such cases a comprehensive survey of the locality to be developed will probably have been made, and will be available for the purposes of the scheme. This, and all other information, will probably be available from government sources. Careful attention should, however, be given to the local habits and customs, which may well determine the form of the facilities to be provided.

The second part of the investigation is usually carried out on paper with the aid of maps, together with information that has been collected. It may often be possible to frame a preliminary report obtained by this means prior to the prosecution of a detailed level survey of the ground. Once the proposals (or alternative proposals) have been made and lines of sewer decided, level, and sometimes topographical and subsoil, surveys of the site have to be made.

Systematic design

There are several ways in which sewerage systems may be designed; the following may appeal to those who have not had the practice which makes it possible for them to see at once the obvious course, or to those who always prefer to apply some form of system. It should also be of use to the expert when dealing with complicated systems.

When the position of outfall or treatment works has been decided, or the positions of the several outfalls required for the scheme, these points of discharge are marked tentatively on a large-scale map. Working sewerage plans are usually prepared on large-scale maps; the designer lays over the map a sheet of detail or tracing paper and, in blue pencil, indicates the subsidiary sewers draining the various existing or proposed roads. Then, with the aid of the levels shown on the map or added thereto from his level survey, he finds the direction of fall of each road and marks an arrow-head on each sewer showing the natural direction of flow. When this has been done for the whole of the area, the general trend of drainage can be seen at a glance and, unless the country is difficult to drain, it should then be necessary only to connect together the low ends of all lengths of sewer with main sewers leading the flow to a point of outfall. This the designer may do with red pencil and, by comparing the proportion of red pencil to blue pencil on the plan, he will be able to see the proportion of unproductive sewer into which no side connections from houses or drains will discharge.

The next step is to estimate the rate of flow expected and to find the sizes of sewer required to accommodate the flows and the gradients necessary to effect the velocities essential for self-cleansing conditions. To do this the engineer must plot sections and calculate gradients.

Advertisement for tenders: the contract documents

Once an authority has decided to go ahead with a scheme and all necessary sanctions have been obtained, the engineer is instructed to obtain tenders from contractors or else to put the work in hand by direct labour.

Before calling for tenders for the construction of works, a set of documents is prepared which includes the general conditions of contract, the specification, the bill of quantities, the schedule of prices and the contract drawings. All these papers are so prepared as to be read together and to serve as the legal agreement binding both parties to the contract. The general conditions of contract are usually stereotyped and not varied except in such circumstances as call for the alteration of clauses. They frequently consist of a printed document with spaces left for the insertion of the names of the contractors and their sureties, date, amount of tender, etc.

The specification is a description of the classes of materials and workmanship to be used in the execution of the contract, together with more detailed descriptions of the method of construction of the individual components of the work. Both the general conditions and the specification as generally in use at the present time by public authorities and consulting engineers are the result of experience, over a long period of time, and are so framed as to safeguard the client against the possible ways in which contractors may default.

A specification is a legal document and therefore should be carefully worded. It should make clear not only to the contractor, but to any unbiased arbitrator, the exact details of the works required and the manner in which they are intended to be executed. Any specification which is merely descriptive, or which can be interpreted in more ways than one, fails in its purpose. Thus it will be appreciated that the art of specification writing is acquired as a result of practice and intelligent study, and is one which well deserves the attention of engineers.

A major part of specification writing consists in copying from previous specifications and takes the form of scissors and paste work, relevant clauses being inserted and new clauses written whenever work of a kind not done before has to be described. By writing specifications in this manner a good deal of time is saved, for a specification is often lengthy and would require a lot of work if it were to be prepared afresh in full detail for every contract. But it is important not only that the writer reads every clause carefully before he inserts it but that he understands what he is reading and has full knowledge of the class of work described. When writing new clauses, or revising those existing, he should be very careful that the wording cannot be misread and does not rely on the punctuation for its meaning.

Standard specifications are very useful to specification writers who, in requiring any article to comply with the appropriate prescribed standard, such as a British Standard Specification, say all that need be said without entering into detail. On the whole, standard specifications are in line with current standards of workmanship and materials, and are preferred by local-government engineers, but not invariably so, as at times the interests of purchasers tend to give place too much to those of the manufacturers.

The engineer should, whenever referring to a standard specification, have read the specification and know what he is specifying. He should be careful not to refer to an obsolete or withdrawn specification or to one that is of very limited scope.

The bill of quantities is a detailed bill based on a quantity survey made by measuring the quantities of all materials and labours shown on the drawings, and setting out every item so that all may be separately priced by the contractor. For each part of the works proposed there is an individual bill giving the quantities of site clearance, excavation, concrete foundations, brickwork, and so forth, with all special labours described and related to the specification. The totals of each bill are carried to a summary, for addition, with the inclusion of various overheads and a percentage to cover contingencies. These make up the grand total of the tender, which is the price for which the contractor is prepared to carry out the work, assuming the quantities of actual work agree with those measured up from the drawings. The details in the bill of quantities should agree with the specification and drawings throughout. Every clause in the specification should be covered by an item in the bill of quantities. It should be appreciated that engineering bills of quantities are of necessity divided under headings according to localities of works or structures, whilst architects' bills are divided under classes of labours: excavator, concretor, bricklayer, etc.

The schedule of prices, which usually follows the general summary page of the bills of quantities, is a schedule giving the contractor's charges for various materials likely to be used in large quantity during the construction of works and for the principal labour rates including profit; also certain main items of completed work. The rates charged under this schedule must be in exact agreement with those charged for the same work as given in the bill of quantities, otherwise the engineer may object to the tender. The schedule of prices may also include classes of material and labour not mentioned in the bill of quantities.

The contract drawings consist of a number of drawings which should be enumerated in the specification and which should agree with the specification and bill of quantities in every detail. These drawings should be sufficiently complete that with their aid, and with the aid of the specification, the contractor will be able to carry out the works without any further guidance, although drawings showing variations or added details may be issued at later stages during the progress of the contract.

When the general conditions, specification, bill of quantities, schedule of prices and contract drawings have been prepared they are printed in sufficient numbers for copies to be available for all persons tendering, for all interested officers of the authority, and so forth. The general conditions, specification, bill of quantities and schedule of prices are frequently bound in one volume, while the drawings are issued separately, or bound together in another volume. Tenders are then advertised in local newspapers and in certain trade journals well known as advertising media for public-works contracts. Contractors wishing to tender apply for copies of the documents, on which they pay a deposit which is returnable on receipt of a bona-fide tender. Tenders are forwarded to the clerk or engineer to the authority under sealed cover, and opened in committee on a prearranged date.

The above is, in brief, the procedure in Britain as regards small sewer and sewage works construction. Where there are proposals for large contracts involving mechanical and electrical equipment, the general construction contract may be advertised with clauses to cover the installation of mechanical and/or electrical equipment which are the subject of separate contracts with mechanical and/or electrical contractors. Tenders for pumps, together with all ancillary electrical equipment, are often invited from selected firms, and tenders are also invited for specific items of sewage-works equipment such as sedimentation-tank sludging mechanisms, trickling-filter distributors, activated-sludge equipment, sludge-digestion tank equipment, etc.

Methods for invitations for tenders have of recent years not always been the same in Britain, and they do vary from the above in America. In Canada bidding procedures generally fall into two categories, as described below. The former method is at present the more usual but the latter, being much more satisfactory for the engineer and the authority, is gaining more acceptance.

(1) Specifications are prepared by the engineers and, where equipment is involved, the specifications describe features which are mechanical or electrical specific requirements of the project but allow competitive bidding by firms manufacturing equipment of the type.

The project is then advertised and major suppliers, mechanical and electrical contractors, as well as general contractors, usually take out documents.

Interested suppliers tender to the mechanical contractor, who in turn tenders to the general contractor, who submits his tenders to the authority. In some cases a general contractor may have his own mechanical and electrical contracting division.

(2) The engineer prepares specifications which are issued to major equipment suppliers for competitive tender on a pre-selection basis.

Contract drawings are then adjusted to suit the selected equipment and, when issued, direct the general contractor to purchase particular equipment from a named supplier, or alternatively permit direct purchase of this equipment by the authority for installation by the general and mechanical contractors.

In the latter case the authority enters into contracts with major equipment suppliers in addition to the general contractor, and payments are made direct.

Engineering quantities

Although it is not intended here to deal in detail with the compilation of Bills of Quantities, it behoves every engineer to familiarize himself with what has tended to become, in Great Britain at any rate, the art of the quantity surveyor. This art, developed to tackle the intricacies of the items of work carried out by the various trades employed in the building industry, reaches its heights in complicated tower structures or office blocks. The civil engineer with his pipelines and straightforward concrete structures has no need of many of the 'labours' described in such detail in bills dealing with intricate brick structures, or carpenters and joiners work. Thus, the Institution of Civil Engineers (ICE) as far back as 1933 produced their *Standard Method of Measurement of Civil Engineering Quantities*, intended to encourage young civil engineers to prepare bills of quantities in a standardized form, without becoming involved in the intricacies of the building surveyor's efforts.

The wide use of this document proves its success, although over the years engineers and contractors have striven to find alternative methods more suited to both sides. The outcome was the introduction in 1977 of the *Civil Engineering Standard Method of Measurement* which, besides attempting to bring civil engineering quantities into the 'computer age', also introduces a new concept, in the form of what are known as 'method-related charges', with the object of relating payments made more closely to capital outlay, thus avoiding the incidence of claims and rendering the contract sum a more accurate reflection not only of the measured work, but of the organization in terms of plant and labour that the contractor must provide. It is intended that these charges may account for up to 40% of the contract value and that their introduction will avoid wide variations in the pricing of common items such as excavation, or concrete construction, brought about by difficult conditions, such as waterlogged ground or particular types of construction. Based as they are on the anticipated performance of machines, mixing plant, etc., their measurement involves careful recording of working times of plant, etc., which formerly were considered almost entirely a matter for the contractor who had to carry out the excavation, or complete the structure at agreed times and possibly with little recorded by way of measurement until a particular stage was reached. A long period of consultation between representatives of the ICE and CIRIA (Construction Industry Research and Information Association) have resulted in the new CESMM, (Civil Engineering Standard Method of Measurement) as described by Martin Barnes in *Measurement in Contract Control*[7] and its companion volume *Examples of CESMM*,[8] and contractors are now busy familiarizing themselves with the new method, and if it is seen to be achieving its purpose no doubt it will come to be adopted in other parts of the world. One thing is already clear: the method has no place for the 'descriptive item' which has often conveyed quite satisfactorily in words something that would otherwise have required a separate drawing! Thus pressure on the drawing office is increased. Another point of some consequence is that unless quantity survey staff are actually present on site and recording working times and conditions throughout a contract, they may be hard put to arrive at a fair evaluation of the method-related items which will have to be settled direct between the engineer and the contractor. It may be that the present vogue for employing quantity surveyors on civil engineering work will lose its popularity, and the task of measurement in its entirety revert to the engineer. If so, in the author's view, this will be no bad thing!

References

1. Symp. on *Storm Sewage Overflows*, Int. of Civil Engnrs. Lndn. (1967).
2. Rep. of the J. Comm. on Field Res. into Drainage Problems, May 1954. Instn. of Public Hlth. Engnrs.
3. *Building Regulations*. HMSO (1976).
4. Code of Practice 301. Br. Standards Inst. Lndn. (1971).
5. *Location of Underground Utilities*, Manual of Practice No. 14. Am. Soc. of Civil Engnrs.
6. *Civil Engineering Standard Method of Measurements*. Instn. Civil Engnrs. Thomas Telford Ltd, Lndn. (1976).
7. Barnes, M., *Measurement in Contract Control*. Thomas Telford Ltd, Lndn. (1977).
8. Barnes, M. *Examples of CESMM*. Thomas Telford Ltd, Lndn. (1977).

CHAPTER 2
Flow in Sewers

Hydraulics will not be discussed generally or at length in this book: it is assumed that the engineer has had a grounding in this subject and, if not, he can refer to textbooks. There are, however, special applications of hydrodynamics to sewerage problems and in this respect advances have been made which cannot be found in student textbooks.

The engineer concerned with sewer design has to make reasonably accurate calculations without waste of time and, in this connection, some advice can be given. At low velocities and in restricted channels laminar flow occurs; the water slides over the surface of the channel, impelled by gravity but retarded by the friction of the surface, and the several layers of water slide over one another. In larger channels and at higher velocities, instead of sliding in laminar flow, the water rolls down the channel in eddies: this is turbulent flow. In virtually all calculations of flow of clean water and of sewage the dimensions of channel and the practical working velocities are such that the flow is turbulent. There is one exception: sewage sludge is sufficiently viscous to flow in laminar condition at velocities that would be practicable for sewage; therefore the critical point at which there is a change from turbulent to laminar flow needs to be known when sludge-flow calculations are being made.

There is an extensive literature in the form of papers, monographs and books devoted to flow in pipes and channels. Many formulae have been devised; some of merely historical interest and, incidentally, notoriously inaccurate; some more reasonably accurate but unnecessarily complicated; and some which are applicable to fluids of all densities and viscosities but which may be of moderate accuracy only when applied to any specific liquid. The most accurate formulae are those which have been prepared to express the results of numerous experiments and these, for water or sewage, are almost invariably in the form

$$V = Km^{\alpha}i^{\beta}$$

where: V = velocity in metres per second;

m = hydraulic mean depth in metres; i.e. the cross-sectional area of flow divided by the 'wetted perimeter' of contact between the water and its bed;

i = the fall in metres divided by the length in metres;

K = a coefficient

Whereas in terms of Imperial measurements the units used for evaluating the terms V, m and i were commonly all in feet, metrication in SI (Systeme International) units implies the use of millimetres in calculating m, the hydraulic mean depth, although V would be in metres, and i will not be affected whether reckoned in feet or metres, the introduction of millimetres has to be compensated for by adjustment in the coefficients formerly associated with well-known formulae using this format, as will be realized when formulae (1) and (2) are quoted in metric units. For simplicity in calculation formula (3), on which Table 2 is based, has been expressed in slightly different units.

In all hydraulic experiments with which the author has been concerned, it has been found that results of careful work, when plotted to logarithmic ordinates and abscissae, have formed straight lines, except when the velocity has been reduced below the critical point. Then the rule has changed and two straight lines at different slopes, one for laminar and one for turbulent flow, have met at the critical point. Some experimenters say that the lines are slightly curved or that there is a transition curve between laminar and turbulent flow, but the author has never seen any indication of such curvature.

The fact that flow data form straight lines when plotted on logarithmic paper makes determination of the constant and indices in formulae of the above kind very easy, and this has been done for pipes and culverts constructed of various materials to different standards of workmanship. Very smooth and regular pipes which have constant diameter, straightness and well-made, widely spaced joints are polyvinyl chloride pipes, pitch-impregnated fibre pipes and asbestos-cement pipes. For all of these the value of K may be taken as 242, α as 0.700, and β as 0.577.

In engineering practice, however, materials and standards of workmanship vary; also water mains become encrusted and sewers coated with a slime. For this reason, design of water mains and sewers is usually in accordance with a selected formula which is considered to be a fair representation of the conditions of flow in pipes and culverts of the usual materials and in average condition after some years of use. One of these, Manning's formula,[1] has been largely used in America and to some extent in Great Britain. It reads

$$V = \frac{0.01}{n} m^{0.667} i^{0.500} \tag{1}$$

where: n = coefficient of roughness,
m = hydraulic mean depth in millimetres
V = velocity in metres per second

This formula is reasonably accurate if the correct coefficient of roughness is selected, but this often is not so, for the reason that the original coefficients were determined with regard to flow in open channels and, as will be explained, this made the calculated discharges for pipes and culverts nearly 20% too high.

Crimp and Bruges' formula has perhaps been more used than any other in Great Britain, probably because of the availability of the excellent Crimp and

Bruges Tables.[2] This reads

$$V = 0.834 \, m^{0.666} \, i^{0.5} \tag{2}$$

It is the equivalent of Manning's formula with the value of n being slightly less than 0.012 in the latter.

Because of the unsuitable value of α in Manning's and Crimp and Bruges' formulae, these give discharges which are too pessimistic for very small pipes as used in domestic plumbing, and somewhat too optimistic for very large culverts constructed of brickwork and concrete. But they are accurate enough for sewers of moderate size.

Perhaps the most accurate formula for concrete-pipe sewer design is Scobey's formula[3] on which the tables in the Appendix are based. The author's formula from which Table 2 and Escritts *Tables of Metric Hydraulic Flow*[4] were prepared) agrees with Scobey's formula very well over the range for which the latter applies, but also gives good figures for pipes and culverts from 10 mm to 9 m diameter. This formula was derived by taking the average of a number of formulae for flow in clay pipes, cast-iron pipes, brick culverts and concrete tunnels and making very slight adjustments for the sake of simplicity. In metric notation, for circular pipes, this becomes

$$V = \frac{26.738 D^{0.62}}{I^{0.5}} \tag{3}$$

where: V = velocity of flow in metres per minute;
$\quad\quad\quad D$ = diameter in millimetres
$\quad\quad\quad I$ = length divided by fall

Another formula that is now widely used, following research by Ackers at the Hydraulics Research Station,[5,6] is the Colebrook–White formula, or uniform flow equation, in which velocity of flow is dependent on pipe bore, kinematic viscosity, gradient and surface roughness. This roughness depends on length of pipes, surface texture, types of joints, deposited grit in the invert, adherent slime and grease on walls, and deviation from longitudinal/straightness.

Bland, of the Clay Pipe Development Association, quotes further experiments carried out at Wimpey's laboratory and at the Hydraulic Research Centre, and finally quotes the following values for K_s, the roughness coefficient:

Pipe material	K_s (mm)
Concrete	6.0
Asbestos cement	6.0
U-PVC	1.5
Clay	1.5

These figures were established as a result of 2 years of experiments with 225 mm bore pipes with sewage flowing through them. Maximum velocity 0.78 m/s. The growth, and later sloughing-off, of slime have a considerable effect on these figures. Bland has prepared design tables,[7] calculated from the Colebrook–White formula in the following form:

$$V = -2\sqrt{(2gDi)} \log\left(\frac{K_s}{3.7D} + \frac{2.51v}{D\sqrt{(2gDi)}}\right)$$

where: A = cross-sectional area of flow;
D = bore of pipeline;
g = gravitational acceleration;
Q = discharge;
V = mean velocity of flow;
i = hydraulic gradient;
K_s = linear measure of roughness;
v = kinematic velocity;
and $Q = AV$.

NOTE. In applying this formula the initial minus sign will be cancelled by the negative logarithm which follows.

In calculating flows in sewers it is usual to neglect the velocity head lost in entering the top end of the pipe, for this is very small when velocities are low. But velocity head can be considerable at high velocities and then allowance for it should be made, particularly in the case of a pipe discharging from a tank, where water would have to accelerate from stationary to the velocity of flow in the pipe.

This can be calculated by the formula:

$$H = \frac{V^2}{2g} \qquad (4)$$

where: H = velocity head in metres;
V = velocity in pipe in metres per second;
g = acceleration under gravity = 9.80665 metres per second per second.

Surcharge

A surcharged sewer is one which is under pressure, the hydraulic gradient being above the crown of the sewer and therefore above the water level except at manholes, where the water rises to the level of the hydraulic gradient. A surcharged sewer can discharge considerably more than might be expected because the hydraulic gradient is usually curved, being steepest at the bottom end of each length of sewer. Suppose a length of sewer of constant gradient and diameter has no flow coming in at the top end but is receiving flow from many laterals equally distributed throughout its length, and suppose the surcharge is such that the pipe is filled just to crown level at the top end and not surcharged above the crown at the bottom end, then the discharge at the bottom end will not be the discharge as calculated according to the hydraulic gradient, but $\sqrt{3}$ times this figure, because the curve of the hydraulic gradient will be three times the gradient of the invert at that point.

In sewerage systems there is inflow at the top end of each length of sewer except the highest length, and surcharge backs up the bottom end of each length.

Table 2 Flow in circular pipes and culverts, based on the formula derived by L. B. Escritt

(Provided by courtesy of the drainage design support group of United Computing, Epsom, Surrey)

(a) Diameters 100–450 mm

	100 mm		150 mm		225 mm		300 mm		375 mm		450 mm	
Length/fall	Velocity (m/min)	Flow (l/s)	Velocity (m/min)	Flow (l/s)	Velocity (m/min)	Flow (l/s)	Velocity (m/min)	Flow (l/s)	Velocity (m/min)	Flow (l/s)	Velocity (m/min)	Flow (l/s)
5	207.7	27.2	267.1	78.7	—	—	—	—	—	—	—	—
6	189.6	24.8	243.8	71.8	—	—	—	—	—	—	—	—
7	175.6	23.0	225.7	66.5	290.3	192.4	—	—	—	—	—	—
8	164.2	21.5	211.2	62.2	271.5	180.0	—	—	—	—	—	—
9	154.8	20.3	199.1	58.7	256.0	169.7	—	—	—	—	—	—
10	146.9	19.2	188.9	55.6	242.9	161.0	290.3	342.1	—	—	—	—
11	140.1	18.3	180.1	53.1	231.6	153.5	276.8	326.2	—	—	—	—
12	134.1	17.6	172.4	50.8	221.7	147.0	265.0	312.3	—	—	—	—
13	128.8	16.9	165.7	48.8	213.0	141.2	254.6	300.0	292.4	538.3	—	—
14	124.1	16.3	159.6	47.0	205.3	136.1	245.3	289.1	281.7	518.8	—	—
15	119.9	15.7	154.2	45.4	198.3	131.4	237.0	279.3	272.2	501.2	—	—
16	116.1	15.2	149.3	44.0	192.0	127.3	229.5	270.4	263.5	485.2	295.1	782.4
17	112.7	14.8	144.9	42.7	186.3	123.5	222.6	262.4	255.7	470.8	286.3	759.0
18	109.5	14.3	140.8	41.5	181.0	120.0	216.4	255.0	248.5	457.5	278.2	737.6
19	106.6	14.0	137.0	40.4	176.2	116.8	210.6	248.2	241.8	445.3	270.8	718.0
20	103.9	13.6	133.6	39.3	171.7	113.8	205.3	241.9	235.7	434.0	263.9	699.8
21	101.4	13.3	130.3	38.4	167.6	111.1	200.3	236.1	230.0	423.6	257.6	682.9
22	99.0	13.0	127.3	37.5	163.7	108.5	195.7	230.6	224.7	413.8	251.6	667.2
23	96.9	12.7	124.5	36.7	160.1	106.1	191.4	225.6	219.8	404.7	246.1	652.6
24	94.8	12.4	121.9	35.9	156.8	103.9	187.4	220.8	215.2	396.2	240.9	638.8
25	92.9	12.2	119.5	35.2	153.6	101.8	183.6	216.3	210.8	388.2	236.1	625.9

Diameter

Table 2 (*Continued*)

Length/fall	100 mm Velocity (m/min)	100 mm Flow (l/s)	150 mm Velocity (m/min)	150 mm Flow (l/s)	225 mm Velocity (m/min)	225 mm Flow (l/s)	300 mm Velocity (m/min)	300 mm Flow (l/s)	375 mm Velocity (m/min)	375 mm Flow (l/s)	450 mm Velocity (m/min)	450 mm Flow (l/s)
26	91.1	11.9	117.1	34.5	150.6	99.8	180.0	212.1	206.7	380.7	231.5	613.8
27	89.4	11.7	114.9	33.9	147.8	98.0	176.7	208.2	202.9	373.5	227.1	602.3
28	87.8	11.5	112.9	33.3	145.1	96.2	173.5	204.4	199.2	366.8	223.1	591.4
29	86.3	11.3	110.9	32.7	142.6	94.5	170.5	200.9	195.7	360.4	219.2	581.1
30	84.8	11.1	109.0	32.1	140.2	92.9	167.6	197.5	192.5	354.4	215.5	571.4
31	83.4	10.9	107.3	31.6	137.9	91.4	164.9	194.3	189.3	348.6	212.0	562.1
32	82.1	10.8	105.6	31.1	135.8	90.0	162.3	191.2	186.3	343.1	208.6	553.2
33	80.9	10.6	104.0	30.6	133.7	88.6	159.8	188.3	183.5	337.9	205.5	544.8
34	79.7	10.4	102.4	30.2	131.7	87.3	157.4	185.5	180.8	332.9	202.4	536.7
35	78.5	10.3	101.0	29.7	129.8	86.0	155.2	182.8	178.2	328.1	199.5	529.0
36	77.4	10.1	99.5	29.3	128.0	84.8	153.0	180.3	175.7	323.5	196.7	521.6
37	76.4	10.0	98.2	28.9	126.3	83.7	150.9	177.8	173.3	319.1	194.0	514.5
38	75.4	9.9	96.9	28.5	124.6	82.6	148.9	175.5	171.0	314.9	191.5	507.7
39	74.4	9.7	95.6	28.2	123.0	81.5	147.0	173.2	168.8	310.8	189.0	501.1
40	73.4	9.6	94.4	27.8	121.4	80.5	145.1	171.0	166.7	306.9	186.6	494.8
41	72.5	9.5	93.3	27.5	119.9	79.5	143.4	168.9	164.6	303.1	184.3	488.8
42	71.7	9.4	92.2	27.2	118.5	78.6	141.6	166.9	162.7	299.5	182.1	482.9
43	70.8	9.3	91.1	26.8	117.1	77.6	140.0	165.0	160.8	296.0	180.0	477.3
44	70.0	9.2	90.0	26.5	115.8	76.7	138.4	163.1	158.9	292.6	177.9	471.8
45	69.2	9.1	89.0	26.2	114.5	75.9	136.8	161.3	157.1	289.3	175.9	466.5
46	68.5	9.0	88.1	25.9	113.2	75.1	135.3	159.5	155.4	286.2	174.0	461.4
47	67.8	8.9	87.1	25.7	112.0	74.3	133.9	157.8	153.8	283.1	172.2	456.5
48	67.0	8.8	86.2	25.4	110.8	73.5	132.5	156.1	152.2	280.2	170.4	451.7
49	66.4	8.7	85.3	25.1	109.7	72.7	131.1	154.5	150.6	277.3	168.6	447.1
50	65.7	8.6	84.5	24.9	108.6	72.0	129.8	153.0	149.1	274.5	166.9	442.6

51	438.2	165.3	271.8	147.6	151.5	128.5	71.3	107.5	24.6	83.6	8.5	65.0
52	434.0	163.7	269.2	146.2	150.0	127.3	70.6	106.5	24.4	82.8	8.4	64.4
53	429.9	162.1	266.6	144.8	148.6	126.1	69.9	105.5	24.2	82.0	8.4	63.8
54	425.9	160.6	264.1	143.5	147.2	124.9	69.3	104.5	23.9	81.3	8.3	63.2
55	422.0	159.2	261.7	142.1	145.9	123.8	68.6	103.6	23.7	80.5	8.2	62.6
56	418.2	157.7	259.4	140.9	144.6	122.7	68.0	102.6	23.5	79.8	8.1	62.1
57	414.5	156.3	257.1	139.6	143.3	121.6	67.4	101.7	23.3	79.1	8.1	61.5
58	410.9	155.0	254.9	138.4	142.0	120.5	66.8	100.8	23.1	78.4	8.0	61.0
59	407.4	153.7	252.7	137.2	140.8	119.5	66.3	100.0	22.9	77.8	7.9	60.5
60	404.0	152.4	250.6	136.1	139.7	118.5	65.7	99.1	22.7	77.1	7.9	60.0
61	400.7	151.1	248.5	135.0	138.5	117.5	65.2	98.3	22.5	76.5	7.8	59.5
62	397.5	149.9	246.5	133.9	137.4	116.6	64.7	97.5	22.3	75.9	7.7	59.0
63	394.3	148.7	244.5	132.8	136.3	115.6	64.1	96.8	22.2	75.2	7.7	58.5
64	391.2	147.5	242.6	131.8	135.2	114.7	63.6	96.0	22.0	74.7	7.6	58.1
65	388.2	146.4	240.8	130.7	134.2	113.9	63.1	95.3	21.8	74.1	7.5	57.6
66	385.2	145.3	238.9	129.8	133.2	113.0	62.7	94.5	21.7	73.5	7.5	57.2
67	382.3	144.2	237.1	128.8	132.2	112.1	62.2	93.8	21.5	73.0	7.4	56.7
68	379.5	143.1	235.4	127.8	131.2	111.3	61.7	93.1	21.3	72.4	7.4	56.3
69	376.8	142.1	233.7	126.9	130.2	110.5	61.3	92.5	21.2	71.9	7.3	55.9
70	374.1	141.1	232.0	126.0	129.3	109.7	60.8	91.8	21.0	71.4	7.3	55.5
71	371.4	140.1	230.4	125.1	128.4	108.9	60.4	91.1	20.9	70.9	7.2	55.1
72	368.8	139.1	228.7	124.2	127.5	108.2	60.0	90.5	20.7	70.4	7.2	54.7
73	366.3	138.1	227.2	123.4	126.6	107.4	59.6	89.9	20.6	69.9	7.1	54.4
74	363.8	137.2	225.4	122.5	125.7	106.7	59.2	89.3	20.5	69.4	7.1	54.0
75	361.4	136.3	224.1	121.7	124.9	106.0	58.8	88.7	20.3	69.0	7.0	53.6
76	359.0	135.4	222.6	120.9	124.1	105.3	58.4	88.1	20.2	68.5	7.0	53.3
77	356.6	134.5	221.2	120.1	123.3	104.6	58.0	87.5	20.1	68.1	6.9	52.9
78	354.4	133.6	219.8	119.4	122.5	103.9	57.6	87.0	19.9	67.6	6.9	52.6
79	352.1	132.8	218.4	118.6	121.7	103.3	57.3	86.4	19.8	67.2	6.8	52.3
80	349.9	132.0	217.0	117.9	120.9	102.6	56.9	85.9	19.7	66.8	6.8	51.9
81	347.7	131.1	215.7	117.1	120.2	102.0	56.6	85.3	19.6	66.4	6.8	51.6
82	345.6	130.3	214.3	116.4	119.5	101.4	56.2	84.8	19.4	66.0	6.7	51.3
83	343.5	129.6	213.1	115.7	118.7	100.8	55.9	84.3	19.3	65.6	6.7	51.0
84	341.5	128.8	211.8	115.0	118.0	100.2	55.5	83.8	19.2	65.2	6.6	50.7

Table 2 (*Continued*)

	100 mm		150 mm		225 mm		300 mm		375 mm		450 mm	
Length/fall	Velocity (m/min)	Flow (l/s)	Velocity (m/min)	Flow (l/s)	Velocity (m/min)	Flow (l/s)	Velocity (m/min)	Flow (l/s)	Velocity (m/min)	Flow (l/s)	Velocity (m/min)	Flow (l/s)
85	50.4	6.6	64.8	19.1	83.3	55.2	99.6	117.3	114.3	210.5	128.0	339.4
86	50.1	6.6	64.4	19.0	82.8	54.9	99.0	116.6	113.7	209.3	127.3	337.5
87	49.8	6.5	64.0	18.9	82.3	54.6	98.4	116.0	113.0	208.1	126.5	335.5
88	49.5	6.5	63.7	18.8	81.9	54.3	97.9	115.3	112.4	206.9	125.8	333.6
89	49.2	6.4	63.3	18.7	81.4	54.0	97.3	114.7	111.7	205.7	125.1	331.7
90	49.0	6.4	63.0	18.5	81.0	53.7	96.8	114.0	111.1	204.6	124.4	329.9
91	48.7	6.4	62.6	18.4	80.5	53.4	96.2	113.4	110.5	203.5	123.7	328.1
92	48.4	6.3	62.3	18.3	80.1	53.1	95.7	112.8	109.9	202.4	123.1	326.3
93	48.2	6.3	61.9	18.2	79.6	52.8	95.2	112.2	109.3	201.3	122.4	324.5
94	47.9	6.3	61.6	18.1	79.2	52.5	94.7	111.6	108.7	200.2	121.7	322.8
95	47.7	6.2	61.3	18.1	78.8	52.2	94.2	111.0	108.2	199.1	121.1	321.1
96	47.4	6.2	61.0	18.0	78.4	52.0	93.7	110.4	107.6	198.1	120.5	319.4
97	47.2	6.2	60.6	17.9	78.0	51.7	93.2	109.8	107.0	197.1	119.8	317.8
98	46.9	6.1	60.3	17.8	77.6	51.4	92.7	109.3	106.5	196.1	119.2	316.1
99	46.7	6.1	60.0	17.7	77.2	51.2	92.3	108.7	105.9	195.1	118.6	314.5
100	46.5	6.1	59.7	17.6	76.8	50.9	91.8	108.2	105.4	194.1	118.0	313.0
105	45.3	5.9	58.3	17.2	74.9	49.7	89.6	105.6	102.9	189.4	115.2	305.4
110	44.3	5.8	56.9	16.8	73.2	48.5	87.5	103.1	100.5	185.1	112.5	298.4
115	43.3	5.7	55.7	16.4	71.6	47.5	85.6	100.9	98.3	181.0	110.1	291.8
120	42.4	5.6	54.5	16.1	70.1	46.5	83.8	98.7	96.2	177.2	107.7	285.7
125	41.5	5.4	53.4	15.7	68.7	45.5	82.1	96.8	94.3	173.6	105.6	279.9
130	40.7	5.3	52.4	15.4	67.4	44.6	80.5	94.9	92.5	170.2	103.5	274.5
135	40.0	5.2	51.4	15.1	66.1	43.8	79.0	93.1	90.7	167.1	101.6	269.3
140	39.3	5.1	50.5	14.9	64.9	43.0	77.6	91.4	89.1	164.0	99.8	264.5
145	38.6	5.1	49.6	14.6	63.8	42.3	76.2	89.8	87.5	161.2	98.0	259.9

Diameter

150	255.5	96.4	158.5	86.1	88.3	74.9	41.6	62.7	14.4	48.8	5.0	37.9
155	251.4	94.8	155.9	84.7	86.9	73.7	40.9	61.7	14.1	48.0	4.9	37.3
160	247.4	93.3	153.4	83.3	85.5	72.6	40.2	60.7	13.9	47.2	4.8	36.7
165	243.6	91.9	151.1	82.1	84.2	71.5	39.6	59.8	13.7	46.5	—	—
170	240.0	90.5	148.9	80.8	83.0	70.4	39.0	58.9	13.5	45.8	—	—
175	236.6	89.2	146.7	79.7	81.8	69.4	38.5	58.1	13.3	45.1	—	—
180	233.3	88.0	144.7	78.6	80.6	68.4	37.9	57.2	13.1	44.5	—	—
185	230.1	86.8	142.7	77.5	79.5	67.5	37.4	56.5	12.9	43.9	—	—
190	227.0	85.6	140.8	76.5	78.5	66.6	36.9	55.7	12.8	43.3	—	—
195	224.1	84.5	139.0	75.5	77.5	65.7	36.5	55.0	12.6	42.8	—	—
200	221.3	83.5	137.2	74.5	76.5	64.9	36.0	54.3	12.4	42.2	—	—
205	218.6	82.4	135.6	73.6	75.6	64.1	35.6	53.6	12.3	41.7	—	—
210	216.0	81.4	133.9	72.7	74.6	63.3	35.1	53.0	12.1	41.2	—	—
215	213.4	80.5	132.4	71.9	73.8	62.6	34.7	52.4	12.0	40.7	—	—
220	211.0	79.6	130.9	71.1	72.9	61.9	34.3	51.8	11.9	40.3	—	—
225	208.6	78.7	129.4	70.3	72.1	61.2	33.9	51.2	11.7	39.8	—	—
230	206.4	77.8	128.0	69.5	71.3	60.5	33.6	50.6	11.6	39.4	—	—
235	204.1	77.0	126.6	68.8	70.6	59.9	33.2	50.1	11.5	39.0	—	—
240	202.0	76.2	125.3	68.0	69.8	59.3	32.9	49.6	11.4	38.6	—	—
245	199.9	75.4	124.0	67.3	69.1	58.6	32.5	49.1	11.2	38.2	—	—
250	197.9	74.6	122.8	66.7	68.4	58.1	32.2	48.6	11.1	37.8	—	—
255	196.0	73.9	121.5	66.0	67.7	57.5	31.9	48.1	11.0	37.4	—	—
260	194.1	73.2	120.4	65.4	67.1	56.9	31.6	47.6	10.9	37.0	—	—
265	192.2	72.5	119.2	64.8	66.5	56.4	31.3	47.2	10.8	36.7	—	—
270	190.5	71.8	118.1	64.2	65.8	55.9	31.0	46.7	10.7	36.3	—	—
275	188.7	71.2	117.0	63.6	65.2	55.4	30.7	46.3	10.6	36.0	—	—
280	187.0	70.5	116.0	63.0	64.6	54.9	30.4	45.9	10.5	35.7	—	—
285	185.4	69.9	115.0	62.4	64.1	54.4	30.2	45.5	—	—	—	—
290	183.8	69.3	114.0	61.9	63.5	53.9	29.9	45.1	—	—	—	—
295	182.2	68.7	113.0	61.4	63.0	53.4	29.6	44.7	—	—	—	—
300	180.7	68.1	112.1	60.9	62.5	53.0	29.4	44.3	—	—	—	—
305	179.2	67.6	111.1	60.4	61.9	52.6	29.1	44.0	—	—	—	—
310	177.7	67.0	110.2	59.9	61.4	52.1	28.9	43.6	—	—	—	—
315	176.3	66.5	109.4	59.4	60.9	51.7	28.7	43.3	—	—	—	—

Table 2 (Continued)

	Diameter											
	100 mm		150 mm		225 mm		300 mm		375 mm		450 mm	
Length/fall	Velocity (m/min)	Flow (l/s)	Velocity (m/min)	Flow (l/s)	Velocity (m/min)	Flow (l/s)	Velocity (m/min)	Flow (l/s)	Velocity (m/min)	Flow (l/s)	Velocity (m/min)	Flow (l/s)
320	—	—	—	—	42.9	28.5	51.3	60.5	58.9	108.5	66.0	114.9
325	—	—	—	—	42.6	28.2	50.9	60.0	58.5	107.7	65.5	173.6
330	—	—	—	—	42.3	28.0	50.5	59.5	58.0	106.8	65.0	172.3
335	—	—	—	—	42.0	27.8	50.2	59.1	57.6	106.0	64.5	171.0
340	—	—	—	—	41.6	27.6	49.8	58.7	57.2	105.3	64.0	169.7
345	—	—	—	—	41.3	27.4	49.4	58.2	56.8	104.5	63.5	168.5
350	—	—	—	—	41.1	27.2	49.1	57.8	56.3	103.8	63.1	167.3
355	—	—	—	—	40.8	27.0	48.7	57.4	55.9	103.0	62.6	166.1
360	—	—	—	—	40.5	26.8	48.4	57.0	55.5	102.3	62.2	164.9
365	—	—	—	—	40.2	26.6	48.0	56.6	55.2	101.6	61.8	163.8
370	—	—	—	—	39.9	26.5	47.7	56.2	54.8	100.9	61.4	162.7
375	—	—	—	—	39.7	26.3	47.4	55.9	54.4	100.2	61.0	161.6
380	—	—	—	—	39.4	26.1	47.1	55.5	54.1	99.6	60.5	160.5
385	—	—	—	—	39.1	25.9	46.8	55.1	53.7	98.9	60.2	159.5
390	—	—	—	—	38.9	25.8	46.5	54.8	53.4	98.3	59.8	158.5
395	—	—	—	—	38.6	25.6	46.2	54.4	53.0	97.7	59.4	157.5
400	—	—	—	—	38.4	25.5	45.9	54.1	52.7	97.0	59.0	156.5
405	—	—	—	—	38.2	25.3	45.6	53.8	52.4	96.4	58.6	155.5
410	—	—	—	—	37.9	25.1	45.3	53.4	52.1	95.9	58.3	154.6
415	—	—	—	—	37.7	25.0	45.1	53.1	51.7	95.3	57.9	153.6
420	—	—	—	—	37.5	24.8	44.8	52.8	51.4	94.7	57.6	152.7
425	—	—	—	—	37.3	24.7	44.5	52.5	51.1	94.2	57.3	151.8
430	—	—	—	—	37.0	24.5	44.3	52.2	50.8	93.6	56.9	150.9
435	—	—	—	—	36.8	24.4	44.0	51.9	50.5	93.1	56.6	150.1
440	—	—	—	—	36.6	24.3	43.8	51.6	50.3	92.5	56.3	149.2

445	148.4	56.0	92.0	50.0	51.3	43.5	24.1	36.4	—	—	—	—
450	147.5	55.6	91.5	49.7	51.0	43.3	24.0	36.2	—	—	—	—
455	146.7	55.3	91.0	49.4	50.7	43.0	23.9	36.0	—	—	—	—
460	145.9	55.0	90.5	49.1	50.4	42.8	23.7	35.8	—	—	—	—
465	145.1	54.7	90.0	48.9	50.2	42.6	23.6	35.6	—	—	—	—
470	144.4	54.4	89.5	48.6	49.9	42.3	—	—	—	—	—	—
475	143.6	54.2	89.1	48.4	49.6	42.1	—	—	—	—	—	—
480	142.8	53.9	88.6	48.1	49.4	41.9	—	—	—	—	—	—
485	142.1	53.6	88.1	47.9	49.1	41.7	—	—	—	—	—	—
490	141.4	53.3	87.7	47.6	48.9	41.5	—	—	—	—	—	—
495	140.7	53.1	87.2	47.4	48.6	41.3	—	—	—	—	—	—
500	140.0	52.8	86.8	47.1	48.4	41.1	—	—	—	—	—	—
510	138.6	52.3	85.9	46.7	47.9	40.6	—	—	—	—	—	—
520	137.2	51.8	85.1	46.2	47.4	40.3	—	—	—	—	—	—
530	135.9	51.3	84.3	45.8	47.0	39.9	—	—	—	—	—	—
540	134.7	50.8	83.5	45.4	46.6	39.5	—	—	—	—	—	—
550	133.4	50.3	82.8	44.9	46.1	39.1	—	—	—	—	—	—
560	132.2	49.9	82.0	44.5	45.7	38.8	—	—	—	—	—	—
570	131.1	49.4	81.3	44.2	45.4	38.4	—	—	—	—	—	—
580	129.9	49.0	80.6	43.8	44.9	38.1	—	—	—	—	—	—
590	128.8	48.6	79.9	43.4	44.5	37.8	—	—	—	—	—	—
600	127.8	48.2	79.2	43.0	44.2	37.5	—	—	—	—	—	—
610	126.7	47.8	78.6	42.7	43.8	37.2	—	—	—	—	—	—
620	125.7	47.4	78.0	42.3	43.4	36.9	—	—	—	—	—	—
630	124.7	47.0	77.3	42.0	43.1	36.6	—	—	—	—	—	—
640	123.7	46.7	76.7	41.7	42.8	36.3	—	—	—	—	—	—
650	122.8	46.3	76.1	41.3	42.4	36.0	—	—	—	—	—	—
660	121.8	45.9	75.6	41.0	42.1	35.7	—	—	—	—	—	—
670	120.9	45.6	75.0	40.7	—	—	—	—	—	—	—	—
680	120.0	45.3	74.4	40.4	—	—	—	—	—	—	—	—
690	119.1	44.9	73.9	40.1	—	—	—	—	—	—	—	—
700	118.3	44.6	73.4	39.8	—	—	—	—	—	—	—	—
710	117.5	44.3	72.8	39.6	—	—	—	—	—	—	—	—
720	116.6	44.0	72.3	39.3	—	—	—	—	—	—	—	—

Table 2 (*Continued*)

| | | Diameter | | | | | | | | | |
| Length/fall | 100 mm | | 150 mm | | 225 mm | | 300 mm | | 375 mm | | 450 mm | |
	Velocity (m/min)	Flow (l/s)	Velocity (m/min)	Flow (l/s)	Velocity (m/min)	Flow (l/s)	Velocity (m/min)	Flow (l/s)	Velocity (m/min)	Flow (l/s)	Velocity (m/min)	Flow (l/s)
730	—	—	—	—	—	—	—	—	39.0	71.8	43.7	115.8
740	—	—	—	—	—	—	—	—	38.8	71.4	43.4	115.0
750	—	—	—	—	—	—	—	—	38.5	70.9	43.1	114.3
760	—	—	—	—	—	—	—	—	38.2	70.4	42.8	113.5
770	—	—	—	—	—	—	—	—	38.0	69.9	42.5	112.8
780	—	—	—	—	—	—	—	—	37.7	69.5	42.3	112.1
790	—	—	—	—	—	—	—	—	37.5	69.1	42.0	111.3
800	—	—	—	—	—	—	—	—	37.3	68.6	41.7	110.6
810	—	—	—	—	—	—	—	—	37.0	68.2	41.5	110.0
820	—	—	—	—	—	—	—	—	36.8	67.8	41.2	109.3
830	—	—	—	—	—	—	—	—	36.6	67.4	41.0	108.6
840	—	—	—	—	—	—	—	—	36.4	67.0	40.7	108.0
850	—	—	—	—	—	—	—	—	36.2	66.6	40.5	107.3
860	—	—	—	—	—	—	—	—	35.9	66.2	40.2	106.7
870	—	—	—	—	—	—	—	—	35.7	65.8	40.0	106.1
880	—	—	—	—	—	—	—	—	35.5	65.4	39.8	105.5
890	—	—	—	—	—	—	—	—	—	—	39.6	104.9
900	—	—	—	—	—	—	—	—	—	—	39.3	104.3

(b) Diameter 525–900 mm

	Diameter											
	525 mm		600 mm		675 mm		750 mm		825 mm		900 mm	
Length/fall	Velocity (m/min)	Flow (l/s)	Velocity (m/min)	Flow (l/s)	Velocity (m/min)	Flow (l/s)	Velocity (m/min)	Flow (l/s)	Velocity (m/min)	Flow (l/s)	Velocity (m/min)	Flow (l/s)
32	229.6	828.5	249.4	1175.6	268.3	1600.5	286.4	2109.4	—	—	—	—
33	226.1	815.9	245.6	1157.6	264.2	1576.1	282.0	2077.2	—	—	—	—
34	222.7	803.8	241.9	1140.5	260.3	1552.7	277.8	2046.4	294.8	2626.9	—	—
35	219.5	792.2	238.5	1124.0	256.5	1530.4	273.8	2016.9	290.5	2589.1	—	—
36	216.4	781.1	235.1	1108.3	252.9	1509.0	270.0	1988.7	286.4	2552.8	—	—
37	213.5	770.5	231.9	1093.2	249.5	1488.5	266.3	1961.7	282.6	2518.1	—	—
38	210.7	760.3	228.9	1078.8	246.2	1468.7	262.8	1935.7	278.8	2484.8	294.3	3121.0
39	208.0	750.5	225.9	1064.8	243.0	1449.8	259.4	1910.7	275.2	2452.7	290.5	3080.7
40	205.3	741.1	223.1	1051.5	240.0	1431.6	256.2	1886.7	271.8	2421.8	286.8	3041.9
41	202.8	732.0	220.3	1038.5	237.0	1414.0	253.0	1863.5	268.4	2392.1	283.3	3004.6
42	200.4	723.2	217.7	1026.1	234.2	1397.1	250.0	1841.2	265.2	2363.5	279.9	2968.6
43	198.0	714.7	215.1	1014.1	231.4	1380.7	247.1	1819.7	262.1	2335.8	276.6	2933.9
44	195.8	706.6	212.7	1002.5	228.8	1364.9	244.2	1798.9	259.1	2309.1	273.5	2900.4
45	193.6	698.7	210.3	991.3	226.2	1349.7	241.5	1778.8	256.2	2283.3	270.4	2868.0
46	191.5	691.0	208.0	980.5	223.8	1334.9	238.9	1759.3	253.4	2258.4	267.5	2836.6
47	189.4	683.7	205.8	970.0	221.4	1320.7	236.3	1740.5	250.7	2234.2	264.6	2806.3
48	187.4	676.5	203.6	959.8	219.1	1306.8	233.8	1722.3	248.1	2210.8	261.8	2776.9
49	185.5	669.6	201.5	950.0	216.8	1293.4	231.4	1704.6	245.5	2188.2	259.1	2748.4
50	183.7	662.8	199.5	940.4	214.6	1280.4	229.1	1687.5	243.1	2166.2	256.5	2720.8
51	181.8	656.3	197.5	931.2	212.5	1267.8	226.9	1670.9	240.7	2144.8	254.0	2694.0
52	180.1	650.0	195.6	922.2	210.5	1255.6	224.7	1654.7	238.3	2124.1	251.6	2668.0
53	178.4	643.8	193.8	913.4	208.5	1243.7	222.5	1639.0	236.1	2104.0	249.2	2642.7
54	176.7	637.8	192.0	904.9	206.5	1232.1	220.5	1623.8	233.9	2084.4	246.8	2618.1
55	175.1	632.0	190.2	896.7	204.6	1220.8	218.5	1609.0	231.7	2065.4	244.6	2594.2
56	173.5	626.3	188.5	888.6	202.8	1209.9	216.5	1594.5	229.7	2046.8	242.4	2570.9

Table 2 (*Continued*)

Length/fall	525 mm		600 mm		675 mm		750 mm		825 mm		900 mm	
	Velocity (m/min)	Flow (l/s)	Velocity (m/min)	Flow (l/s)	Velocity (m/min)	Flow (l/s)	Velocity (m/min)	Flow (l/s)	Velocity (m/min)	Flow (l/s)	Velocity (m/min)	Flow (l/s)
57	172.0	620.8	186.9	880.8	201.0	1199.2	214.6	1580.5	227.6	2028.8	240.3	2548.3
58	170.5	615.4	185.2	873.2	199.3	1188.8	212.7	1566.8	225.7	2011.2	238.2	2526.2
59	169.1	610.2	183.7	865.8	197.6	1178.7	210.9	1553.5	223.8	1994.1	236.2	2504.7
60	167.7	605.1	182.1	858.5	195.9	1168.9	209.2	1540.5	221.9	1977.4	234.2	2483.7
61	166.3	600.1	180.6	851.4	194.3	1159.2	207.4	1527.8	220.1	1961.1	232.3	2463.3
62	164.9	595.2	179.2	844.5	192.7	1149.9	205.8	1515.4	218.3	1945.3	230.4	2443.3
63	163.6	590.5	177.7	837.8	191.2	1140.7	204.1	1503.3	216.5	1929.8	228.5	2423.9
64	162.3	585.9	176.3	831.2	189.7	1131.7	202.5	1491.5	214.8	1914.6	226.7	2404.9
65	161.1	581.3	175.0	824.8	188.2	1123.0	200.9	1480.0	213.2	1899.8	225.0	2386.3
66	159.9	576.9	173.7	818.6	186.8	1114.5	199.4	1468.8	211.6	1885.4	223.3	2368.1
67	158.7	572.6	172.4	812.4	185.4	1106.1	197.9	1457.8	210.0	1871.3	221.6	2350.4
68	157.5	568.4	171.1	806.4	184.0	1098.0	196.5	1447.0	208.4	1857.5	220.0	2333.1
69	156.3	564.2	169.8	800.6	182.7	1090.0	195.0	1436.5	206.9	1844.0	218.4	2316.1
70	155.2	560.2	168.6	794.8	181.4	1082.2	193.6	1426.2	205.4	1830.7	216.8	2299.5
71	154.1	556.2	167.4	789.2	180.1	1074.5	192.3	1416.1	204.0	1817.8	215.3	2283.2
72	153.0	552.4	166.3	783.7	178.9	1067.0	190.9	1406.2	202.6	1805.1	213.8	2267.3
73	152.0	548.6	165.1	778.3	177.6	1059.7	189.6	1396.6	201.2	1792.7	212.3	2251.7
74	151.0	544.8	164.0	773.0	176.4	1052.5	188.3	1387.1	199.8	1780.6	210.9	2236.5
75	150.0	541.2	162.9	767.9	175.2	1045.5	187.1	1377.8	193.5	1768.7	209.5	2221.5
76	149.0	537.6	161.8	762.8	174.1	1038.6	185.8	1368.7	197.1	1757.0	208.1	2206.9
77	148.0	534.1	160.8	757.8	173.0	1031.8	184.6	1359.8	195.9	1745.5	206.7	2192.5
78	147.0	530.7	159.7	753.0	171.8	1025.2	183.4	1351.1	194.6	1734.3	205.4	2178.4
79	146.1	527.3	158.7	748.2	170.7	1018.7	182.3	1342.5	193.4	1723.3	204.1	2164.5
80	145.2	524.0	157.7	743.5	169.7	1012.3	181.1	1334.1	192.2	1712.5	202.8	2151.0
81	144.3	520.8	156.8	738.9	168.6	1006.0	180.0	1325.8	191.0	1701.9	201.6	2137.7

Diameter

82	2124.6	200.3	1691.5	189.8	1317.7	178.9	999.8	167.6	734.4	155.8	517.6	143.4
83	2111.7	199.1	1681.3	188.7	1309.7	177.8	993.8	166.6	729.9	154.9	514.5	142.5
84	2099.1	197.9	1671.2	187.5	1301.9	176.8	987.9	165.6	725.6	153.9	511.4	141.7
85	2086.8	196.8	1661.4	186.4	1294.2	175.7	982.0	164.6	721.3	153.0	508.4	140.9
86	2074.6	195.6	1651.7	185.3	1286.7	174.7	976.3	163.7	717.1	152.1	505.4	140.0
87	2062.6	194.5	1642.2	184.3	1279.3	173.7	970.7	162.7	713.0	151.2	502.5	139.2
88	2050.9	193.4	1632.8	183.2	1272.0	172.7	965.2	161.8	708.9	150.4	499.6	138.4
89	2039.3	192.3	1623.6	182.2	1264.8	171.7	959.7	160.9	704.9	149.5	496.8	137.7
90	2028.0	191.2	1614.6	181.2	1257.8	170.8	954.4	160.0	701.0	148.7	494.0	136.9
91	2016.8	190.2	1605.7	180.2	1250.9	169.8	949.1	159.1	697.1	147.9	491.3	136.1
92	2005.8	189.1	1596.9	179.2	1244.0	168.9	943.9	158.2	693.3	147.1	488.6	135.4
93	1995.0	188.1	1588.3	178.2	1237.3	168.0	938.9	157.4	689.6	146.3	486.0	134.7
94	1984.3	187.1	1579.8	177.3	1230.7	167.1	933.8	156.5	685.9	145.5	483.4	133.9
95	1973.9	186.1	1571.5	176.3	1224.2	166.2	928.9	155.7	682.3	144.7	480.9	133.2
96	1963.6	185.1	1563.3	175.4	1217.8	165.3	924.1	154.9	678.7	144.0	478.4	132.5
97	1953.4	184.2	1555.2	174.5	1211.5	164.5	919.3	154.1	675.2	143.2	475.9	131.9
98	1943.4	183.2	1547.3	173.6	1205.3	163.7	914.6	153.3	671.7	142.5	473.4	131.2
99	1933.6	182.3	1539.4	172.7	1199.2	162.8	910.0	152.5	668.3	141.8	471.1	130.5
100	1923.9	181.4	1531.7	171.9	1193.2	162.0	905.4	151.8	665.0	141.1	468.7	129.9
105	1877.5	177.0	1494.8	167.7	1164.5	158.1	883.6	148.1	649.0	137.7	457.4	126.7
110	1834.4	173.0	1460.4	163.9	1137.7	154.5	863.3	144.7	634.0	134.5	446.9	123.8
115	1794.0	169.2	1428.3	160.3	1112.7	151.1	844.3	141.5	620.1	131.6	437.1	121.1
120	1756.3	165.6	1398.2	156.9	1089.3	147.9	826.5	138.5	607.1	128.8	427.9	118.6
125	1720.8	162.2	1370.0	153.7	1067.3	144.8	809.8	135.7	594.8	126.2	419.2	116.2
130	1687.4	159.1	1343.4	150.7	1046.5	142.1	794.1	133.1	583.2	123.7	411.1	113.9
135	1655.8	156.1	1318.3	147.9	1027.0	139.4	779.2	130.6	572.0	121.4	403.4	111.8
140	1626.0	153.5	1294.5	145.3	1008.5	136.9	765.2	128.3	562.0	119.2	396.1	109.8
145	1597.7	150.6	1272.0	142.7	990.9	134.5	751.9	126.0	552.2	117.2	389.2	107.8
150	1570.9	148.1	1250.6	140.3	974.3	132.3	739.3	123.9	543.0	115.2	382.7	106.0
155	1545.3	145.7	1230.3	138.0	958.4	130.1	727.2	121.9	534.1	113.3	376.5	104.3
160	1521.0	143.4	1210.9	135.9	943.3	128.1	715.8	120.0	525.7	111.5	370.5	102.7
165	1497.7	141.2	1192.4	133.8	928.9	126.1	704.9	118.1	517.7	109.8	364.9	101.1
170	1475.6	139.1	1174.8	131.8	915.2	124.3	694.4	116.4	510.0	108.2	359.5	99.6
175	1454.3	137.1	1157.9	129.9	902.0	122.5	684.4	114.7	502.7	106.6	354.3	98.2

Table 2 (*Continued*)

	Diameter											
	525 mm		600 mm		675 mm		750 mm		825 mm		900 mm	
Length/fall	Velocity (m/min)	Flow (l/s)	Velocity (m/min)	Flow (l/s)	Velocity (m/min)	Flow (l/s)	Velocity (m/min)	Flow (l/s)	Velocity (m/min)	Flow (l/s)	Velocity (m/min)	Flow (l/s)
180	96.8	349.3	105.2	495.7	113.1	674.8	120.8	889.4	128.1	1141.7	135.2	1434.0
185	95.5	344.6	103.7	488.9	111.6	665.7	119.1	877.3	126.4	1126.1	133.4	1414.5
190	94.2	340.0	102.3	482.4	110.1	656.8	117.5	865.7	124.7	1111.2	131.6	1395.7
195	93.0	335.6	101.0	476.2	108.7	648.4	116.0	854.5	123.1	1096.9	129.9	1377.7
200	91.8	331.4	99.8	470.2	107.3	640.2	114.6	843.7	121.5	1083.1	128.3	1360.4
205	90.7	327.3	98.5	464.5	106.0	632.4	113.2	833.4	120.0	1069.8	126.7	1343.7
210	89.6	323.4	97.4	458.9	104.7	624.8	111.8	823.4	118.6	1057.0	125.2	1327.6
215	88.6	319.6	96.2	453.5	103.5	617.5	110.5	813.8	117.2	1044.6	123.7	1312.1
220	87.6	316.0	95.1	448.3	102.3	610.4	109.2	804.5	115.9	1032.7	122.3	1297.1
225	86.6	312.5	94.1	443.3	101.2	603.6	108.0	795.5	114.6	1021.1	120.9	1282.6
230	85.6	309.0	93.0	438.5	100.1	597.0	106.8	786.8	113 3	1010.0	119.6	1268.6
235	84.7	305.7	92.0	433.8	99.0	590.6	105.7	778.4	112 1	999.2	118.3	1255.0
240	83.8	302.5	91.1	429.3	98.0	584.4	104.6	770.2	110.9	988.7	117.1	1241.9
245	83.0	299.4	90.1	424.9	97.0	578.4	103.5	762.3	109.8	978.6	115.9	1229.1
250	82.1	296.4	89.2	420.6	96.0	572.6	102.5	754.7	108.7	968.7	114.7	1216.8
255	81.3	293.5	88.3	416.4	95.0	567.0	101.5	747.2	107.6	959.2	113.6	1204.8
260	80.5	290.7	87.5	412.4	94.1	561.5	100.5	740.0	106.6	949.9	112.5	1193.1
265	79.8	287.9	86.7	408.5	93.2	556.2	99.5	733.0	105.6	940.9	111.4	1181.8
270	79.0	285.2	85.9	404.7	92.4	551.0	98.6	726.2	104.6	932.2	110.4	1170.8
275	78.3	282.6	85.1	401.0	91.5	546.0	97.7	719.5	103.6	923.7	109.4	1160.2
280	77.6	280.1	84.3	397.4	90.7	541.1	96.8	713.1	102.7	915.4	108.4	1149.7
285	76.9	277.6	83.6	393.9	89.9	536.3	96.0	706.8	101.8	907.3	107.4	1139.6
290	76.3	275.2	82.8	390.5	89.1	531.7	95.1	700.7	100.9	899.4	106.5	1129.7
295	75.6	272.9	82.1	387.2	88.4	527.1	94.3	694.7	100.1	891.8	105.6	1120.1
300	75.0	270.6	81.5	383.9	87.6	522.7	93.5	688.9	99.2	884.3	104.7	1110.8

305	74.4	268.4	80.8	380.8	86.9	518.4	92.8	683.2	98.4	877.1	103.9	1101.6
310	73.8	266.2	80.1	377.7	86.2	514.2	92.0	677.7	97.6	870.0	103.0	1092.7
315	73.2	264.1	79.5	374.7	85.5	510.1	91.3	672.3	96.8	863.0	102.2	1084.0
320	72.6	262.0	78.9	371.7	84.8	506.1	90.6	667.0	96.1	856.3	101.4	1075.5
325	72.0	260.0	78.3	368.9	84.2	502.2	89.9	661.9	95.3	849.6	100.6	1067.2
330	71.5	258.0	77.7	366.1	83.5	498.4	89.2	656.9	94.6	843.2	99.9	1059.1
335	71.0	256.1	77.1	363.3	82.9	494.7	88.5	651.9	93.9	836.9	99.1	1051.1
340	70.4	254.2	76.5	360.6	82.3	491.0	87.9	647.1	93.2	830.7	98.4	1043.4
345	69.9	252.3	76.0	358.0	81.7	487.5	87.2	642.4	92.5	824.6	97.7	1035.8
350	69.4	250.5	75.4	355.5	81.1	484.0	86.6	637.8	91.9	818.7	97.0	1028.4
355	68.9	248.8	74.9	352.9	80.5	480.5	86.0	633.3	91.2	812.9	96.3	1021.1
360	68.4	247.0	74.4	350.5	80.0	477.2	85.4	628.9	90.6	807.3	95.6	1014.0
365	68.0	245.3	73.8	348.1	79.4	473.9	84.8	624.6	90.0	801.7	94.9	1007.0
370	67.5	243.7	73.3	345.7	78.9	470.7	84.2	620.3	89.4	796.3	94.3	1000.2
375	67.1	242.0	72.9	343.4	78.4	467.5	83.7	616.2	88.8	791.0	93.7	993.5
380	66.6	240.4	72.4	341.1	77.9	464.5	83.1	612.1	88.2	785.7	93.1	986.9
385	66.2	238.9	71.9	338.9	77.3	461.4	82.6	608.1	87.6	780.6	92.4	980.5
390	65.8	237.3	71.4	336.7	76.8	458.5	82.0	604.2	87.0	775.6	91.9	974.2
395	65.3	235.8	71.0	334.6	76.4	455.6	81.5	600.4	86.5	770.7	91.3	968.0
400	64.9	234.3	70.5	332.5	75.9	452.7	81.0	596.6	85.9	765.9	90.7	961.9
405	64.5	232.9	70.1	330.4	75.4	449.9	80.5	592.9	85.4	761.1	90.1	956.0
410	64.1	231.5	69.7	328.4	75.0	447.1	80.0	589.3	84.9	756.5	89.6	950.1
415	63.7	230.1	69.3	326.4	74.5	444.4	79.5	585.7	84.4	751.9	89.0	944.4
420	63.4	228.6	68.8	324.5	74.1	441.8	79.1	582.2	83.9	747.4	88.5	938.8
425	63.0	227.3	68.4	322.6	73.6	439.2	78.6	578.8	83.4	743.0	88.0	933.2
430	62.6	226.0	68.0	320.7	73.2	436.6	78.1	575.4	82.9	738.7	87.5	927.8
435	62.3	224.7	67.6	318.8	72.8	434.1	77.7	572.1	82.4	734.4	87.0	922.4
440	61.9	223.4	67.3	317.0	72.4	431.6	77.2	568.9	81.9	730.2	86.5	917.2
445	61.6	222.2	66.9	315.2	71.9	429.2	76.8	565.6	81.5	726.1	86.0	912.0
450	61.2	220.9	66.5	313.5	71.5	426.8	76.4	562.5	81.0	722.1	85.5	906.9
455	60.9	219.7	66.1	311.8	71.1	424.5	76.0	559.4	80.6	718.1	85.0	901.9
460	60.6	218.5	65.8	310.1	70.8	422.1	75.5	556.3	80.1	714.2	84.6	897.0
465	60.2	217.3	65.4	308.4	70.4	419.9	75.1	553.3	79.7	710.3	84.1	892.2
470	59.9	216.2	65.1	306.7	70.0	417.6	74.7	550.4	79.3	706.5	83.7	887.4

Table 2 (Continued)

	Diameter											
	525 mm		600 mm		675 mm		750 mm		825 mm		900 mm	
Length/fall	Velocity (m/min)	Flow (l/s)	Velocity (m/min)	Flow (l/s)	Velocity (m/min)	Flow (l/s)	Velocity (m/min)	Flow (l/s)	Velocity (m/min)	Flow (l/s)	Velocity (m/min)	Flow (l/s)
475	59.6	215.0	64.7	305.1	69.6	415.4	74.3	547.5	78.9	702.8	83.2	882.7
480	59.3	213.9	64.4	303.5	69.3	413.3	73.9	544.6	78.4	699.1	82.8	878.1
485	59.0	212.8	64.1	302.0	68.9	411.1	73.6	541.8	78.0	695.5	82.4	873.6
490	58.7	211.7	63.7	300.4	68.6	409.0	73.2	539.0	77.5	692.0	81.9	869.1
495	58.4	210.7	63.4	298.9	68.2	406.9	72.8	536.3	77.2	688.5	81.5	864.7
500	58.1	209.6	63.1	297.4	67.9	404.9	72.5	533.6	76.9	685.0	81.1	860.4
510	57.5	207.5	62.5	294.5	67.2	400.9	71.7	528.4	76.1	678.3	80.3	851.9
520	57.0	205.5	61.9	291.6	66.6	397.0	71.0	523.3	75.4	671.7	79.5	843.7
530	56.4	203.6	61.3	288.9	65.9	393.3	70.4	518.3	74.7	665.3	78.8	835.7
540	55.9	201.7	60.7	286.2	65.3	389.6	69.7	513.5	74.0	659.1	78.1	827.9
550	55.4	199.8	60.2	283.6	64.7	386.1	69.1	508.8	73.3	653.1	77.3	820.4
560	54.9	198.1	59.6	281.0	64.1	382.6	68.5	504.2	72.6	647.3	76.7	813.0
570	54.4	196.3	59.1	278.5	63.6	379.2	67.9	499.8	72.0	641.6	76.0	805.8
580	53.9	194.6	58.6	276.1	63.0	375.9	67.3	495.5	71.4	636.0	75.3	798.9
590	53.5	193.0	58.1	273.8	62.5	372.7	66.7	491.2	70.8	630.6	74.7	792.1
600	53.0	191.3	57.6	271.5	62.0	369.6	66.1	487.1	70.2	625.3	74.1	785.4
610	52.6	189.8	57.1	269.2	61.4	366.6	65.5	483.1	69.6	620.2	73.4	779.0
620	52.2	188.2	56.7	267.1	60.9	363.6	65.1	479.2	69.0	615.1	72.9	772.7
630	51.7	186.7	56.2	264.9	60.5	360.7	64.5	475.4	68.5	610.2	72.3	766.5
640	51.3	185.3	55.8	262.9	60.0	357.9	64.0	471.7	67.9	605.5	71.7	760.5
650	50.9	183.8	55.3	260.8	59.5	355.1	63.5	468.0	67.4	600.8	71.1	754.6
660	50.6	182.4	54.9	258.8	59.1	352.4	63.1	464.5	66.9	596.2	70.6	748.9
670	50.2	181.1	54.5	256.9	58.6	349.8	62.6	461.0	66.4	591.8	70.1	743.3
680	49.8	179.7	54.1	255.0	58.2	347.2	62.1	457.6	65.9	587.4	69.6	737.8
690	49.4	178.4	53.7	253.2	57.8	344.7	61.7	454.3	65.4	583.1	69.1	732.4

700	49.1	177.1	53.3	251.3	57.4	342.2	61.2	451.0	65.0	578.9	68.6	727.2
710	48.7	175.9	52.9	249.6	57.0	339.8	60.8	447.8	64.5	574.8	68.1	722.0
720	48.4	174.7	52.6	247.8	56.6	337.4	60.4	444.7	64.1	570.8	67.6	717.0
730	48.1	173.5	52.2	246.1	56.2	335.1	60.0	441.6	63.6	566.9	67.1	712.1
740	47.7	172.3	51.9	244.5	55.8	332.8	59.6	438.6	63.2	563.1	66.7	707.2
750	47.4	171.1	51.5	242.8	55.4	330.6	59.2	435.7	62.8	559.3	66.2	702.5
760	47.1	170.0	51.2	241.2	55.1	328.4	58.8	432.8	62.3	555.6	65.8	697.9
770	46.8	168.9	50.8	239.6	54.7	326.3	58.4	430.0	61.9	552.0	65.4	693.3
780	46.5	167.8	50.5	238.1	54.3	324.2	58.0	427.2	61.5	548.4	65.0	688.9
790	46.2	166.8	50.2	236.6	54.0	322.1	57.6	424.5	61.1	545.0	64.5	684.5
800	45.9	165.7	49.9	235.1	53.7	320.1	57.3	421.9	60.8	541.5	64.1	680.2
810	45.6	164.7	49.6	233.7	53.3	318.1	56.9	419.3	60.4	538.2	63.7	676.0
820	45.4	163.7	49.3	232.2	53.0	316.2	56.6	416.7	60.0	534.9	63.3	671.9
830	45.1	162.7	49.0	230.8	52.7	314.3	56.2	414.2	59.7	531.7	63.0	667.8
840	44.8	161.7	48.7	229.4	52.4	312.4	55.9	411.7	59.3	528.5	62.6	663.8
850	44.5	160.8	48.4	228.1	52.1	310.5	55.6	409.3	59.0	525.4	62.2	659.9
860	44.3	159.8	48.1	226.8	51.8	308.7	55.2	406.9	58.6	522.3	61.9	656.0
870	44.0	158.9	47.8	225.5	51.5	307.0	54.9	404.5	58.3	519.3	61.5	652.3
880	43.8	158.0	47.6	224.2	51.2	305.2	54.6	402.2	57.9	516.3	61.1	648.5
890	43.5	157.1	47.3	222.9	50.9	303.5	54.3	400.0	57.6	513.4	60.8	644.9
900	43.3	156.2	47.0	221.7	50.6	301.8	54.0	397.7	57.3	510.6	60.5	641.3
910	43.1	155.4	46.8	220.4	50.3	300.1	53.7	395.6	57.0	507.8	60.1	637.8
920	42.8	154.5	46.5	219.2	50.0	298.5	53.4	393.4	56.7	505.0	59.8	634.3
930	42.6	153.7	46.3	218.1	49.8	296.9	53.1	391.3	56.4	502.3	59.5	630.9
940	42.4	152.9	46.0	216.9	49.5	295.3	52.8	389.2	56.1	499.6	59.2	627.5
950	42.1	152.1	45.8	215.8	49.2	293.7	52.6	387.1	55.8	497.0	58.9	624.2
960	41.9	151.3	45.5	214.6	49.0	292.2	52.3	385.1	55.5	494.4	58.5	620.9
970	41.7	150.5	45.3	213.5	48.7	290.7	52.0	383.1	55.2	491.8	58.2	617.7
980	41.5	149.7	45.1	212.4	48.5	289.2	51.8	381.2	54.9	489.3	57.9	614.6
990	41.3	149.0	44.8	211.3	48.2	287.8	51.5	379.2	54.6	486.8	57.7	611.5
1000	41.1	148.2	44.6	210.3	48.0	286.3	51.2	377.3	54.4	484.4	57.4	608.4
1050	40.1	144.6	43.5	205.2	46.8	279.4	50.0	368.2	53.0	472.7	56.0	593.7
1100	39.3	141.3	42.5	200.5	45.8	273.0	48.8	359.8	51.8	461.8	54.7	580.1
1150	38.3	138.2	41.6	196.1	44.8	267.0	47.8	351.9	50.7	451.7	53.5	567.3

Table 2 (Continued)

| Length/fall | Diameter | | | | | | | | | | | |
| | 525 mm | | 600 mm | | 675 mm | | 750 mm | | 825 mm | | 900 mm | |
	Velocity (m/min)	Flow (l/s)	Velocity (m/min)	Flow (l/s)	Velocity (m/min)	Flow (l/s)	Velocity (m/min)	Flow (l/s)	Velocity (m/min)	Flow (l/s)	Velocity (m/min)	Flow (l/s)
1200	37.5	135.3	40.7	192.0	43.8	261.4	46.8	344.5	49.6	442.2	52.4	555.4
1250	36.7	132.6	39.9	188.1	42.9	256.1	45.8	337.5	48.6	433.2	51.3	544.2
1300	36.0	130.0	39.1	184.4	42.1	251.1	44.9	330.9	47.7	424.8	50.3	533.6
1350	—	—	38.4	181.0	41.3	246.4	44.1	324.8	46.8	416.9	49.4	523.6
1400	—	—	37.7	177.7	40.6	242.0	43.3	318.9	45.9	409.4	48.5	514.2
1450	—	—	37.0	174.6	39.9	237.8	42.5	313.4	45.1	402.2	47.6	505.2
1500	—	—	36.4	171.7	39.2	233.8	41.8	308.1	44.4	395.5	46.8	496.7
1550	—	—	35.8	168.9	38.5	230.0	41.2	303.1	43.7	389.1	46.1	488.7
1600	—	—	—	—	37.9	226.3	40.5	298.3	43.0	382.9	45.3	481.0
1650	—	—	—	—	37.4	222.9	39.9	293.8	42.3	377.1	44.7	473.6
1700	—	—	—	—	36.8	219.6	39.3	289.4	41.7	371.5	44.0	466.6
1750	—	—	—	—	36.3	216.4	38.7	285.2	41.1	366.1	43.4	459.9
1800	—	—	—	—	35.8	213.4	38.2	281.2	40.5	361.0	42.8	453.5
1850	—	—	—	—	—	—	37.7	277.4	40.0	356.1	42.2	447.3

(c) Diameter 975–1350 mm

	Diameter											
	975 mm		1050 mm		1125 mm		1200 mm		1275 mm		1350 mm	
Length/fall	Velocity (m/min)	Flow (l/s)	Velocity (m/min)	Flow (l/s)	Velocity (m/min)	Flow (l/s)	Velocity (m/min)	Flow (l/s)	Velocity (m/min)	Flow (l/s)	Velocity (m/min)	Flow (l/s)
59	248.2	3089.1	259.8	3751.1	271.2	4494.2	282.3	5322.2	293.1	6238.4	—	—
60	246.1	3063.2	257.7	3719.7	268.9	4456.6	279.9	5277.7	290.6	6186.2	—	—
61	244.1	3038.0	255.5	3689.1	266.7	4419.9	277.6	5234.3	288.2	6135.3	—	—
62	242.1	3013.4	253.5	3659.2	264.6	4384.2	275.4	5191.9	285.9	6085.6	—	—
63	240.2	2989.4	251.5	3630.0	262.4	4349.2	273.2	5150.5	283.6	6037.1	—	—
64	238.3	2966.0	249.5	3601.6	260.4	4315.1	271.0	5110.1	281.4	5989.8	291.6	6957.5
65	236.4	2943.1	247.6	3573.8	258.4	4281.8	268.9	5070.7	279.2	5943.5	289.3	6903.7
66	234.6	2820.7	245.7	3546.6	256.4	4249.2	266.9	5032.1	277.1	5898.3	287.1	6851.2
67	232.9	2898.8	243.8	3520.0	254.5	4217.4	264.9	4994.4	275.0	5854.1	284.9	6799.9
68	231.2	2877.4	242.0	3494.0	252.6	4186.3	262.9	4957.5	273.0	5810.9	282.8	6749.7
69	229.5	2856.5	240.3	3468.6	250.8	4155.8	261.0	4921.5	271.0	5768.7	280.8	6700.6
70	227.8	2836.0	238.6	3443.8	249.0	4126.0	259.1	4886.2	269.1	5727.3	278.8	6652.6
71	226.2	2816.0	236.9	3419.4	247.2	4096.9	257.3	4851.7	267.2	5686.8	276.8	6605.6
72	224.7	2796.3	235.2	3395.6	245.5	4068.3	255.5	4817.9	265.3	5647.2	274.9	6559.6
73	223.1	2777.1	233.6	3372.3	243.8	4040.4	253.8	4784.8	263.5	5608.4	273.0	6514.5
74	221.6	2758.3	232.0	3349.4	242.2	4013.0	252.0	4752.3	261.7	5570.4	271.1	6470.3
75	220.1	2739.9	230.5	3327.0	240.5	3986.1	250.4	4720.5	259.9	5533.1	269.3	6427.0
76	218.7	2721.8	228.9	3305.0	238.9	3959.8	248.7	4689.4	258.2	5496.6	267.5	6384.6
77	217.2	2704.0	227.5	3283.5	237.4	3934.0	247.1	4658.8	256.5	5460.8	265.8	6343.0
78	215.8	2686.6	226.0	3262.4	235.9	3908.7	245.5	4628.9	254.9	5425.7	264.1	6302.2
79	214.5	2669.6	224.6	3241.7	234.4	3883.9	243.9	4599.5	253.3	5391.2	262.4	6262.2
80	213.1	2652.8	223.1	3221.3	232.9	3859.6	242.4	4570.6	251.7	5357.4	260.8	6222.9
81	211.8	2636.4	221.8	3201.4	231.5	3835.7	240.9	4542.3	250.1	5324.2	259.2	6184.4
82	210.5	2620.3	220.4	3181.8	230.0	3812.2	239.4	4514.5	248.6	5291.7	257.6	6146.6
83	209.2	2604.5	219.1	3162.6	228.7	3789.2	238.0	4487.3	247.1	5259.7	256.0	6109.5

Table 2 (Continued)

Diameter

Length/fall	975 mm Velocity (m/min)	Flow (l/s)	1050 mm Velocity (m/min)	Flow (l/s)	1125 mm Velocity (m/min)	Flow (l/s)	1200 mm Velocity (m/min)	Flow (l/s)	1275 mm Velocity (m/min)	Flow (l/s)	1350 mm Velocity (m/min)	Flow (l/s)
84	208.0	2588.9	217.8	3143.7	227.3	3766.5	236.6	4460.5	245.6	5228.3	254.5	6073.0
85	206.8	2573.6	216.5	3125.2	225.9	3744.3	235.2	4434.2	244.2	5197.4	253.0	6037.2
86	205.6	2558.6	215.2	3106.9	224.6	3722.5	233.8	4408.3	242.8	5167.1	251.5	6001.9
87	204.4	2543.6	214.0	3089.0	223.3	3701.0	232.5	4382.9	241.4	5137.4	250.1	5967.4
88	203.2	2529.4	212.8	3071.4	222.1	3679.9	231.1	4357.9	240.0	5108.1	248.6	5933.4
89	202.1	2515.1	211.6	3054.1	220.3	3659.2	229.8	4333.4	238.6	5079.3	247.2	5899.9
90	200.9	2501.1	210.4	3037.1	219.6	3638.8	228.5	4309.2	237.3	5051.0	245.9	5867.1
91	199.8	2487.3	209.2	3020.4	218.4	3618.8	227.3	4285.5	236.0	5023.2	244.5	5834.7
92	198.7	2473.8	208.1	3003.9	217.2	3599.1	226.0	4262.1	234.7	4995.8	243.2	5802.9
93	197.7	2460.5	207.0	2987.7	216.0	3579.7	224.8	4239.2	233.4	4968.9	241.9	5771.6
94	196.6	2447.3	205.9	2971.8	214.9	3560.6	223.6	4216.5	232.2	4942.4	240.6	5740.9
95	195.6	2434.4	204.8	2956.1	213.7	3541.8	222.4	4194.3	231.0	4916.3	239.3	5710.6
96	194.6	2421.7	203.7	2940.7	212.6	3523.3	221.3	4172.4	229.8	4890.6	238.1	5680.8
97	193.6	2409.2	202.7	2925.5	211.5	3505.1	220.1	4150.8	228.6	4865.3	236.8	5651.4
98	192.6	2396.9	201.6	2910.5	210.4	3487.1	219.0	4129.6	227.4	4840.5	235.6	5622.5
99	191.6	2384.7	200.6	2895.8	209.4	3469.5	217.9	4108.7	226.3	4815.9	234.4	5594.0
100	190.6	2372.8	199.6	2881.3	208.3	3452.1	216.8	4088.1	225.1	4791.8	233.2	5566.0
105	186.0	2315.6	194.8	2811.8	203.3	3368.9	211.6	3989.6	219.7	4676.3	227.6	5431.8
110	181.8	2262.4	190.3	2747.2	198.6	3291.4	206.7	3897.8	214.6	4568.8	222.4	5306.9
115	177.8	2212.6	186.1	2686.8	194.3	3219.1	202.2	3812.2	209.9	4468.4	217.5	5190.3
120	174.0	2166.0	182.2	2630.2	190.2	3151.3	197.9	3731.9	205.5	4374.3	212.9	5081.0
125	170.5	2122.3	178.5	2577.1	186.3	3087.6	193.9	3656.5	201.4	4285.9	208.6	4978.4
130	167.2	2081.1	175.1	2527.1	182.7	3027.7	190.2	3585.5	197.4	4202.7	204.6	4881.7
135	164.1	2042.0	171.8	2479.8	179.3	2971.1	186.6	3518.5	193.8	4124.1	200.7	4790.4
140	161.1	2005.4	168.7	2435.1	176.1	2917.6	183.2	3455.1	190.3	4049.8	197.1	4704.1
145	158.3	1970.5	165.7	2392.8	173.0	2866.8	180.1	3395.0	187.0	3979.4	193.7	4622.3

150	155.6	1937.4	163.0	2352.5	170.1	2818.6	177.0	3337.9	183.8	3912.5	190.4	4544.6
155	153.1	1905.9	160.3	2314.3	167.3	2772.8	174.2	3283.6	180.8	3848.9	187.3	4470.7
160	150.7	1875.8	157.8	2277.8	164.7	2729.1	171.4	3231.9	178.0	3788.3	184.4	4400.3
165	148.4	1847.2	155.4	2243.1	162.2	2687.5	168.8	3182.6	175.3	3730.4	181.6	4333.1
170	146.2	1819.8	153.1	2209.8	159.8	2647.6	166.3	3135.4	172.7	3675.2	178.9	4268.9
175	144.1	1793.7	150.9	2178.0	157.5	2609.5	163.9	3090.3	170.2	3622.3	176.3	4207.5
180	142.1	1768.6	148.8	2147.6	155.3	2573.0	161.6	3047.1	167.8	3571.6	173.8	4148.6
185	140.2	1744.5	146.7	2118.3	153.2	2538.0	159.4	3005.6	165.5	3523.0	171.5	4092.2
190	138.3	1721.4	144.8	2090.3	151.1	2504.4	157.3	2965.8	163.3	3476.3	169.2	4038.0
195	136.5	1699.2	142.9	2063.3	149.2	2472.1	155.3	2927.5	161.2	3431.5	167.0	3985.9
200	134.8	1677.8	141.1	2037.4	147.3	2441.0	153.3	2890.7	159.2	3388.3	164.9	3935.7
205	133.1	1657.2	139.4	2012.4	145.5	2411.0	151.4	2855.3	157.2	3346.7	162.9	3887.4
210	131.5	1637.4	137.7	1988.3	143.7	2382.2	149.6	2821.1	155.3	3306.7	161.0	3840.9
215	130.0	1618.2	136.1	1965.0	142.1	2354.3	147.9	2788.1	153.5	3268.0	159.1	3796.0
220	128.5	1599.7	134.6	1942.5	140.4	2327.4	146.2	2756.2	151.8	3230.6	157.3	3752.6
225	127.1	1581.9	133.1	1920.8	138.9	2301.4	144.5	2725.4	150.1	3194.5	155.5	3710.7
230	125.7	1564.6	131.6	1899.8	137.4	2276.2	143.0	2695.6	148.4	3159.6	153.8	3670.1
235	124.4	1547.8	130.2	1879.5	135.9	2251.9	141.4	2666.8	146.9	3125.8	152.1	3630.8
240	123.0	1531.6	128.8	1859.8	134.5	2228.3	140.0	2638.9	145.3	3093.1	150.6	3592.8
245	121.8	1515.9	127.5	1840.8	133.1	2205.5	138.5	2611.8	143.8	3061.4	149.0	3556.0
250	120.6	1500.7	126.2	1822.3	131.7	2183.3	137.1	2585.5	142.4	3030.6	147.5	3520.2
255	119.4	1485.9	125.0	1804.3	130.4	2161.8	135.8	2560.1	141.0	3000.7	146.1	3485.6
260	118.2	1471.5	123.8	1786.9	129.2	2140.9	134.5	2535.3	139.6	2971.8	144.7	3451.9
265	117.1	1457.6	122.6	1769.9	128.0	2120.6	133.2	2511.3	138.3	2943.6	143.3	3419.2
270	116.0	1444.0	121.5	1753.5	126.8	2100.9	132.0	2487.9	137.0	2916.2	141.9	3387.4
275	115.0	1430.8	120.4	1737.5	125.6	2081.7	130.7	2465.2	135.8	2889.6	140.6	3356.4
280	113.9	1418.0	119.3	1721.9	124.5	2063.0	129.6	2443.1	134.5	2863.7	139.4	3326.3
285	112.9	1405.5	118.2	1706.7	123.4	2044.8	128.4	2421.6	133.4	2838.4	138.2	3297.0
290	111.9	1393.3	117.2	1691.9	122.3	2027.1	127.3	2400.6	132.2	2813.8	137.0	3268.5
295	111.0	1381.5	116.2	1677.5	121.3	2009.9	126.2	2380.2	131.1	2789.9	135.8	3240.6
300	110.1	1369.9	115.2	1663.5	120.3	1993.1	125.2	2360.3	130.0	2766.6	134.7	3213.5
305	109.2	1358.7	114.3	1649.8	119.3	1976.7	124.1	2340.8	128.9	2743.8	133.6	3187.1
310	108.3	1347.6	113.4	1636.4	118.3	1960.7	123.1	2321.9	127.9	2721.6	132.5	3161.3

Table 2 (*Continued*)

| | Diameter | | | | | | | | | | | |
| | 975 mm | | 1050 mm | | 1125 mm | | 1200 mm | | 1275 mm | | 1350 mm | |
Length/fall	Velocity (m/min)	Flow (l/s)	Velocity (m/min)	Flow (l/s)	Velocity (m/min)	Flow (l/s)	Velocity (m/min)	Flow (l/s)	Velocity (m/min)	Flow (l/s)	Velocity (m/min)	Flow (l/s)
315	107.4	1336.9	112.5	1623.4	117.4	1945.0	122.2	2303.4	126.8	2699.9	131.4	3136.1
320	106.6	1326.4	111.6	1610.7	116.4	1929.8	121.2	2285.3	125.8	2678.7	130.4	3111.5
325	105.7	1316.2	110.7	1598.2	115.6	1914.9	120.3	2267.7	124.9	2658.0	129.4	3087.5
330	104.9	1306.2	109.9	1586.1	114.7	1900.3	119.4	2250.4	123.9	2637.8	128.4	3064.0
335	104.1	1296.4	109.0	1574.2	113.8	1886.1	118.5	2233.6	123.0	2618.0	127.4	3041.0
340	103.4	1286.8	108.2	1562.6	113.0	1872.2	117.6	2217.1	122.1	2598.7	126.5	3018.6
345	102.6	1277.5	107.5	1551.2	112.2	1858.5	116.7	2201.0	121.2	2579.8	125.6	2996.6
350	101.9	1268.3	106.7	1540.1	111.3	1845.2	115.9	2185.2	120.3	2561.3	124.7	2975.1
355	101.2	1259.3	105.9	1529.2	110.6	1832.2	115.1	2169.7	119.5	2543.2	123.8	2954.1
360	100.5	1250.6	105.2	1518.6	109.8	1819.4	114.3	2154.6	118.6	2525.5	122.9	2933.5
365	99.8	1242.0	104.5	1508.1	109.0	1806.9	113.5	2139.8	117.8	2508.2	122.1	2913.4
370	99.1	1233.5	103.8	1497.9	108.3	1794.7	112.7	2125.3	117.0	2491.1	121.3	2893.6
375	98.4	1225.3	103.1	1487.9	107.6	1782.7	112.0	2111.1	116.3	2474.5	120.4	2874.3
380	97.8	1217.2	102.4	1478.1	106.9	1770.9	111.2	2097.2	115.5	2458.1	119.7	2855.3
385	97.2	1209.3	101.7	1468.4	106.2	1759.3	110.5	2083.5	114.7	2442.1	118.9	2836.7
390	96.5	1201.5	101.1	1459.0	105.5	1748.0	109.8	2070.1	114.0	2426.4	118.1	2818.4
395	95.9	1193.9	100.4	1449.7	104.8	1736.9	109.1	2056.9	113.3	2411.0	117.4	2800.5
400	95.3	1186.4	99.8	1440.6	104.2	1726.0	108.4	2044.0	112.6	2395.9	116.6	2783.0
405	94.7	1179.0	99.2	1431.7	103.5	1715.4	107.7	2031.4	111.9	2381.1	115.9	2765.8
410	94.1	1171.8	98.6	1423.0	102.9	1704.9	107.1	2019.0	111.2	2366.5	115.2	2748.8
415	93.6	1164.8	98.0	1414.4	102.3	1694.6	106.4	2006.8	110.5	2352.2	114.5	2732.2
420	93.0	1157.8	97.4	1405.9	101.6	1684.4	105.8	1994.8	109.8	2338.2	113.8	2715.9
425	92.5	1151.0	96.8	1397.6	101.0	1674.5	105.2	1983.0	109.2	2324.4	113.1	2699.9
430	91.9	1144.3	96.3	1389.5	100.5	1664.7	104.6	1971.5	108.6	2310.8	112.5	2684.2
435	91.4	1137.7	95.7	1381.5	99.9	1655.2	104.0	1960.1	107.9	2297.5	111.8	2668.7

440	1131.2	90.9	1373.6	95.2	1645.7	99.3	1948.9	103.4	2284.4	107.3	2653.5	111.2
445	1124.8	90.4	1365.8	94.6	1636.4	98.7	1937.9	102.8	2271.5	106.7	2638.5	110.6
450	1118.5	89.9	1358.2	94.1	1627.3	98.2	1927.1	102.2	2258.9	106.1	2623.8	110.0
455	1112.4	89.4	1350.8	93.6	1618.4	97.7	1916.5	101.6	2246.4	105.5	2609.4	109.3
460	1106.3	88.9	1343.4	93.1	1609.5	97.1	1906.1	101.1	2234.2	105.0	2595.1	108.7
465	1100.3	88.4	1336.2	92.6	1600.9	96.6	1895.8	100.5	2222.1	104.4	2581.2	108.2
470	1094.5	87.9	1329.0	92.1	1592.3	96.1	1885.7	100.0	2210.3	103.8	2567.4	107.6
475	1088.7	87.5	1322.0	91.6	1583.9	95.6	1875.7	99.5	2198.6	103.3	2553.8	107.0
480	1083.0	87.0	1315.1	91.1	1575.7	95.1	1866.0	99.0	2187.2	102.8	2540.5	106.5
485	1077.4	86.6	1308.3	90.6	1567.5	94.6	1856.3	98.5	2175.8	102.2	2527.4	105.9
490	1071.9	86.1	1301.6	90.2	1559.5	94.1	1846.8	97.9	2164.7	101.7	2514.5	105.4
495	1066.5	85.7	1295.0	89.7	1551.6	93.6	1837.5	97.5	2153.8	101.2	2501.7	104.8
500	1061.1	85.3	1288.5	89.3	1543.8	93.2	1828.3	97.0	2143.0	100.7	2489.2	104.3
510	1050.7	84.4	1275.8	88.4	1528.6	92.2	1810.2	96.0	2121.9	99.7	2464.7	103.3
520	1040.5	83.6	1263.5	87.5	1513.8	91.4	1792.8	95.1	2101.3	98.7	2440.8	102.3
530	1030.7	82.8	1251.5	86.7	1499.5	90.5	1775.8	94.2	2081.4	97.8	2417.7	101.3
540	1021.1	82.0	1239.9	85.9	1485.5	89.6	1759.2	93.3	2062.1	96.9	2395.2	100.4
550	1011.8	81.3	1228.6	85.1	1472.0	88.8	1743.2	92.5	2043.2	96.0	2373.3	99.5
560	1002.7	80.6	1217.6	84.3	1458.8	88.0	1727.5	91.6	2024.9	95.1	2352.1	98.6
570	993.8	79.8	1206.8	83.6	1445.9	87.3	1712.3	90.8	2007.1	94.3	2331.3	97.7
580	985.2	79.2	1196.4	82.9	1433.4	86.5	1697.5	90.0	1989.7	93.5	2311.1	96.8
590	976.9	78.5	1186.2	82.2	1421.2	85.8	1683.0	89.3	1972.8	92.7	2291.5	96.0
600	968.7	77.8	1176.3	81.5	1409.3	85.0	1669.0	88.5	1956.3	91.9	2272.3	95.2
610	960.7	77.2	1166.6	80.8	1397.7	84.3	1655.2	87.8	1940.1	91.1	2253.6	94.4
620	952.9	76.6	1157.1	80.2	1386.4	83.7	1641.8	87.1	1924.4	90.4	2235.4	93.7
630	945.3	75.9	1147.9	79.5	1375.3	83.0	1628.7	86.4	1909.1	89.7	2217.5	92.9
640	937.9	75.4	1138.9	78.9	1364.6	82.3	1616.0	85.7	1894.1	89.0	2200.1	92.2
650	930.7	74.8	1130.1	78.3	1354.0	81.7	1603.5	85.0	1879.5	88.3	2183.2	91.5
660	923.6	74.2	1121.5	77.7	1343.7	81.1	1591.3	84.4	1865.2	87.6	2166.6	90.8
670	916.7	73.6	1113.1	77.1	1333.7	80.5	1579.4	83.8	1851.2	87.0	2150.3	90.1
680	909.9	73.1	1104.9	76.5	1323.8	79.9	1567.7	83.1	1837.6	86.3	2134.5	89.4
690	903.3	72.6	1096.9	76.0	1314.2	79.3	1556.3	82.5	1824.2	85.7	2118.9	88.8
700	896.8	72.0	1089.0	75.4	1304.8	78.7	1545.2	81.9	1811.1	85.1	2103.7	88.2
710	890.5	71.5	1081.3	74.9	1295.5	78.2	1534.2	81.4	1798.3	84.5	2088.9	87.5

Table 2 (*Continued*)

| | Diameter | | | | | | | | | | | |
| | 975 mm | | 1050 mm | | 1125 mm | | 1200 mm | | 1275 mm | | 1350 mm | |
Length/fall	Velocity (m/min)	Flow (l/s)	Velocity (m/min)	Flow (l/s)	Velocity (m/min)	Flow (l/s)	Velocity (m/min)	Flow (l/s)	Velocity (m/min)	Flow (l/s)	Velocity (m/min)	Flow (l/s)
720	71.0	884.3	74.4	1073.8	77.6	1286.5	80.8	1523.5	83.9	1785.8	86.9	2074.3
730	70.6	878.2	73.9	1066.4	77.1	1277.7	80.2	1513.1	83.3	1773.5	86.3	2060.1
740	70.1	872.3	73.4	1059.2	76.6	1269.0	79.7	1502.8	82.8	1761.5	85.7	2046.1
750	69.6	866.4	72.9	1052.1	76.1	1260.5	79.2	1492.8	82.2	1749.7	85.2	2032.4
760	69.1	860.7	72.4	1045.1	75.6	1252.2	78.6	1482.9	81.7	1738.2	84.6	2019.0
770	68.7	855.1	71.9	1038.3	75.1	1244.0	78.1	1473.2	81.1	1726.8	84.1	2005.8
780	68.3	849.6	71.5	1031.7	74.6	1236.0	77.6	1463.8	80.6	1715.7	83.5	1992.9
790	67.8	844.2	71.0	1025.1	74.1	1228.2	77.1	1454.5	80.1	1704.8	83.0	1980.3
800	67.4	838.9	70.6	1018.7	73.6	1220.5	76.7	1445.4	79.6	1694.2	82.5	1967.9
810	67.0	833.7	70.1	1012.4	73.2	1212.9	76.2	1436.4	79.1	1683.7	82.0	1955.7
820	66.6	828.6	69.7	1006.2	72.7	1205.5	75.7	1427.6	78.6	1673.4	81.5	1943.7
830	66.2	823.6	69.3	1000.1	72.3	1198.2	75.3	1419.0	78.1	1663.3	81.0	1932.0
840	65.8	818.7	68.9	994.1	71.9	1191.1	74.8	1410.5	77.7	1653.3	80.5	1920.4
850	65.4	813.9	68.5	988.3	71.5	1184.1	74.4	1402.2	77.2	1643.6	80.0	1909.1
860	65.0	809.1	68.1	982.5	71.0	1177.2	73.9	1394.0	76.8	1634.0	79.5	1898.0
870	64.6	804.4	67.7	976.8	70.6	1170.4	73.5	1386.0	76.3	1624.6	79.1	1887.0
880	64.3	799.9	67.3	971.3	70.2	1163.7	73.1	1378.1	75.9	1615.3	78.6	1876.3
890	63.9	795.4	66.9	965.8	69.8	1157.1	72.7	1370.3	75.5	1606.2	78.2	1865.7
900	63.5	790.9	66.5	960.4	69.4	1150.7	72.3	1362.7	75.0	1597.3	77.7	1855.3
910	63.2	786.6	66.2	955.1	69.1	1144.4	71.9	1355.2	74.6	1588.5	77.3	1845.1
920	62.8	782.3	65.8	949.9	68.7	1138.1	71.5	1347.8	74.2	1579.8	76.9	1835.0
930	62.5	778.1	65.4	944.8	68.3	1132.0	71.1	1340.5	73.8	1571.3	76.5	1825.2
940	62.2	773.9	65.1	939.8	67.9	1125.9	70.7	1333.4	73.4	1562.9	76.1	1815.4
950	61.8	769.8	64.8	934.8	67.6	1120.0	70.3	1326.4	73.0	1554.7	75.7	1805.8
960	61.5	765.8	64.4	929.9	67.2	1114.2	70.0	1319.4	72.7	1546.6	75.3	1796.4

970	61.2	761.9	64.1	925.1	66.9	1108.4	69.6	1312.6	72.3	1538.6	74.9	1787.1
980	60.9	758.0	63.8	920.4	66.5	1102.7	69.3	1305.9	71.9	1530.7	74.5	1778.0
990	60.6	754.1	63.4	915.7	66.2	1097.1	68.9	1299.3	71.5	1522.9	74.1	1769.0
1000	60.3	750.3	63.1	911.1	65.9	1091.6	68.6	1292.8	71.2	1515.3	73.8	1760.1
1050	58.8	732.3	61.6	889.2	64.3	1065.3	66.9	1261.6	69.5	1478.8	72.0	1717.7
1100	57.5	715.4	60.2	868.7	62.8	1040.8	65.4	1232.6	67.9	1444.8	70.3	1678.2
1150	56.2	699.7	58.9	849.6	61.4	1018.0	63.9	1205.5	66.4	1413.0	68.8	1641.3
1200	55.0	685.0	57.6	831.7	60.1	996.5	62.6	1180.1	65.0	1383.3	67.3	1606.8
1250	53.9	671.1	56.5	814.9	58.9	976.4	61.3	1156.3	63.7	1355.3	66.0	1574.3
1300	52.9	658.1	55.4	799.1	57.8	957.4	60.1	1133.8	62.4	1329.0	64.7	1543.7
1350	51.9	645.8	54.3	784.2	56.7	939.5	59.0	1112.6	61.3	1304.2	63.5	1514.9
1400	50.9	634.2	53.3	770.0	55.7	922.6	57.9	1092.6	60.2	1280.7	62.3	1487.6
1450	50.1	623.1	52.4	756.7	54.7	906.6	56.9	1073.6	59.1	1258.4	61.3	1461.7
1500	49.2	612.6	51.5	743.9	53.8	891.3	56.0	1055.5	58.1	1237.2	60.2	1437.1
1550	48.4	602.7	50.7	731.8	52.9	876.8	55.1	1038.4	57.2	1217.1	59.2	1413.8
1600	47.7	593.2	49.9	720.3	52.1	863.0	54.2	1022.0	56.3	1198.0	58.3	1391.5
1650	46.9	584.1	49.1	709.3	51.3	849.8	53.4	1006.4	55.4	1179.7	57.4	1370.2
1700	46.2	575.5	48.4	698.8	50.5	837.3	52.6	991.5	54.6	1162.2	56.6	1349.9
1750	45.6	567.2	47.7	688.8	49.8	825.2	51.8	977.2	53.8	1145.5	55.8	1330.5
1800	44.9	559.3	47.0	679.1	49.1	813.7	51.1	963.6	53.1	1129.4	55.0	1311.9
1850	44.3	551.7	46.4	669.9	48.4	802.6	50.4	950.5	52.3	1114.1	54.2	1294.1
1900	43.7	544.4	45.8	661.0	47.8	792.0	49.7	937.9	51.6	1099.3	53.5	1276.9
1950	43.2	537.3	45.2	652.5	47.2	781.7	49.1	925.8	51.0	1085.1	52.8	1260.4
2000	42.6	530.6	44.6	644.3	46.6	771.9	48.5	914.1	50.3	1071.5	52.2	1244.6
2050	42.1	524.1	44.1	636.4	46.0	762.4	47.9	902.9	49.7	1058.3	51.5	1229.3
2100	41.6	517.8	43.6	628.7	45.5	753.3	47.3	892.1	49.1	1045.7	50.9	1214.6
2150	41.1	511.7	43.0	621.4	44.9	744.5	46.8	881.7	48.6	1033.4	50.3	1200.4
2200	40.6	505.9	42.6	614.3	44.4	736.0	46.2	871.6	48.0	1021.6	49.7	1186.7
2250	40.2	500.2	42.1	607.4	43.9	727.8	45.7	861.8	47.5	1010.2	49.2	1173.4
2300	39.7	494.8	41.6	600.8	43.4	719.8	45.2	852.4	46.9	999.2	48.6	1160.6
2350	39.3	489.5	41.2	594.4	43.0	712.1	44.7	843.3	46.4	988.5	48.1	1148.2
2400	38.9	484.3	40.7	588.1	42.5	704.7	44.3	834.5	46.0	978.1	47.6	1136.2
2450	38.5	479.4	40.3	582.1	42.1	697.4	43.8	825.9	45.5	968.1	47.1	1124.5
2500	38.1	474.6	39.9	576.3	41.7	690.4	43.4	817.6	45.0	958.4	46.6	1113.2

Table 2 (Continued)

Diameter

Length/fall	975 mm		1050 mm		1125 mm		1200 mm		1275 mm		1350 mm	
	Velocity (m/min)	Flow (l/s)	Velocity (m/min)	Flow (l/s)	Velocity (m/min)	Flow (l/s)	Velocity (m/min)	Flow (l/s)	Velocity (m/min)	Flow (l/s)	Velocity (m/min)	Flow (l/s)
2550	37.7	469.9	39.5	570.6	41.3	683.6	42.9	809.6	44.6	948.9	46.2	1102.2
2600	37.4	465.3	39.1	565.1	40.9	677.0	42.5	801.7	44.1	939.8	45.7	1091.6
2650	37.0	460.9	38.8	559.7	40.5	670.6	42.1	794.1	43.7	930.8	45.3	1081.2
2700	36.7	456.6	38.4	554.5	40.1	664.4	41.7	786.8	43.3	922.2	44.9	1071.2
2750	36.4	452.5	38.1	549.4	39.7	658.3	41.3	779.6	42.9	913.8	44.5	1061.4
2800	36.0	448.4	37.7	544.5	39.4	652.4	41.0	772.6	42.5	905.6	44.1	1051.9
2850	35.7	444.5	37.4	539.7	39.0	646.6	40.6	765.8	42.2	897.6	43.7	1042.6
2900	35.4	440.6	37.1	535.0	38.7	641.0	40.3	759.1	41.8	889.8	43.3	1033.6
2950	35.1	436.9	36.7	530.5	38.4	635.6	39.9	752.7	41.4	882.2	42.9	1024.8
3000	34.8	433.2	36.4	526.0	38.0	630.3	39.6	746.4	41.1	874.9	42.6	1016.2
3050	34.5	429.6	36.1	521.7	37.7	625.1	39.3	740.2	40.8	867.7	42.2	1007.8
3100	34.2	426.2	35.8	517.5	37.4	620.0	38.9	734.2	40.4	860.6	41.9	999.7
3150	34.0	422.8	35.6	513.4	37.1	615.1	38.6	728.4	40.1	853.8	41.6	991.7
3200	—	—	—	—	36.8	610.2	38.3	722.7	39.8	847.1	41.2	983.9

(d) Diameters 1425–1800 mm

Diameter

	1425 mm		1500 mm		1575 mm		1650 mm		1725 mm		1800 mm	
Length/fall	Velocity (m/min)	Flow (l/s)	Velocity (m/min)	Flow (l/s)	Velocity (m/min)	Flow (l/s)	Velocity (m/min)	Flow (l/s)	Velocity (m/min)	Flow (l/s)	Velocity (m/min)	Flow (l/s)
86	260.1	6915.3	268.5	7910.0	276.7	8988.5	284.8	10153.7	292.8	11407.9	300.6	12753.5
87	258.6	6875.4	266.9	7864.4	275.1	8936.7	283.2	10095.1	291.1	11342.1	298.9	12680.0
88	257.1	6836.3	265.4	7819.6	273.6	8885.8	281.6	10037.6	289.4	11277.5	297.2	12607.7
89	255.7	6797.8	263.9	7775.5	272.0	8835.8	280.0	9981.1	287.8	11214.0	295.5	12536.7
90	254.2	6759.9	262.5	7732.2	270.5	8786.5	278.4	9925.5	286.2	11151.5	293.9	12466.8
91	252.8	6722.6	261.0	7689.6	269.0	8738.1	276.9	9870.8	284.6	11090.0	292.2	12398.1
92	251.5	6686.0	259.6	7647.7	267.6	8690.5	275.4	9817.0	283.1	11029.6	290.7	12330.6
93	250.1	6650.0	258.2	7606.4	266.1	8643.6	273.9	9764.1	281.6	10970.1	289.1	12264.1
94	248.8	6614.5	256.8	7565.9	264.7	8597.6	272.4	9712.0	280.1	10911.6	287.5	12198.7
95	247.5	6579.6	255.5	7525.9	263.3	8552.2	271.0	9660.7	278.6	10854.1	286.0	12134.3
96	246.2	6545.2	254.1	7486.7	261.9	8507.5	269.6	9610.3	277.1	10797.4	284.5	12071.0
97	244.9	6511.4	252.8	7448.0	260.6	8463.6	268.2	9560.6	275.7	10741.6	283.1	12008.6
98	243.6	6478.1	251.5	7409.9	259.2	8420.3	266.8	9511.7	274.3	10686.6	281.6	11947.1
99	242.4	6445.3	250.2	7372.3	257.9	8377.6	265.5	9463.6	272.9	10632.5	280.2	11886.7
100	241.2	6413.0	249.0	7335.4	256.6	8335.6	264.1	9416.1	271.5	10579.2	278.8	11827.1
105	235.4	6258.4	243.0	7158.6	250.4	8134.7	257.8	9189.2	265.0	10324.3	272.1	11542.0
110	230.0	6114.5	237.4	6994.0	244.7	7947.7	251.8	8977.9	258.9	10086.9	265.8	11276.7
115	224.9	5980.1	232.2	6840.3	239.3	7773.0	246.3	8780.6	253.2	9865.2	260.0	11028.8
120	220.2	5854.2	227.3	6696.3	234.3	7609.4	241.1	8595.7	247.9	9657.5	254.5	10796.6
125	215.7	5735.9	222.7	6561.0	229.5	7455.6	236.3	8422.0	242.9	9462.3	249.4	10578.4
130	211.5	5624.5	218.4	6433.6	225.1	7310.8	231.7	8258.5	238.1	9278.6	244.5	10373.0
135	207.6	5519.4	214.3	6313.3	220.9	7174.2	227.3	8104.1	233.7	9105.2	239.9	10179.1
140	203.8	5420.0	210.4	6199.5	216.9	7044.9	223.2	7958.1	229.5	8941.1	235.6	9995.7
145	200.3	5325.7	206.8	6091.7	213.1	6922.4	219.4	7819.7	225.5	8785.6	231.5	9821.9
150	196.9	5236.2	203.3	5989.3	209.5	6806.0	215.7	7688.2	221.7	8637.9	227.6	9656.8

Table 2 (Continued)

Length/fall	1425 mm Velocity (m/min)	1425 mm Flow (l/s)	1500 mm Velocity (m/min)	1500 mm Flow (l/s)	1575 mm Velocity (m/min)	1575 mm Flow (l/s)	1650 mm Velocity (m/min)	1650 mm Flow (l/s)	1725 mm Velocity (m/min)	1725 mm Flow (l/s)	1800 mm Velocity (m/min)	1800 mm Flow (l/s)
155	193.7	5151.0	200.0	5891.9	206.1	6695.3	212.2	7563.3	218.1	8497.4	223.9	9499.7
160	190.7	5069.9	196.8	5799.1	202.9	6589.9	208.8	7444.1	214.7	8363.6	220.4	9350.1
165	187.8	4992.5	193.8	5710.6	199.8	6489.3	205.6	7330.4	211.4	8235.9	217.0	9207.4
170	185.0	4918.5	191.0	5626.0	196.8	6393.1	202.6	7221.8	208.2	8113.9	213.8	9071.0
175	182.3	4847.8	188.2	5545.0	194.0	6301.1	199.7	7117.9	205.3	7997.1	210.7	8940.4
180	179.8	4780.0	185.6	5467.5	191.3	6213.0	196.9	7018.4	202.4	7885.3	207.8	8815.4
185	177.3	4714.9	183.1	5393.1	188.7	6128.5	194.2	6922.9	199.6	7778.0	205.0	8695.4
190	175.0	4652.5	180.6	5321.6	186.2	6047.3	191.6	6831.2	197.0	7675.0	202.3	8580.3
195	172.7	4592.4	178.3	5253.0	183.8	5969.3	189.2	6743.0	194.4	7575.9	199.6	8469.5
200	170.5	4534.7	176.1	5186.9	181.5	5894.2	186.8	6658.2	192.0	7480.6	197.1	8363.0
205	168.5	4479.0	173.9	5123.3	179.2	5821.9	184.5	6576.5	189.6	7388.8	194.7	8260.4
210	166.4	4425.4	171.8	5061.9	177.1	5752.1	182.3	6497.7	187.4	7300.4	192.4	8161.5
215	164.5	4373.6	169.8	5002.7	175.0	5684.9	180.1	6421.7	185.2	7215.0	190.1	8066.0
220	162.6	4323.6	167.9	4945.5	173.0	5619.9	178.1	6348.3	183.1	7132.5	188.0	7973.8
225	160.8	4275.3	166.0	4890.3	171.1	5557.1	176.1	6277.4	181.0	7052.8	185.9	7884.7
230	159.0	4228.6	164.2	4836.8	169.2	5496.4	174.2	6208.8	179.0	6975.7	183.8	7798.5
235	157.3	4183.4	162.4	4785.1	167.4	5437.6	172.3	6142.4	177.1	6901.1	181.9	7715.1
240	155.7	4139.6	160.7	4735.0	165.7	5380.6	170.5	6078.1	175.3	6828.9	180.0	7634.3
245	154.1	4097.1	159.1	4686.4	164.0	5325.4	168.8	6015.7	173.5	6758.8	178.1	7556.0
250	152.5	4055.9	157.5	4639.3	162.3	5271.9	167.1	5955.3	171.7	6690.9	176.3	7480.1
255	151.0	4016.0	155.9	4593.6	160.7	5220.0	165.4	5896.6	170.0	6625.0	174.6	7406.4
260	149.6	3977.2	154.4	4549.2	159.2	5169.5	163.8	5839.6	168.4	6561.0	172.9	7334.8
265	148.2	3939.5	153.0	4506.1	157.6	5120.5	162.3	5784.3	166.8	6498.8	171.3	7265.3
270	146.8	3902.8	151.5	4464.2	156.2	5072.9	160.8	5730.5	165.2	6438.3	169.7	7197.7
275	145.4	3867.2	150.1	4423.4	154.8	5026.6	159.3	5678.1	163.7	6379.5	168.1	7132.0

280	144.1	3832.5	148.8	4383.7	153.4	4981.5	157.9	5627.2	162.3	6322.3	166.6	7068.0
285	142.9	3798.7	147.5	4345.1	152.0	4937.6	156.5	5577.6	160.8	6266.6	165.1	7005.8
290	141.6	3765.8	146.2	4307.5	150.7	4894.9	155.1	5529.3	159.4	6212.3	163.7	6945.1
295	140.4	3733.8	145.0	4270.8	149.4	4853.2	153.8	5482.1	158.1	6159.5	162.3	6886.0
300	139.3	3702.5	143.8	4235.1	148.2	4812.6	152.5	5436.4	156.8	6107.9	161.0	6828.4
305	138.1	3672.1	142.6	4200.2	146.9	4773.0	151.2	5391.7	155.5	6057.6	159.6	6772.2
310	137.0	3642.3	141.4	4166.2	145.8*	4734.3	150.0	5348.0	154.2	6008.6	158.3	6717.3
315	135.9	3613.3	140.3	4133.0	144.6	4696.6	148.8	5305.4	153.0	5960.7	157.1	6663.8
320	134.8	3585.0	139.2	4100.6	143.5	4659.8	147.7	5263.8	151.8	5914.0	155.8	6611.5
325	133.8	3557.3	138.1	4068.9	142.4	4623.8	146.5	5223.1	150.6	5868.3	154.6	6560.5
330	132.8	3530.2	137.1	4038.0	141.3	4588.6	145.4	5183.4	149.5	5823.7	153.5	6510.6
335	131.8	3503.8	136.0	4007.8	140.2	4554.2	144.3	5144.6	148.3	5780.1	152.3	6461.8
340	130.8	3477.9	135.0	3978.2	139.2	4520.6	143.3	5106.6	147.3	5737.4	151.2	6414.1
345	129.9	3452.6	134.0	3949.2	138.2	4487.8	142.2	5069.5	146.2	5695.7	150.1	6367.5
350	128.9	3427.9	133.1	3920.9	137.2	4455.6	141.2	5033.1	145.1	5654.8	149.0	6321.8
355	128.0	3403.7	132.1	3893.2	136.2	4424.1	140.2	4997.6	144.1	5614.9	148.0	6277.2
360	127.1	3379.9	131.2	3866.1	135.3	4393.3	139.2	4962.7	143.1	5575.7	146.9	6233.4
365	126.2	3356.7	130.3	3839.5	134.3	4363.1	138.3	4928.6	142.1	5537.4	145.9	6190.6
370	125.4	3334.0	129.4	3813.5	133.4	4333.5	137.3	4895.2	141.2	5499.9	144.9	6148.6
375	124.6	3311.7	128.6	3788.0	132.5	4304.5	136.4	4862.5	140.2	5463.1	144.0	6107.5
380	123.7	3289.8	127.7	3763.0	131.6	4276.1	135.5	4830.4	139.3	5427.0	143.0	6067.2
385	122.9	3268.4	126.9	3738.5	130.8	4248.2	134.6	4798.9	138.4	5391.7	142.1	6027.6
390	122.1	3247.3	126.1	3714.4	130.0	4220.9	133.8	4768.0	137.5	5357.0	141.2	5988.9
395	121.4	3226.7	125.3	3690.8	129.1	4194.1	132.9	4737.8	136.6	5323.0	140.3	5950.8
400	120.6	3206.5	124.5	3667.7	128.3	4167.8	132.1	4708.1	135.8	5289.6	139.4	5913.5
405	119.8	3186.6	123.7	3645.0	127.5	4142.0	131.3	4678.9	134.9	5256.9	138.5	5876.9
410	119.1	3167.1	123.0	3622.7	126.7	4116.7	130.5	4650.3	134.1	5224.7	137.7	5841.0
415	118.4	3148.0	122.2	3600.8	126.0	4091.8	129.7	4622.2	133.3	5193.1	136.8	5805.7
420	117.7	3129.2	121.5	3579.3	125.2	4067.4	128.9	4594.6	132.5	5162.1	136.0	5771.0
425	117.0	3110.8	120.8	3558.2	124.5	4043.4	128.1	4567.5	131.7	5131.7	135.2	5737.0
430	116.3	3092.6	120.1	3537.4	123.8	4019.8	127.4	4540.9	130.9	5101.8	134.4	5703.5
435	115.6	3074.8	119.4	3517.1	123.0	3996.6	126.6	4514.7	130.2	5072.4	133.7	5670.6
440	115.0	3057.3	118.7	3497.0	122.3	3973.9	125.9	4485.0	129.4	5043.4	132.9	5638.3
445	114.3	3040.0	118.0	3477.3	121.7	3951.5	125.2	4463.7	128.7	5015.0	132.2	5606.6

Table 2 (*Continued*)

Length/fall	Diameter											
	1425 mm		1500 mm		1575 mm		1650 mm		1725 mm		1800 mm	
	Velocity (m/min)	Flow (l/s)	Velocity (m/min)	Flow (l/s)	Velocity (m/min)	Flow (l/s)	Velocity (m/min)	Flow (l/s)	Velocity (m/min)	Flow (l/s)	Velocity (m/min)	Flow (l/s)
450	113.7	3023.1	117.4	3457.9	121.0	3929.5	124.5	4438.8	128.0	4987.1	131.4	5575.3
455	113.1	3006.5	116.7	3438.9	120.3	3907.8	123.8	4414.3	127.3	4959.6	130.7	5544.6
460	112.5	2990.1	116.1	3420.1	119.7	3886.5	123.2	4390.3	126.6	4932.6	130.0	5514.4
465	111.9	2974.0	115.5	3401.7	119.0	3865.6	122.5	4366.6	125.9	4906.0	129.3	5484.7
470	111.3	2958.1	114.8	3383.6	118.4	3844.9	121.8	4343.3	125.2	4879.8	128.6	5455.4
475	110.7	2942.5	114.2	3365.7	117.8	3824.7	121.2	4320.4	124.6	4854.1	127.9	5426.6
480	110.1	2927.1	113.6	3348.1	117.1	3804.7	120.6	4297.9	123.9	4828.7	127.2	5398.3
485	109.5	2912.0	113.1	3330.8	116.5	3785.0	119.9	4275.6	123.3	4803.8	126.6	5370.4
490	109.0	2897.1	112.5	3313.8	115.9	3765.7	119.3	4253.8	122.7	4779.2	125.9	5342.9
495	108.4	2882.4	111.9	3297.0	115.3	3746.6	118.7	4232.2	122.0	4755.0	125.3	5315.9
500	107.9	2868.0	111.3	3280.5	114.8	3727.8	118.1	4211.0	121.4	4731.2	124.7	5289.2
510	106.8	2839.7	110.3	3248.2	113.6	3691.1	117.0	4169.5	120.2	4684.6	123.4	5237.1
520	105.8	2812.3	109.2	3216.8	112.5	3655.4	115.8	4129.2	119.1	4639.3	122.3	5186.5
530	104.8	2785.6	108.2	3186.3	111.5	3620.8	114.7	4090.1	117.9	4595.3	121.1	5137.4
540	103.8	2759.7	107.1	3156.7	110.4	3587.1	113.7	4052.1	116.3	4552.6	120.0	5089.6
550	102.8	2734.5	106.2	3127.8	109.4	3554.3	112.6	4015.1	115.3	4511.0	118.9	5043.1
560	101.9	2710.0	105.2	3099.8	108.4	3522.5	111.6	3979.0	114.7	4470.5	117.8	4997.9
570	101.0	2686.1	104.3	3072.5	107.5	3491.4	110.6	3944.0	113.7	4431.2	116.8	4953.8
580	100.1	2662.9	103.4	3045.9	106.6	3451.2	109.7	3909.8	112.7	4392.8	115.8	4910.9
590	99.3	2640.2	102.5	3019.9	105.7	3431.7	108.7	3876.6	111.8	4355.4	114.8	4869.1
600	98.5	2618.1	101.6	2994.7	104.8	3403.0	107.8	3844.1	110.8	4319.0	113.8	4828.4
610	97.7	2596.5	100.8	2970.0	103.9	3375.0	106.9	3812.5	109.9	4283.4	112.9	4788.6
620	96.9	2575.5	100.0	2946.0	103.1	3347.7	106.1	3781.6	109.0	4248.7	112.0	4749.9
630	96.1	2555.0	99.2	2922.5	102.2	3321.0	105.2	3751.5	108.2	4214.9	111.1	4712.0
640	95.3	2535.0	98.4	2899.6	101.4	3295.0	104.4	3722.1	107.3	4181.8	110.2	4675.1

650	4639.0	109.3	4149.5	106.5	3693.3	103.6	3269.5	100.7	2877.2	97.7	2515.4	94.6
660	4603.7	108.5	4118.0	105.7	3665.2	102.8	3244.6	99.9	2855.3	96.9	2496.3	93.9
670	4569.2	107.7	4087.1	104.9	3637.8	102.0	3220.3	99.1	2833.9	96.2	2477.6	93.2
680	4535.5	106.9	4057.0	104.1	3610.9	101.3	3196.6	98.4	2813.0	95.5	2459.3	92.5
690	4502.5	106.1	4027.4	103.4	3584.7	100.6	3173.3	97.7	2792.5	94.8	2441.4	91.8
700	4470.2	105.4	3998.6	102.6	3559.0	99.8	3150.6	97.0	2772.5	94.1	2423.9	91.2
710	4438.6	104.6	3970.3	101.9	3533.8	99.1	3128.3	96.3	2752.9	93.4	2406.8	90.5
720	4407.7	103.9	3942.6	101.2	3509.2	98.4	3106.5	95.6	2733.7	92.8	2390.0	89.9
730	4377.4	103.2	3915.6	100.5	3485.1	97.8	3085.2	95.0	2715.0	92.2	2373.6	89.3
740	4347.7	102.5	3889.0	99.8	3461.4	97.1	3064.2	94.3	2696.5	91.5	2357.5	88.7
750	4318.6	101.8	3863.0	99.1	3438.3	96.5	3043.7	93.7	2678.5	90.9	2341.7	88.1
760	4290.1	101.1	3837.5	98.5	3415.6	95.8	3023.7	93.1	2660.8	90.3	2326.2	87.5
770	4262.2	100.5	3812.5	97.9	3393.3	95.2	3004.0	92.5	2643.5	89.7	2311.1	86.9
780	4234.8	99.8	3788.0	97.2	3371.5	94.6	2984.6	91.9	2626.5	89.2	2296.2	86.4
790	4207.9	99.2	3763.9	96.6	3350.1	94.0	2965.7	91.3	2609.8	88.6	2281.6	85.8
800	4181.5	98.6	3740.3	96.0	3329.1	93.4	2947.1	90.7	2593.5	88.0	2267.3	85.3
810	4155.6	98.0	3717.2	95.4	3308.5	92.8	2928.8	90.2	2577.4	87.5	2253.3	84.7
820	4130.2	97.4	3694.4	94.8	3288.3	92.2	2910.9	89.6	2561.6	86.9	2239.5	84.2
830	4105.2	96.8	3672.1	94.2	3268.4	91.7	2893.3	89.1	2546.2	86.4	2226.0	83.7
840	4080.7	96.2	3650.2	93.7	3248.9	91.1	2876.1	88.5	2531.0	85.9	2212.7	83.2
850	4056.7	95.6	3628.6	93.1	3229.7	90.6	2859.1	88.0	2516.0	85.4	2199.6	82.7
860	4033.0	95.1	3607.5	92.6	3210.9	90.1	2842.4	87.5	2501.3	84.9	2186.8	82.2
870	4009.8	94.5	3586.7	92.1	3192.4	89.6	2826.0	87.0	2486.9	84.4	2174.2	81.8
880	3986.9	94.0	3566.3	91.5	3174.2	89.0	2809.9	86.5	2472.8	83.9	2161.8	81.3
890	3964.4	93.4	3546.2	91.0	3156.3	88.5	2794.1	86.0	2458.8	83.5	2149.6	80.8
900	3942.4	92.9	3526.4	90.5	3138.7	88.0	2778.5	85.5	2445.1	83.0	2137.7	80.4
910	3920.6	92.4	3507.0	90.0	3121.4	87.6	2763.2	85.1	2431.7	82.5	2125.9	80.0
920	3899.3	91.9	3487.9	89.5	3104.4	87.1	2748.2	84.6	2418.4	82.1	2114.3	79.5
930	3878.3	91.4	3469.1	89.0	3087.7	86.6	2733.4	84.2	2405.4	81.6	2102.9	79.1
940	3857.6	90.9	3450.6	88.6	3071.2	86.2	2718.8	83.7	2392.5	81.2	2091.7	78.7
950	3837.2	90.4	3432.4	88.1	3055.0	85.7	2704.4	83.3	2379.9	80.8	2080.6	78.3
960	3817.2	90.0	3414.4	87.6	3039.0	85.3	2690.3	82.8	2367.5	80.4	2069.8	77.8
970	3797.4	89.5	3396.8	87.2	3023.3	84.8	2676.4	82.4	2355.3	79.9	2059.1	77.4
980	3778.0	89.1	3379.4	86.7	3007.9	84.4	2662.7	82.0	2343.2	79.5	2048.6	77.0

Table 2 (*Continued*)

| Length/fall | Diameter | | | | | | | | | | | |
| | 1425 mm | | 1500 mm | | 1575 mm | | 1650 mm | | 1725 mm | | 1800 mm | |
	Velocity (m/min)	Flow (l/s)	Velocity (m/min)	Flow (l/s)	Velocity (m/min)	Flow (l/s)	Velocity (m/min)	Flow (l/s)	Velocity (m/min)	Flow (l/s)	Velocity (m/min)	Flow (l/s)
990	76.7	2038.2	79.1	2331.3	81.6	2649.2	84.0	2992.6	86.3	3362.3	88.6	3758.9
1000	76.3	2028.0	78.7	2319.7	81.2	2636.6	83.5	2977.6	85.9	3345.4	88.2	3740.1
1050	74.4	1979.1	76.8	2263.8	79.2	2572.4	81.5	2905.9	83.8	3264.8	86.0	3649.9
1100	72.7	1933.6	75.1	2211.7	77.4	2513.3	79.6	2839.1	81.9	3189.8	84.1	3566.0
1150	71.1	1891.1	73.4	2163.1	75.7	2458.0	77.9	2776.7	80.1	3119.6	82.2	3487.6
1200	69.6	1851.3	71.9	2117.5	74.1	2406.3	76.3	2718.2	78.4	3054.0	80.5	3414.2
1250	68.2	1813.9	70.4	2074.8	72.6	2357.7	74.7	2663.3	76.8	2992.3	78.9	3345.2
1300	66.9	1778.6	69.1	2034.5	71.2	2311.9	73.3	2611.6	75.3	2934.2	77.3	3280.2
1350	65.6	1745.4	67.8	1996.4	69.8	2268.7	71.9	2562.7	73.9	2879.3	75.9	3218.9
1400	64.5	1713.9	66.5	1960.5	68.6	2227.8	70.6	2516.6	72.6	2827.4	74.5	3160.9
1450	63.3	1684.1	65.4	1926.4	67.4	2189.0	69.4	2472.8	71.3	2778.2	73.2	3105.9
1500	62.3	1655.8	64.3	1894.0	66.3	2152.3	68.2	2431.2	70.1	2731.5	72.0	3053.7
1550	61.3	1628.9	63.2	1863.2	65.2	2117.3	67.1	2391.7	69.0	2687.1	70.8	3004.1
1600	60.3	1603.2	62.2	1833.8	64.2	2083.9	66.0	2354.0	67.9	2644.8	69.7	2956.8
1650	59.4	1578.8	61.3	1805.8	63.2	2052.1	65.0	2318.1	66.8	2604.4	68.6	2911.6
1700	58.5	1555.4	60.4	1779.1	62.2	2021.7	64.1	2283.7	65.9	2565.8	67.6	2868.5
1750	57.7	1533.0	59.5	1753.5	61.3	1992.6	63.1	2250.9	64.9	2528.9	66.6	2827.2
1800	56.8	1511.6	58.7	1729.0	60.5	1964.7	62.3	2219.4	64.0	2493.5	65.7	2787.7
1850	56.1	1491.0	57.9	1705.4	59.7	1938.0	61.4	2189.2	63.1	2459.6	64.8	2749.7
1900	55.3	1471.2	57.1	1682.9	58.9	1912.3	60.6	2160.2	62.3	2427.0	64.0	2713.3
1950	54.6	1452.3	56.4	1661.1	58.1	1887.6	59.8	2132.3	61.5	2395.7	63.1	2678.3
2000	53.9	1434.0	55.7	1640.2	57.4	1863.9	59.1	2105.5	60.7	2365.6	62.3	2644.6
2050	53.3	1416.4	55.0	1620.1	56.7	1841.0	58.3	2079.7	60.0	2336.6	61.6	2612.2
2100	52.6	1399.4	54.3	1600.7	56.0	1819.0	57.6	2054.8	59.3	2308.6	60.8	2580.9
2150	52.0	1383.1	53.7	1582.0	55.3	1797.7	57.0	2030.7	58.6	2281.6	60.1	2550.7

2200	51.4	1367.3	53.1	1563.9	54.7	1777.2	56.3	2007.5	57.9	2255.5	59.4	2521.5
2250	50.8	1352.0	52.5	1546.4	54.1	1757.1	55.7	1985.1	57.2	2230.3	58.8	2493.4
2300	50.3	1337.2	51.9	1529.5	53.5	1738.1	55.1	1963.4	56.6	2205.9	58.1	2466.1
2350	49.8	1322.9	51.4	1513.2	52.9	1719.5	54.5	1942.4	56.0	2182.3	57.5	2439.7
2400	49.2	1309.0	50.8	1497.3	52.4	1701.5	53.9	1922.1	55.4	2159.5	56.9	2414.2
2450	48.7	1295.6	50.3	1482.0	51.8	1684.1	53.4	1902.3	54.9	2137.3	56.3	2389.4
2500	48.2	1282.6	49.8	1467.1	51.3	1667.1	52.8	1883.2	54.3	2115.8	55.8	2365.4
2550	47.8	1270.0	49.3	1452.6	50.8	1650.7	52.3	1864.7	53.8	2095.0	55.2	2342.1
2600	47.3	1257.7	48.8	1438.6	50.3	1634.8	51.8	1846.7	53.3	2074.8	54.7	2319.5
2650	46.9	1245.8	48.4	1425.0	49.9	1619.3	51.3	1829.1	52.7	2055.1	54.2	2297.5
2700	46.4	1234.2	47.9	1411.7	49.4	1604.2	50.8	1812.1	52.3	2036.0	53.7	2276.1
2750	46.0	1222.9	47.5	1398.8	48.9	1589.5	50.4	1795.6	51.8	2017.4	53.2	2255.3
2800	45.6	1211.9	47.1	1386.3	48.5	1575.3	49.9	1779.5	51.3	1999.3	52.7	2235.1
2850	45.2	1201.3	46.6	1374.0	48.1	1561.4	49.5	1763.8	50.9	1981.7	52.2	2215.4
2900	44.8	1190.9	46.2	1362.1	47.7	1547.9	49.1	1748.5	50.4	1964.5	51.8	2196.2
2950	44.4	1180.7	45.8	1350.6	47.2	1534.8	48.6	1733.6	50.0	1947.8	51.3	2177.5
3000	44.0	1170.8	45.5	1339.3	46.9	1521.9	48.2	1719.1	49.6	1931.5	50.9	2159.3
3050	43.7	1161.2	45.1	1328.2	46.5	1509.3	47.8	1705.0	49.2	1915.6	50.5	2141.5
3100	43.3	1151.8	44.7	1317.5	46.1	1497.1	47.4	1691.2	48.8	1900.1	50.1	2124.2
3150	43.0	1142.6	44.4	1307.0	45.7	1485.2	47.1	1677.7	48.4	1884.9	49.7	2107.3
3200	42.6	1133.7	44.0	1296.7	45.4	1473.5	46.7	1664.6	48.0	1870.2	49.3	2090.8
3250	42.3	1124.9	43.7	1286.7	45.0	1462.2	46.3	1651.7	47.6	1855.7	48.9	2074.6
3300	42.0	1116.4	43.3	1276.9	44.7	1451.0	46.0	1639.1	47.3	1841.6	48.5	2058.8
3350	41.7	1108.0	43.0	1267.4	44.3	1440.2	45.6	1626.9	46.9	1827.8	48.2	2043.4
3400	41.4	1099.8	42.7	1258.0	44.0	1429.5	45.3	1614.9	46.6	1814.3	47.8	2028.3
3450	41.1	1091.8	42.4	1248.9	43.7	1419.2	45.0	1603.1	46.2	1801.1	47.5	2013.6
3500	40.8	1084.0	42.1	1239.9	43.4	1409.0	44.6	1591.6	45.9	1788.2	47.1	1999.1
3550	40.5	1076.3	41.8	1231.1	43.1	1399.0	44.3	1580.4	45.6	1775.6	46.8	1985.0
3600	40.2	1068.8	41.5	1222.6	42.8	1389.3	44.0	1569.4	45.3	1763.2	46.5	1971.2
3650	39.9	1061.5	41.2	1214.2	42.5	1379.7	43.7	1558.6	44.9	1751.1	46.1	1957.6
3700	39.7	1054.3	40.9	1205.9	42.2	1370.4	43.4	1548.0	44.6	1739.2	45.8	1944.4
3750	39.4	1047.2	40.7	1197.9	41.9	1361.2	43.1	1537.6	44.3	1727.6	45.5	1931.4
3800	39.1	1040.3	40.4	1190.0	41.6	1352.2	42.8	1527.5	44.0	1716.2	45.2	1918.6
3850	38.9	1033.5	40.1	1182.2	41.4	1343.4	42.6	1517.5	43.8	1705.0	44.9	1906.1

Table 2 (*Continued*)

	Diameter											
	1425 mm		1500 mm		1575 mm		1650 mm		1725 mm		1800 mm	
Length/fall	Velocity (m/min)	Flow (l/s)	Velocity (m/min)	Flow (l/s)	Velocity (m/min)	Flow (l/s)	Velocity (m/min)	Flow (l/s)	Velocity (m/min)	Flow (l/s)	Velocity (m/min)	Flow (l/s)
3900	38.6	1026.9	39.9	1174.6	41.1	1334.8	42.3	1507.8	43.5	1694.0	44.6	1893.8
3950	38.4	1020.4	39.6	1167.1	40.8	1326.3	42.0	1498.2	43.2	1683.3	44.4	1881.8
4000	38.1	1014.0	39.4	1159.8	40.6	1318.0	41.8	1488.8	42.9	1672.7	44.1	1870.0
4050	37.9	1007.7	39.1	1152.6	40.3	1309.8	41.5	1479.6	42.7	1662.4	43.8	1858.4
4100	37.7	1001.5	38.9	1145.6	40.1	1301.8	41.3	1470.6	42.4	1652.2	43.5	1847.1
4150	37.4	995.5	38.7	1138.7	39.8	1293.9	41.0	1461.7	42.1	1642.2	43.3	1835.9
4200	37.2	989.5	38.4	1131.9	39.6	1286.2	40.8	1452.9	41.9	1632.4	43.0	1825.0
4250	37.0	983.7	38.2	1125.2	39.4	1278.6	40.5	1444.4	41.6	1622.8	42.8	1814.2
4300	36.8	978.0	38.0	1118.6	39.1	1271.2	40.3	1435.9	41.4	1613.3	42.5	1803.6
4350	36.6	972.3	37.8	1112.2	38.9	1263.8	40.0	1427.7	41.2	1604.0	42.3	1793.2
4400	36.4	966.8	37.5	1105.9	38.7	1256.6	39.8	1419.5	40.9	1594.9	42.0	1783.0
4450	36.2	961.3	37.3	1099.6	38.5	1249.6	39.6	1411.5	40.7	1585.9	41.8	1773.0
4500	36.0	956.0	37.1	1093.5	38.3	1242.6	39.4	1403.7	40.5	1577.1	41.6	1763.1
4550	35.8	950.7	36.9	1087.5	38.0	1235.8	39.2	1395.9	40.3	1568.4	41.3	1753.4

Nevertheless it is not uncommon for a considerable length of sewer, starting from the top, to be surcharged so as to accommodate 50% more than the theoretical capacity without any flooding occurring.

Flow in open channels and partly filled culverts

All early formulae have been used indiscriminately for pipes, open channels and partly filled culverts. However, turbulent flow in open channels is different from that in fully charged pipes or culverts for as soon as there is a free surface of water in contact with the air, waves form and these dissipate energy. The difference between the flow calculated according to the formula $V = Km^{\alpha}i^{\beta}$ does not appear to be great in the case of pipes that are very smooth, straight and regular, but study of the researches referred to in the footnote to Table 3 showed that the hydraulic mean depth was not the cross-sectional area divided by the wetted perimeter, but averaged, with remarkable accuracy, the cross-sectional area divided by the sum of the wetted perimeter and one-half the width of the water-to-air surface. Thus, it would appear that the dissipation of energy by waves was directly related to the surface over which this dissipation could take place. The values in Table 3 were therefore calculated to the author's formula, using this new assumption.

In the past, tables and diagrams were prepared showing proportional flows in partly-filled circular pipes and culverts and, on the basis of these theoretical calculations, it was thought that at a depth of flow of 0.94 of the diameter, the discharge would be about 1.0757 times the flowing-full discharge. For the reasons just described this has been proved a fallacy, as is shown by Table 3.

What has been learnt by studying partly filled circular pipes and culverts can be applied to open channels because the proportional flow, as calculated by the

Table 3 Proportional cross-sectional areas, velocities and discharges of partly-filled circular pipes and culverts relative to the running-full conditions[8-13]

Proportional depth	Proportional area	Proportional* velocity	Proportional* discharge
1.0	1.000	1.000	1.000
0.9	0.9480	1.0394	0.9854
0.8	0.8576	1.0189	0.8738
0.7	0.7477	0.9765	0.7302
0.6	0.6265	0.9173	0.5747
0.5	0.5000	0.8425	0.4213
0.4	0.3735	0.7517	0.2808
0.3	0.2523	0.6427	0.1621
0.2	0.1424	0.5102	0.0727
0.1	0.0520	0.3373	0.0175

* Based on numerous tests on partly-filled small pipes and large sewers, and now accepted as more accurate than previous figures.

author's formula with the new assumption, agreed with the averages of the recorded flows very closely for very varied shapes of cross-section of flow. It can thus be said that the rule can be applied to any open channel or partly filled culvert of reasonable shape.

Calculations often have to be made for flows in channels of rectangular or trapezoidal cross-section at sewage works and these can occupy time because there are few tables or diagrams available. A method given herein willy apply to most cases. Table 4 gives the depths and widths of rectangular channels in terms of diameters of circular culverts of similar discharges. This can be used in conjunction with Table 2 or the table now reproduced in Appendix II of this volume (pp. 481–518).

Table 4 Depths and widths of rectangular channels in terms of diameters of circular culverts of equal discharge

Depth	Width	Depth divided by width
1.900	0.475	4.0
1.758	0.502	3.5
1.609	0.536	3.0
1.451	0.580	2.5
1.367	0.608	2.25
1.281	0.641	2.0
1.190	0.680	1.75
1.095	0.730	1.5
0.993	0.794	1.25
0.940	0.835	0.125
0.884	0.884	1.0
0.837	0.930	0.9
0.788	0.986	0.8
0.733	1.048	0.7
0.684	1.140	0.6
0.626	1.253	0.5
0.596	1.324	0.45
0.564	1.410	0.4
0.530	1.515	0.35
0.494	1.648	0.3
0.456	1.824	0.25
0.435	1.935	0.225
0.414	2.068	0.2
0.391	2.232	0.175
0.366	2.440	0.15
0.339	2.713	0.125
0.325	2.885	0.1125
0.309	3.092	0.1
0.296	3.292	0.09
0.282	3.530	0.08

NOTE. To use Table 4 decide the most suitable ratio of depth to width and gradient of the proposed channel, then, by use of Table 2 or the Table in Appendix II, find the diameter of the circular pipe laid at this gradient which will give the required discharge. This figure, multiplied by the particular constants in the table opposite the ratio selected, gives the required dimensions.

References

1. Manning, R. On the flow of water in open channels and pipes. *Trans. Inst. Civil Engnrs. of Ireland*, Vol. XX, 1891.
2. Crimp, W. S. and Bruges, C. E. *Crimp and Bruges Tables and Diagrams for Designing Sewers and Water Mains, 1897*. Revised by Bruges, W. E., 1936. Metric edn, 1969.
3. Scobey, F. C. The flow of water in concrete pipes. *U.S. Dept. Agric., Bull.*, **852**, 1920.
4. Escritt, L. B. and V. P. *Escritts Tables of Metric Hydraulic Flow*. George Allen & Unwin, London, 1970.
5. Ackers, P. *The Roughness of Pitch-fibre Sewer Pipes*. Hydraulics Research Station. Nov. 1959.
6. *Charts for the Hydraulic Design of Channels and Pipes*, 4th edn (metric), HMSO, 1978.
7. Bland, C. E. G. *Design Tables for Determining the Flow Capacity of Vitrified Clay Pipes*. Clay Pipe Development Assoc., London.
8. Yarnell, and Woodward, S. M. *The Flow of Water in Drain Tile*, Bull. No. 854, U.S. Dept. of Agriculture, 1920.
9. Wilcox, E. R. *A Comparative Test of the Flow of Water in 8-inch Concrete and Vitrified Clay Sewer Pipe*. Bull. No. 27, Univ. of Wash. Engnrg. Expl. Stn., 1924.
10. Cosens, K. W., Sewer pipe roughness coefficients. *Sewage and Industrial Wastes 26* pt. 1, p. 42, Jan. 1954.
11. Pomeroy, R. D., Flow Velocities in Small Sewers, *J. Wat. Pollution Control Fed.*, Sept 1967.
12. Escritt, L. B., Modified theory for turbulent flow, *Engg*, Apr. 1972.
13. Escritt, L. B., *Public Health Engineering Practice*, Vol. II, Appendix III, Macdonald & Evans, London, 1972.

CHAPTER 3
Sanitary or Soil Sewerage

The flow in a sanitary sewerage system which has been designed to take domestic and trade waste, but from which surface water is excluded, is made up of domestic waste, trade waste, infiltration of subsoil water by leakage into the sewers and drains to sewers, leakage of rain water into inspection chambers and manholes, and illicit discharge of surface water. All of these components are very varied in different localities, largely uncertain and liable to change. The engineer has to make an estimate of the future flows in, say 30 years (Great Britain) or 40 years or more (United States), and design accordingly.

The main part of the flow is made up of the water supply which is very similar in quantity from day to day or hour to hour as the sanitary sewage flow. It is not, however, quite the same, for sometimes as much as 40% and often 25% of the water supply does not reach the sewers. Whereas the water supply rises to a sharp peak during the morning, of about twice the average for the day, the time of concentration of the sewerage system tends to round off the flow curve, slightly reducing the height of the peak and making it come later.

Nevertheless, the water demand is a very valuable clue to the sanitary-sewage flow and therefore information from the water authority as to flows per head of population or per house served is very useful. The water authority's estimates of future flows by virtue of increase in population and increase per head of population should also be ascertained.

Water demands are greatest in large cities and industrial areas. For example, in Chicago the demand some years ago was in the region of 1000 litres per head of population per day. This is a high figure compared with the supply of the Metropolitan Water Board, London, England, which in 1964/65 was 268 litres per head of population per day.

The average water supply to 67 cities in Massachusetts was 238 litres per head per day and the average flow of domestic sewage of 15 cities in the United States was 240 litres per head per day.

Imhoff, Müller and Thistlethwayte[1] give flows in litres per head per day, as about 246–473 for the United States; 93–189 for Australia; 93–170 for Great Britain; and 82–189 for Germany. The average daily flow of sewage to the

West-Southwest Sewage-treatment Plant, Chicago, in 1971 was 1059 litres per head per day, excluding by-passes, as against 317 for Beckton Sewage-treatment works, London, England, in the 1964 design calculations (again excluding discharge of storm water). In Canada 377 litres per head per day is an accepted design figure.

In Great Britain it was found, by gauging the flow from a new housing estate in an area where there could be no infiltration, that the discharge from that estate was 122 litres per head of population per day: this is considerably less than some authorities were allowing at that time in the design of sewers for new housing estates. For long it had been British practice to allow up to 68 litres per head per day for small villages, 113 litres for medium-sized provincial towns, and upwards of 142 litres for large towns. Records for 1936–38 showed that the largest groups of water authorities were supplying about 125 litres per head per day including both domestic and trade demands, while at the same time the largest groups of sewage-treatment works were receiving average dry-weather flows about 147 litres per head of population per day in dry weather. Water demands and sewage flows are increasing but there is a tendency to overestimate in the design of works. Figures published by the National Water Council in 1980, put average household consumption of water still at only 120 litres per head per day, although this is somewhat at variance with the figures attributed to the Dept. of the Environment (see Figure 2) for the years 1976 and 1980. These, of course, include trade as well as domestic consumption, but assuming that unmetered supplies are almost entirely domestic a resulting average of about 200 litres/head for 1976, and 225 litres/head for 1980 is obtained. In terms of overall consumption the figure of 125 litres/head for 1936, quoted above, plus 10% increase for every 4 years since, would equate with the figures given in the Table.

Estimating future population for determining future water demands can be done by plotting the population from previous government censuses, or other reliable sources, for as many years back as records are available, and then attempting to extrapolate the curve to give predictions for up to 30 or 40 years ahead. Populations in undeveloped areas tend to increase logarithmically until either the development extends beyond the boundaries of the catchment concerned or the population tends to exceed the available food supply; then the rate of increase reduces until finally there is no increase at all. It is best to plot population curves on logarithmic paper of the kind on which years can be plotted to an arithmetic scale but population to a logarithmic scale. The curve so produced often gives a good suggestion of how the population is increasing or otherwise; but the estimate made in this manner should be checked against other factors such as the amount of land remaining available for development, and political or planning proposals. After the future population has been decided, adjustments according to the water authority's estimates of future water demand per head of population should be made to give a reasonable estimate of the future water demand. Estimates made in this way may be very good but unforeseen events may upset them.

Figure 2   Water consumption by water authority areas (Source: DOE)

Megalitres per day

	Total		Metered[1]		Unmetered		Average daily per capita consump- tion[2]	
	1976[3]	1980	1976[3]	1980	1976[3]	1980	1976[3]	1980
United Kingdom	17,263		5,893		11,370		310	
North West	2,404	2,474	943	865	1,461	1,609	344	358
Northumbrian	971	1,049	525	546	446	503	364	398
Severn-Trent	2,018	2,260	654	660	1,364	1,600	247	276
Yorkshire	1,200	1,367	383	375	817	992	265	302
Anglian	1,384	1,550	528	519	856	1,031	292	315
Thames	3,262	3,612	765	685	2,497	2,927	280	313
Southern	1,032	1,167	332	331	700	836	274	305
Wessex	735	803	288	285	447	518	326	349
South West	346	397	93	97	253	300	253	284
Welsh	1,071	1,196	410	394	661	802	356	396
England and Wales	14,423	15,875	4,921	4,757	9,502	11,118	293	322
Scotland[4]	2,246		782		1,464		432	
Northern Ireland	594	686	190	188	404	498	429	493

(1) Metered supplies include some non-potable (unfit for drinking) water; about 10 per cent on average.
(2) Litres per head per day.
(3) Figures distorted by effects of the drought in 1976.
(4) Figures for Scotland refer to the financial year beginning 1 April. Data for 1980/81 are not available.

Future industrial demands are even more speculative. Present demand can be determined from water authority's receipts for metered flow or a sewerage authority's records of discharge of wastes. But the tendency is for new industries to use large quantities of water, and it is hard to estimate what these will be in the future.

In the United States industrial flows are usually estimated in terms of flow per hectare per day. The average for 15 towns was 18.75 m^3 per hectare per day: the maximum figure for one of the towns was 23 m^3 per hectare per day. This would appear to average in the region of 312 litres per head of population per day, but the maximum figure was about 75% higher than this.

Infiltration of subsoil water is very varied indeed. At one English town with which the writer was concerned infiltration accounted for 50% of the total average dry-weather flow: in another it was a negative value. It is possible that infiltration to old sewerage systems may well, on the average, add 40% to the dry-weather flow: it is hard to obtain accurate figures. In the United States, in spite of the great variability of infiltration, it was found that for new sewers it averaged about 181 litres per head per day and, for old sewers, 380 litres.

Leakage of more than 15 litres per day per m² of interior surface of sewer is considered excessive.

Having estimated the future flows of domestic sewage, trade waste and infiltration, and expressed these in terms of litres per head of population per day, the peak rates have to be determined because rate of flow varies throughout the day and from day to day throughout the year, for which facts there are several causes. The sanitary sewers need to be constructed of such capacity as to be able to take the maximum flow that may occur at any time, and it is international practice that when this flow occurs there is no surcharge in a separate sanitary sewer.

The peak daily water demand, and therefore the flow of sanitary sewage in the sewers of a town, is usually twice the average for the day. On certain days of the week there is more flow than on others; in certain months of the year (e.g. July in England) more water than the average is demanded. The combined result of this irregularity of water demand is that on one day in the year the flow is about $1\frac{1}{2}$ times the average. By taking the peak daily flow on the day of

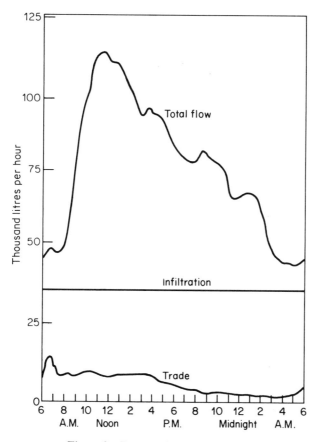

Figure 3   Dry-weather flow diagram

maximum water demand we find that the peak flow of sanitary sewage is about 3 times the average. To the estimated sewage flow has to be added the estimated future infiltration: this is fairly steady from day to day but is a little more in wet than in dry weather. In the United States and Canada it is considered that peak flows from separate sewerage systems rarely exceed $2-2\frac{1}{2}$ times the average.

Another distinction between British and US practice is that whilst the former base all their calculations on dry-weather flow (DWF), which will not normally include any allowance for infiltration, the latter refer to 'average' flow which will include average increments to cover infiltration, the effects of rainfall, and any trade waste flow. Assessment of peak daily flows by British engineers is built up by adding an allowance for infiltration to a multiple of DWF (commonly 3 times DWF), together with a further allowance for trade waste flow where necessary.

Suppose an engineer is constructing sewers for a developed area and he has estimated that the domestic flow will be 240 litres per head of population per day and the infiltration 382 litres per head per day. If the domestic flow is multiplied by three and to this is added the infiltration, the total flow will be 1102 litres per head of population per day or about 4.6 times the average flow of sanitary sewage. This is in the range of figures commonly used in sanitary-sewerage design. Imhoff, Müller and Thistlethwayte, in their international study of sewage disposal, give as the basis of sewer-capacity design an allowance of $3\frac{1}{2}-5$ times the average DWF. In Great Britain the crude but practical rule is to allow not less than 4 times DWF and, whenever there is any doubt as to what future flows may be, up to 6 times DWF: several consulting engineers normally work to the latter figure, which is particularly appropriate for small systems where the time of concentration of flow is short.

In the United States several rules have been used, and Babbitt[2] quoted two formulae, one of which gives the figure of 5 times DWF and the other 4.5 for sewers draining populations of not more than 1000 persons and reduced quantities for larger populations. Figures in WPCF Manual of Practice No. 9[3] and British experience suggest that peak rates of flow in purely sanitary sewers average in the region of 15 times DWF divided by the sixth root of the population. Such rules, while not without value, should be used with caution because the peak rate of flow in sanitary sewers does not appear to be directly related to population but to the proportion of infiltration and the time of concentration of the sewerage system. A large proportion of infiltration (as in the foregoing calculation) can give a peak flow of 4.6 times DWF and this could be applied to large and small populations with little modification. When, however, the time of concentration is very long, e.g. 6 hours, it has the effect of rounding the angular curve; but it does not reduce the daily peak rate of sanitary flow (excluding infiltration) by more than about 14%.

To continue with the example case: if the peak flow of sanitary sewage including infiltration is 1104 litres per head of population per day and it is ascertained that there will be, on the average, 4.43 persons per house, the peak flow per house will be 4890 litres per house per day or 3.4 litres per minute per house.

Table 5   Sanitary sewer design calculation

Locality (manhole to manhole)	Number of houses served	Flow (1/min)	Gradient (1 in:)	Diameter (mm)	Capacity (1/min)
E5 to E4	65	220	120	150	1019
E4 to E3	98	333	120	150	1019
E3 to E2	133	452	135	150	962
E2 to E1	176	609	100	150	1116
E1 to B2	200	680	90	150	1178
B7 to B6	20	68	145	150	928
B6 to B5	91	300	180	150	832
B5 to B4	118	401	180	150	832
B4 to B3	202	686	165	150	869
B3 to B2	282	968	165	200	1849
B2 to B1	491	1670	145	200	1974
B1 to outgo	577	1960	145	200	1974

The flow from each part of the estate to each part of the sewerage system can then be found by counting the number of houses and multiplying by this figure. This is illustrated in Table 5, which relates to Figure 4.

Some engineers, observing that according to the sewer calculations the top ends of sewers receive very little flow in proportion to their capacities, calculate the gradients necessary to bring velocities up to self-cleansing at the peak daily rate of flow. An example of where such a method might be used is the length of sewer from B7 to B6 in Table 5 which, running full, would discharge 928 litres

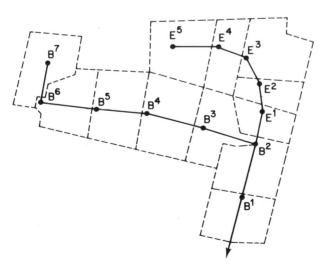

Figure 4   Drainage area diagram (used in example sewer-design calculations)

per minute but, according to the calculation, would be receiving 68 litres per minute. On this argument the proportional discharge would be about 0.0727 giving a proportional velocity of 0.5102, or only 0.435 m per second at the peak rate of flow (see Table 3). This would not be a self-cleansing condition. In fact, the top ends of sewers receive from time to time flushes of flow due to the simultaneous discharges of the several sanitary appliances draining thereto. As already mentioned, a single water closet produces a flush of about 90.6 litres per minute: a closet plus one tap would give 100.2 litres per minute. In a group of houses there will be occasions when several appliances will discharge simultaneously, producing a good self-cleansing flush. Table 6, which is calculated to a more modest rule than that used by designers of indoor sanitation (described in Chapter 1) gives some indication of the magnitude of flushes of sewage likely to occur in small housing estates. According to this table, the sewer length B7 to B6 would receive flushes of 276 litres per minute, which would produce a velocity of more than 0.607 metres per second and self-cleansing conditions.

Table 6   Simultaneous discharge of sanitary appliances

Number of houses served	Simultaneous discharge (litres/minute)
1	102
2	128
3	146.5
4	161
5	174
6	187.5
8	202
10	220
12	233.5
15	250.5
20	276
25	297.5
30	314
40	348
50	376

Table 7 is based on the American Standards Institute National Plumbing Code, with an appropriate conversion from US gallons to litres per minute. Having totalled the fixture units, the flow per house is obtained by multiplying the total units by the number of persons per house (usually between 3 and 4) and then adjusting to litres per minute.

Table 7 Fixture-units per fixture or group*

Fixture type	Fixture-unit value as load factors
1 bathroom group consisting of tank-operated water closet, lavatory, and bathtub or shower stall	6
Bathtub† (with or without overhead shower)	2
Bidet	3
Combination sink-and-tray	3
Combination sink-and-tray with food-disposal unit	4
Dental unit or cuspidor	1
Dental lavatory	1
Drinking fountain	$\frac{1}{2}$
Dishwasher, domestic	2
Floor drains	1
Kitchen sink, domestic	2
Kitchen sink domestic, with food waste grinder	3
Lavatory	1
Lavatory	2
Lavatory, barber, beauty parlor	2
Lavatory, surgeon's	2
Laundry tray (1 or 2 compartments)	2
Shower stall, domestic	2
Showers (group) per head	3
Sinks	
Surgeon's	3
Flushing rim (with valve)	8
Service (trap standard)	3
Service (P trap)	2
Pot, scullery, etc.	4
Urinal, pedestal, syphon jet, blowout	8
Urinal, wall lip	4
Urinal stall, washout	4
Urinal trough (each 2-ft section)	2
Wash sink (circular or multiple), each set of faucets	2
Water closet, tank-operated	4
Water closet, valve-operated	8

* From United States of America Standards Institute National Plumbing Code, USASI A40.8–1955.
† A shower head over a bathtub does not increase the fixture value.
Note: For a continuous or semicontinuous flow into a drainage system, such as from a pump, pump ejector, air-conditioning equipment, or similar device, fixture unit shall be allowed for every 2 liters/min of flow. *Source:* ASCE Manual No. 37.

References

1. Imhoff, K., Müller, W. J. and Thistlethwayte, D. K. *Disposal of Sewage and other Water-borne Wastes.* Butterworth, London (1956).
2. Babbit, H. E. *Sewerage and Sewage Treatment*, John Wiley & Sons, N.Y. 7th edn. (1953).
3. Man. of Practice No. 9, Design & Constr. of Sanitary & Storm Sewers, Jt. Comm. Wat. Pollution Contr. Fed. & Am. Soc. Civ. Engnrs. (1969).

CHAPTER 4
Surface-water Sewerage

Surface-water sewers are designed and constructed to carry away the run-off due to rainfall on roofs and paved surfaces or, where the roofed and paved areas are a comparatively small proportion of the total, they may be made to take run-off from unpaved areas. The calculations for their sizes are not so simple as those required for separate sanitary sewers for which estimates for average and peak flows may be easily made. The difficulty with surface-water sewers is that it cannot be said that any particular intensity of rainfall will have to be catered for so many times during the financial or structural life of the sewer for, although there are rainfall statistics that can be used in design, they relate to averages only. An engineer may design a system of sewers for one locality and eventually learn that, over the course of the years, they have never had to accommodate a storm nearly as great as that for which they were constructed, whereas, in another locality, a system constructed to the same standard of design may be seriously overloaded within months of completion of work.

The engineer must, therfore, accept the position that on the one hand he may be accused of extravagant design and on the other of designing sewers to too low a standard. If he attempts to make the sewers capable of taking without risk of flooding the most intense storm likely to occur, say, once in a hundred years, he would certainly be extravagant. If, on the other hand, his sewers are so small as to involve even moderate flooding once a year, they are certainly inadequate. Thus it has come about that there are recognized standards of design based on many decades of experience in different countries, and which are considered reasonably economic in view of the particular local conditions.

In the past engineers found by experience that paved and roofed surfaces varied in degree of impermeability and also that the intensity of rainfall that could be safely assumed varied with the size of the area to be drained. Whereas a high intensity of rainfall would be required for a small building site, a much lower intensity could be allowed for in calculations for a drainage area of several square miles. Thus, some engineers prepared scales of rainfall intensity related to gross area drained: E. J. Silock[1] wrote as follows:

As an example of practical working, the following table may be quoted, giving the basis on which sewers have for many years been designed in a North Country city with an average rainfall of 914 mm, and nearly all streets being covered with impervious paving.

Table 8   Flat-rate rainfall intensities per hectare

Drainage area (ha)	Rainfall per hour (mm)	Drainage area (ha)	Rainfall per hour (mm)	Drainage area (ha)	Rainfall per hour (mm)
8.08	25.4	30.3	11.8	141	8.87
9.30	24.3	40.4	11.4	151	8.62
10.52	22.8	50.5	11.2	161.4	8.36
11.73	21.6	60.6	10.9	171.4	8.11
12.95	20.3	70.8	10.7	181.5	7.85
14.30	19.0	80.8	10.4	191.5	7.60
15.35	17.8	91.0	10.2	200.8	7.35
16.6	16.5	100.8	9.9	212.0	7.10
17.8	15.3	110.8	9.6	222.0	6.85
19.0	13.9	121.0	9.4	232.0	6.60
20.2	12.7	131.0	9.1	242.0	6.34

The figures in Table 8 were understood to have been originated by Thomas Hewson for use in areas of very dense paving in Leeds. Similar rainfall intensities were allowed for on the continent; for example, the maximum rainfall intensity used in the nineteenth-century sewerage of Berlin was 22 mm/h, of which one-third was considered to enter the sewers; in the sewerage of Dresden a maximum of 18 mm/h was allowed. Of more recent years flat-rate intensities of 25, 19, or even as little as 13 mm/h of rainfall have been allowed for on impermeable surfaces over large housing estates in Britain, and no reports of inadequacy have been recorded. For small estates and building sites it has been general British practice to allow 25 mm of rain per hour over the roofed and paved surfaces, often assuming that these were 100% impermeable, while rain falling on garden land, etc., was assumed to soak away.

The change from this practice came about on the discovery that the greatest run-off liable to result from a storm of a particular frequency of occurrence was caused by one having a duration equal to the time of concentration, i.e. the time taken for water to flow from the most distant part of the catchment to the point at which flow was to be known, and from this originated the time-of-concentration methods.

Time of concentration

Thus, former practices became replaced by these time-of-concentration methods, in particular the 'rational' method in America and the very similar Lloyd-Davies method in England. The two main methods developed more or less independently on each side of the Atlantic. In Britain, Lloyd-Davies[2] found empirically that, of the storms of equal frequency of occurrence, the one that had a duration equal to the time of concentration of the drainage area produced the greatest run-off in that area. To do this he used early designs of autographic rain gauges and flow gauges. On the basis of this argument he produced the Lloyd-Davies formula which, as used in Great Britain at the present day, reads

$$Q = 0.166 \, ApR \tag{5}$$

where: 0.166 = a constant which converts mm rainfall per hour multiplied by area in hectares to m^3/min;

Q = run-off in m^3/min;

Ap = impervious area or area drained in hectares impermeability factor;

R = intensity of rainfall in mm/h.

In the original Lloyd-Davies formula the rainfall intensity R did not denote a storm of any particular frequency of occurrence but what Lloyd-Davies considered a reasonable intensity to allow for sewer construction and his suggested 'working curve' was according to the Birmingham formula

$$R = \frac{1016}{t + 20} \tag{6}$$

where: R = intensity of rainfall (mm/h)

t = time of concentration (or duration of storm) in minutes.

The Ministry Departmental Committee on Rainfall and Run-off[3], perhaps mistakenly, adjusted this formula by restricting its use to times of concentration between 20 and 100 min and, for 5–20 min, substituting

$$R = \frac{762}{t + 10} \tag{7}$$

These two formulae became known as the Ministry of Health Standard Curve and were found to be a rough representation of British storms liable to occur about once every 15 months.

British engineers, by making no allowance for change of impermeability during rainfall, tended to over-estimate impermeability of surfaces and also to use questionable methods to allow for irregular distribution of impermeable area resulting, in the author's view, in the over-sizing of sewers from about 1932 to after the Second World War when there came a change in opinion as will be described.

"Rational" method

The American 'rational' method can be expressed by the formula

$$Q = 0.166 \, CiA \tag{8}$$

where: Q = run-off in m^3/min;

0.166 = a constant to convert mm of rainfall per hour $\times$ hectares to m^3/min;

c = a run-off coefficient representing the ratio of run-off to rainfall*;

i = rainfall in mm/h;

A = drainage area in hectares

There have been several ways in which this formula has been used. In some crude methods c was given a fixed value as ascertained by comparing rainfall to run-off from specific areas; i was determined by a crude formula for intensity of rainfall according to time of concentration; and A was the gross area of the catchment in question. Such methods were liable to lead to error, for it is now well known that the run-off coefficient is a factor that varies not only for type of surface but very greatly as rainfall proceeds after start of storm and, to a lesser extent, according to the average intensity of storms for the particular locality. Rainfall intensity varies inversely as a function of duration and inversely as a function of frequency of occurrence. Thus it has been necessary to produce scales giving values of c and i for various localities. Furthermore, in the author's view, the drainage area is best expressed as the impermeable area and not the gross area.

Rainfall intensity

In 1891 Talbot expressed the opinion that storms of any particular frequency of occurrence could be represented by the formula

$$R = \frac{a}{t + b} \tag{9}$$

and, in 1905 Kuichling suggested values on this basis.[5]

The disadvantages of formulae of this type are that they do not accurately express the data and they cannot be extrapolated without the risk of involving gross error. For example, the Ministry of Health Standard Formulae are applicable to the limits already stated and should not be used beyond these limits. There are occasions, for example, in the calculation of storage for storm water in tanks or ponds, when the use of formulae of this kind would lead to very misleading results. Their use in theoretical investigations, moreover, can produce nonsensical findings.[6] (Unless the power of t is less than unity, a storage tank of finite capacity and without outlet can hold all the rainfall that drains into it in geological time.) For this reason a more accurate expression of rainfall statistics is really desirable.

---

* In the words of Wat. Pollution Control Fed. Manual of Practice No. 9, *Design and Construction of Sanitary and Storm Sewers*,[4] 'this factor c must represent a host of variables including infiltration, ground slope, ground cover, surface and depression storage, evaporation, transpiration, antecedent precipitation, shape of drainage area, overland flow velocity, etc.' Some authorities use a value of c which includes all the factors mentioned and yet make further allowances, forgetting that these allowances have already been made.

In 1935 E. G. Bilham published a discussion[7] on the classification of heavy rainstorms in which he gave a formula that can be expressed

$$R = \frac{267.7 \ N^{0.2817}}{t^{0.7183}} \ \frac{152.4}{t} \tag{10}$$

where: R = mm of rainfall per hour;
N = number of years between storms of this magnitude;
t = duration of storm in minutes.

Table 9 gives values calculated to Bilham's formula.

Bilham's data were collected by finding the shortest intervals of time in which specified amounts of rain fell. This did not mean that the measured 'storms' included any rainfall before or after the period of time applicable to the specified intensity. Also, a storm was considered as one storm even if there was a temporary cessation of rainfall between the beginning and end. This investigation was applied to 10 years' records.

Table 9 Probable British rainfall in mm per hour according to Bilham's formula

Duration (min)	Once-a-year storm	Once-in-2-years storm	Once-in-3-years storm	Once-in-4-years storm	Once-in-5-years storm	Once-in-10-years storm	Once-in-15-years storm
5	51.5	72.0	84.4	94.0	102.0	131.0	150.0
5.5	51.0	68.0	79.5	88.6	96.0	126.0	140.0
6	48.5	64.4	76.0	83.8	90.9	116.0	136.0
6.5	46.0	61.2	71.6	80.5	86.3	110.0	126.0
7	43.4	58.7	68.2	76.0	82.4	104.5	120.0
7.5	42.7	56.3	65.4	72.6	76.5	100.0	115.8
8	41.0	54.0	62.8	69.8	75.5	95.9	109.8
8.5	39.6	52.1	60.5	67.1	72.5	92.0	105.8
9	38.2	50.1	58.2	64.5	69.8	88.8	101.4
9.5	37.1	48.6	56.2	62.5	67.5	85.6	97.7
10	35.9	47.0	54.6	60.6	65.3	82.6	94.7
11	33.9	44.3	51.0	56.8	61.4	77.6	88.8
12	32.2	41.9	48.5	53.5	58.0	73.3	88.4
13	30.6	39.8	46.1	51.0	55.0	68.2	78.3
14	29.3	38.0	43.8	48.5	52.5	66.0	75.0
15	28.1	36.4	42.0	46.4	50.1	63.1	71.5
16	27.0	34.9	40.2	44.5	47.8	62.8	68.7
17	26.0	33.6	38.7	42.7	46.0	58.2	66.0
18	25.1	32.4	37.6	41.2	44.4	55.7	63.6
19	24.3	31.3	35.9	39.7	42.7	53.5	61.3
20	23.6	30.2	34.8	38.4	41.4	51.9	58.3
21	22.8	29.3	33.6	37.2	40.0	50.2	57.2
22	22.2	28.4	32.7	36.0	38.8	48.7	55.3
23	21.6	27.6	31.8	34.8	37.6	47.2	53.6
24	20.9	26.8	30.8	34.0	36.6	45.8	52.0

Duration (min)	Once-a-year storm	Once-in-2-years storm	Once-in-3-years storm	Once-in-4-years storm	Once-in-5-years storm	Once-in-10-years storm	Once-in-15-years storm
25	20.4	26.1	30.1	33.4	35.6	44.6	50.7
26	19.9	25.5	29.3	32.2	34.6	43.5	49.3
27	19.5	24.8	28.5	31.4	33.8	42.3	48.2
28	19.0	24.3	27.7	30.6	33.0	41.0	47.0
29	18.8	23.5	27.2	30.0	32.2	40.3	45.8
30	18.2	23.2	26.6	29.3	31.5	39.4	44.8
32	17.4	22.2	25.5	28.0	30.2	37.6	42.8
34	16.8	21.4	24.4	26.9	29.0	36.5	41.1
36	16.2	20.6	23.6	25.9	27.8	34.8	39.5
38	15.6	19.8	22.8	25.0	26.8	33.6	38.0
40	15.1	19.2	22.0	24.2	26.0	32.4	37.5
42	14.6	18.6	21.3	23.4	25.1	31.3	35.5
44	14.2	18.0	20.6	22.7	24.3	30.3	34.4
46	13.8	17.5	20.0	22.0	23.6	29.4	33.4
48	13.4	17.0	19.4	21.4	22.9	28.6	32.4
50	13.1	16.5	18.9	20.7	22.3	27.8	31.5
55	12.3	15.5	17.7	19.5	20.9	26.0	29.5
60	11.6	14.7	16.7	18.4	19.7	24.5	27.8
65	11.0	13.9	15.8	17.4	19.1	21.2	26.3
70	10.5	13.2	15.1	16.5	17.7	22.0	25.0
75	10.0	12.6	14.4	15.8	16.9	21.0	23.8
80	9.6	12.1	13.8	15.1	16.2	20.1	22.8
85	9.2	11.6	13.2	14.5	15.5	19.3	21.8
90	8.9	11.1	12.7	13.9	14.9	18.5	21.0
95	8.6	10.7	12.2	13.4	14.4	17.8	20.2
100	8.3	10.4	11.8	13.0	13.9	17.2	19.5
110	7.8	9.7	11.1	12.1	13.0	16.1	18.2
120	7.3	9.2	10.4	11.4	12.2	15.2	17.2
130	6.9	8.7	9.9	10.8	11.6	14.3	16.2
140	6.6	8.3		10.3	11.0	13.6	15.4
150	6.3	7.2	9.4	9.8	10.5	13.0	14.7
160	6.0	7.5	8.6	9.4	10.1	12.4	14.0
170	5.8	7.2	8.2	9.0	9.6	11.9	13.5
180	5.6	7.0	7.9	8.6	9.3	11.4	12.9
190	5.4	6.7	7.6	8.3	8.9	11.0	12.4
200	5.2	6.5	7.3	8.0	8.6	10.6	12.0
210	5.0	6.3	7.1	7.8	8.3	10.3	11.6
220	4.9	6.1	6.9	7.5	8.1	10.0	11.2
230	4.7	5.9	6.7	7.3	7.8	9.7	10.9
240	4.6	5.7	6.5	7.1	7.6	9.3	10.6
250	4.5	5.6	6.3	6.9	7.4	9.1	10.3
260	4.3	5.4	6.1	6.7	7.2	8.8	10.0
270	4.2	5.3	6.0	6.5	7.0	8.6	9.7
280	4.1	5.1	5.8	6.4	6.8	8.4	9.5
290	4.0	5.0	5.7	6.2	6.7	8.2	9.2

From the data obtained and expressed by Bilham's formula it is possible to estimate the number of times in a period of years on which so many millimetres of rain are liable to fall in so many minutes, and curves can be plotted for the once-a-year storm, once-in-5-years storm, once-in-10-years storm, etc. The description, however, of a curve as the 'once-a-year' curve does not mean that storms complying with the curve will occur on the average only once a year. If a particular point on the curve is considered it can be implied that, on average, a storm of that intensity and duration will occur once a year. But the rainfall station will also record other storms complying with the curve. For example, a storm of about 3.5 mm/h lasting for 10 min may be recorded, but also on the same or another occasion there may be a precipitation of about 25 mm/h lasting for 20 min. Thus, when the whole curve is considered, the number of 'once-a-year' storms that occur during the year complying with the curve, must exceed one, but just how many cannot be said at present, for this matter has not been investigated.

After many years of using the Ministry of Health Standard Curve, British engineers have come to accept Bilham's formula, using it for storms of occurrence of once a year or once in 3 years, according to their preference, and for this purpose it served very well.

With the aid of new intensity rain gauges a re-examination was made of Bilham's work, and the results published by the Meteorological Office.[8,9] As a result of these investigations, two basic recommendations were made. The first was, in effect, to require that the number of years between occurrence of storms as found by Bilham's formula should be multiplied by 0.9. This would mean that the constant in Bilham's formula, as expressed above, should be altered from 267.7 to 275.8. The second adjustment was one which made no difference except for rainfall durations of less than 1 min and except for storms more frequent than once a year. The writers of the above publication considered that the former change was too small compared with the variations between the results as recorded at different stations to justify amendment of published tables, and that the second adjustment could be ignored. (The figures in Table 9 are to the original formula.)

The storm intensities in Great Britain are probably the lowest in Europe and intensities generally increase farther towards the east of the continent.

In the United States the lowest storm intensities are in the extreme west and are comparatively low in the region of the boundary between the United States and Canada from Michigan to Maine. Intensities increase progressively from west to east and then from north to south, becoming a maximum on the coast of Texas, Louisiana and Florida where intensities are more than four times as much as experienced in Washington and parts of Oregon, Idaho and Nevada. The average intensity for the whole of the United States is experienced on a belt running through Texas, Kansas, Nebraska, South Dakota, North Dakota, Minnesota, Wisconsin, Ohio and Pennsylvania.

Meyer[10] compiled a series of formulae of Talbot's type applicable to five regional groups in the United States and storm frequencies of once in 1, 2, 5, 10,

25, 50 and 100 years. While these figures have some irregularities, they showed that, when storms of long duration were considered, the intensity of rainfall varied approximately as $N^{0.2}$, where N equals the number of years between occurrences of the storm in question, and inversely as $t^{0.8}$, t being the duration of rainfall in minutes. To make storms of short duration agree this formula a constant would have to be added to the appropriate value of t.

Yarnell[11] compiled rainfall intensities and frequencies for 211 United States Weather Bureau stations based on records available through 1933, and prepared outline maps for the different intensities, frequencies and durations. These figures were studied by Kirpich.[12] Spot checks between figures worked from Meyer's compilation and Kirpich's study of Yarnell's work agree very well apart from minor irregularities. Unlike Meyer's figures, the figures which Kirpich gave suggested that whereas the power of N was about 0.2 for most areas, it was 0.15 where the intensity of the 60 min storm occurring once in 10 years was more than 70 mm/h.

Many authorities have studied Chicago and New York rainfall, the storm intensities of which are about the average for the whole of the United States. Schafmayer's figures for Chicago[13] agree with the above studies very well and, among several formulae, Bleich[14] gave the formula for the 2-year storm near New York City

$$R = \frac{106}{(t + 9)^{0.813}} \tag{11}$$

and for the 10-year storm near New York City

$$R = \frac{162}{(t + 12)^{0.795}} \tag{12}$$

On the basis of the above data, the author suggests a general formula for the United States:

$$R = \frac{94xN^{0.2}}{(t + 8.5)^{0.8}} \tag{13}$$

where: R = intensity of rainfall in mm per hour
 x = a constant for the locality (see Table 11)
 N = number of years between storms of intensity R
 t = duration of rainfall in minutes.

From this formula Table 10 has been prepared.

The value of x in Table 11 was determined for the various localities in the United States according to a map, based on Yarnell's figures, giving the intensities of once-in-10-years storms having a duration of 60 min. According to this, x was unity where the rainfall was 51 mm/h and proportionately higher or lower elsewhere.

The figures in Table 10 could be applied to all parts of the United States, subject to the accuracy of the information on which the values of x were determined. Where, however, there is reliable local information on storms this should not be ignored.

Table 10   Average US storms liable to occur once in 5 years (multiply these intensity figures by the appropriate factor *x* of Table 11 according to the locality)

Duration (min)	Intensity (mm/h)	Duration (min)	Intensity (mm/h)	Duration (min)	Intensity (mm/h)
11	121	32	67	110	28
12	116	34	65	120	27
13	111	36	62	130	25
14	107	38	60	140	24
15	104	40	58	150	23
16	100	42	56	160	21
17	97	44	55	170	20.5
18	94	46	53	180	20
19	92	48	51	190	19
20	89	50	50	200	18
21	87	55	47	210	17.5
22	84	60	44	220	17.75
23	82	65	42	230	16.25
24	80	70	39	240	15.75
25	78	75	38	250	15.25
26	76	80	36	260	14.75
27	75	85	34	270	14.37
28	73	90	33	280	14.0
29	71	95	32	290	13.5
30	70	100	30	300	13.25

For storms of different frequencies of occurrence, multiply the above intensities by the following factors

Years between occurrence	Factor
1	0.725
2	0.833
3	0.903
10	1.149

The WPCF (Wat. Pollution Control Fed.) Manual of Practice No. 9, *Design and Construction of Sanitary and Storm Sewers* (1960)[4] and subsequent editions, is considered the best book of reference on United States practice in sewer design. (Incidentally, while semi-official codes of British drainage and sewerage practice have been published, there is nothing in Britain to compare with the American book.) The manual contains, among much other useful information, a series of sixteen maps of the United States giving the amounts of rainfall (not expressed as intensities) for storms having durations of 15, 30, 60 and 120 min and frequencies of occurrence of once every 2, 5, 10 and 25 years. These were omitted from the 1970 edition which, however, refers the reader to the US

Table 11   Factors for use with Tables 10 and 14   79

State	Factor x	Factor y*
Alabama	1.2 to 1.5	1.36 to 1.40
Arizona	0.5 to 0.6	1.0 to 1.08
Arkansas	1.2 to 1.3	1.36
California	0.4 to 0.5	0.91 to 1.0
Colorado	0.6 to 0.9	1.08 to 1.21
Connecticut	1.0	1.26
Delaware	1.1	1.31
Florida	1.5 to 1.7	1.44 to 1.48
Georgia	1.2 to 1.4	1.36 to 1.40
Idaho	0.3 to 0.5	0.91 to 1.0
Illinois	1.0 to 1.1	1.26 to 1.31
Indiana	1.0 to 1.1	1.26 to 1.31
Iowa	1.0 to 1.2	1.31
Kansas	0.9 to 1.2	1.26 to 1.36
Kentucky	1.1	1.31
Louisiana	1.3 to 1.6	1.40 to 1.48
Maine	0.5 to 0.8	1.08 to 1.14
Maryland	1.1 to 1.2	1.31
Massachusetts	0.9 to 1.0	1.21
Michigan	0.7 to 0.9	1.14 to 1.21
Minnesota	0.8 to 1.1	1.21 to 1.26
Mississippi	1.2 to 1.5	1.36 to 1.44
Missouri	1.0 to 1.2	1.31
Montana	0.4 to 0.7	0.91 to 1.14
Nebraska	0.8 to 1.2	1.14 to 1.31
Nevada	0.3 to 0.4	0.91
New Hampshire	0.6 to 0.8	1.08 to 1.21
New Jersey	1.0 to 1.1	1.26 to 1.31
New Mexico	0.6 to 0.8	1.08 to 1.21
New York	0.6 to 1.0	1.08 to 1.26
N. Carolina	1.2 to 1.3	1.36
N. Dakota	0.7 to 1.0	1.14 to 1.26
Ohio	0.9 to 1.1	1.26
Oklahoma	0.9 to 1.3	1.26 to 1.36
Oregon	0.3 to 0.4	0.91
Pennsylvania	0.9 to 1.1	1.21 to 1.26
Rhode Is.	1.0	1.26
S. Carolina	1.2 to 1.4	1.36 to 1.40
S. Dakota	0.8 to 1.1	1.14 to 1.31
Tennessee	1.2	1.33
Texas	0.7 to 1.6	1.14 to 1.48
Utah	0.4 to 0.6	1.00
Vermont	0.6 to 0.8	1.08 to 1.21
Virginia	1.1 to 1.3	1.31 to 1.36
Washington	0.3 to 0.4	0.91
W. Virginia	1.0 to 1.1	1.26 to 1.31
Wisconsin	0.9 to 1.0	1.21 to 1.26

* y has been based on the cube root of depth of rain liable to fall in 1 hour, once in 10 years. For Great Britain according to Bilham's Formula as adjusted by the Meteorological Office, this is almost unity.

Weather Bureau's Technical Paper No. 40 (6). Anyone with skill in the preparation of logarithmic graphs should have no difficulty in deriving from these maps a formula for any locality similar to formula (13) with appropriate values for the constants and indices.

The figures in Table 10 could be applied in all parts of the United States; where, however, there is reliable local information on rainstorms this should be ignored. Rainfall data are available in the United States Weather Bureau Technical Paper No. 40.

Formulae of the type

$$R = \frac{a}{(t + b)n}$$

are used in Australia for the once-a-year storm, and *Australian Rainfall and Run-off*, the Report of the Stormwater Standards Committee of the Institution of Engineers, Australia, contains intensity data for the whole of the continent and procedures for flood estimation. It includes maps of each state giving isopleths for the constants and indices of the formula.

Impermeability

Run-off to sewers during rainfall is that amount of rainfall which runs off the surface and finds its way to the sewers. All surface water that soaks away into the ground, is evaporated, otherwise lost, or too delayed for it to affect peak run-off, is excluded. At the end of a storm of the same duration as the time of concentration, the flow in the sewer reaches a peak and thereafter tails off. It is this peak which has to be known for sewer design and, for practical purposes, the flow before or after the peak does not need to be considered.

At one time the selected intensity of rainfall was multiplied by the area and an appropriate impermeability factor of the type shown in Tables 12 and 13. But it became evident that impermeability increased during rainfall very greatly indeed, particularly near the beginning of the storm, and only after a long time did the run-off coefficient become equal to the ultimate impermeability factor. Accordingly curves of run-off coefficient against duration of rainfall were prepared for

Table 12   Ultimate impermeability factors according to Fruhling

Type of surface	Impermeability factor
Metal, glazed tile and slate roofs	0.95
Ordinary tile and roofing papers	0.90
Asphalt and other smooth and dense pavements	0.85 to 0.90
Closely jointed wood or stone-block pavements	0.80 to 0.85
Block pavements with wide joints	0.50 to 0.70
Cobblestone pavements	0.40 to 0.50
Macadam roadways	0.25 to 0.45

Table 13   Ultimate impermeability factors according to Kuichling and Bryant

Type of surface	Impermeability factor
Watertight roof surfaces	0.70 to 0.95
Asphalt pavements in good order	0.86 to 0.90
Stone, brick and wood-block pavements with tightly cemented joints	0.75 to 0.85
Same with open or uncemented joints	0.50 to 0.70
Inferior block pavements with open joints	0.40 to 0.50
Macadam roadways	0.25 to 0.60
Gravel roadways and walks	0.15 to 0.30
Parks, gardens, lawns, meadows, depending on surface, slope and character of subsoil	0.05 to 0.25
Wooded areas depending as before	0.01 to 0.20

roofed and paved areas (and also for previous areas) for various localities in America, including London, Ontario, Louisville, St Louis, Kansas City, Decatur, Grand Rapids, Peoria, and Montreal. These curves had a general similarity of form but the impermeability was greater in those areas that had very heavy storms and lesser where storms were moderate.

For many years, as has been mentioned, sewers in Great Britain were designed on the 'ultimate' impermeability factor, i.e. the factor as determined by comparing total rainfall with total run-off over a long period, say 1 year. But engineers of experience had become aware that surface-water and combined sewers were being designed to unnecessarily large capacities. On the instigation of W. A. M. Allan, Divisional Engineer, Main Drainage, Chief Engineer's Department, London County Council, field experiments were made by the Road Research Laboratory, Department of Scientific and Industrial Research, at various sites and the collected data published in confidential reports.[15-17] While a proportion of the data published in these reports appeared to be reliable enough, the theory of sewer design proposed by the Laboratory, in the author's view, is not, but as a result of Government-department publicity, has come to be used by too many engineers not sufficiently versed in the subject to detect the fallacies.[18]

The author spent a considerable time in analysing these data and found that impermeability changed during rainfall very greatly with time and, to a lesser extent, with quantity of rain. On the basis of this work he devised a formula

$$I = i\frac{\left(t - 2 + 4.6052 \log_{10} \frac{2}{t}\right)}{t} \tag{14}$$

where: I = average impermeability during rainfall
i = ultimate impermeability
t = duration of rainfall in minutes

He also found that the average ultimate impermeability i for roofed and paved areas and also for densely built-up areas, was 0.8. This factor, used as i in the above formula, gave the scale of run-off coefficients in Table 14. Furthermore, the most consistent results were obtained when the impermeability of permeable areas such as garden land was excluded, indicating that the run-off from these surfaces tended to arrive at the sewers too late to affect the peak run-off except in special cases; for example where a steeply sloping bank of earth drained down to a road or other paved surface.

A curve prepared from Table 14 was found to be very similar in form to all United States and Canadian curves inspected; also, that by multiplying the values in Table 14 by the cube root of the once-in-10-years storm of 1 h duration for the appropriate American locality and dividing it by the cube root of the once-in-10-years British storm of 1 h duration, suitable curves for any part of the United States could be found. Factor y in Table 11 multiplied by the values in Table 14 will give such curves.

Table 14   Run-off coefficients for roofs and pavements of developed areas in Great Britain

Duration of storm (min)	Run-off coefficient	Duration of storm (min)	Run-off coefficient	Duration of storm (min)	Run-off coefficient
11	0.406	32	0.611	110	0.727
12	0.427	34	0.620	120	0.732
13	0.446	36	0.627	130	0.736
14	0.463	38	0.634	140	0.740
15	0.478	40	0.640	150	0.743
16	0.492	42	0.646	160	0.746
17	0.504	44	0.651	170	0.749
18	0.515	46	0.656	180	0.751
19	0.526	48	0.661	190	0.753
20	0.535	50	0.665	200	0.755
21	0.545	55	0.674	210	0.757
22	0.553	60	0.682	220	0.759
23	0.560	65	0.690	230	0.760
24	0.567	70	0.696	240	0.761
25	0.574	75	0.701	250	0.763
26	0.581	80	0.706	260	0.764
27	0.586	85	0.710	270	0.765
28	0.592	90	0.715	280	0.766
29	0.597	95	0.718	290	0.767
30	0.602	100	0.721	300	0.768

The reasons for choosing the cube root of the ratio of American storms to British storms were, first, that in the research data impermeability varied approximately as the cube root of the intensity when storms of equal duration were compared; secondly, that the rule fits the American published data very well; thirdly, that research by Meek showed that impermeability varied as the cube root of rainfall intensity.[19]

It will be noted that whereas rainfall intensities for storms of short duration are very high, run-off factors are very low and it is found that when the rainfall intensity is multiplied by the appropriate run-off factor when the time of concentration is less than about 11 min, a lower run-off results than applicable to the 11-min (Bilham) storm. In such a circumstance the 11-min storm produces a greater run-off than any storm of shorter duration and therefore, for all areas having times of concentration of 11 min or under, a flat rate of rainfall and a fixed run-off factor appropriate to the 11-min time of concentration should be used in Britain. In other words, for these short times of concentration the rational method is a waste of time and a flat-rate method is all that is needed. This fact simplifies the surface-water sewerage design for the majority of schemes.

Table 15 shows how the maximum run-off results from the US storm of 14 min duration and that little harm would be done if all schemes with not more than 20 min time of concentration were designed on a flat rate.

Table 15 Minimum time of concentration for use with 'rational' method in the United States

Duration (min)	Rainfall (mm/h)	Run-off coefficient	Product of rainfall and run-off coefficient
11	121	0.406	49.13
12	116	0.427	49.53
13	112	0.446	49.95
14	107	0.463	49.50
15	104	0.478	49.70
16	100	0.492	49.20
17	97	0.504	48.88
18	94	0.515	48.41
19	92	0.526	48.39
20	89	0.535	47.61

Selection of the design storm

If, for a drainage area having a time of concentration of 15 min or more, the paved areas are multiplied by the appropriate factor for rainfall likely to occur say, once in 5 years, by the appropriate run-off coefficient and by a factor to convert mm of rainfall per hour to cumecs per hectare, a fair estimate will be made of the run-off likely to occur once in 5 years and, if the sewer were made of such a size as to take this flow at the available gradient, it would momentarily run just full on the average once every five years. But what would happen on the occasion of, say, the much more intense storm liable to occur once in thirty years? Should not the engineer design on, say, the once-in-30 or even once-in-50-years storm? Experience has shown that the use of data from such rare storms in Britain is neither necessary nor desirable for, although storms of such intensity are recorded at rainfall stations, they are spot records only and do not apply to large areas such as run-off catchments. Storms of very high intensity cover small areas only, which may be smaller than the catchment in question and often will not coincide with the catchment. An investigation made by the Greater London Council into drainage areas in Middlesex showed that the most intense rainfall likely to occur in 30 years was not that due to the once-in-30-years storm but to the storm likely to occur once in 9 years according to Bilham's formula.

An investigation was made by D. J. Holland[20] into the areas covered by storms in the category used for sewer design, and plottings from his formula were given in the Appendix to Hydrological Memorandum No. 33. The plotted figures suggest that allowance could be made (if desired), according to area of the circle enclosing the catchment, by multiplying the rainfall according to Bilham's formula by an appropriate factor. It would appear that the figures shown in the table below could be used in design calculations.

Area of circle enclosing catchment (hectares)	Factor
61	0.95
324	0.9
889	0.85
1497	0.8
2529	0.75
3885	0.7
5261	0.65

When rainfall run-off becomes greater than the capacity of a sewer, surcharge occurs, and this increases the discharge of the sewer as described in Chapter 2. There is a common fallacy that the storage capacity of the sewerage system always permits the run-off to the sewers to exceed the discharge capacity of the sewers. In fact the rational method may be considered a means by which the effect of the storage capacity may be estimated. The sewage stored in a system of sewers is the difference between the run-off to the sewers and what has run out of the sewers. Both can be calculated and, at the end of the time of concentration, the amount of surface water in the sewerage system is just about one-half of the total storm water that has run into the sewers, the other half having already been discharged. The amount of water stored in a properly designed storm-water sewer is that which is held back by the resistance to flow due to fall available as allowed for in the design calculations. The total volume of the sewers, i.e. cross-sectional area multiplied by length, can be occupied as storage capacity only if the bottom end is plugged to cause the sewage to back up, or if the run-off to the sewers is so increased as to surcharge the system right up to crown level throughout. In the latter case, outflow is increased, not reduced.

Taking everything into account, if a sewerage system in America is designed to accommodate the storm liable to occur once in 5 years, if the run-off factor has been accurately estimated, and if there is not a marked increase of building development in the area, the sewerage system should not surcharge so much as to cause floods, over a very much longer period of time than 5 years, because increase of discharge due to surcharge, the limits of areas covered by storms of high intensity, and the fact that sewers usually discharge rather more than the calculated amount, in all, to sufficient available capacity over and above the design capacity, to allow for storms of greater intensity than the once-in-5-years storm.

If a sewerage system is designed, as the author has suggested for Great Britain, on the basis of the once-in-3-years Bilham storm, the rainfall would be, for an 11-min time of concentration, 50. 7 mm/h. If moderate surcharge increases the discharge of the sewer by 73 %, that discharge will be the equivalent of 89 mm/h or slightly more than the once-in-15-years storm, and if the limitation of the Bilham storm to less than the area of the catchment permits a rainfall 26 % higher, or an intensity of 112 mm/h, the sewer should accommodate something more than the once-in-30-years storm. This suggests that the once-in-3-years storm is adequate provision in Great Britain for the small sewer at the top ends of a system.

If the once-in-3-years storm is applied to a large sewer at the bottom end of a system and the time of concentration is 1 h, the Bilham storm would be 16.75 mm/h. In such a case it would not be unreasonable to expect permissible surcharge to increase discharge by about $41\frac{1}{2}$ %, making the sewer capable of taking 23.6 mm/h, or the run-off due for the once-in-9-years storm which, as in the case of the Middlesex investigations, could well mean that the once-in-30-years storm could be accommodated.

There is much uncertainty here, and therefore the engineer must use his discretion, bearing in mind recognized practice. At one time in Great Britain the storm that occurred approximately once in 15 months was generally accepted, and proved excessive because of the tendency at the time to over-estimate the impermeable area or the run-off factor. The writer suggested that the once-in-3-years storm was desirable if a more accurate run-off coefficient were used. In the United States the once-in-5-years storm appears to be most frequently adopted and may well be justified: there is so very small a difference between intensities of, say, once-in-3-years and once-in-5-years storms, according to the US Weather Bureau figures.

Designing a flat-rate scheme

Many sewerage schemes which engineers have to design are for areas such as housing estates or small villages with times of concentration of under 15 min. As shown in Table 14 the run-off factor reduces more rapidly than the rainfall intensity increases when the duration of rainfall is reduced below 14 min. In such a circumstance it is not the storm equal in duration to the time of concentration, but the storm liable to last for 14 min that produces the greatest run-off and, therefore, in the United States the run-off factor for rainfall intensity applicable to 14 min should be used. Design is then a very simple matter, for all that is necessary is to determine the roofed and paved area draining to each length of sewer and find the run-off by multiplying the paved and roofed area in hectares by 0.166, the run-off coefficient and the rainfall in mm/h applicable to 14 min duration, to give the run-off in m³/min. The diameter of sewer can then be found with the aid of Table 2.

Suppose that a housing estate in the north of the State of Illinois has to be sewered on the separate system. In this case factor x is unity and, assuming that the once-in-5-years storm is to be used, the storm intensity for 14 min (Table 10) is 107 mm/h. Factor y is 1.26 and for 14 min time of concentration the run-off coefficient, according to Table 14, is 0.463. The run-off to sewers will then be:

$$0.166 \times 107 \times 1.26 \times 0.463 = 10.37 \text{ m}^3/\text{ha}$$

The roofed and/or paved areas draining to each length of sewer should be ascertained by measuring, on the ground or on proposal drawings, a sample area of sufficient size, calculating the proportion of paved to gross area, and applying this to the whole. The calculation for the estate then becomes as shown in Table 16 (for which Figure 4 may be used).

Table 16   Flat-rate surface-water sewer-design calculation

Location (manhole to manhole)	Increment roofed and paved area (ha)	Roofed and paved area (ha)	Run-off (m³/min)	Gradient (1 in:)	Diameter of sewer (mm)	Discharge capacity (m³/min)
E5 to E4	0.28	0.28	2.89	425	300	3.34
E4 to E3	0.12	0.40	4.21	500	375	5.52
E3 to E2	0.15	0.56	5.83	360	375	6.51
E2 to E1	0.18	0.74	7.70	670	450	7.70
E1 to B2	0.08	0.81	8.50	700	530	11.30
B7 to B6	0.09	0.09	0.96	120	150	1.02
B6 to B5	0.24	0.33	3.23	405	300	3.40
B5 to B4	0.17	0.50	5.24	325	300	6.85
B4 to B3	0.35	0.85	8.83	465	450	9.26
B3 to B2	0.37	1.22	12.74	500	525	13.37
B2 to B1	0.14	2.17	22.74	640	675	22.88
B1 to outfall	0.25	2.42	25.35	640	750	30.16

A time of concentration calculation

Suppose a system of large sewers has to be constructed in the same locality in Illinois but, in this case, to pick up surface-water flows from a number of areas each of which individually could have been designed on a flat rate. Again using Figure 4, but this time considering that the scale is different and that the points B1, B2, etc., are those at which diameters and gradients of main sewers change, the calculation becomes as in Table 17. It will be observed that in this case the only flat-rate figures are for the sewers E5 to E4 and B7 to B6. The remaining calculations have to be made on the run-off coefficients and the rainfall intensities

Table 17 Surface-water Sewer-design calculation by the 'rational' method

Locality	Increment paved and/or roofed area (ha)	Sum of paved and/or roofed areas (ha)	Time of concentration (min)	Run-off coefficient (Table 13 × 1.26)	Rainfall (mm/h) (Table 9)	Run-off (m³/min)	Gradient (length divided by fall)	Diameter (mm)	Discharge capacity (l/s) (from Table 2)	Velocity (m/min)	Length (m)	Increment time of concentration (nearest min)
E5 to E4	0.68	0.68	9	0.583*	107.4*	7.14	590	450	128.8	48.6	458	9
E4 to E3	0.62	1.31	15	0.602	103.9	13.65	700	600	251.3	53.3	312	6
E3 to E2	0.38	1.69	21	0.687	86.6	16.71	1000	675	279.4	46.8	276	6
E2 to E1	0.47	2.15	27	0.739	74.7	19.77	1100	750	359.8	48.8	282	6
E1 to B2	0.70	2.85	34	0.781	64.8	24.67	1300	825	424.8	47.7	335	7
B7 to B6	0.59	0.59	10	0.583*	107.4*	6.23	590	450	128.8	48.6	512	10
B6 to B5	0.42	1.01	22	0.697	84.3	9.91	800	525	165.7	45.9	568	12
B5 to B4	1.22	2.28	34	0.781	64.8	19.26	1350	750	324.8	44.1	544	12
B4 to B3	1.16	3.44	46	0.827	53.1	25.15	1300	825	424.8	47.7	583	12
B3 to B2	1.08	4.52	58	0.860	44.2	28.80	1000	825	484.4	54.4	570	12
B2 to B1	0.96	8.34	69	0.878	38.6	47.01	1900	1125	782.0	47.2	518	11
B1 to outgo	0.73	9.07	80	0.890	36.1	48.43	2350	1200	843.3	44.1	489	11

* Flat-rate for times of concentration below 15 min.

Note: Calculation in this table are by slide rule. Nearest figures in tables are taken.

88

dependent on the several times of concentration of the parts of the sewerage system draining to each point in question. The time of concentration is not known until the sewer has been designed, the velocity determined, and the increment time of concentration calculated according to the increment of length. Therefore the procedure is one of trial and error. A first attempt is made, the time of concentration found and, if this gives run-off and rainfall coefficients that produce too high or too low a run-off for the diameter of sewer chosen, the diameter is changed and the process repeated until the appropriate size is found.

It will be noted that where two lengths of sewer join, as at B2, the time of concentration is taken as the longer of the two, and this, with the increments of time of concentration, is used until another main line is met (not shown in the present calculation) and the greater time of concentration again selected.

Time-area graph and tangent method

The methods that were used in the 1930s for finding the effect of the shape of catchment involved the preparation of a time–area graph, or of tabular calculation which served the same purpose. Of those in which the time–area graph was used, one of the easiest was Riley's tangent method.[21]

A time–area graph is a curve showing the area contributing flow at all times after start of rainfall to the point at which flow is desired to be known. It is constructed in the following manner.

In preparing the time–area graph for a large drainage area it is usual to assume that the distribution of impervious area in each component drainage area is even, and that therefore the time–area curves of the component areas are shown as straight lines rising from zero hectare to total hectares during the time of concentration. The time–area curve for the whole area is the sum curve of these components.

Figure 6 is the time-area curve relating to point Y for the drainage area shown in Figure 5. It is constructed as follows: immediately on commencement of a

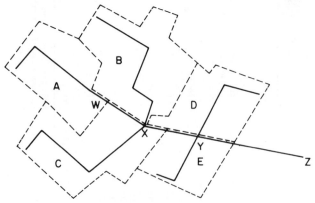

Figure 5   Drainage area diagram (used in preparation of
time–area graph)

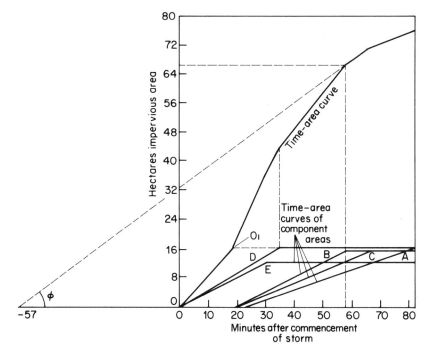

Figure 6 Time–area graph and tangent method

storm area D begins to contribute to point Y where the sewer from area D connects to the main sewer, and after 35 min the whole of area D is contributing. Thus the time–area curve for D is a straight line rising from 0 min, 0 ha to 35 min, 16.5 ha. According to one of the conventions which are unavoidable, it is assumed that the storm in question is one of infinite duration and that area D continues to contribute indefinitely. Therefore the curve continues from 35 min, 16.5 ha as a straight line showing the impervious area contributing as 16.5 ha up to the maximum period of time under consideration. If the storm in the calculation turned out to be one of short duration a time would arrive when area D would cease to contribute and the time–area graph should slope down again to 0 ha, but to make allowance for such a contingency would render the calculation extremely complicated and the advantage in accuracy obtained thereby would not be justified.

The time–area curve for area E is drawn in the same manner as that for area D. But areas B and C do not contribute until 18.4 min after commencement of storm, and therefore the time–area curves in each case commence at 0 ha, 18.4 min instead of at 0 ha, 0 min. In the case of area B the time of concentration is 40 min and the impervious area 15.78 ha, and so the curve slopes up to the point 40 plus 18.4 min (58.4 min), 15.78 ha. The curve for area C is found similarly. Area A commences to contribute 4 min after area B, or 22.4 min after commencement of storm, and ceases to contribute after its time of concentration of 60 min, or 82.4 min after commencement of storm. This is the time that determines the limit of the graph.

Having drawn the component time–area graphs a sum is drawn by adding them together with the aid of dividers. This curve should end, in the case in question, at 82.4 min, 76.4 ha.

Tangent method

The tangent method is applicable to rainfall formulae of the type

$$R = \frac{a}{t + b}$$

and it cannot be applied to any other type of formula. This is because it is based on the argument:

(1) $Q = C \tan \varphi$;
(2) the greater the value of φ the greater the value of Q;
(3) therefore the greatest angle can be produced by drawing a line tangential to the time–area graph from point located at $-$'b' min from any part of the time–area curve, denotes the maximum run-off from the area concerned.

If, in the design of the sewerage system, a formula not of the above type has been used, it is necessary to transform it to the required type, and this means that the values of a and b must be found. If in the case illustrated in Figure 6 it is assumed that the work in question is in Illinois, that $x =$ unity and $y = 1.26$, then the run-off per hectare for 82 min time of concentration will be:

$$36 \times 0.706 \times 1.26 = 32 \text{ mm/h}$$

and, for 14 min time of concentration, will be:

$$107 \times 0.463 \times 1.26 = 62 \text{ mm/h},$$

according to the nearest figures in Tables 10 and 14. Then, using 82 and 14 as the approximate limits of time of concentration, the values of a and b are found as follows:

$$R = \frac{a}{t + b}$$

$$32 = \frac{a}{82 + b} \quad \text{and} \quad 62 = \frac{a}{14 + b}$$

$$\therefore \quad \frac{62a}{82 + b} = \frac{32a}{14 + b}$$

On cancelling out of a, b becomes 57 (and $a = 175$).

Having found that the value of b is 57, measure backwards on the time scale this amount from what appears to be the most concave point on the time–area graph (see Figure 6). More than one attempt may have to be made to ascertain which is the most concave point giving the largest angle of φ. In the example it is found that the effective time of concentration is 58.4 minutes, at which time 66.8 ha would be contributing. The rainfall run-off per hectare would then be

$$47 \times 0.674 \times 1.26 = 39 \text{ mm/h}$$

and the run-off for the 66.8 ha would be

$$0.166 \times 39 \times 66.8 = 430 \text{ m}^3/\text{min.}$$

The run-off from the whole area, due to the rainfall applicable to the total time of concentration of 82.4 min, would be

$$0.166 \times 36 \times 0.706 \times 1.26 \times 76.5 = 406 \text{ m}^3/\text{min.}$$

Thus, the effect of using the tangent method is to find that, in the example, the run-off from part only of the drainage area is about 9% more than the run-off from the whole area.

The author's 'modified tangent method', which requires the construction of a transparent overlay of curved lines, can be applied to any rainfall formula in which the power of t is less than unity. It is described in Vol. II of *Public Health Engineering Practice*.[22]

Storm-water storage

The capacity required in ponds or watercourses to accommodate the excess surface water, due to rainfall, over and above the acceptable discharge, frequently needs to be known.

The author's formula is:

$$C = \left[(0.166\Delta pk)^{1/n} \times n\left(\frac{1-n}{p}\right)^{1-n/n} \right] - nPtc \qquad (15)$$

where : C = required storage capacity in m^3;
    k = the constant in the rainfall formula $R = k/t^n$;
    Ap = area in hectares of roofed and/or paved areas multiplied by ultimate impermeability factor;
    n = power of t in formula $R = k/t^n$;
    P = rate of outgo from storage in m^3/min;
    tc = time of concentration of the sewerage system in min.

(Copas and the author, working separately, arrived at the same general conclusion). In practical surface-water storage problems, times of concentration are nearly always very long and, therefore, the factor 8.5 in Formula 13 becomes negligible. Thus, the rainfall formula for the United States becomes $R = (37.1xN^{0.2})/(t^{0.8})$ and substituting $37.1xN^{0.2}$ for k and 0.8 for n in formula 15, the formula for storage becomes

$$C = \frac{297Ap^{1.25}x^{0.25}}{p^{0.25}} - 0.8\ ptc \qquad (16)$$

where: x = a constant for the locality, in Table 11 (p. 79);
    N = number of years.

The last part of the formula $(-nPtc)$ is an approximation to allow for the storage in the sewerage system. This is applicable only if tc is less than duration of the storm that requires the maximum storage. It is usual, however, to neglect this part of the formula.

The equivalent storage formula for Great Britain, based on Bilham's rainfall formula, is approximately

$$C = \frac{88Ap^{1.4}N^{0.4}}{p^{0.4}} - \frac{5ptc}{7} \tag{17}$$

Storm overflows

At one time it was considered permissible to discharge to watercourses untreated storm water in excess of some specified rate of flow without treatment. This became a cause of serious nuisance because storm water from combined sewers can be very foul and the quantity large. There came a change of opinion and sections 30 and 31 of the Public Health Act, 1936, applicable to England and Wales, read

30.   Nothing in this Part of this Act shall authorise a local authority to construct or use any public or other sewer, or any drain or outfall, for the purpose of conveying foul water into any natural or artificial stream, watercourse, canal, pond or lake.

31.   A local authority shall so discharge their functions under the foregoing provisions of this Part of this Act as not to create a nuisance.

Following this some British engineers, with the approval of the appropriate Ministry, provided storm tanks to store, settle and return as much flow as considered practicable to the sewers after the storm. Despite the above clauses there are still many storm overflows for crude sewage causing pollution.

The Technical Committee on Storm Overflows in the formulation of its 1970 Report,[23] so far as Great Britain is concerned has raised the generally accepted overflow rate of 6 × DWF, to about 9 × DWF, or higher, if any quantity of trade waste is present. They deduce three formulae for calculation of overflow rate. The normal one, Formula A, reads as follows:

$$Q = \text{DWF} + 1360P + 2E$$

where: Q = overflow weir setting in litres per day;
$\quad\quad P$ = population;
$\quad\quad E$ = volume of industrial effluent discharged in 24 h, in litres;
$\quad$ DWF = average daily flow in dry weather including industrial effluent and an allowance for filtration, in litres per day.

Whilst such requirements present no real difficulty in new construction it is their application to existing systems, many of which will be either 'combined' or 'partially separate' with storm overflows at lesser or greater intervals, where difficulties will quickly arise and may result in the need for duplication, or replacement of the whole of the existing system from the point where the first storm overflow occurs, downwards to the point of outfall. Such considerations, added to the problems of 'ageing' sewers, make it clear that even in the highly developed countries much work still remains for sewerage engineers, in fact it must be said that until it *is* done, the standards of the rivers will show little sign of improvement.

At a symposium held at the Institution of Civil Engineer, in 1967[24] ideas were put forward for tackling such problems, one of which was to convert 'combined' to 'separate' systems by introducing a small-diameter pipe system within the existing large diameter sewers, and to connect to it the foul sewage from built-up areas, blocks of flats, etc., roof and ground water continuing to find their way to the old sewers, thus avoiding the heavy cost of duplicating such systems through-out. In the United States,[24] Chicago is developing an elaborate system of under-ground storage for storm water, and New Orleans,[24] with its built-up area mostly below the river level of the Mississippi, relies on fourteen major pumping stations, of average capacity 70 cumec, to beat the 1965 disaster caused by Hurricane Betsy, and plans further improvements to make the city safe. These were due for completion by 1980.

In Holland and Belgium[24] a rather different approach to the storm water prob-lem has developed. Precipitation data recorded at the Royal Meteorological Institute De Bilt, Utrecht in the centre of the Netherlands, from 1938 to 1948 is used as the basis of calculation—average precipitation being 780 mm in 510 h. Until 10 years ago, storm overflows in the Netherlands were required to operate so as to discharge flow diluted by rainwater—thus up to 5 × DWF had to be carried by the sewerage system. Over the 10 years (to 1967) interest has centred on frequency of operation rather than degree of dilution or duration of operation. Overflows should be located on the outskirts of the district served—available static storage capacity being that part of the system between the top of the lowest overflow and the highest water level in the pumping sump. Dynamic storage is the amount that can be accommodated in those parts of the system higher than the overflows, without any sewage being discharged over them; usually a negligible quantity.

Disposal of surface water by soakaway

In certain localities the subsoil is of so permeable a nature that, before any building development, the whole of the rainfall soaked into the ground. In such districts there are no ditches or natural streams except, perhaps, winter-bournes which flow only when the water table rises above ground level in the wettest season, or small streams taking surface drainage from the silt in the bottoms of valleys. In these circumstances surface-water drainage should continue to be soaked into the ground, for if a surface-water sewerage system is constructed it will result in either the delivery of large storm flows to tiny local streams quite incapable of taking them, or the need for the construction of expensive sewers to transport the flows out of the district and discharge them to larger streams which, again, may be of inadequate capacity. The need for soakaway systems as against surface-water sewerage can be ascertained by study of geological maps and memoirs, or by looking at the site itself. In cases of doubt, experimental soakaways may be dug and tested.

Where soakaways are the alternative to either separate or combined systems, it is usual for developers of building land to provide them within the curtilages to take the drainage of roofs and yards, and for the estate developer, local authority, or highway authority to provide soakaways to take the drainage from roads.

In the latter case it is usual to construct these road soakaways at the same intervals as sewer manholes and to connect from one to another with drains of small capacity. These drains receive flow from laterals serving gullies or street inlets and to which other surface-water discharges may be made. They also serve the purpose of providing overflow from one soakaway to another so that should one fail to take the flow it will be relieved by the next. It is common practice to construct an extra-large soakaway at the bottom of any line of drain or, if possible, have a drain leading away to some natural watercourse not too distant.

Soakaways need to have storage capacity to hold the excess of inflow over the rate of soakage. In practice it has been found that a capacity equal to x inches of rain over the paved area is adequate, x being of the appropriate value for the locality as given in Table 11. If it is found that soakaways generally need to have capacities of as much as twice the above amount, it may be assumed that the subsoil is not suitable for soakaway drainage.

The capacity of a soakaway is the volume that can be occupied by the rainfall, and this must all be above the natural level of subsoil water because any capacity below that level is already occupied. Soakaways should be constructed with proper walls provided with openings to let the water out. They should not be filled with rubble because not only does this rob them of capacity; it is also more expensive than the construction of walls.

Recent developments

Growing concern about the environment, coupled with the rapidly increasing number of computers available for such studies, have produced a considerable volume of work over the last few years, as witness the various international and other symposia starting with the one held at the Institution of Civil Engineers, in London, in 1967 (already mentioned; see p. 93), which produced the information used by the Technical Committee on Storm Overflows, in formulating its long-awaited Report of 1970 (to which reference is also made on p. 93).

Since about that time it is probably true that most storm water run-off calculations in Great Britain have been based on the TRRL Method, as described in RN 35(1).[25] That this method had shortcomings was confirmed by the NERC Flood Studies Report, of 1975,[26] and the publication in 1976 of RN 35(2)[27]; coupled with these, the Meteorological office has advised that the design storm should be the 50 percentile storm of such duration as gives the peak flow in the pipeline being considered. This implies that engineers should use the storm of duration $2 \times tc$.[28] Work by Hepworth[29] indicates that complex design procedure arising from the ever-increasing mass of data becoming available may well not be required, if what he terms the 'direct' method of design is employed. Briefly, this consists in using the data from the second half of an incremental storm, on the ground that only during this period does the storm water run-off reach its peak intensity so far as sewer flow is concerned. In effect this agrees with the Meteorological Office advice, to use a duration of $2 \times tc$, as already quoted.

Papers presented at the Southampton Conference of 1978 by Ackers,[30] and by American and Canadian engineers, suggest that despite the detailed information given in Report No. EPA 670/27–75 017 (SWMM)[31] and other recent pronouncements,[32] no clear guidance is yet available so that most design work in the US is still based on the rational method. For any but large population centres, therefore, the methods described in this chapter—coupled with the refinements just suggested in regard to time of concentration as related to storm duration—will continue to serve all practical purposes. Indeed even large schemes, unless designed on the basis of accurate rainfall records as to duration and intensity, may well not improve on these precepts, as witness the findings of Crockett and Dugdale[33] when, after a detailed study of hydrographs applied to the sewer system of Edinburgh, it was decided that 'the sizing of the interceptor sewer system should be based on the simple summation of the inputs'.

References

1. Silcock, E. J. Sanitary Engng. *Kempe's Engrs. Year Bk.* (1935). Morgan Bros. (Publishers) Ltd and Crosby Lockwood & Son, London.
2. Lloyd-Davies, D. E. The elimination of storm water from sewerage systems. *Proc. ICE*, (Instn Civ. Engnrs.) **CLXIV** (1906).
3. Min. of Hlth. Depl. Committee on Rainfall and Run-off. *J. Instn. Munic. Engnrs.* **LVI** (22), 1172.
4. Man. of Practice No. 9, *Design and Constr. of Sanitary and Storm Sewers.* Jt. Comm. Wat. Pollution Contr. Fed. and Am. Soc. Civ. Engnrs. (1969).
5. Kuichling, *Trans. Am. Soc. Civ. Engnrs.*, **54**, 192 (1905).
6. Copas, B. A. Storm water storage calculations. *J. Instn. Public Hlth. Engnrs.*, **LVI**, Pt. 3 (July 1957) (Discussion).
7. Bilham, E. G. Classification of heavy falls in short periods. *British Rainfall* (1925).
8. Hydrological Memorandum No. 33. The Meteorological Office (Nov. 1964).
9. Memorandum No. 33, The Meteorological Office (Aug. 1968).
10. Meyer, A. F. *Elements of Hydrology.* 2nd edn. (1928).
11. Yarnell, D. L. *Rainfall Intensity Frequency Data,* US Dept. of Agric., Misc. Pub. 204, Wash. DC (1935).
12. Kirpich, P. Z. Hydrology. *Handbook of Appl. Hydraulics*, 2nd edn. (1952).
13. Schafmayer, A. J. Rainfall intensities and frequencies, *Proc. Am. Soc. Civ. Engnrs.* (Feb. 1937).
14. Bleich, S. D. *Proc. Am. Soc. Civ. Engnrs.* **60**, 157 (1904).
15. Watkins, L. H. Research Note No. RN/2361/LHW, B.W.9. Transport & Road Research Laboratory.

16. Watkins, L. H. Research Note No. RN/2362/LHW, B.W.11. Transport & Road Res. Lab.
17. Watkins, L. H. Research Note No. RN/2366/LHW, B.W.14. Transport & Road Res. Lab.
18. Escritt, L. B. *Design of Surface Water Sewers*, C. R. Books, Lond. (1964).
19. Meek, J. B. L. Sewerage with special reference to rainfall. Engnrg. Conf. at Instn. Civ. Engnrs. (1928).
20. Holland, D. J. The Cardington rainfall experiment. *Meteorological Magazine*, **96**, 193–202 (July 1967).
21. Riley, D. W. Notes on calculating the flows in surface water sewers. *J. Instn. Munic. Engnrs.* **LVIII** (20), 1483.
22. Escritt, L. B. *Public Health Engineering Practice*, Vol. II, Macdonald & Evans, London, 1972.
23. Min. of Local Government and Housing—Final Report of the Techl. Comm. on Storm Overflows. HMSO, 1970.
24. Symposium on Storm Sewage Overflows, Instn. Civ. Engnrs. (1967).
25. Transp. and Rd. Res. Lab. A guide for Engineers to the design of storm water systems. Road Note No. 35. HMSO (1963).
26. Natural Environment Res. Coun. *Flood Studies Rep.* (1975).
27. Transp. and Rd. Res. Lab. Rd. Note No. 35, 2nd edn. HMSO (1976).
28. Price, A. J. and Howard, R. A. An engineering evaluation of storm profiles for the design of urban drainage systems in the UK. Proc. Int. Conf. on Urban Storm Drainage, Southampton, p. 2. Pentech Press 1(978).
29. Hepworth, R. Direct design of surface water sewers. Proc. Int. Conf. on Urban Storm Drainage, Southampton, pp. 193–206. Pentech Press, 1978.
30. Ackers, P. Urban Drainage. The effect of sediment on performance and design criteria. Proc. Int. Conf. on Urban Storm Drainage, Univ. of Southampton, pp. 535–545 (Apr. 1978).
31. Metcalf and Eddy (Inc.) 1975. Univ. of Florida and Wat. Resources Inc. Storm Water Management Model—Users Manual, Version III. US Environmental Prot. Agency.
32. Smisson, R. P. A review of the storm water drainage of new developments. *J. Instn. Public Hlth. Engnrs.* **8** (1), (April, 1980).
33. Crockett, A. S. and Dugdale, J. Development of Edinburgh's sewage disposal scheme. *Proc. Instn. Civil Engnrs.*, **60** (Feb. 1976).

CHAPTER 5
Sewage, Sludge, and Effluent Pumping

It is common practice for engineers, when ordering pumping plant, to provide the manufacturer with basic data as to output required and maximum static head, and to leave the detailed arrangement of plant and pipework in his hands, concentrating only on the design of a building to house it. This approach, in the author's view, has sometimes led to troubles which might well have been avoided had the engineer coupled his knowledge of exact conditions under which he expected the plant to operate, with a fuller understanding of the type of plant and the arrangement of it most likely to achieve his purpose. This chapter therefore treats the matter in more detail than usual, in the hope of repairing past deficiencies in this respect.

Pumping is to be avoided unless it is necessary or shows a distinct advantage in overall economy, because mechanical installations are more liable to break down than simple sewerage structures and because, unlike the sewers, pumping plant will not pass more than its rated capacity under conditions of surcharge. Storm-water pumping is still less desirable than sanitary-sewage pumping, because it involves the installation of large pumps and motors to deal with heavy storm flows that occur at infrequent intervals; consequently, capital costs, kVA charges and overheads can be disproportionately high.

The types of pumps used for sewage include

(1)   reciprocating pumps
    (a)   lift pumps
    (b)   force pumps

(2)   centrifugal pumps
    (a)   volute pumps
    (b)   axial-flow pumps
    (c)   mixed-flow pumps

(3)   air-operated devices
    (a)   sewage ejectors
    (b)   low-lift air lifts

98

To the above list should be added the inclined-screw type of low-lift high-discharge pump which is a revival of the ancient invention by Archimedes (287–212 BC). This consists of a long screw in an inclined rotating cylinder. It is coming to be recognized as an economic alternative to centrifugal pumps for dealing with crude sewage. It does, however, require more space than a centrifugal-pump installation. The 'Mono' pump, which may be called a miniature high-speed version of the screw pump, has also become popular for certain duties.

Reciprocating pumps

Reciprocating pumps, at one time most commonly used, have been very largely superseded by centrifugal pumps for most purposes owing to the improvements in design of the latter, particularly with regard to efficiency, low cost, and small space required.

There are several classifications of reciprocating pumps—one of which divides them into those which have pistons moving in cylinders, and plunger pumps, i.e. those having plungers working through glands into vessels. The latter are preferable for delivering against high heads or when dealing with sewage or other liquids containing detritus.

Three-throw pumps, having three pistons or plungers moving out of phase with each other in three separate cylinders, or vessels, with their own inlet and outlet reflux valves, are generally preferred because they cause less change of velocity in the rising main than either single- or two-throw pumps: the use of a fourth phase gives little added advantage. An air vessel is necessary on the delivery side of a reciprocating pump, to absorb the shock caused by the varying velocity in the rising main, which would otherwise cause slamming of reflux valves with possibility of damage. An air vessel may also be provides on the suction side.

Reciprocating pumps are useful for delivering to very high heads, and have the advantage of passing a constant quantity at constant rotational speed regardless of the head. Except at low heads the horsepower can be considered as being almost in direct proportion to the head and the quantity, and efficiency is good, varying from 60% to 85% for small and large pumps, respectively, when the manometric head is high. At low heads efficiency is low because the comparatively high friction of the working parts absorbs an appreciable part of the total power input. Disadvantages of reciprocating pumps are high capital cost and the large amount of space occupied. They also require comparatively large-diameter suction and delivery pipework and valves, adding further to the capital cost.

The delivery of a reciprocating pump is the product of the capacity of the cylinder multiplied by the number of effective strokes per minute, less about 5% for leakage past the piston, reflux valves, etc.

While reciprocating pumps will handle sewage and sludge, it is generally considered advisable to protect them against large solids by the use of screens.

Centrifugal pumps

Centrifugal pumps are broadly of four types:

(1) volute pumps
(2) axial-flow pumps
(3) mixed-flow pumps
(4) turbine pumps.

Volute pumps are true centrifugal pumps working on the principle that a rotating impeller throws water outwards into a volute casing where the velocity of the water is converted into pressure. The water is fed into the eye of the pump (the space between the inlet ends of the vanes) and is thrown outwards by centrifugal force between the blades or vanes which are so curved that, at prescribed distances from the centre, they form angles varying according to design theory, or a compromise between the theoretical optimum and some practical consideration such as convenience in manufacture. The liquid leaves the edges of the vanes at a velocity which depends on their circumferential velocity, their angle, and other factors which cannot be determined with accuracy by theory.

For any particular design of centrifugal pump, broadly, the quantity delivered varies directly as the speed of rotation, the head varies as the square of the speed of rotation and, because the input horsepower depends on the quantity delivered multiplied by the head and divided by the efficiency, the horsepower varies as the cube of the revolutionary speed.

The speed of rotation needs to be such as to develop sufficient velocity in the water to be converted to the manometric head to be overcome. But at any

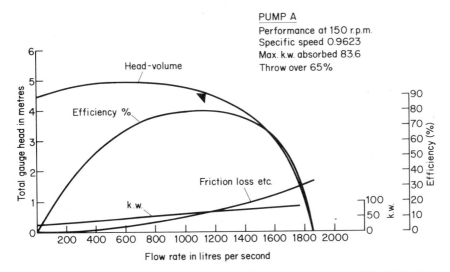

Figure 7   Characteristic curve for volute centrifugal pump (Courtesy NEI-APE Ltd., W. H. Allan)

100

particular speed of rotation, the delivery will vary, as the head is changed, according to a curve known as the characteristic curve of the pump. At any particular speed, also, the delivery will be greatest at the lowest head and will reduce as the head is increased until finally a manometric head is reached at which there will be no flow at all. To deliver against a higher head than this would require a higher speed of rotation characteristic curves for the most widely used types of centrifugal pumps are given in Figures 7, 7a, and 7b.

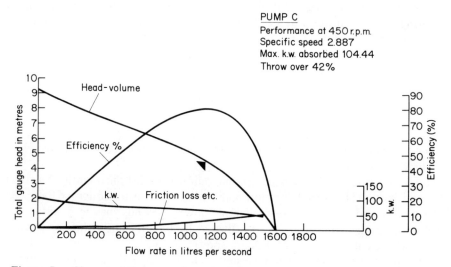

Figure 7a   Characteristic curve for mixed-flow pump (Courtesy NEI-APE Ltd., W. H. Allan)

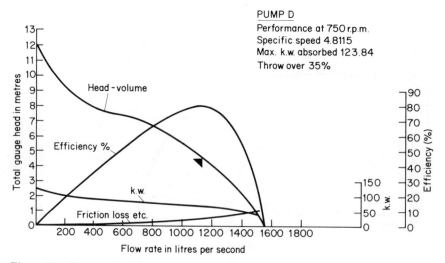

Figure 7b   Characteristic curve for axial-flow pump (Courtesy NEI-APE Ltd., W. H. Allan)

As head is varied, so efficiency varies, there being a maximum efficiency at some intermediate position on the characteristic curve. The combination of these factors results in horsepower variations at different heads, but not in accordance with the head. In some kinds of pump the horsepower is greatest at the lowest flow and highest head, and in others the horsepower reaches its peak as the flow increases and the head is reduced. The latter characteristic is undesirable for installations where head varies considerably, for it can cause overloading of electric motors at low heads unless they are made disproportionately large relative to normal running load.

Specific speed

The performance of a centrifugal pump is expressed by its specific speed, which can be calculated by the formula

$$N_s = \frac{3.65NQ^{0.5}}{H^{0.75}} \tag{19}$$

where: N_s = specific speed;
N = revolutions per min;
Q = delivery in m^3/s;
H = manometric head in metres, at maximum efficiency.

Specific speed is calculated in America in different units from those used in Great Britain prior to metrication. Thus, when comparing specific speeds calculated in different countries, it is necessary to know the units and constants that were used, for the value of N_s found for a pump in one country, will vary from that found for the same pump elsewhere.

At the time of writing no metrication committee recommendation has been made on specific speed: the above formula was provided to the author by a well-known international firm of pump manufacturers.

The specific speed is used by pump manufacturers in calculations for the empiric design of pumps; but knowledge of the specific speed is also useful to the engineer, for it tells him broadly the type of pump referred to (whether axial-flow, mixed-flow or volute) and gives him some indication of the probable form of its characteristic and the extent to which it will tolerate negative pressure on the suction side without cavitation. For a clear exposition on this subject and the design of pumps in general, the reader is advised to consult books such as Addison's 'Centrifugal and other Rotadynamic Pumps'[1]

Cavitation

Should a pump suck against a suction head greater than can be lifted, a partial vacuum containing water vapour is formed in the suction pipe or in the pump, or in both. If water were unable to vaporize, a pump should theoretically be capable of sucking against a total head equal to the atmospheric pressure of air; but vaporization occurs at lower pressures than this, subject to temperature, the figures being as in Table 17. In practice it is usual to allow for a suction head of not more than two-thirds of the figures in Table 18 for clean, or virtually clean, water.

Table 18   Effect of temperature on suction head

Degrees Centigrade	Metres head
15.5	10.0
37.8	9.6
60.0	8.1
82.2	5.1
100.0	0.0

Cavitation takes place because, owing to the suction remaining constant at a level determined by the vaporization negative pressure of the water, the flow in the suction pipe remains steady, whereas the pump may vary momentarily in speed, causing the water to fall away from the pump working parts, then catch them up, and when this occurs it strikes a blow which can do damage. Cavitation can be detected by a rattling noise, and its effect in the course of time is to cause deterioration of the impeller by pitting or honeycombing.

Impellers

There are several forms of impeller, designed for high efficiency on the one hand, or unchokability with sewage or sludge on the other, or to give a particular shape of characteristic curve. For high efficiency the vanes are shaped according to theory and are seldom fewer than six or more than ten in number: the number reduces for liquids containing solids, but not necessarily with a marked reduction of efficiency, if the pumps are large.

Impellers are described as 'shrouded' (enclosed by circular plates front and back); 'semi-shrouded' (with plate at the back, away from the inlet); or 'open' (not shrouded). Shrouded impellers are more expensive than non-shrouded impellers because the arrangement interferes with machining of the vanes and involves more handwork in manufacture. Non-shrouded impellers are suited to low speeds.

Success in the design of an unchokable pump for sewage or sludge is achieved by having an impeller without leading edges to the vanes, but in the form of one vane, or two parallel vanes which continue right across the pump, passing near the eye tangentially. The reason for this is that pieces of fibrous matter such as paper, string and rag tend to fold over on each side of these edges and remain in position, held by the flow, until a mass of fibrous matter builds up and chokes the eye of the pump. When specifying unchokable pumps it is necessary to insist on the desired type of impeller; otherwise manufacturers have been known to offer, for the purposes of pumping sewage or sludge, impellers with few vanes but with leading edges, and trouble has resulted.

The efficiency of volute centrifugal pumps can be very high but unchokable volute pumps, because of their special simple impellers, can only be of comparatively low efficiency.

Volute centrifugal pumps

Volute centrifugal pumps are useful for middle ranges of head, i.e. from 6 m to about 37 m (if single-stage) and are suitable for water or (if not too small in size) for screened sewage. Special unchokable volute pumps of 75 mm inlet diameter or upwards can deal with crude sewage, sewage sludge, etc.

Centrifugal pumps can be simple single-entry types with an entry in one direction to the eye. These are described as left-hand or right-hand, according to whether their rotation is anti-clockwise or clockwise respectively as viewed from the inlet branch. Both hands are available, but left-hand pumps are most common. A vertical-spindle left-hand pump rotates clockwise when viewed from above: this fact may be confusing and could lead to incorrect description.

Centrifugal pumps can also have two inlets, one on each side of the axis. This arrangement is most commonly adopted for horizontal pumps of large size.

For lifting against high heads, i.e. more than about 37 m head, multi-stage pumps are available. A two-stage pump has two impellers, the delivery from one passing via ducts to the eye of the second. These are useful for clean water but not for sewage, for there is too much risk of choking unless very fine screening is used. On the rare occasions when sewage pumps have to deliver to very high heads, two ordinary pumps or unchokable pumps can be arranged to run together, the delivery of the first discharging to the suction of the second.

Another variation of the centrifugal pump better known on the Continent and in America, is the offset impeller type, which achieves its unchokable characteristics by having its impeller set back so as to allow solids to by-pass it through the pump casing. It has lately aroused some interest in Great Britain, although its efficiency is lower even than that of the conventional 'unchokable' pump.

Axial-flow pumps

An axial-flow pump is not truly centrifugal, although included under the general title, for the water is not thrown out centrifugally but is pushed through the pump by a rotating impeller very similar in form to a ship's propeller. These pumps are most useful for dealing with large quantities delivered against low heads. One of their main duties is pumping flood water from marshes by lifting maybe only a metre or so from marsh level to some raised watercourse. In the larger sizes they are used for pumping sewage and to a great extent at sewage-treatment works for the recirculation of effluent in recirculation percolating-filter schemes, or for dealing with returned activated sludge.

Axial-flow pumps may be of special design fitted with variable-pitch impellers; the pitch can be varied so as to adjust efficiencies at various duties.

The characteristic curve of most axial-flow pumps is convenient in that the horsepower does not increase on reduction of head but, if anything, tends to fall off because the discharge does not increase disproportionately. The efficiencies of some axial-flow pumps are very high, approaching 90%. Axial-flow pumps are vary liable to cavitation and therefore suction lift should be the minimum possible.

Mixed-flow pumps

Mixed-flow pumps fall between volute centrifugal pumps and axial-flow pumps in almost every respect. The water is pushed partly outwards and partly forwards. They are useful for a range of heads overlapping, and generally between, those of volute and axial-flow pumps. The delivery varies more or less inversely as the head, and the horsepower remains fairly constant, not rising with high heads or with low heads. This property makes the pumps very useful for many purposes where head varies. Efficiencies can be high, often exceeding 85%. They are liable to cavitation and suction lift should be limited.

Turbine pumps

Turbine pumps are useful for pumping clean water to very high heads. They are unsuitable for sewage.

Types of pumping plant

Pumping sets are of either the horizontal-spindle or vertical-spindle type. Horizontal-spindle pumps are driven by electric motor or prime mover, in the latter case usually through a gear to change the rate of rotation, on the same level of floor as the pump. This arrangement is suitable for many purposes and less costly than vertical-spindle installations, but it necessitates the motor or prime mover being down in the dry well, if the pump is primed by charging from the head of water in the suction well, or if it is up at engine or motor floor level above the water in the suction well, priming has to be by hand-controlled priming gear or automatic control. Priming is an added complication, and automatic priming can fail to act, which is why horizontal pumps are not much favoured for use in automatic pumping stations. Limits on suction lift by cavitation also tend to restrict the use of horizontal sets.

Vertical-spindle pumps are therefore usually installed in sewage-pumping stations, in such a position that they are below the level of the water in the suction well at all times, are thus automatically primed, and are protected against cavitation. The shaft driving the pump continues vertically upwards with a muff—or flexible—coupling at the bottom end connecting to the pump, and a flexible coupling at the top end, connecting to the electric motor or to a gear head, if another form of prime mover is used, transferring rotation about a vertical axis to rotation about a horizontal one. The motor or prime mover is arranged at a level where it is easily accessible and above any risk of flooding (see Figure 8).

Vertical-spindle pumps fall broadly into two types: submersible pumps and pumps for installation in a dry well. Submersible pumps originally consisted of a unit built up of a pump at the bottom submerged in the liquid in the wet

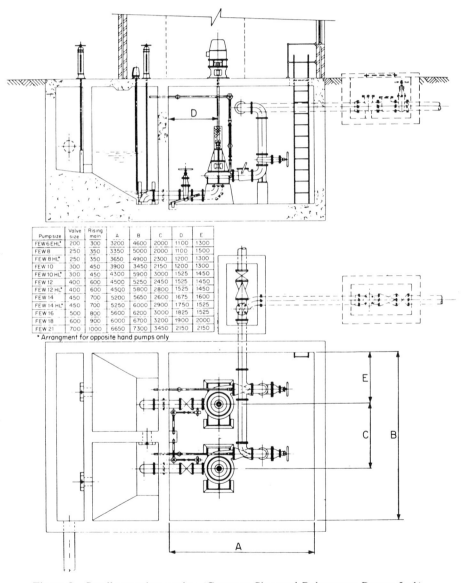

Pump size	Valve size	Rising main	A	B	C	D	E
FEW 6 EHL*	200	300	3200	4600	2000	1100	1300
FEW 8	250	350	3350	5000	2000	1100	1500
FEW 8 HL*	250	350	3650	4900	2300	1200	1300
FEW 10	300	450	3900	3450	2150	1200	1300
FEW 10 HL*	300	450	4300	5900	3000	1525	1450
FEW 12	400	600	4500	5250	2450	1525	1450
FEW 12 HL*	400	600	4500	5800	2800	1525	1450
FEW 14	450	700	5200	5650	2600	1675	1600
FEW 14 HL*	450	700	5250	6000	2900	1750	1525
FEW 16	500	800	5600	6200	3000	1825	1525
FEW 18	600	900	6000	6700	3200	1900	2000
FEW 21	700	1000	6650	7300	3450	2150	2150

* Arrangement for opposite hand pumps only

Figure 8   Small pumping station (Courtesy Sigmund Pulsometer Pumps Ltd.)

well, a motor at the top supported on the motor floor, and an interconnecting tube joining the motor to the pump and containing the rotating shaft. The pump has a suction inlet or bellmouth connected directly to its inlet branch and the rising main is brought up vertically from a bend on its delivery side. Reflux and sluice valves are placed on, or immediately below, motor floor level.

Submersible pumps are comparatively inexpensive and in Great Britain have often been used for unimportant work or plant that is not likely to be in use for a long time. Initially they had the marked disadvantage that when a pump had to be repaired or maintained the whole installation would have been impracticable.

Of recent years, the use of fully protected submersible motors, mounted directly above the pumps, has become increasingly common within the smaller pump size range, and this arrangement avoids to some extent the difficulties with pipework, just described, and moreover does away with the necessity for any form of superstructure, other than to house the electric starters and controls. A typical installation is shown in Figure 9.

Moreover a series of ingenious guides and couplings have been developed which enable the pump and motor to be uncoupled and withdrawn without the necessity for lifting or dismantling any pipework. Motors are fully protected and mounted above the pump, with a short suction pipe immediately underneath it. Thus, in the smaller size ranges, little more than a glorified manhole is required for an automatic pumping installation though a slight aesthetic problem has arisen at times due to the necessity to house the starters and controls in a separate cubicle away from the very damp conditions in the pump well. The kiosks used for this purpose become the only structures above ground, and are an anathema to many town planners! Increasing costs of sewer construction, as compared with that of rising mains in the 75–150 mm range particularly has emphasized these advantages, and many such installations are now to be found both in Great Britain and on the Continent. In America, things may be said to have gone even further, as pressure systems with a pump on every house-connection or, even more complicated, vacuum systems sucking the household discharges to a central station, have been constructed and are the subject of a great deal of interest because of the high cost of conventional sewerage (see Chapter 18).

A dry well pumping station consists normally of two underground chambers: a wet well to contain the water, and a dry well to contain the vertical-spindle pumps, the motor and switch-gear being in the house or superstructure, above. The pump-suctions pass through the watertight walls of the wet and dry wells, and the pumps are set at such a level that they are automatically primed by the level of the water in the well. This is by far the most usual type of installation for sewage pumping and is also largely preferred for many water-supply schemes and other purposes. The cost of pumps and structure is greater than that of some other types of installation such as submersible pumps or horizontal primed pumps set at motor floor level, but it is far the best for practical working and maintenance purposes for the majority of automatic stations. (See Figure 8).

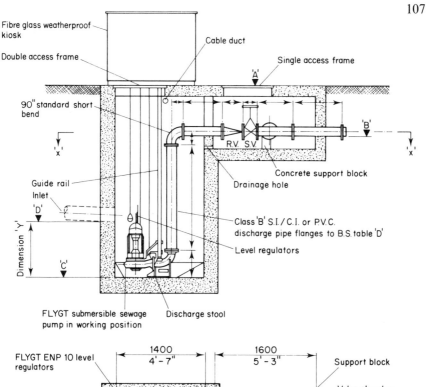

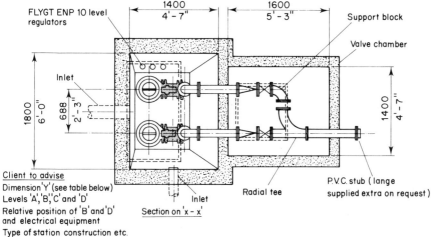

Pumping time = storage+assumed flow			
		Discharge rate	
Dimension 'Y'		STORAGE *	
Metres	Feet	Litres	Gallons
0.900	3'-0"	1500	330
1.200	4'-0"	2300	500
1.500	5'-0"	3000	670
1.800	6'-0"	3800	840

*These figures were calculated on storage capacity between stop and start regulators

The above layout may be varied to suit individual requirements

| Millimetres |
| Imperial units |

Figure 9   Submersible pump installation (Courtesy Flygt Pumps Ltd.)

It may be worth mentioning that the well-known and widely used Sykes Auto-vac pumps, although normally used under arduous conditions on construction sites, can also be applied in permanent installations. They have the ability to overcome suction lift and priming problems through the medium of a vacuum pump continuously driven from the pump shaft. As the latter controls the rate of suction lift, as distinct from that of delivery by the pump, cavitation is only avoided by careful design of the unit.

Other pumps

It is not always realized that in a separate system the average dry-weather flow generated by 1000 population may well be not more than 1 1/s. Thus it becomes difficult on small schemes to equate pump size to delivery requirements, particularly where delivery is direct to the sewage works, where pumping operations at comparatively low rates over long periods are much to be preferred to short intermittent bursts of high-rate pumping. For some years now the Mono pump, which may be described as a miniature high-speed version of the Archimedes screw, has been used for various applications both at the sewage works and as the link between sewers and works. On its own this pump can pass only small-diameter solids (say up to 14 mm) but teamed with a disintegrator unit, installed on the suction side, flows of crude sewage as low as 40 l/min may be pumped through 38 mm diameter rising mains. The normal range of such installations is up to about 5 l/s, but it must not be forgotten that to all intents and purposes these are positive pumps with delivery related to speed and not to pressure. As pressure in small-diameter mains, although normally steady enough under any given set of conditions, can vary noticeably if there is a tendency towards partial airlock or considerable variation in solids content of the disintegrated sewage, automatic sets are usually provided with a pressure relief valve with a return connection to the suction well. If for any reason the relief valve continues to operate for long periods there will, of course, be considerable waste of power, as well as diminution in delivery, not to mention possible repercussions in the form of surcharged sewers! Daily recording of pressure-gauge readings will usually give sufficient warning of untoward conditions, but conditions in the suction well should also be checked, so that overflow from the pressure release valves does not go un-noticed. For such small flows a sump 2 m in diameter will often suffice, but for convenience in operation and maintenance it is well worth providing a conventional structure either wholly or partly over the wet wall analogous to the motor room provided for ordinary centrifugal pumps.

The peristaltic pump

The peristaltic principle derives from the wave of contraction passing along tubular organs, by which their contents is propelled. The last few years have seen the development of improved versions of this type of pump using modern tube materials and better methods of fixing the tubes to the mechanisms. In practice, a series of exterior rollers pass along the tube and either squeeze it flat or push it at an angle at which it kinks and closes (see Figure 10). Although pumps have been designed to deliver up to 2.5 l/s against a negligible head, normal flow is up to 0.25 l/s. It is used for laboratory apparatus, in continuous pumping of effluent flows, and for dosing application. It is easily cleaned, liquids never come in contact with driving or metallic moving parts, and as there is no slip, it is suitable for metering applications.

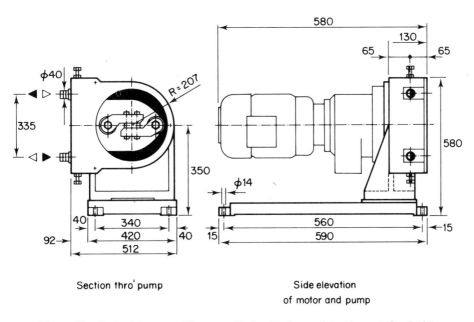

Section thro' pump

Side elevation
of motor and pump

Figure 10   Peristaltic pump (Courtesy Harley Hydrosand Equipment Co. Ltd.)

The principle is being increasingly applied on feed pumps to filter press installations dealing with various industrial slurries, farmyard excrement, and sewage sludges. The accuracy of delivery allows such processes to be automated, the slurries being conditioned, if necessary, by chemicals injected by a separate peristaltic pump. Although generally housed in a prefabricated structure, a trailer chassis is also available, making the unit mobile.

Pumping liquids containing solids

When fluids contain solids the choice is either of an unchokable pump with a comparatively low efficiency, or of a high-efficiency pump which may need to be protected by screens against solids which could choke it. In the case of sewage sludge the unchokable pump is almost always to be preferred because reliability is the first requirement and, as the quantity of sewage sludge is small relative to other liquids to be pumped in most sewerage and sewage-treatment installations, efficiency matters very little. Reciprocating pumps have often been selected for this purpose because it is thought that the sludge would be too thick for a centrifugal pump to handle, but in fact such circumstances are rare.

Unchokable pumps are most commonly used for pumping small flows of crude sewage. In these circumstances the reason for the choice is that it is impracticable to give constant attention to small pumping stations, and although the pumps may be running for long hours every day, consuming an appreciable amount of current during the year, reliability without attention is much more important and economical than mechanical efficiency.

When pumps are installed for dealing with storm water and are likely to run for a comparatively few hours in the year, mechanical efficiency may still not be important and it may be possible to use a pump which does not need protection from solids, but the choice depends very much on circumstances, particularly the size of the units, as there is a limit to the size of unchokable pumps that are obtainable.

In towns having sewers large enough to be entered by men, occasional objects such as motor tyres and builders' waste, dead animals, etc., find their way to pumping stations. Extraordinary solids cause much comment, and not unnaturally difficulties from such causes become exaggerated.

At one time it was not uncommon to provide screens for the protection of small centrifugal pumps. But this practice was found unsatisfactory, for such screens were neglected, rapidly became choked, and, when they had to be cleared, disposal of the screenings was not easy without risk of nuisance. It is now normal practice not to protect small unchokable pumps by screens, but the writer has found some advantage in providing one or two vertical slots 50 mm wide and 0.6 m or more in depth between a manhole on the incoming sewer and the suction well. These slots will intercept large solids and take a long time to choke because there are no bars around which fibrous matter can wrap. If there are two slots, they should be at least 0.3 m apart.

Sewage ejectors

Sewage ejectors are used for delivering small flows of crude sewage against moderate heads. They have the advantage of simplicity of installation, for they can be installed in underground chambers with little or no wet well, the air-compressing plant being placed either underground or at ground level. Their disadvantages are low efficiency (very low at high heads), and higher capital costs on plant as compared with pumps of equal capacity, although this may well be offset by saving on building work.

The capacity of an ejector is stated in terms of the cubic capacity of the cast-iron vessel of the ejector body and this is often taken to be the delivery per minute on the assumption that the ejector will discharge once a minute. In fact, an ejector having a capacity of x m^3 will discharge more than x m^3 a minute, given an adequate supply of air.

Most sewage ejectors work on a simple principle. The sewage gravitates into the vessel from the incoming sewer, or drain, via a non-return valve. As the sewage level rises in the vessel a float, or bell, is lifted which operates an air valve admitting compressed air, under the force of which the sewage is blown out through another non-return valve into the rising main. Then the float closes the air-inlet valve and operates another air valve which releases the pressure, the compressed air in the vessel being discharged, usually to the inlet manhole or sewer.

The cycle of operations takes place in a time dependent on the rate of flow of sewage into the ejector under gravity plus the rate at which it is blown out by the air. The latter is usually rapid, which is advantageous for, provided the air pressure is adequate, it ensures a self-cleansing velocity in the rising main.

Sewage ejectors are usually installed in pairs and mechanically linked so that

Table 19   Power required for compression of air

Head (m)	Pressure* (hg/cm^2)	Cubic metres free air per m^3 of compressed air	Kilowatts per m^3 free air per minute (at 100% efficiency)
0.9	0.09	1.09	0.141
1.2	0.12	1.12	0.185
1.5	0.15	1.15	0.228
1.8	0.18	1.18	0.270
2.1	0.21	1.21	0.312
2.4	0.24	1.24	0.352
2.7	0.27	1.27	0.392
3.0	0.30	1.30	0.430
3.3	0.33	1.32	0.466
3.6	0.36	1.35	0.503
4.2	0.42	1.41	0.573
4.8	0.48	1.47	0.643
5.4	0.54	1.53	0.706
6.0	0.60	1.59	0.771
7.5	0.75	1.74	0.919
9.0	0.90	1.89	1.05
10.5	1.05	2.03	1.18
12.0	1.20	2.18	1.30
13.5	1.35	2.33	1.41
15.0	1.50	2.48	1.50
18.0	1.80	2.77	1.69
21.0	2.10	3.07	1.86
24.0	2.40	3.36	2.00
27.0	2.70	3.65	2.14
30.0	3.00	3.96	2.28

* These figures, multiplied by 98,066.5, give newtons per square metre or pascals.

each in turn will discharge to the rising main while the other is filling. Thus any storage capacity on the incoming side is rendered unnecessary. When two ejectors are so installed it is usual to consider the working capacity of the station as being the output of one ejector, although it will usually be slightly greater, provided the air supply is adequate.

The quantity of air required for a sewage ejector can be found by the formula

$$C = Q\frac{(H + 10.3323)}{10.3323} \tag{20}$$

where: C = m^3 free air per minute;

H = total manometric head in metres;

Q = discharge of sewage in m^3/min.

Pressure should be arranged to be 40% higher than the theoretical requirement.

The ancillary equipment of an ejector station is a small air compressor capable of delivering a sufficient volume of air which, when compressed, will be at least equal in m^3/min to the maximum volume of sewage to be ejected, and at a pressure greater than the maximum manometric head required when the flow is discharged during an appropriate fraction of the time occupied by the whole cycle. The compressor delivers to an air receiver which can be of quite small capacity, but is often made large so as to permit the maximum rate of flow of sewage temporarily to exceed the capacity of the compressor. The compressor is stopped and started when the pressure in the air vessel rises above, or falls below, predetermined limits.

At least one form of ejector is capable of 'sucking' as well as 'blowing' the sewage, thus avoiding the need for anything more than a small storage chamber below ground, with the ejector plant situated in a conventional house at ground level. These plants give reliable automatic service for small duties on 'separate' systems.

Another combination tried with some success by the author on village schemes is an ejector for dry-weather flow, and a centrifugal pump for higher duties in times of rainfall. The combination takes advantage of the fact that in order to suit the characteristics of the ejector, the rising main would probably be 100 mm diameter and thus capable of dealing with pumped flows up to say 10 l/s, or even more. The ejector sump would be arranged to overflow into the pump well at a predetermined level. Both the pump and the ejector would normally be housed in the 'dry well' with the pump motor and all electrical gear located in a motor room above.

Low-lift air lifts

An air lift consists of a vertical pipe having its greater part below water level. A jet of air is blown into the lower end, filling the column of liquid with bubbles and so reducing its density below that of the liquid in the well outside it. This causes the mixture of air and water in the pipe to rise.

There are three types of air lift. In the first the pipe supplying the air passes down inside the air lift, not reaching quite to the bottom. In the second the air pipe is carried down outside, and parallel with, the rising main and turned into it near the bottom. The third has the rising main inside a larger, longer air pipe (see Figure 11).

The formula applicable to air lifts for deep wells cannot be applied to those used at sewage works which have submergences in the region of 3.0–4.5 m. This is largely due to the relatively high importance of velocity head in the case of short pipes. On the basis of tests made by Activated Sludge Ltd, with 305 mm and 380 mm internal-diameter tubes about 5.64 m long, the author derived the

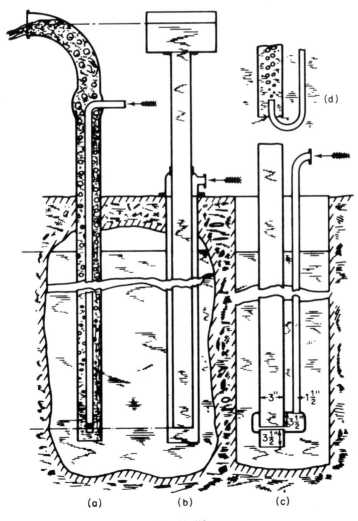

Figure 11   Air lift pumps

Table 20 Optimum efficiency of low-lift air lifts

Submergence ratio (H/h)	Percentage efficiency	Submergence ratio (H/h)	Percentage efficiency
1	19.1	4	27
1.5	21.15	4.5	27.8
2	22.8	5	28.6
2.5	24	5.5	29.2
3	25.2	6	29.9
3.5	26		

Table 21 Performance of low-lift air lifts: approximate cubic metres of air per cubic metre of water lifted

Submergence ratio	Velocity of water in metres per second									
	0.30	0.45	0.60	0.75	0.90	1.05	1.20	1.35	1.50	1.65
1.5	4.54	3.98	3.58	3.30	3.16	3.16	3.32	3.60	4.03	4.60
2.0	3.18	2.78	2.50	2.30	2.22	2.22	2.32	2.50	2.82	3.20
2.5	2.40	2.10	1.89	1.74	1.67	1.67	1.75	1.90	2.13	2.42
3.0	1.91	1.68	1.50	1.39	1.33	1.33	1.39	1.51	1.69	1.93
3.5	1.58	1.38	1.24	1.15	1.09	1.09	1.15	1.25	1.39	1.59
4.0	1.33	1.17	1.05	0.96	0.93	0.93	0.97	1.06	1.18	1.34
4.5	1.15	1.01	0.90	0.83	0.80	0.80	0.84	0.92	1.02	1.16
5.0	1.01	0.88	0.79	0.73	0.70	0.70	0.74	0.80	0.90	1.02
5.5	0.90	0.78	0.70	0.65	0.62	0.62	0.65	0.71	0.79	0.90
6.0	0.80	0.70	0.63	0.58	0.56	0.56	0.58	0.63	0.71	0.82

information given in Tables 20 and 21. In the experiments the maximum efficiency occurred at the (water only) velocity of 1 m/s (see Table 21) at which the efficiency of the air lift (excluding that of the compressor, etc.) would be as in Table 20. This velocity is much lower than considered economic for deep-well air lifts.

An air lift should have free discharge at the top. If it has not, not only must the head over the point of discharge be added to the lift, but it must also be deducted from the submergence.

The degree of pressure in the air supply requires to be only slightly in excess of that necessary to blow the air down to the submergence of the air pipe, i.e. about 0.3 bar per 0.3 m of depth of submergence. Thus the necessary power can be calculated from Table 19.

Planning pumping-station layout

In most instances, particularly in the case of automatic sewage-pumping stations of large capacity, it is best to have as many of the pumps as possible similar in size, hand of rotation, and all details, in order that any pump may take over the duties of another, and all component parts be interchangeable. This may seem obvious; nevertheless perhaps the majority of pumping stations have been designed with pumps of various sizes intended to operate in succession; first a small one cutting in, then a large one cutting in and the small one cutting out, then the small one cutting in again, and so on. In the author's view this system of operation, being dependent on an elaborate system of relays and switches, is unnecessarily liable to break down.

One of the few exceptions to this rule is when a sewage-pumping station may require to have DWF pumps delivering a controlled quantity to sewage-treatment plant and much larger pumps dealing with all storm water. Unless the duties required are within the scope of single pumps, such an installation may be considered as two pumping stations, one for DWF, having pumps all of one size, and the other for storm-water, having pumps all of another size.

As mentioned previously (e.g. p. 108), rates of flow pumped direct to the treatment works are worthy of careful attention and should, if possible, fit in with maximum rates of flow permissible through the settlement tanks or any by-pass to storm overflow or storm water tanks. Rates up to say 6 × DWF have often been adequately covered by a small pump and a large one. With centrifugal pumps it will usually be of no advantage to run the two together, as with the bigger pump operating little or no delivery will be obtained from the smaller one under the conditions commonly obtaining. Pumps of the same size working together will take equal shares of the duty, but unless the rising main is remarkably short the combined output will be substantially below that of two pumps operating singly. This, of course, is due to the unavoidable increase in head which could only be overcome by arranging for each pump to deliver through a separate rising main. Two rising mains laid in the same trench may, in fact, be well worth considering to overcome velocity and friction loss problems.

Assuming that several pumps all of one size are to be used, the designer has to decide whether to have a large number of small pumps or a smaller number of larger pumps. This choice is not altogether arbitrary and can, at times, be strictly limited by circumstances. In a very small station it is virtually a rule that there shall be one duty pump capable of dealing with the maximum flow to be accommodated, plus one stand-by. In a larger station it might suffice to divide the maximum rate of flow to be passed by the minimum rate occurring once a day to find the number of duty pumps and add one or more stand-by units.

It is often thought that pumping stations having a small number of large pumps cost less than pumping installations of the same total capacity but having a large number of small pumps. Examination of tenders shows that there is little, if any, difference when the whole of the equipment is taken into account. In a pumping station the pumps themselves represent only a small part of the total cost. The cost of electric motors of any particular kind increases with KW rating, roughly as the fourth root of the power of the individual motor, but types change with size, offsetting this. The suction and delivery valves, however, are by no means a small part of the total cost and the larger they are the more costly they are in proportion to the rate of flow that passes through them. Where there are several pumps the cost of stand-by units is reduced, but the dry well and motor house may be a little larger. Taking all these, and some other factors, into account there is very little difference in the cost of pumping stations of the same duty regardless of the sizes of the individual units; in fact, a very close approximation to the cost of a pumping station can be found merely by multiplying the total pumping capacity, including stand-by, by a constant dependent on the quality of the installation, the product being in dollars or pounds.

There is sometimes an advantage in having several small pumps in order to reduce the degree of change of rate of flow when any pump cuts in or out; the advantage of this can be exaggerated: often changes of rate do not matter at all.

In automatically operated stations the number of pumps may be limited by the operating depth in the suction well, and its ability to suit a number of 'cut-in' and 'cut-out' levels of the float switch controls. Where rod floats are used a separation in level of, say, 150 mm or rather more is necessary to avoid the conditions known as 'hunting' where pumps are continually cutting in and out, due, say, to turbulent conditions in the suction well. Thus, in the somewhat rare cases where a large-diameter sewer serves as a suction well, the designer would probably be forced to use a small number of large pumps.

Having decided on the number of pumps and the capacity of an individual pump, it is next necessary to determine its dimensions and then the dimensions of the various components of valves and pipework, the amount of space required round and between pumps for access and maintenance purposes, etc., and so arrive at the size of the station as a whole.

When pumps are of comparatively unusual design (e.g. some classes of axial-flow pumps) it may be necessary to obtain information from a manufacturer. However by far the larger number of pumps are likely to be of the single-entry volute centrifugal or mixed-flow type, and in many pumping stations will be of the vertical-spindle type. While the pumps of all manufacturers vary in detail and performance to some extent, and a large-diameter pump running at a slow speed can have the same duty as a small pump running at a higher rate of rotation, examination of several makers' catalogues indicates that, broadly, there is little difference in size in the majority of instances between one make and another. Usually, the larger the capacity of the pump, the slower its rotation of speed and the greater its dimensions, while increase of head calls for higher speed of rotation with some slight reduction in dimensions. Bearing this in mind the author has deduced the following formula for predetermining the sizes of pumps likely to be required in pumping stations

$$D = \frac{KQ^{0.5}}{H^{0.25}} \tag{21}$$

where: D = diameter of suction branch in mm;
$\quad\quad\quad Q$ = delivery of volute centrifugal or mixed-flow pump in m³/min;
$\quad\quad\quad H$ = manometric head in m;
$\quad\quad\quad K$ = a constant.

The value of K was found to vary between a maximum of about 180 and a minimum of about 130, the average being about 156. The writer has used the last figure in predetermining the sizes of pumping stations for some considerable time and has found it adequate for the purpose.

Particularly in the case of small-size pumps of the volute type, the delivery branch is sometimes slightly smaller than the suction branch, but this is not a rule and, for drawing purposes, it can be assumed to be the same size. It is, however, as well to allow extra length for a taper-pipe connection in the delivery pipework.

From the size of the suction branch the remaining dimensions of the pump, for drawing office purposes, can be estimated by the following approximations, where D = nominal internal diameter of pump suction branch in mm:

(1)   distance back from face of flange of volute pump delivery
      to a plane passing through axis of pump. $D + 280$ mm
(2)   distance from axis of pump to axis of delivery $D + 178$ mm
(3)   distance from face of flange of suction to a plane passing
      through axis of delivery $D + 100$ mm

The minimum distance between the axes of any two vertical pumps can be established according to the following rule:

Distance between axes of pumps: $2.50D + 1280$ mm

Having found the sizes, dimensions, and positions of the pumps relative to one another, their general position in the dry well of the pumping stations will be fixed by the dimensions of the pipework and the incorporated valves.

The dimensions of sluice valves and reflux valves can be found for drawing purposes by the following rule, where D = nominal internal diameter of sluice valve or reflux valve in mm

(1)   distance between flanges of sluice valves $30D^{0.5}$ mm
(2)   height of sluice valve measured from axis of pipe to cap
      of spindle $2D + 380$ mm
(3)   distance between flanges of reflux valves
      (In large sizes this last rule gives excessive lengths be-
      cause the type of reflux valve changes, but the rule still
      proves useful). $2D + 150$ mm

The dimensions of cast-iron pipe fittings are best obtained from any standard specification available in the country concerned; also there are several easily obtained, very useful, non-standard fittings, listed in some countries with full dimensions and particulars, in the catalogues of the leading manufacturers.

Pumping-station pipework

The suction pipes of a pumping station of the dry-well type consist of the inlet end from the wet well, the isolating sluice valve and the connection to the pump. In the case of very small sewage-pumping stations the inlet may be in the form of one or two slots about 50 mm wide in the narrow direction, but having an area at least equal to that of the pipe, but tending to exclude solids that might cause chokage. In a larger station having unchokable pumps the inlet may consist of the bare end of a pipe, not reduced to a slot and not in the form of a bellmouth, for bellmouths, unless very large, can concentrate balls of fibrous material and cause chokage. Very large sewage pumps and all pumps for water free from solids should have bellmouths to reduce head losses at entry. Many inlets are made to turn downwards in the suction well so as to draw as low as possible. This arrangement has the effect of lifting the level of the pump relative to the suction well, which can set the pumps inconveniently high if they are of vertical-spindle type and not double inlet.

When suctions are turned down towards the floor they should not be set too low, otherwise free flow may be impeded; if the installation is for sewage or other liquids containing suspended solids, however, the suction should be near enough to the floor to pick up the solids. The lowest water level in the suction well must be above the entry to the suction pipe by not less than the velocity head of entry, otherwise air will be drawn in by vortex formation.

Suction pipes should continue from the dry well at a steady rise to the pump. Portions of the crown may be level; but in no case should any change of diameter or of level be allowed to cause an air pocket in the crown. Generally, the diameter should be large so as to prevent cavitation and, as a rule, velocities not exceeding 1.5 m/s for centrifugal, and 1 m/s for reciprocating pumps, are desirable. To drop velocity much below the latter figure for any pump lifting solids may also create problems.

The sluice valves on suctions should be so arranged that they may be operated from motor floor level and in all cases above any possible flood level.

As a means of pumping out the dry well, should it become partly flooded, and also for dealing with water spilled on the floor during maintenance, auxiliary suctions should be provided on the main suctions, between the sluice valves and the pumps, and controlled by their own sluice valves. If practicable such suctions should branch from the bottom of the main suction, not the sides, in order that the main suction can be completely emptied on to the floor when necessary. Although suctions are of small size, being no larger than necessary to avoid chokage, their sluice valve spindles should also be carried above flood level for obvious reasons.

In large pumping stations small automatically operated cellar-emptying pumps are usually provided to keep the dry wells free from any water which may be spilled on the floor or leak in through the walls. These are in addition to the auxiliary suctions, which are still useful for pump-emptying or in an emergency.

If the pump is of vertical-spindle type and single-entry, the suction pipe finishes in the form of a 'medium' 90° bend which turns up and connects to the inlet branch of the pump. In the case of pumps dealing with sewage, sludge, or other liquids containing solids, this bend should have an access for cleaning or be easily removable: however, such bends are sometimes difficult to remove because of their position under the pump and between the concrete stools which support the casing. This bend is usually of the same diameter as the inlet branch of the pump and in many instances the whole of the suction pipe may be of the same diameter, provided an unduly high velocity does not result. Any taper to increase the diameter of the suction should be as near the pump as practicable, and generally should be omitted in the case of small pumps dealing with sewage or sludge,

Pump suctions are best arranged above the floor of the dry well. An alternative is to set them in trenches covered by chequer plating, but this method has two disadvantages: it complicates the construction of the floor, and the trenches are liable to become full of water and dirt which, being very difficult to remove, is usually neglected.

The delivery branches of centrifugal pumps are often slightly smaller than the suction branches, and the pipes leading to the rising main can generally be smaller than the suction pipes, for high velocities are not deleterious except that they add to manometric head and, if very excessive, can cause scour at bends. A maximum velocity of 2.5 m/s is usually stipulated. Broadly, high velocities are economical because the relatively high cost of reflux and sluice valves. Table 22 gives economic velocities for pumps on fairly regular duty: the same figures could be used without undue extravagance for storm-water pumps.

The fittings on the delivery of a pump consist of any necessary taper pipe connecting with the delivery branch, a reflux valve, an isolating sluice valve and

Table 22   Economic diameters of pipework and valves

Discharge of one pump (m^3/min)	Economic diameter of suction and delivery pipework and valves (nominal mm)	Approximate economic velocity (m/s)
0.40	100	0.823
0.68	125	0.885
1.08	150	0.975
1.53	175	1.04
2.21	200	1.13
2.97	225	1.22
3.85	250	1.28
6.09	300	1.40
10.65	375	1.52
16.8	450	1.71
24.7	525	1.83
34.4	600	1.95
46.2	675	2.07
60.2	750	2.22
76.3	825	2.32
95.0	900	2.41

such bends and pipework as are required to connect to the rising main. Reflux valves for sewage or sludge should always be on horizontal pipes for if they are placed in vertical pipe lines, solids will fall on the back of the door, causing it to stick open with the danger of slamming and damage to the plant. Reflux valves for water may be in vertical members, but are then of different design from those intended to be placed on the horizontal.

There are several types of reflux valve; some that have flaps freely hung from the top, some that are balanced, and some that have external arms showing the position of the door. The last should be carefully adjusted, using balance weights on the arms, for otherwise the friction of the gland through which the axle passes may cause the door to jam open, and later to slam. Reflux valves for crude sewage or primary sludge should be of the single-door type—and top, not centrally hung—for any bifurcation of the flow is liable to cause rags and string to hang up and interfere with the operation of the valves.

Sluice valves on deliveries, like those on suctions, should have their spindles carried up to motor floor level. They should always be set vertical.

The delivery pipe should connect to the rising main, preferably with a radial feed, but this is not essential. In the case of sewage-pumping stations, each connection should enter the main laterally, not from underneath, for otherwise solids may fall down the branch when flow from other pumps is passing across the top of it.

It is not always practicable to make delivery pipework of a diameter which is best economically, or to make suction pipework exactly the diameter required to avoid cavitation. Delivery pipes and valves may be of larger diameter than the delivery branch of the pump and connect thereto by a taper pipe; but they should not be smaller and, when the pump delivery branch is equal to or greater in diameter than the economic diameter of the delivery pipes, this greater size may be maintained right through to where the delivery pipes connect to the rising main (see Table 22).

Suction pipes are usually larger than delivery pipes when connected to large pumps if they are sized so as to give a velocity of 1.22 m/s. In the case of small pumps, however, an economic velocity for both delivery and suction pipes is less than this, so the suction pipes should be no smaller in diameter than the delivery pipes.

Pipe joints and supports

All the pipework inside a pumping station, as contrasted with pipework laid in the ground, should have flanged joints, for the pressure developed in a rising main can often blow open the various kinds of socket joints, particularly when pressures are great and diameters are large. (The ultimate strength of a caulked lead joint is said to be 14 kg/cm^2 of lead in contact with the spigot.) Some flexible joints for cast-iron pipes have no resistance to internal pressure that is worth taking into account.

Joints for flanged pipes are usually made with rubber caulking rings. (There is a British Standard for these rings, but it relates to the consistency of the rubber only and the Classes A, B, C, D, E, and F are for rubber hardness and bear no relation to the similar classes for cast-iron pipes.) The flanges of pipes can be bolted together either with a narrow rubber ring placed inside the ring of bolts, or a wide ring the full breadth of the flange, perforated with the bolt holes. The latter type is generally preferable, for those rings which are not perforated are very difficult to place in position when making the joints of large-diameter pipes.

When designing pipework it is necessary to consider the possibility of taking down the installation, re-erecting it, and tightening up joints. If two ends of a straight line of flanged pipes are built into the two walls of a building, the pipework cannot be taken down and re-erected unless one socket joint, or expansion joint, or a pipe with flanges at an angle to the axis is inserted. On the other hand if there is a right-angle bend in a line of pipe, a socket joint or expansion joint should not be necessary and could be a danger unless the pipework were anchored against movement. Although pump manufacturers usually manage to correct designers' errors in these respects the design should be correct in the first instance; the author has seen cases where work has had to be done to correct faults of this kind after the pumping stations have been put into service.

It is almost invariably possible to design pumping station pipework using standard or stock-size flanged fittings throughout, apart from a number of special template-length straight pipes or cut pipes and expansion or socket joints, and in designing pumping station pipework the best results are obtained by strictly adhering to standard fittings and avoiding awkward special castings. The writer prefers also to specify that (in Britain) British Standard fittings and pipes shall be used as far as practicable, although this may not be popular with pump manufacturers, for the makers of pumps frequently obtain their pipework from foundries who make all the castings to the dimensions shown on manufacturers' drawings without regard to whether they are standard or not. It will be appreciated that replacement of any non-standard pipe could be very inconvenient, should damage occur.

Suctions, deliveries, and rising mains inside buildings need to be supported by concrete stools from the floor, or by wall brackets, beams, etc., so that there is no undue stress put on the pipes themselves. Horizontal cast-iron pipes should be supported where necessary on concrete stools or piers, or by brackets from the walls. Concrete stools should support one-quarter of the circumference of the pipe and be at least 300 mm thick in the direction of the axis of the pipe and be at least 75 mm wider than the internal diameter.

Where pipes pass through walls it is usual to provide puddle-flanges, either specially cast-on or of the bolted-on type, to reduce the risk of leakage through the walls along the face of the pipe. Leakage in this position is quite common if great care is not taken in placing the concrete. Generally, all such concrete around pipes and in the walls of wet and dry wells should be vibrated. It is good practice also to order pipes for 'building-in', to be uncoated externally, so as to secure better adhesion with the concrete.

Where the rising main leaves a pumping station it usually changes by means of a standard flange and socket pipe, from flanged to socketed piping, the latter being much less costly than flanged piping and perfectly satisfactory when laid in trench and supported by concrete to the face of the excavation at all bends and junctions. The bottom end of the rising main generally should have a wash-out pipe and valve and, if there are two rising mains, each wash-out may discharge to the suction well. If there is only one rising main the suction well should be large enough to hold the contents of the rising main if the wash-out is connected to it.

Like the general layout of pumps etc., pipework should be simple and orderly. It is desirable that the headstocks of suction and delivery sluice valves should form two straight lines and each headstock, by its position, clearly indicate to which pump it relates. Auxiliary suctions do not need headstocks: they may be operated by key via a surface box in the motor-room floor so as not to occupy too much space.

When inviting tenders it is as well to specify that manufacturers shall adhere as closely as practicable to the drawings, otherwise they may greatly distort the scheme, adding complications and untidy and unsightly features.

Valves

There are several kinds of sluice valves and reflux or non-return valves, and each type should be used in correct circumstances. Sluice valves are of two types: first, those for burying under the ground on rising mains have internal screws working in the liquid but protected from the earth outside; secondly, valves for use indoors have external screws which are easily accessible for lubrication.

Sluice valves for sewage, sludge, and the like, unless buried in the ground, should have access doors or plugs arranged at the bottom in the position where detritus is likely to settle and prevent proper closing. When sluice valves are a very great distance below the point from which they are operated, they may have, in lieu of rotating spindles, pull-and-push spindles (usually of square cross-section) operated by screws in the headstocks, for long tortional spindles spring too much and, when a man gives a turn to a handwheel, it may go back half a turn when it is released between turns.

Large valves and penstocks can take a surprising time to open and the work can be very laborious. A rough estimate of the time to open a valve can be found by allowing 5 min per 900 cm² of waterway through the valve. If the time taken to open a valve as estimated in this way will be too long in the circumstances, or if the valve is one which will have to be opened and closed frequently, it may be necessary to have a power-operated valve which will open in a much shorter time.

The sluice valves of the duty pumps of automatic sewage pumping stations should always be left open so as to permit flow to take place should the pump start. If a reciprocating pump is started against a closed delivery valve it may cause immediate damage before being thrown out by overload. If a centrifugal pump is run against a closed delivery valve it will absorb the horsepower applicable to the closed-valve conditions according to its characteristic curve, converting the whole of the energy to heat, so that in a very short time the water in the pump will be raised to boiling point, producing steam, again with the possibility of causing damage. There is also the risk of the pump seizing up. The risk of damage by pumping against closed valves is one reason for not having electrical gear for valve closing: a valve that is laborious to close is not likely to be closed unnecessarily.

The main trouble experienced with reflux valves is slamming due to their closing too late. This is most particularly troublesome in sewage and sludge-pumping stations, where the valves are likely to get stuck open by solid matter being packed behind the door. If a reflux valve sticks open at the cessation of pumping, water will rush backwards from the rising main, through the pump and into the suction well, eventually clearing the dirt from behind the door of the reflux valve. Then the valve will close very violently, bringing the long column of water in the rising main suddenly to a stop, and developing immense pressures which in some instances can easily break open joints, burst the lids off hatch boxes, or split the rising main often a long distance from the pumping station.[2]

While the proper selection and installation of a reflux valve will help to obviate such troubles, additional precautions may be advisable, particularly if rising mains are long. An air vessel similar to, but larger than, that normally installed on the delivery side of a reciprocating pump will considerably reduce the shock. If the manometric head is not great, the lower end of the rising main can be connected to a stand-pipe over which water can be forced by the shock and discharged to the suction well. The risk of damage to fittings was well described by B. Smith in a paper presented to the Institution of Public Health Engineers.[3]

Pumping-power and cost calculations

The purpose of a pump is to develop pressure in a fluid and if the fluid is a liquid, not a gas, the pressure is most frequently required for lifting it from one level to another, so as to produce a static head.

The work that has to be done by the pump is this dead lift or static head plus a number of losses of energy which are classed under velocity head and friction head. The last is in part made up of friction in the rising main and other pipes plus a number of velocity-head losses which are all classed as part of the friction head.

The total head against which the pump has to deliver, made up of all these components, is known as the manometric head because it can be measured by manometer or pressure gauge. If the fluid is water, the head is best expressed in terms of metres head of water, which can be converted into bar by multiplying by 10.

The dead lift is the distance vertically from the water level in the suction well to the highest level to which the water has to be pumped. The highest level may be the level in a reservoir at the delivery end or the crown of the rising main at its farthest end; but it should be borne in mind that the latter is not always the case when the rising main has a free discharge, for sometimes the highest point on a rising main is intermediate in its length and it may be that the manometric head required to deliver to this point is greater than that which would be needed to deliver to the farthest end. What is the case can be determined by drawing a section of the rising main and drawing on the hydraulic gradient from the outlet end and from any high intermediate point.

As the level in the suction well usually varies, a decision has to be made which level shall be used in head calculations. This must depend on circumstances. In many instances the level at the delivery end also varies during pumping.

The velocity head is the head required to accelerate the water from stationary to the velocity in the pipework. It is calculated by formula (4), p. 26.

The friction head is the loss in the suction and delivery pipework, valves, and in the rising main. The friction head in a pipe system can be calculated by the appropriate formula for all straight lengths of suction and delivery pipes and rising main. To this should be added the losses due to bends, junctions, and sluice and reflux valves. The loss of head or friction in the pump itself is calculated as part of the pump's efficiency, and appears in the characteristic curve of the pump.

The loss of head through reflux valves follows the velocity-head formula only when the velocity exceeds about 2.5 m/s. At lower velocities the head also is increased by the partial closing of the flap. The figures in Table 23 are representative.

Table 23 Loss of head through reflux valves

Velocity (m/s)	Loss of head (m)
0.6	0.110
0.9	0.117
1.2	0.128
1.5	0.143
1.8	0.158
2.1	0.176
2.4	0.206

Power requirements

The required power in kilowatts to drive a pump for water or sewage is found by the formula

$$\text{kW} = \frac{QH}{6.1183} \times \frac{100}{\text{pump efficiency } (\%)} \tag{22}$$

where: Q = m^3/min;

$\quad H$ = manometric head in metres;

$\quad$ kW = *mechanical* load on motor in kilowatts.

In this the metres head is the total manometric head and the pump efficiency is the estimated efficiency for the type and size of pump. kW × h running gives kilowatt-hours, on which the charge for electricity is usually based.

$$\text{kilowatts} = \frac{100}{\text{power factor } (\%)}$$

$$= \text{kilovolt-amps, on which are based kVA charges.}$$

Charges for electricity are mainly based on kilowatt-hours consumed. There is frequently an additional annual charge for kilovolt-amps and sometimes for maximum actual peak demand recorded during the period.

Table 24　Efficiencies and power factors of electric motors (experimental values)

Kilowatts at 100% motor efficiency	Efficiency (%)			Power factor (%)		
	Full load	$\frac{3}{4}$-load	$\frac{1}{2}$-load	Full load	$\frac{3}{4}$-load	$\frac{1}{2}$-load
1.5	74	73	70	80	75	65
2.25	78	78	76	81	76	66
3.0	81	81	79	82	78	69
7.5	85	85	83	85	81	72
15.0	88	89	88	89	87	80
30.0	89	89	87	90	88	82
75.0	91	92	91	88	85	79
150.0	93	$93\frac{1}{2}$	93	89	87	81

Capacities of generators, transformers, and cables to carry the power necessary for the station depend on the kVA in the case of alternating current, although the horsepower at the generating station will depend on the kW. (In the case of direct current there is no power-factor calculation.) Unity power factor alternating current motors are obtainable at extra cost, or the power factor of an installation may be adjusted by the use of condensers.

To ascertain the annual running cost of a pumping station, the average daily flows must be found and the running time of one or more pumps in hours determined so as to give the number of units consumed in a day. This, multiplied by $365\frac{1}{4}$, gives annual units from which the total cost of units may be obtained. It should be remembered that in the case of a sewage pumping station the average daily flow will invariably be greater than the average DWF by some amount, even if the station is supposed to pump sanitary sewage only. The cost of the units plus the kVA charge and/or any maximum load charge gives the cost for power for the year.

When using internal combustion engines the following figures give a rough guide to fuel costs: 0.30 kg of crude oil produces 1 kilowatt-hour, and 0.9 m³ of sewage sludge gas produces 1 kilowatt-hour in normal dual-fuel engines.

To find the total annual cost of running the station, to the cost of power must be added the labour and maintenance costs, rates on the building, etc., and repayment of loan on the capital cost of the installation according to the current rate of interest and the period over which the loan is paid in equal instalments.

Large pumping stations, fully attended day and night for manual operation, involve three shifts of 8 h plus an additional shift for holiday and sick relief, costing several thousands of dollars or pounds a year per station. Unless the station is very large, a shift consists of one driver and one cleaner.

When automatic pumping stations are so designed that the probability of serious failure is negligibly small, there is no need for them to be manned. All that is necessary is for each pumping station to be visited about once a week (or more frequently according to size) for general tests, inspection, and overhaul, cleaning, etc., with possibly more frequent inspection of electrodes at sewage pumping stations, should this prove necessary. To be economical, an authority needs to have a sufficient number of pumping stations or similar establishments for a travelling maintenance gang of at least two men to visit at least one station a day.

Economic size of rising main

As a general rule it has often been assumed that rising mains for sanitary sewage that are in use for several hours a day shall have diameters that give velocities in the region of 0.8 m/s.

Contrary to a not-uncommon misconception, the economic diameter of a rising main does not depend on the length of the main unless the length is so short as to be negligible. The reason for this is that the major variable costs of a pumping scheme are the capital cost of the rising main and the running costs of electric power or fuel. The differences of horsepower do not have a very great effect on the cost of the pumping plant: an increase of head adds to the cost of electric motors and slightly to that of switch-gear, but it may slightly *reduce* the cost of the pumps themselves. Thus, in making comparisons between schemes with large- and small-diameter rising mains, the cost of the pumping plant can usually be neglected.

It is perhaps most often desirable to know the economic diameter of a rising main taking the flow from a large storm-water pumping station or a station dealing with flow from a combined sewerage scheme. The method of finding the economic velocity is to find by trial the scheme which gives the lowest overall annual cost, including total annual charge for power, e.g. unit and kVA charges for electricity, plus annual repayment of loan on rising main.

Automatic sewage pumping stations

Small automatic electric sewage pumping stations have been constructed for very many years where installations were too small to justify constant attendance and where an occasional breakdown would not be a serious matter. Practice has been to provide pumps in duplicate, one duty and the other stand-by, at one time invariably float-operated, later by pneumatic devices, floats, or electrodes.

The extension of automatic operation to larger installations has proceeded with some caution, but there have been few failures or serious troubles (usually due to designers' errors). Now, stations of the largest type can be rendered entirely automatic, saving a great cost on the day and night shift control that manually-operated stations require.

The differences between modern and earlier pumping stations, apart from the increase of size to which automation has been applied, are improvements resulting from experience and a high degree of reliability dependent not on elaboration but on simple, correct design.

The need for simplicity deserves reiteration. At a time when automation is a topical subject there is some tendency for engineers to put in automatic gear and warning devices even in manned stations, thereby not only adding to cost but introducing a risk that the operational staff, relying on the machine, will lose their alertness.

There are some rules which must be adhered to if large automatic stations are to be absolutely reliable in all reasonable circumstances. First there should be not fewer than two entirely separate sources of electricity supply, and these should be so connected that each would be immediately and automatically available should the other fail. In several instances two or more sources of electricity have been provided but have been connected to a main switchboard in such a manner that should one fail the other can be switched on in place. This arrangement is unsatisfactory and could prove disastrous at a storm-water pumping station, the failure of which could result in a flood before an attendant could arrive to change over the supply. While change-over can be automatic, this means an additional complication with added risk of breakdown.

The alternative is to have both sources connected in the following manner: the first, third, and fifth pumps to cut in; and the second to the second, fourth, and sixth pumps. This will ensure that on failure of one source of supply one-half of the pumping power will be available until an attendant can arrive.

The next and most important rule is that each pump with its motor, electric gear, starting floats, or electrodes, etc., and all appurtenances should be entirely separate from the others and capable of functioning on its own, regardless of almost any disaster that could happen to the remainder. As previously mentioned, a not-uncommon practice was to provide a small pump to cut in first then, as flow increased, for a larger pump to cut in and the smaller one to cut out until further flow brought it in again. This meant an electrical interconnection of the pumps and, unavoidably, the danger that, should a fault develop in the relay system, the whole pumping station could be put out of commission, although the individual pumps, motors, and starts might still be perfect and capable of operation had they been separately controlled.

Pump sets are mechanically independent in so far as each has its own suction pipe with isolating sluice valve, delivery with reflux valve and sluice valve, but all may connect into a common rising main. There will then be no danger of the mechanical failure of one affecting another except by the failure of a sluice valve or breakage of the pipework and rising main. In most large stations it is well worth having two rising mains with pumps capable of connection by valve adjustment to either, as may be desired. A common pipe may run through the pumping station with the delivery of the various pumps connecting into it and its two ends leading away as separate rising mains. In this case each of the rising mains should be isolated from the common pipe by a sluice valve and above each sluice valve should be a wash-out capable of emptying each rising main to the suction well.

With the foregoing arrangements carefully carried out, there is very little risk indeed of a serious breakdown and therefore constant attendance becomes unnecessary. In fact, should a plant attendant be present there would seldom be anything he could do, until such time as a gang of fitters could be brought to the station.

There is, however, a further precaution that can be taken, and that is to install an alarm system giving information to some remote control room which is constantly manned, e.g. at a manned pumping station or sewage-treatment works, or via the public telephone system to the house of a responsible officer. Failure or partial failure of a station as described would be indicated by pumps not cutting-in in the predetermined order, or by the requisite number of pumps not running at the indicated water level in the suction well. These facts can be detected by electrical gear, interpreted, and transmitted.

Automatic control of pumps

The first method of automatic control by water level to be used was by means of floats. These were arranged in several ways, some better than others.

In one, the float controlling any particular pump was suspended on a stranded non-corrodable wire which passed round a drum and had a counter-balance. The float had to be inside a vertical cast-iron tube or be surrounded by a cage of guides, otherwise it might be carried away by the turbulence of the water. Another method, more generally useful, was to have floats with holes through the centre, through which a rod passed. The float moved up and down the rod until it came against adjustable stops, when it caused the rod to move downwards or upwards as the case might be, switching pumps off or on respectively. Such floats did not necessarily require any protection against currents or eddies in the suction well, but needed to be so located that they were not too violently held by the flow against the rod, preventing sliding motion.

These methods, particularly the latter, are still in use and can be very satisfactory. Some troubles may be experienced: generally floats for sewage should not be so arranged that they come in contact with settled sludge at the bottom of the well or the float tube and get stuck. There is always some tendency for rags to twine round rods or wire, but this is seldom a serious trouble.

When flexible wires are used they are often brought up vertically from the float, cast over a pulley, and carried horizontally through the wall of the wet well and into the pumping station. This can be a cause of trouble. To some extent foul odours can pass through the hole where the wire goes, but a greater danger is the possibility of water flowing from the wet well to the dry well, flooding it, if for any reason the pumps fail.

To avoid complications, rods with floats on them are brought straight up to the float-operated switch: any attempts at angular motion by quadrants or similar devices are undesirable complications. The float switches employed usually require to be moved 75 mm to turn them on or off, a fact that should be kept in mind when determining tolerances of cut-in and cut-out water levels.

Floats are made of copper or other non-ferrous metal that will not corrode too rapidly in water or sewage, or preferably of suitable ceramic material. They are generally flattish, of such a weight that they normally float half-submerged so as to be equally effective in pulling down and pushing up, and large enough for their displacement to be easily sufficient to operate the switch.

The control of pumps by electrodes has largely displaced the use of floats at British pumping stations of all sizes. Electrode rods are suspended vertically so as to dip into the suction well at various levels. The cut-in electrode of any pump operates a relay starting the pump when the water level rises sufficiently to touch the end of the electrode. The cut-out electrode stops the pump when the water loses contact with its lower end. Thus, by the use of two electrodes (for cut-in and cut-out respectively), set at different levels, a pump may be arranged to pump from one level down to another. The electrodes are supplied with low-voltage alternating current and installations are available in two grades of sensitivity.

Electrodes are not without their troubles. They require regular maintenance, mainly cleaning. If they are too long they may sway in the turbulence of the suction well and eventually their fixings may become damaged, for electrodes cannot always be stayed satisfactorily even by insulated stays in a wet well, where short-circuiting can take place over a damp surface. The rod type of float-switch gear needs a minimum clearance of some 200 mm between rods, and between rods and the walls of the well. Lacking this, short-circuiting can occur. Sluggish operation can also occur due to heavy accumulation of grease, etc. on the rods themselves, which need frequent cleaning. Various other types of gear have become available, notably the pear-shaped plastic float, housing a mercury switch, which operates when the water level rises, and raises the float to a horizontal position. There are also systems depending on slight variations in air pressure in submerged tubes fed with air from a small compressor. A rise in water level produces an increase in air pressure in the tube, which at a predetermined figure operates a pressure switch, thereby starting or stopping the pump. This device is in fact one of the more sensitive and accurate ones available.

In America these pneumatic devices are much favoured, but according to the WPCF Manual of Practice[4] pneumatic types are considered most trouble-free, floats next to these, and electrodes most troublesome of all. Pneumatic controls are now more commonly used in Great Britain.

Size of suction well

In water-supply schemes the size of the suction well is seldom of any importance. Often the pumps suck directly from reservoirs, which are then suction wells of very large size. In other cases the suction well may be made large at comparatively little cost, for there is no risk of trouble by settlement of solids.

In sewage-pumping stations large suction wells tend to become sedimentation tanks, in which sludge settles at the bottom and scum accumulates at the surface, setting up a problem of frequent cleansing, which may have to be manual or, at best, require an arrangement of the pumping equipment and shaping of the suction well which will permit periodical complete pumping out of all solids.

The problem is reduced or eliminated by making the suction well of as small a size as practicable and carefully shaping it, in order that the turbulence of the entry of the sewage and the pumping out shall prevent sludge and scum accumulation.

Opinions vary considerably as to how the bottom of the suction well should be formed so as to minimize settlement of silt and accumulation of scum. One method which is generally satisfactory is to have the suctions of the several pumps suitably spaced apart and drawing from a channel one side, or preferably both sides, of which slope outwards at about 60° to the horizontal till they meet the walls. To completely prevent silting, the suctions of the pumps should be close together, but this is impracticable, for if sufficiently close for this purpose they cause mutual hydraulic interference between the pumps.

Another form of suction well is one which has a long channel with a benching sloping down to it at an angle (about 1 in 6) which will cause sludge to run off moderately well provided that the lowest cut-out level of the pumps is sufficiently low to draw the water below the benching. The advantages of this design are that it gives a surface on which men can stand for any maintenance purposes, and that the capacity between cut-in and cut-out levels (required by the calculations which follow) can be secured in a moderate depth with less difficulty than would be the case in the former design. The disadvantage is that the benching does not keep clean well enough for all purposes unless a high degree of turbulence can be induced in the well by the incoming flow.

Turbulence can be caused by permitting the liquid to fall some distance on entry—the distance, of course, increasing as the water level drops below the invert of the incoming pipe. But for turbulence to be effective in preventing settlement, the capacity of the well must be small in order that the body of water to be stirred up is not too great. It is in this respect that there has been some considerable change in the design of automatic sewage-pumping stations as compared with those which were constructed some 50 years ago. It is here also that proper design of suction well in relation to pumping plant has to be considered.

The extent to which the capacity of a suction well can be reduced in size is limited by the ability of the electric starting gear to tolerate frequent starts. In the case of any one pump, the most frequent starts occur when the rate of flow into the well is one-half of the delivery capacity of the pump. Suppose the pump can empty the well from its cut-in level to its cut-out level in 1 min; then, when the rate of inflow is one-half the pumping rate, it will pump the well out in 2 min, after which it will stop and the inflow will refill the well in another 2 min, making a total cycle of 4 min from start to start. When the rate of flow is greater than one-half the pumping rate the pump will take a longer time to empty the well, but the refilling rate will not be much quicker than before; and should the incoming rate be less than one-half the pumping rate, the time of refilling will be increased to a greater extent than the reduction of time of pumping out. These facts can easily be ascertained by anyone working out a few examples.

On the basis of the foregoing it can be said that the most frequent cutting-in of an automatic pump is four times the time taken for that pump to empty the well from its cut-in level to its cut-out level. Therefore if the pump is to start once every 4 min, the capacity between the cut-in and cut-out levels of that pump must be the equivalent of 1 min delivery of that pump.

It is necessary to state in the specification the number of starts per hour, and to make it clear that there is no limit on the duration of this rate of starting. Frequent-duty starters capable of 40 starts per hour could reasonably be called for where only 15 starts per hour are expected: this is a practice which the writer has used for many years with satisfaction.

Where there are several automatic pumps the capacity between cut-in and cut-out level of each pump can be calculated without any regard to the others, provided that the cut-in and cut-out levels are arranged so that the first pump to cut in is the last to cut out, as given in the following example

No. 3 cuts in at 750 mm above assumed datum,
No. 2 cuts in at 600 mm above assumed datum,
No. 1 cuts in at 450 mm above assumed datum,
No. 3 cuts out at 300 mm above assumed datum,
No. 2 cuts out at 150 mm above assumed datum,
No. 1 cuts out at 0.0 mm above assumed datum,

The distance between cut-in levels should not be too small, for more than one reason. First, pumps do not deliver *any* flow until they have attained the speed of rotation necessary to overcome static head. This may take approximately the times (in seconds) given in Table 25. A number of pumping stations have been designed with all the cut-out levels alike. This is not recommended, for it would

Table 25   Starting times of pumps (s)

Type of pump	40 kW	110 kW	225 kW	375 kW	750 kW
Centrifugal	20	30	40	60	90
Reciprocating	20	40	60	60	90

be liable to lead to trouble by causing two pumps to cut out at once, thereby approximately halving the cycle of start to restart.

Pumping station without suction well

There are rare occasions in which pumping stations can be designed without *ad hoc* suction wells. Simple examples are such as drawing from a large reservoir or from the open dyke of a land-drainage scheme. A comparatively tricky problem is designing an automatic station in which a sewer or culvert of very large diameter has to serve as suction well so as to avoid the considerable expense of constructing a suction well of adequate capacity to control the starting and stopping of the pumps. The first fully automatic installation to be based on the theory described herein was the sewage-pumping station at Crossness (London) Sewage-treatment Works designed by the author. This pumped directly from two low-level and one high-level sewers each of 3.5 m diameter and had a total pumping capacity, including stand-by, of 1.3 million m³/day. After 5 years operation it was reported that all pumping units had functioned satisfactorily with few minor defects. In this circumstance, the number of pumps that can be installed for automatic operation is limited by the number of cut-in and cut-out levels that can be arranged between the lowest water level in the culvert and the crown of the culvert.

Sewers or circular culverts are normally laid to a gradient; accordingly the amount of water in a culvert between two levels is a slice taken through a cylinder and the volume per unit of depth will vary according to depth above the invert. The storage in a circular sewer can be found by the following formula, which is near enough for practical purposes

$$C = \frac{\pi D^2}{4} \times \frac{A_1 + 4A_2 - 5A_3}{6} \times (H_1 - H_2)i \qquad (23)$$

where: C = storage capacity in cubic metres;
$\quad D$ = diameter of sewer in metres;
$\quad A_1$ = proportional area applicable to H_1 (see Table 3, p. 59);
$\quad A_2$ = proportional area applicable to $(H_1 + H_2)/2$;
$\quad A_3$ = proportional area applicable to H_2;
$\quad H_1$ = depth over invert in metres at cut-in level;
$\quad H_2$ = depth over invert in metres at cut-out level;
$\quad i$ = inclination or length in metres in which sewer falls 1 m.

The highest cut-in level must be well below the crown of the sewer, otherwise high surge pressures may develop on the suction side of the pumps and put the station out of action.

The calculation is of course complicated if the sewer is not circular or of even diameter or gradient throughout the length in which storage can take place.

Automatic diesel stations

In the absence of an electric supply it is possible to have an automatic diesel-driven pumping station. Electric gear transmits from the floats or electrodes when the water is at cut-in level and the diesel engine is turned over by compressed air to start it. The diesel then runs until the water has dropped to cut-out level. The diesel needs to run for at least 15 min so as to give adequate time for compressing the air for the next start and recharging the batteries of the electric gear. This means that a large suction well is necessary. The method is unavoidably more complicated and less reliable than automatic electric pumping, and for this reason it is not often used.

Remote-controlled unattended pumping stations

Intermediate between pumping stations which are manned day and night by three shifts or at any times when they are required to operate, and those stations which are entirely automatic, are unattended installations manually controlled from a distance. The control can be by private land-line or by telephone. Where the control is by land-line, starting and stopping signals are sent out initiated by pushbuttons on a control board. These operate relays at the receiving end and set in motion any necessary sequence of starting operations. A signal is then sent back automatically to the control board showing that the plant has started or stopped, or indicating any cause of failure to do so.

One method of using telephone communication is to install electric gear so that any part of a pumping plant may be started or stopped by dialling the telephone number reserved for this purpose and then dialling a coded instruction.

Open-air stations

Some economy in construction of automatic or remote-controlled stations is effected by arranging for a greater part of the pumping plant to be in the open air instead of in a pump- or motor-house. It is quite practicable for weather-proof motors and starters and weather-proofing over the crane to be provided, the appropriate classes of motors and starters being used to withstand all weather conditions. Alternatively, the pumps and motors may be in the open and associated electric gear in some other pumping station or building not necessarily close at hand.

Storm-water storage

A pumping station for combined or surface-water sewerage can be arranged to store storm water in its suction well or in a tank or tank-sewer discharging thereto, so that the maximum rate of pumping can be considerably less than the peak rate of flow into the suction well. For calculating the required storage capacity, formula (16) or (17) pp. 91–2 should be used as appropriate. As the pumping rate enters the formula, the storage capacity must be provided *above* the highest cut-in level of the pumps, any capacity below this level being used only for the purpose of controlling the starting and stopping of the pumps, as already described.

The building and ancillary works

The building consists of the wet well, the dry well (if a dry well station), and any necessary screen chambers and manholes below ground level and (usually) above ground level, the motor room and ancillary rooms including men's accommodation, store, and any necessary substation.

Below-ground structures were formerly of brickwork or mass concrete. At the present time reinforced concrete is much more often used, although mass concrete can be employed quite economically in some circumstances. All below-ground work must be watertight. This was formerly achieved by the grouting or flushing-up of brickwork and the rendering of concrete and, later, by bitumen tanking, an expensive and not necessarily reliable method.

At the present time watertight construction can be secured in either mass or reinforced concrete by thoroughly vibrating the concrete during construction and entirely cutting away the surface at all construction joints before placing further concrete.

Mass concrete has two advantages over reinforced concrete; first it has much less tendency to develop shrinkage cracks: secondly it is heavy enough to resist flotation. Further it can often compete against reinforced concrete in terms of economy.

As a compromise between mass and reinforced concrete, for small stations, particularly where ground conditions are difficult, a 'caisson' type of structure, with circular or oval mass concrete walls about 0.5 m thick cast at ground level over a metal and reinforced concrete 'cutting ring', can be sunk 'lift' by 'lift' to the desired depth. A fairly heavy reinforced floor slab is then formed *in situ*, followed by a reinforced concrete division wall between wet and dry wells, and finally a reinforced roof slab probably forming the floor to the motor room. A grab type of excavator is used to take out the bulk of the excavation, but hand-trimming may be necessary to ensure reasonably accurate downward movement of the cutting edge. This type of structure is favoured by many contractors, owing to the savings to be made by avoiding a heavily timbered excavation. Where unstable ground exists to considerable depth, the 'caisson' method can be applied to bring quite complicated reinforced concrete structures to rest on *in-situ* piles previously formed up to foundation depth. The placing of the floor slab may be difficult in very soft ground, and excavation requires to be carefully controlled, otherwise distortion of the pile heads can take place due to 'running' of silty material under the cutting edge, as the walls descend.

The introduction of reinforced concrete has lightened the weight of structures and thereby augmented the risk of the flotation of buildings by upward water pressure. Care should be taken in design that not only is the entire completed structure heavy enough to withstand flotation under the influence of the highest ground water level, but also that no part of the below-ground structure shall float during construction. Flotation can be resisted by making the work heavy enough in itself to be equal to the weight of the water it displaces in the excavation. This may be easy with a mass concrete construction but extravagant with reinforced concrete. A reinforced concrete substructure can be held down by extending the floor outwards beyond the walls to give ledges on which refilled earth can rest, adding its weight to that of the structure. Any building founded on piles is almost certainly safe against flotation.

The suction well is usually of the same length or width as the dry well, as deep as may be necessary to accommodate all the cut-in and cut-out levels of the pumps and wide enough, with the other dimensions, to give the required storage capacity. It should preferably have access from *outside* the building only, for men to enter by ladder or stairway extending to the bottom and, except in small stations, to a gangway above highest water level. The access should be large enough for the suction pipework to be lifted in and out. If the pumps are controlled by electrodes it is advantageous for the suction well to be partly under the motor room so that the heads of the electrodes can be mounted on the motor floor and not in an external manhole or in the open air. The suction well must be ventilated sufficiently but not so as to resist incoming flow. The interior surfaces of the well should be smooth and, with good supervision and suitable specification, adequate smoothness can be attained without the use of rendering or granolithic flooring, etc.

The dry-well floor should slope slightly and evenly to a drainage channel formed of half-round channel laid at a grade to a drainage sump. From this, auxiliary suctions or a cellar-emptying pump can take away any water that seeps into the dry well or is discharged on to the floor when pumps are taken down, or leaks from glands, etc. Small drainage pipes should lead from main glands to this channel. Stools for supporting pumps are anchored to the floor with bars in the case of a reinforced concrete structure. If the dry well is deep, it is economical to build intermediate floors at intervals between the bottom floor and the motor floor above. These should be open near to or at the centre, to permit the lifting in and out of pumps, valves, pipework etc., during construction and maintenance. The remaining strips of the intermediate floors against the long wall and the end walls form stiffening beams which can be held apart by reinforced concrete struts or steel joists to take the thrust of the earth on the outside or the water pressure from the wet well. There should be handrails (at least 1 m high) around the openings, and ladders or stairways should terminate at each floor. The floors can serve as supports for bearings on vertical driving shafts, pipework and auxiliary-hoist girders.

The motor room accommodates the electric motors or prime movers, headstocks for valves etc., electric starting gear, and main switchboard. Its floor should have openings to permit the erection and maintenance of the machinery below. These should be covered with (usually) 10 mm steel or aluminium chequered plating secured to cast-iron rebated frames with non-ferrous countersunk set-screws, and supported at all joints and, where necessary because of the span, by removable rolled-steel sections. Whereas below-ground floors are often finished with smoothly screeded concrete, motor-room floors really deserve a good finish such as quarry tiles or terrazzo. Where the latter is used, a border of a different colour from the main surface improves the appearance. Either tiles or terrazzo may be swept up at the walls to form a skirting and render cleaning easy.

The height of the motor room depends on the height of the traveller crane or other hoisting arrangements and clearance space above. The doors to the motor room must be large enough to give easy access for the largest machinery, e.g. to allow electric motors to be brought in right way up (but not necessarily with the exciters in position): this may call for a removable fanlight (transom). The floor should be able to take the heaviest piece of machinery considered as a live load. In a large station the floor can be made strong enough for a loaded vehicle to be able to enter.

Formerly large pumping stations had pitched roofs with trusses, where the spans were large, or collar-beam construction for small buildings. These were very satisfactory, but went out of fashion. For many years flat reinforced-concrete roofs rendered waterproof with bitumen sheeting or asphalt were popular, but in a number of instances it was found that these expanded and contracted with temperature changes and tended to crack off from the walls and move laterally relative to the rest of the building.

In rural areas much stress tends to be put on harmonizing the appearance of a pumping station with its surrounds. A flat roof is seldom suited to such conditions and so pitched roofs have often to be used, and sometimes local stone for the walls and rustic tiles on the roof to match those of nearby cottages. Often concealing all but one side of the building by lowering the floor level a little and creating an embankment round the other three sides is the most economical way of obscuring such structures.

At the present time many alternative roofing materials are available. In the selection of untried materials and the detailing of their use, great care should be taken to avoid any risk of leakage that could lead to damage to electric gear.

Screen houses are generally similar to motor rooms, with overhead cranes, but they should be fly-proof.

The men's accommodation of a manned station includes an office, a mess-room, a clothes-locker room, a store, and lavatories. In a small station some of these offices may be combined. The lavatories should contain a urinal, water closet, basins, and sink. There should be one or more showers, and basins should have surgeon's taps, wrist- or pedal-operated. The type of hot-water system depends on circumstances, but in any case the lavatories call for overhead water-storage tanks and space to accommodate them. The messroom requires a sink with drinking-water tap and a suitable cooking stove.

Pumping stations that are not manned still preferably include men's accommodation, in particular lavatories, for the visiting maintenance gangs, although not on the same scale as required for manned stations.

Ventilation

Ventilation of the above-ground structure of most pumping stations is usually best effected by natural convection by ventilator openings or windows at high and low level. Natural ventilation may be adequate also for the dry well if it is not deep and has fair-sized openings between it and the ground floor. On the other hand, deep dry wells of large pumping stations can become pockets in which exhaust fumes and other objectionable gases may collect, and this calls for mechanical ventilation. Mechanical ventilation of dry wells should always be by extraction, not injection, the trunking reaching down to lowest floor level so as to remove heavy gases and the discharge being delivered to the open air: blowing in fresh air is not a certain means of removing foul gas.

The primary purpose of ventilation of the wet well of a pumping station is to prevent development of negative or positive pressure on the falling and rising of the water level in the well. As a general rule it suffices for this purpose to supply a ventilating pipe of such a size that a negative or positive pressure of more than 12 mm water gauge shall not develop during the most rapid fall or rise of the water surface.

Such a ventilating pipe will serve also to ventilate the sewer or drain discharging to the well, in which case its size should also be related to this secondary requirement. In Britain the use of a rainwater pipe (leader) as a ventilating shaft is an offence against the Public Health Act, 1936 (Section 40).

Heating and lighting

Heating and lighting for pumping stations have to be considered with regard to the type of pumping station, its size, and intended operation.

Small pumping stations and storm-water stations or other installations which may be at rest for long periods require heating and ventilation to prevent corrosion of metal or freezing of any water pipes. It is usual to provide heaters inside electric motors and starter panels, but thermostatically controlled heating of the space generally is desirable.

When one or more pumps will be running continuously, the efficiency losses of the electric motors should be taken into account in working out heating requirements, for the whole of the motor efficiency loss is converted to heat and this heats the air. In large stations the amount of such waste heat can be easily sufficient to provide all heating required in the building in the coldest weather, or can be made so by attention to heat insulation, the avoidance of excessive ventilation, and the avoidance of unnecessary windows.

The matter of artificial as against natural lighting is, in these days, receiving more serious consideration than formerly. In many industrial buildings it is an economy to have artificial light throughout the day and night so as to avoid loss of heat through windows. In the case of unattended pumping stations which are visited for a few hours in the week for maintenance purposes this is particularly true, and the amenity value of natural lighting ceases to be of real importance.

There is a second reason for avoiding windows at unattended stations. If the buildings have no means of access except by the door and this is made of steel or other strong material, vandalism, not only causing expensive damage but possible serious breakdown of an installation, can largely be obviated. It has been known for the greater part of maintenance cost to be for repair of windows broken by vandals!

When windows are provided they should not be made unnecessarily large but generally in the region of about one-sixth to one-seventh of the floor area and should be located for easy cleaning, if possible without the aid of portable ladders. High-level windows forming a clerestory can be satisfactory if accessible from a gallery and/or from the flat roofs of outbuildings.

Cranes and hoisting arrangements

Pumping plant requires maintenance in the form of replacement of worn impellers, glands, etc., necessitating that each pump, the shafting above it, and, in some instances, the electric motors, have to be taken down and the parts lifted out.

In small pumping stations all that is necessary is a steel joist arranged over the pump or line of pumps, to which can be fixed hoisting gear. At one time manual chain tackle was used for this purpose, and with two sets of chain tackle the heaviest components of the installation could be lifted, moved laterally, and lowered to the floor. It is now usual except at small stations to provide portable electric hoists that can be hung on the joists which in all instances should be at a sufficient height to give ample space above the electric motors or other machinery to allow for the working depth of the hoist or chain tackle and for lifting the largest component of the machinery over the motors and any other obstacles on the floor.

In pumping stations of moderate and large size travelling cranes are usually provided. These move longitudinally on rails supported on beams along the walls of the motor room and a crab moves laterally across the travelling girders. To prevent cross-winding when in longitudinal motion the wheel-base of the travelling girder should not be less than one-fifth of the span between the rails.

Cranes of this kind can be made to lift by electric power, or the crab may move by electric power and the girders move by electric power: alternatively, any of these motions can be manually operated by continuous chains hanging down to motor-floor level. Whether or not electric power is provided wholly or in part depends on the size of the installation. Electricity is supplied either by plug and socket (there being a trailing cable and a number of socket points), or by contacts on bare wires fixed along one side of the building. The former arrangement has some inconvenience; the latter an element of danger.

A travelling crane should be capable of lifting the heaviest individual component (usually the electric motor) in the station. Cranes are often included in the pump contract, the pump manufacturer being asked to specify the weight of the heaviest component and the weight that the crane will lift. But there are occasions when the crane may have to lift some item not in the pump contract, in which case the weight must be specified and the contract for the crane may be let separately.

A not very common, but convenient, arrangement is to have a walkway at crane-rail level, giving access to the crane and, for cleaning purposes, to any clerestory windows. A walkway across the beam of the crane is an added convenience, for not only does it give access to the crane; it also makes easy the maintenance of electric light fittings which are often difficult of access in some pumping stations. In this connection adequate headroom above the crane should be allowed, apart from that necessary for the machinery itself. As in the case of simple hoists, the highest position of the hook of the crane should be studied from the point of view of lifting the largest pieces of machinery over obstacles on the motor floor.

The crane hook should be capable of reaching right down through openings left by the removal of chequer plating to the floor of the dry well. However, there may in large stations be positions where the crane hook cannot lift such objects as sluice valves, reflux valves, or their parts except at an awkward angle. In this case, auxiliary hoists should be fixed below the motor floor in the required positions, either chain tackle or portable electric hoists being provided.

Fuel oil store

If pumps are driven by diesel engine, a store for fuel oil should be provided of sufficient capacity to tide over strikes or other causes of temporary interference with supply.

Preferably there should be two oil stores, each of such a minimum capacity that a full load can be taken from a tank-vehicle when any one tank is nearly empty.

These oil tanks are not, in all countries, under the same rigorous control as petrol stores, but it is usual for precautions to be taken against risk of fire or leakage. Any large tank or tanks should be out of doors, well away from buildings or combustible structures, and away from each other. Each tank should be set in a pit or surrounded by a wall so that should the tank burst or leak, the oil will not escape to waste to the surrounding ground or to the sewers. Naturally, such surrounding pit will collect rainwater. This can be pumped out by a small cellar-emptying pump controlled by electrodes, not a float. On the rise of water level, an electrode will set the pump running, whereas a rise of oil level due to the bursting or leakage of the tank would not affect the electrode owing to the low electrical conductivity of the oil.

Oil-storage tanks are filled by tank-vehicles having their own pumps for pumping the oil into the storage tank at a moderate elevation, and flexible hoses with couplings for connecting to the tank inlets.

Specification of pumping machinery

Specifications for pumping machinery are simple compared with those for general building contracts and a comprehensive bill of quantities is unusual, for measured work is normally carried out by the building contractor who should wait on the machinery contractor under terms mentioned in both contracts. In all cases, however, even when only small plants are concerned, detailed specification is desirable and in most instances the main components of works should be separately priced. In particular, full information should be given on the requirements of the plant, for annoyance and delay are caused when the manufacturer is wrongly or inadequately advised on the required duties of pumps or ratings of motors. The main heads of information that should be supplied, even when ordering or specifying a small plant, are:

(1) Locality of the proposed pumping station.
(2) The number of pumps of each kind required.
(3) The minimum quantity of sewage that has to be delivered by each pump when working alone and when working in conjunction with other pumps.
(4) The nature of the liquid to be pumped, including specific gravity, whether sewage, surface water, or sludge, hot or cold, and if corrosive.
(5) The total lift, including friction in the rising main, from lowest (or other working) level in suction well to discharge point at far end of rising main. This is sometimes calculated by the engineer, but pump manufacturers prefer to be also informed on the highest and lowest levels in suction wells and the length, diameter, and material of the rising main. If any part of the rising main is higher than the discharge point the pump manufacturer should be informed of the level and distance from the pumping station. In addition to supplying this information, the engineer may also specify the maximum head against which the pumps shall deliver, assuming future conditions different from those immediately envisaged.
(6) The type of pumps, where situated, and whether self-priming.
(7) The source of power—whether steam, petrol, oil, gas, or electricity: if electric power is to be used, the type of electric current available, whether AC or DC, the voltage and, if AC, the frequency and whether single- or three-phase, and the name and address of the supply authority.
(8) Type of motors required.
(9) Type of starter, and whether hand or automatic.
(10) If pumps are not electrically operated, the type of prime mover should be specified in detail.

It is really the responsibility of the engineer who designs the pumping station and specifies the plant to determine for himself the manometric head to which the pump will have to deliver in the circumstances of the specified duty. To ask pump manufacturers to quote for pumps capable of delivering a set quantity against a dead lift of so many feet, plus the friction losses through the station pipework and so many feet length of rising main of such and such diameter and material is an unfair requirement, for it places on the pump manufacturer responsibility for matters over which he has no control, and may, perhaps, not have been adequately informed. All a pump manufacturer can guarantee is that he can make a centrifugal pump capable of delivering a set quantity against a total manometric head within certain tolerances. as most of this may be friction loss, there is the possibility of different assessments being made, leading to serious misunderstandings where performance is being checked!

While it can be stated that a centrifugal pump shall be capable of delivering a specific quantity against a specific total manometric head, it is not reasonable to require that the pump shall also deliver a different quantity against a different head, for this will presuppose a particular shape of characteristic curve. An inexperienced engineer might get hold of some manufacturer's characteristic curve of a pump having the required delivery at the required head and, based on this curve, ask for one or more duties at different heads. By so doing, he would rule out most, or perhaps all, competitors of the maker of the pump to which that particular curve related.

It is, however, reasonable to require a specified delivery at a specified head and require that the characteristic curve shall be of such shape that, for example, it does not cause overload of the electric motor by delivering much greater quantities at a lower head, provided, of course, that the type of pump likely to be available for the duty concerned will be capable of having such a characteristic.

Very commonly it is specified that the pump shall deliver the required quantity and have its maximum efficiency when the water level in the suction well is at its lowest, the level at the delivery end at its highest (if it varies), and that the maximum number of duty pumps are discharging through the rising main; in other words, when the manometric head is the maximum possible. This rule does not always apply: there are, for example, times when the pumping station should be so designed that it never delivers more than a limited quantity. Take, for instance, the case where a DWF pumping station serves the purpose not only of lifting the sewage but controlling a quantity of, say, 3 × DWF to the sewage works for treatment, allowing storm flow to pass elsewhere. In this case the specified quantity of 3 × DWF should occur when all duty pumps are running and the water level in the suction well is at the highest, not lowest, cut-in level. Again, in this instance, the lowest charge for power would be secured if the maximum efficiency occurred at a time when one DWF were being pumped and the level in the suction well were half-way between the cut-in and cut-out levels applicable to that circumstance.

Erection and testing of centrifugal pumps

Small to moderate-size horizontal centrifugal pumps with suction diameters of 450 mm and downwards usually have cast-iron bed-plates for pump and motor; where larger pumps and motors are installed, these are frequently mounted on separate concrete beds.

Horizontal pumps should be secured to a solid floor which, if not of reinforced concrete, should be at least 600 mm thick. The pump or the bed-plate for pump and engine should be carefully levelled on iron packing and grouted in position before the holding-down bolts are tightened, and care should be taken not to move the bed-plate when tightening the bolts. Whenever practicable, the pump and motor should be taken off the bed-plate before it is levelled and the shafts between coupling and engine carefully set in line: the flexible coupling cannot be expected to take up any error of alignment.

In bolting-up flanged pipework care must be taken not to put any stress on the pump so as to move it out of line, also the alignment of the coupling should be checked and corrected if necessary after the pipes have been bolted up.

Vertical-spindle centrifugal pumps are most often mounted on concrete stools built to the floor. These have to withstand turning moments and any unbalanced thrusts or shocks, and accordingly should be adequately strong, preferably being reinforced in continuity with the floor should the latter be reinforced. Whilst the long driving shafts often associated with this type of pump at one time required considerable skill in alignment, the common use of the Hardy-Spicer or similar flexible coupling has obviated these difficulties and may even compensate for structural deviation, due, say, to caisson sinking.

Before starting a pump the watertightness of the pipework and, in particular, the airtightness of the suction pipework must be checked, and also that all pump glands have been packed. Then the lubrication system should be checked over. Glands should not be too tightly screwed in the first instance: stuffing boxes may at first run hot, but in time they should settle down and run cool. Glands of pumps should weep slightly to show that the packing is not dry. This weeping should be collected by suitable catching arrangements and drainpiping.

Pumps that are automatically primed or hand-primed must be primed according to the procedure required for the particular installation. Those that are self-priming by being below water level will, if they have automatic air valves, fill with water on the opening of the reflux valve on the suction side. Any hand air cocks should be opened to let air out of the casing or any other castings.

A pump should never be started with the suction sluice valve closed. But in the first starting the delivery valve may be closed for a short time to check the pressure developed by the pump when it has reached normal running speed: then it should be opened progressively before the pump has run long enough to heat the water. The opening should not be too rapid if the rising main is empty, otherwise too great a discharge of water against negligible head might overload the motor.

If after this procedure the pump fails to deliver, the plant should be shut down and again checked, in particular the airtightness of the suction pipework, stuffing boxes, and other places where air could leak into the pump.

All the foregoing work is best carried out by the pump manufacturer's staff and the manufacturer should be made responsible for the erection and any fault that may develop. This is not the testing of the pumps, which should follow when the manufacturer is satisfied that the plant is ready for testing.

Pump tests for performance

Pump tests are carried out at the manufacturer's works to ascertain that the pumps and associated equipment perform in accordance with the specification, and again after installation on site to make sure that the plant has been properly installed and is in working order.

The main items to be ascertained during the works test are that the pump will deliver the quantity specified against the specified head when running at the speed stated by the manufacturer in his tender, that the said quantity will be delivered against the said head, that the power required by the pump is as stated in the manufacturer's tender, and that the efficiencies of the pump and the plant as a whole are in accordance with the tender.

Manufacturers of pumps have, at their works, tanks to hold water, and accurate weirs or other measuring devices to determine rates of flow. The pump on test sucks from the tank and delivers into a stilling chamber from which the water flows over the gauging weir and discharges back into the tank, the water thus being pumped in a circle and not to waste. The manometric head on the pump is varied during the test by first pumping through open pipework and then, stage by stage, closing the sluice valve on the delivery until the pump is delivering against a closed valve. At each stage of closing of the valve, measurements are taken of the manometric head, the quantity pumped, the current input to the electric motor (if it is electric plant), and the speed of rotation. From these data curves are plotted of delivery against head, of horsepower, and efficiency.

The rotation of the pump must be measured by a revolution counter or a tachometer that has been calibrated. This is necessary even when electric motors are used because of the slip of the motor. This should be recorded as a percentage of theoretical revolutions per minute.

The head is measured by two manometers, preferably mercury U-tubes set on the suction and delivery pipes respectively, the total manometric head at any time being the sum of the negative pressure in metres measured on the suction manometer, the positive pressure in metres measured on the delivery manometer, and the vertical distance in metres measured between the two manometers.

The delivery is measured by hook gauge, if a weir is installed, and calculated usually by reference to tables. Figure 12 illustrates such a gauge, which would be zeroed to sill-level, the hook then being drawn up to break surface at the selected measuring point, to give an accurate measure of depth over the sill.

The results of a test are plotted straight away on graph paper prepared for

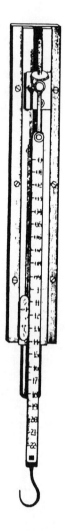

Figure 12   Hook gauge

the purpose, and on which have been previously marked the stated values of delivery and head, so that the experimental characteristic curve of the pump will show any discrepancy of discharge and head and of pump efficiency between the experimental curve and the specified duty, within the tolerances laid down in the specification.

The input of the electric motor is read by watt meter or, in the case of three-phase current, by two watt meters on two of the phases, in which case it is the sum of the readings. The efficiency of the electric motor is usually tested first at the appropriate maker's works. This being known, the overall efficiency of the pump-set can be calculated.

Site tests

Tests should also be required to be repeated after the pumps have been finally installed in the pumping station. These again may take a form similar to those at the works but with such omissions as the engineer may consider permissible or as may be enforced by circumstances. For example, it may not always be practicable at the pumping station to measure quantity delivered. The test at the pumping station is mainly to ascertain that the plant has been properly installed, with no mal-alignments of shafting or other faults liable to lead to breakdown or maintenance troubles.

The temperature of bearings and stuffing boxes should be checked to ascertain that they do not heat up unduly, and observations should be made to ensure that there is no vibration, cavitation, or leakage of air or water.

Just as it is unreasonable to specify that the manufacturer should guarantee that the pump will deliver a specific quantity through a specific length and diameter of rising main, so it is unreasonable to expect that, on a site test, the pump will deliver the required quantity through the rising main: it can only be expected to deliver that quantity when the specified manometric head is observed on the pressure gauges.

In the site test the pump should be started against closed delivery valve (the reverse of the works procedure) if the rising main is empty, or against an open valve if it is full. The delivery valve should then be adjusted until the pump is delivering against the specified manometric head when, provided means for metering are available, the discharge should be recorded. Then the motor should be stopped, upon which it should decelerate rapidly and the reflux valve close with a gentle click. Any violent slamming would call for immediate remedy, and absence of a click would suggest that the valve had not closed at all.

Automatic starting gear, floats, electrodes, etc., should be checked for degree of accuracy of starting with regard to water level and reliability of function. Readings should be taken of all electrical meters, etc., to ascertain that current consumption is not excessive, starting loads not excessive, and power factors not excessive, the last if necessary being adjusted to the requirements of the specification and/or the electricity authority.

References

1. Addison, Prof. H. *Centrifugal and Other Rotodynamic Pumps*. Chapman & Hall (1948).
2. Smith, B. Analysis of surge in pipelines, *Pubk Hlth Eng.* **4** (4) (July 1976).
3. Smith, B. Surge and Pumping Main Equipment, *Publk Hlth Eng*, **7** (2) (Apr. 1979).
4. Wat. Pollution Control Fed. Manual of Practice No. 3. *Design and Construction of Sanitary and Storm Sewers*.

CHAPTER 6
Construction of Sewers

When considering the economics of drainage and sewerage it should be remembered that by far the greater length of drain or sewer is made up of the smallest appropriate diameters. When large sewers have to be laid there may be objections made to the amount of public expenditure because it may seem considerable at the time, but it is usually insignificant compared with the total cost of small drains and sewers that, over the course of years, have been laid in the same locality. Apart from initial cost, the long lengths of small drains and sewers involve the greater part of the maintenance costs and are also responsible for most of the infiltration to the sewerage system. For these reasons the materials, workmanship, and general standards of construction of small sewers deserve careful attention.

Once drains and sewers in Great Britain of even as small an internal diameter as 300 mm, were constructed of radial brickwork jointed in lime mortar or such crude materials as rough masonry in puddled clay. But these forms of construction gave place to socketed pipes of salt-glazed ware or vitreous-enamelled fireclay, the material dependent on what was most easily obtainable locally, and the pipes were jointed in cement mortar. Several sizes were available but during the first half of the present century, it was found that pre-cast concrete pipes with either socket or ogee joints were considerably cheaper than stoneware except in the smaller sizes and, as a result of this competition (in Britain) large-diameter stoneware pipes ceased to be manufactured.

Vitrified clay pipes

Of comparatively recent years a new material described as vitrified clay pipe has been introduced and, having several advantages, has already virtually replaced glazed pipes. Vitrified clay pipe, being dense and of smooth surface, does not need to be, and is not, glazed. It can be made to closer tolerances and in longer lengths than salt-glazed or vitreous-enamelled fireclay pipes and for these reasons alone the new material is to be preferred. A further advantage is that vitrified clay pipes are available with various designs of special joints, several of which permit a good degree of flexibility and all of which ensure proper centring of the spigot in the socket without the need for the precautions that had to be taken when laying stoneware pipes with cement joints.

The new material is available with some slight differences of constitution in America, Britain, and West Germany, but the available sizes are not the same in these countries. In the United States the standard internal diameters in inches are 4, 6, 8, 10, 12, 15, 18, 21, 24, 27, 30, 33, and 36, and there are some non-standard sizes available. In Great Britain since metrication the available (sanitary-drain or sewer) sizes are 100, 150, 225, 300, 375, and 450 mm. These are the British standard sizes. Recent amalgamations of makers have taken place in an attempt to satisfy world demand and meet international competition. In West Germany vitrified clay pipes are made in the following millimetre sizes; 50, 75, 100, 125, 150, 200, 250, 300, 350, 400, 450, 500, 600, 700, 800, 900, 1000, and 1200 (the italics denote sizes which the manufacturers would like to abolish). The normal range of diameters are available in 'standard strength' and 'extra strength', whilst 100 and 150 mm pipes are also made in superstrength. These classifications correspond quite closely to the concrete and asbestos cement categories quoted hereafter.

Pre-cast concrete pipes

Pre-cast concrete pipes are available in most countries in a very large range of diameters and, in view of their lower cost, are largely used in lieu of vitrified clay pipes in diameters that to some extent overlap those of the clay pipes, the change from one material to the other depending on the preferences of the engineer who designs the work.

In Britain the standard sizes are still in 3-in. (75 mm) stages from 6 in. (150 mm) to 48 in. (1200 mm) internal diameter, but larger diameters up to 72 in. (1800 mm) are commonly stocked. 'Nominal' metric sizes are achieved by converting inches to millimeters by 75 mm increments. Perhaps the maximum special order was 84 in. (2100 mm) diameter, as compared with 12 ft 8 in. (3850 mm) as constructed in America.[1] Pre-cast concrete pipes usually have either socket or ogee joints. Whilst both types may be jointed with cement mortar, the rubber-ring, or gasket joint is now almost universal, thus achieving flexibility. There are also special designs for jointing with poured bitumen, or cement grout. An amendment to BS556 (1966) was issued in 1970, whereby the original class 1 pipes were designated 'standard', and new classes L, M, and H (Light, Medium, and Heavy, were introduced to achieve higher beam strenths, to coincide with these for asbestos cement pipes.

Prestressed concrete pressure pipes are useful for large sewers and particularly rising mains, because they can stand considerable internal pressures and in this respect serve as a substitute for cast iron or steel. They have been made in large sizes.

Asbestos-cement pipes

Asbestos-cement pipes (not to be confused with asbestos-cement pressure pipes) are used for sewerage purposes and are available in the nominal diameters (in inches) 4, 5, 6, 7, 8, 9, 10, 12, 15, 18, 21, 24, 27, 30, 33, and 36 or the nominal metric equivalents except that the metric equivalents of 5-in., 7-in., 8-in., and 10-in. are not included in National Building Studies Special Report No. 32, presumably being in disfavour. The metric sizes are, like the English sizes, nominal and do not provide any diameters additional to the imperial sizes.

The pipes are jointed with asbestos-cement sleeves and rubber rings. The procedure is to mount two rubber rings on each pipe end so that one ring is in close proximity to the end of the pipe and the other ring one complete roll distant from the position occupied by the first. With pipes in sizes 4-in. to 9-in. (100–225 mm) diameter this operation is facilitated by use of a taper plug. With pipes in sizes upwards of 9-in. (225 mm) diameter the rings are pulled over the pipe end by hand. Twist should then be removed by inserting a ring locator under the ring and describing one or two revolutions. The ring will then be square to the axis of the pipe and in approximately correct location. The location should be checked by rolling the ring towards the pipe end to ensure that it comes to rest in close proximity to it.

By means of simple leverage the sleeve is pushed over the rubber rings on the end of the pipe about to be laid until the pipe end is protruding slightly. Then the sleeve is pushed over the rubber rings on the adjacent pipe to bring it into the central and final position in which the ends of the sleeves should coincide with location marks on the pipes.

Cast-iron pipes

Cast-iron pipes are used for sewerage in places where future connections are not expected and where the use of this material would be more economical than concrete surround or other protection to vitrified clay, concrete, or asbestos-cement pipes. Cast-iron is a commonly used material for rising mains and inverted siphons. At one time the joints were nearly always of spigot-and-socket type made with poured molten lead caulked, but these have largely been replaced by bolted-gland or more recently rubber-ring joints thus reducing, or entirely avoiding, the use of lead. For indoor work, as in pumping stations, bolted flanged pipes are used because caulked joints have limited resistance to being blown open where pipes are not secured in position and bolted-gland joints have virtually no resistance to such pressures at all.

With the coming of metrication a change is taking place in Britain, and the proposal is that in place of spigot-and-socket pipes, all straight pipes shall have spigots at both ends and be connected by double-socket fittings as in other countries. While in Britain cast-iron sewers are of the same kind of material as used for water mains and are usually British Standard Class 'B', cast-iron drains, while of spigot-and-socket pipes, are made to a different specification and have a different kind of socket.

There are also ductile iron pipes now available in sizes 80–300 mm, made by inoculating molten iron with magnesium so as to give higher strength and more ductility than ordinary grey cast iron without any loss of resistance to corrosion.

Deterioration of cement mortar, concrete, and cast iron as a result of sulphates in the ground

The presence of sulphates, such as Epsom salts, in the ground water necessitates special construction of sewers. Sulphates combine with the lime content of cement, causing cement mortar and concrete to deteriorate. When it is known that sulphates are present (usually there is local information on this subject) stoneware pipes should be laid with special joints. Patent joints incorporating an asphaltic jointing medium are preferred by some but the use of flexible joints overcomes the difficulty.

Sulphates also have the effect of destroying cast-iron pipes. The deterioration in this case, which is known as graphitization, is particularly liable to occur in clay and waterlogged soils in the conditions which favour the growth of anaerobic bacteria. The effect is that the pipe becomes blackened and softened to such an extent that it may be cut with a knife. Sulphates are often very local in their distribution; consequently analyses of the subsoil and subsoil water are advisable when their presence is suspected.

Where graphitization is likely to occur, cast-iron pipes should not be used for gravitational sewers unless protected, and rising mains are best constructed of steel, protected by a substantial external covering of bitumen not containing vegetable fibre, but held on by glass or asbestos fibre.

Effective protection of cast-iron pipes can sometimes be accomplished by measures such as drainage of the ground or filling the trench with material other than clay, so as to eliminate anaerobic conditions, but where long mains are concerned such methods may be impracticable. Cathodic protection with sacrificial zinc electrodes has also been used. The author has observed graphitization attack from the inside of rising mains as well as from the outside.

Asbestos-cement pressure pipes

Whilst iron pipes have for long been almost universally used for sewage pumping mains it is worth remembering that asbestos-cement pressure pipes are manufactured for the European market in four classes corresponding with those for cast-iron pipes. The equivalent classes are made for the same test pressures in each material and the pipes have the same external diameters; but asbestos-cement pipes are thicker than cast-iron or spun-iron pipes and therefore the actual internal bores are smaller. While the majority of actual sizes of cast-iron pipes are greater than the nominal diameters, the actual diameters of asbestos-cement pipes are smaller than the nominal diameters, in order that the external diameters may be kept the same as those of cast-iron pipes for convenience of jointing thereto.

Asbestos-cement pipes are easy to lay and joint, and easy to cut and handle. They cost less than cast-iron in the smaller sizes. For these reasons they are useful for many purposes.

The pipes are jointed by a number of methods, all of which involve rubber joint rings which are pressed against the outside of the barrel by means of circular discs bolted or screwed together. In one design cast-iron jointing flanges are bolted together with wrought-iron bolts in such a manner that they press two rubber joint rings against the outside of the barrel of the pipe and against the edges of a cast-iron collar. An alternative to the cast-iron ring is a somewhat similar design applied to asbestos-cement rings bolted together with bolts which may be completely protected with bitumen. In addition to the above types of joint there is a screw joint which, although it employs the same principle involving two rubber joint rings per joint, does not require the use of metal.

All these joints are flexible, and to such an extent that the minimum number of bends is required.

Asbestos-cement bends are manufactured, but for all other connections such as tees, tapers, angle branches, and junctions to valves, special cast-iron fittings are made. Short-radius iron bends have always been available, and with the coming of ductile iron a complete range of specials is now available for use with asbestos-cement pipes.

Asbestos-cement pipes are laid and jointed in the same manner as are other pipes as regards preparation of the trench bottom and setting out the lines. They are tested in the same manner as cast-iron or steel pipes.

Although initially considered to be immune from sulphate attack it has been found that under certain conditions this is not the case, and careful coating with bitumastic solution, accompanied by the sheathing of metal joints, if these are employed, is advisable if ground conditions are suspect.

Steel tubes

There are several types of steel tubes, or pipes, and joints, including screwed barrel, socketed pipes, pipes suitable for welded jointing, or joints permitting expansion. Steel pipes are quick to lay, and, in normal circumstances, are durable when protected externally by suitable bitumen wrapping. Steel tubes may be purchased with bitumen lining, but this is intended for water supply purposes, and the added cost is not justified for sewage rising mains in which, owing to the lack of oxygen and the protective effect that sewage has for iron, little rusting is likely to occur. The best external protection is bitumen held in place by glass fibre or asbestos spun on as already described. The Johnson coupling joint is often employed with steel pipes, and metal moulds are available for encasing these joints in bituminous compound to complete the protection against aggressive ground conditions, and at the same time retain flexibility.

Unplasticised PVC pipes

The advantage of long-length, light-weight pipes are exemplified by the ever-increasing range of plastic materials, reinforced or unreinforced. It is, of course the very quality of plasticity and therefore liability to distortion that has restricted the use of plastic pipes in sewerage. However, unplasticized varieties such as u-PVC are being used and can be designed to resist loads due to traffic and trench refill by the suitable application of Marston theory. In its unplasticised state PVC is sensitive to sharp points and requires careful bedding on close-grained soil, or requires suitable granular fill. The main criticism of its use has been liability to distortion under load, and this naturally increases rapidly in severity as diameters increase, so that whereas for 100–150 mm pipes there is little cause for concern, with large diameters reliance has to be placed on the refill material for permanent provision of such support. Use of concrete protection, whilst providing an answer to these problems, is not only likely to be expensive but can introduce other difficulties due to consequent rigidity. In Great Britain u-PVC pipes are available up to 600 mm diameter and the criteria for pipeline design may be found in correct adaptation of Marston theory. The joints available, in addition to several versions of rubber ring, include solvent cement, though this material has been known to fail unless very carefully applied. These pipes have more frequently been used as rising mains for sewage and storm water, most often on rural schemes where 100 mm or 150 mm pipes cover a wide field of application. The speed and simplicity of laying has added to the attraction of the combined pump and mascerator installation where diameters down to 25 mm can be used. Their use has not been without problems, as some structural weakness has come to light under conditions of repetitive changes of stress as induced by the starting and stopping of pumps.[2] However, careful attention to these problems by the engineer, and the use of pipes of greater wall thickness than normally required, will probably obviate these troubles in the future.

Other sewer pipes

At least two other types of pipes are now of interest in sewer construction in Great Britain and elsewhere. One is the pitch fibre pipe (BS 2760, 1966, etc.), and the other the reinforced plastic pipe. Both have the advantage of lightness and ease in handling, but in sizes above say 300 mm are suspect due to the possibility of distortion under loads, due to backfill in the latter case and, in the former, loss of strength due to gradual absorption of moisture in waterlogged ground. Whilst the former may be regarded as a rigid pipe, and pipeline design can follow the same lines as for the materials already referred to, the use of the latter requires a different approach.

The development of glass-reinforced plastic pipes has extended the possible range of sizes very considerably. To resist distortion a technique of consolidating the fill on either side of the pipe has been applied with some success, but the

152

strong probability of changes in moisture content, during the year, with considerable effect on the available strength of any natural fill, robs this particular technique of its attraction. The use of large-diameter pipes with much easier handling characteristics obviously has its advantages, but short of the possibility of using some form of soil-cement, only concrete in large quantities would appear to guarantee regular and consistent support to such pipes. The use of glass fibre is not confined to plastic pipes; a type of concrete pipe with a complete glass-fibre wrapping is available. It is commonly tested until the concrete cracks and the pressure is retained only by the glass-fibre wrapping. It has been used both for gravity-flow and pressurized pipelines.

Joints for sewer pipes

Although there may be some who, like the author, are still not averse to the competent use of the mortar joint for clay, or concrete pipes, laid in an accurately bottomed trench without artificial bedding, there is no doubt that over the years this form of joint (due partly to the shortcomings of pipe-jointers, many of whom mixed the mortar to their own specification rather than that laid down, or failed in some other way, to make satisfactory joints, and partly to cracking

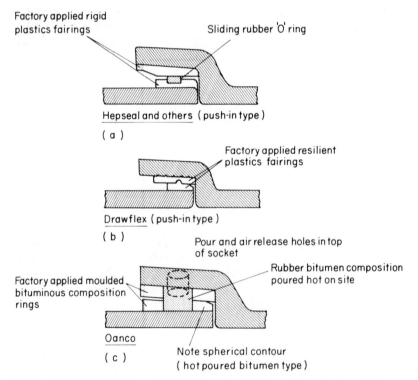

Factory applied rigid plastics fairings

Sliding rubber 'O' ring

Hepseal and others (push-in type)

(a)

Factory applied resilient plastics fairings

Drawflex (push-in type)

(b)

Pour and air release holes in top of socket

Factory applied moulded bituminous composition rings

Rubber bitumen composition poured hot on site

Oanco

(c)

Note spherical contour (hot poured bitumen type)

Figure 13   Flexible Telescopic Joints for non-pressure clayware pipes (*J. Instn. Public Hlth. Engnrs.*)

caused by some form of overstressing due to the resulting rigidity of the pipeline) has been the cause of the high degree of infiltration experienced in many existing systems. It was in an attempt to remedy these troubles that the work of Clarke[3, 4] and others at the Building Research Station in England was done, resulting in the re-examination of pipelines as structures and the introduction of flexibility to limit the build-up of stress to single pipe lengths. Clarke's work is epitomized in National Building Studies Special Report No. 32, sponsored by the Building Research Station.[5 - 7]

This approach was satisfied by joints such as the Cornelius rubber ring, adopted by Stanton at about the same time as the introduction into Great Britain of the spun-iron pipe and since used by them for both iron and concrete pipes. Problems of tolerance in concrete pipes have been countered by the shaping and tapering of the interior of the socket and the exterior of the spigot, whilst the clay pipe manufacturers have turned to plastic material applied to spigots and sockets at works; used either with or without a rubber ring. Asbestos-cement sewer pipes have a rubber ring and sleeve joints such as the 'Turnall' joint. Clarke illustrated practically all the types of joint used in Great Britain in a paper to the Institution of Public Health Engineers[8] and his illustrations are reproduced in Figures 13–20. The description of jointing procedure given

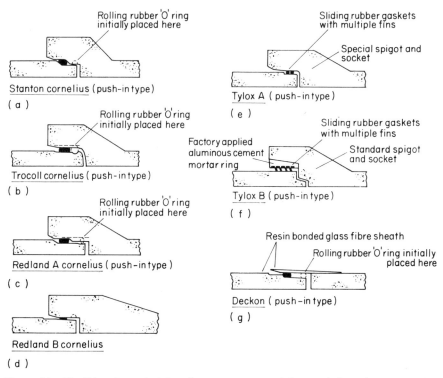

Figure 14   Flexible telescopic joints for non-pressure plain or reinforced concrete pipes
(*Instn. Public Hlth. Engnrs.*)

154

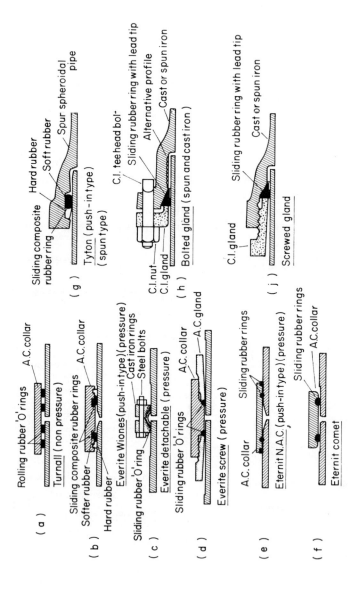

Figure 15 Flexible telescopic joints for asbestos-cement pipes and spun-iron pipes (*J. Instn. Public Hlth. Engnrs.*)

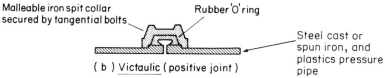

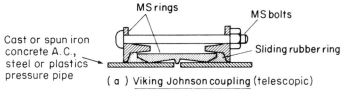

Figure 16   (a) Johnson coupling; (b) Victaulic joint (*Instn. Public Hlth. Engnrs.*)

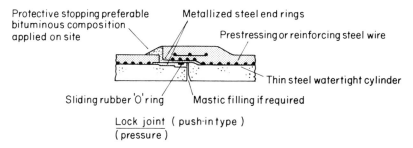

Lock joint   (push-in type)
(pressure)

Figure 17   Flexible joint for reinforced or prestressed concrete pipes (*J. Instn. Public Hlth. Engnrs.*)

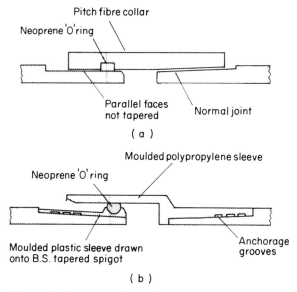

Figure 18   Telescopic joints for pitch fibre pipes (*J. Instn. Public Hlth. Engnrs.*)

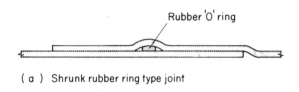

(a) Shrunk rubber ring type joint

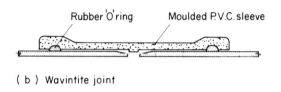

(b) Wavintite joint

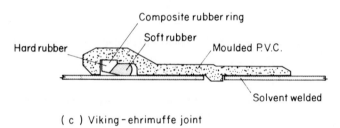

(c) Viking – ehrimuffe joint

Figure 19   Telescopic joints for PVC pressure pipe (*J. Instn. Public Hlth. Engnrs.*)

under the heading 'Asbestos-cement pipes' is meant to apply to the 'Turnall' joint for non-pressure asbestos cement pipes.

The first joint offering some degree of flexibility was that using hot-poured pitch, or asphalt. Several ingenious types are illustrated in late nineteenth-century textbooks, inspired by evidence of similar practices in ancient Meso-potamia. This practice has never become as popular in this country as it still is in Germany, although the Oanco joint (Figure 13c) is popular in mining areas. In about 1890 the Gibault joint appeared in France and the Victaulic joint for cast iron and steel appeared in the early 1920s. The Cornelius joint is now used all over the world and appeared in the US as the lock Joint in 1942.

One of the best known and most universally used joints is the one known in Great Britain as the Viking-Johnson, or Johnson Coupling in the US as the Dresser Coupling, and in France as the Gibault Joint (see Figure 16). It was originally associated with steel tubes, but has been developed and used for concrete, asbestos-cement and plastic pipes. The size range is also very wide, and has been carried down to the water service pipe range with screwed glands to compress the joint rings, instead of bolted flanges. Transfer from plain-ended

Figure 20   Cut-away view of the Viking Johnson coupling (Courtesy Victaulic Co.)

pipes to flanged pipes or fittings is accomplished by the 'flange adaptor' joint where the 'sleeve' becomes a flanged collar and the bolts secure a single rubber ring on one side and a normal flanged joint on the other.

The paper just referred to covers the development of the rubber ring which is also dealt with by Lindley.[9] Satisfactory application of rubber rings in their simplest form has only been achieved after many tests and experiments. Any twisting or distortion during jointing operations will probably lead to failure under test, or more insidiously some time after the pipeline has been in service. Some advantage would seem to be gained by use of the wedge-shaped lead-tipped ring used in the Johnson coupling joint, although any laxity in tightening up the bolts on these joints can also lead to weakness and leakage. Jointing instructions issued by the makers of any particular joint, and the use of the jointing tackle provided by them, is essential if the advantages of flexibility and ease in jointing are to be secured.

It will be obvious that, having gone for flexibility, the use of concrete for bedding or otherwise protecting a pipeline must be done in such a way as to avoid complete rigidity. Thus bedding should be interrupted at each joint,

158

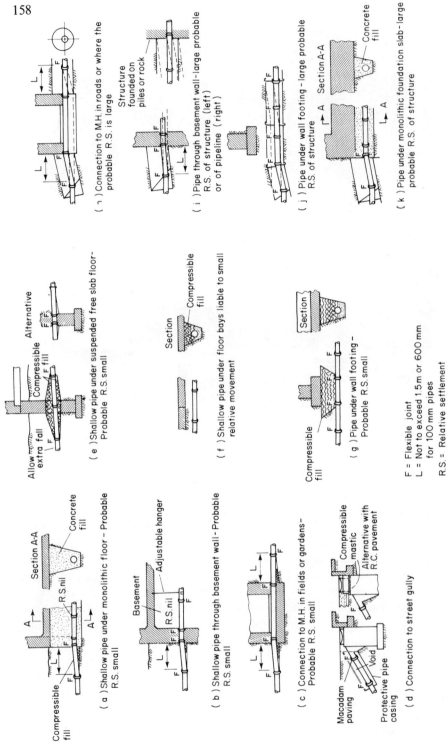

Figure 21   Methods of alleviating differential settlement on pipes passing through or under walls of other structures (*J. Instn. Public Hlth Engnrs.*)

which is not difficult provided pipe lengths remain standard, but takes a little more organizing if junctions require to be positioned at particular points. Although the gaps in the concrete represent some saving in material, the overall cost of the bedding, with incidental items for stop-ends, etc., will probably show some increase over a continuous bed. However, one of the objects behind the research into flexibility and the method of design based on Marston theory,[10, 11, 12] has been to avoid the use of concrete as far as possible by relying on granular bedding to support the pipes even in relatively poor ground, and at the same time provide flexibility for movement at the joints. There has, of course, been considerable controversy over the method of consolidating the material introduced for this purpose. Although with pipes above, say, 450 mm diameter the use of a tamping machine may be practical, for smaller pipes reliance is placed on the addition of extra granular material bringing the finished level up to half pipe, at the time of laying. Dextrous use of the shovel in this operation achieves accurate bedding of the pipe, and a good degree of consolidation at the same time. The latest pronouncements by the Working Party in their 1975 report,[13] emphasize that bedding material should be related to pipe size, the bigger the pipe the coarser the material, and as an improvement on Class B bedding, where granular material is only brought up to half pipe, another class of bedding is recognized which completely surrounds the pipe, and finishes at the same distance above the barrel as the depth of 'bedding' below it i.e. one-sixth of the outside diameter, or 100 mm minimum. The bedding factor for this type of surround is given as 2.2 as against 1.9 for the Class B bedding. The Working Party recognize a higher factor where concrete protection in the form of arching over the pipe is used. Its object is to prevent cracking or deformation of the pipe under heavy loading. The concrete therefore requires to be reinforced, and should also be provided with a reasonable degree of flexibility so as to follow any movement of the pipes. Figure 24 (p. 163) shows the commonest types of concrete protection and bedding for concrete pipes.

Freedom of movement in pipelines implies careful attention to all connections between pipes and rigid structures. Methods of alleviating the effects of differential settlement are illustrated in Figure 21.

Laying clay or concrete pipes

Where rigid joints are to be used, excavation is made in trench to a depth nearly the thickness of the barrel of the pipe below the invert of the pipe to be laid, and of sufficient width for convenience of working. In order that the levels and gradients may be correct, sight rails are set at predetermined heights above drain invert and, with the aid of boning rods, pegs are driven into the ground at the centre of the bottom of the trench, their tops being dead level with the line of invert. To ensure that a straight line is maintained on plan, a line is stretched at the height of the centre of the pipe in such a position that it just fails to touch the pipe collar (if the string touches the collar during pipelaying there is a tendency for the pipes to form a long sweeping curve).

In laying the pipes, the pipelayer carefully removes sufficient earth for the main body of the pipe to rest throughout its length on the ground with its invert at the correct level and true grade. The pipelayer commences work at the lowest end of the trench and works up the gradient. He lays the first pipe, testing the position of its invert by resting a long straight-edge in the invert and on two of the pegs driven into the trench bottom. The first pipe is placed in line, with the collar not touching, but within 3 mm of the string-line and the spigot a measured distance from the string-line. To accommodate the collar, and to make room for jointing, a hole, only just sufficiently large for the purpose, is excavated in the trench bottom. (see Fig. 22)

The next stoneware pipe is then lowered into the trench and, when suspended, struck with a wooden mallet to make certain that it is sound. The spigot is placed in the socket of the first pipe; a ring of tarred yarn is inserted and driven into position so as to raise the spigot to a central position in the socket and so as to raise the spigot to a central position in the socket and so as to prevent cement from entering the pipe. The socket is then filled with cement mortar, which is continued outside the socket in the form of a fillet to a distance of not less than 50 mm from the face of the socket. Before the next pipe is laid any cement that may have entered the pipe through the joint is carefully removed with a wooden scraper. Of late years the laser beam has been applied on major works as a means of positioning pipes for both line and level.

As already described the wide application of Marston theory, both in Great Britain and the USA, has led to a very general use of artificial bedding. Under these circumstances it is usual to rely on machine excavation to a depth below invert equivalent to the desired thickness of bedding and then to deposit the granular material in the trench, first consolidating it to the level of the underside

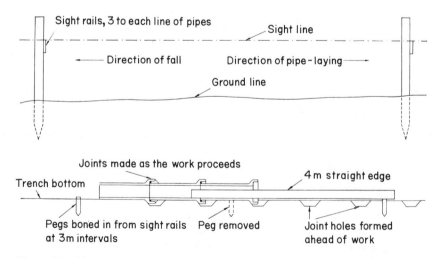

Figure 22   Method of laying socketed pipes (Courtesy Clay Pipe Development Association)

of the pipe barrel, after which the pipe layer and jointer bed the pipe accurately and support it on either side with more granular material.

Experience has shown that rigid joints and granular bedding do not go together, despite some recommendations to this effect. It is recognized, however, that flexible joints can be used where pipes are laid on reasonably good virgin ground. In both cases freedom of movement for the joints must be secured by joint holes or gaps in the consolidated bedding.

Structural protection of pipes

According to the 'Ministry' requirements so long applied in Great Britain, vitrified clay pipes, concrete pipes, and asbestos cement sewer pipes needed protection when laid with less than 4 ft (1.2 m) of cover under roads or less than 3 ft (0.9 m) elsewhere, in which circumstances they would be surrounded with concrete. All pipes laid in heading or with more than 20 ft (6.1 m) of cover would be surrounded by concrete. Pipes laid in open cut with more than 14 ft (4.3 m) cover but less than 20 ft (6.1 m) would be bedded on concrete and haunched with concrete to at least the horizontal diameter of the pipe, any splaying of the concrete to be above that level. All pipes of more than 18 in. (450 mm) internal diameter would be benched and haunched as above described or given such additional protection as required when they are laid very shallow, very deep, or in heading, and generally, large pipes should be surrounded with concrete. Very large concrete pipes must be surrounded with an appropriate thickness of concrete, for they will withstand only slight vertical pressures; the author has seen 1.9 m diameter pipes which, laid on the ground and covered with 305 mm of earth, fail by fracture at the crown, invert and both sides.

In England and Wales the many sewerage schemes carried out as a result of the spreading of the financial burden to whole districts rather than single parishes brought about by the 1936 Public Health Act, and much accelerated by the grants made available under the Rural Water and Sewerage Act of 1944, were all subject to Ministry approval, and all pipelines incorporated in them met the 'Requirements' as just stated. In general it is probably true to say that these simple rules seldom let engineers down, and may still be accepted as representing good practice and be suitable for use as an alternative to the Marston technique. The weakest link in the chain has been the mortar joint, which with its accompanying rigidity has been responsible for a build-up of stress in pipelines that the normal sewer pipes of clay or concrete could not be expected to withstand. Investigation of some of these failures led Clarke and his co-workers to turn to flexibility in joints,[8] thereby restricting the build-up of stresses to single pipes and to give careful attention to the work of Marston and Spangler,[7] whose theories were already widely accepted in the States. The outcome of this work has already been referred to, and has certainly made possible a good deal of economy in sewerage costs, due to wide reliance on granular bedding (see Figure 21), coupled with various grades of crushing strength in the pipes themselves, to avoid the use of concrete wherever possible.

162

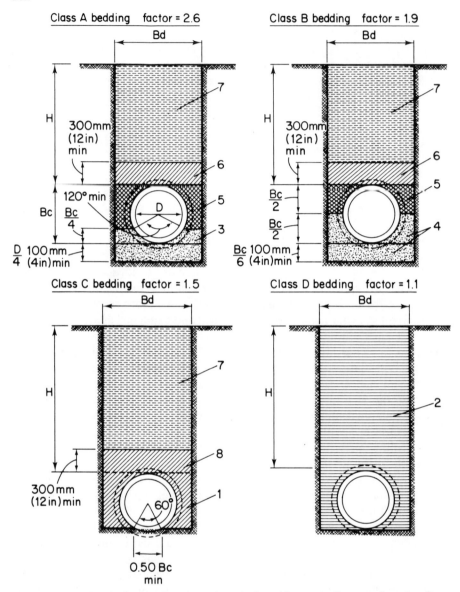

Figure 23 Bedding factors for pipes in trenches (Courtesy Stanton Staveley Spun Concrete)

There is even a vogue for return to bedding of pipes on the natural ground, subject to acceptance of some reduction in ability to withstand loads due to refilled material etc. Bland and Picken, reviewing the position in 1973[14] describe some experimental work on this subject, and question total reliance on Marston theory, which deals with strength available in cross-section without due regard to longitudinal strength. It is felt that design covering both directions might lead to further economies. Theories applied in Germany tend to point in this direction, but real clarification must still be the subject of further research. In general the larger the pipe, the more acute the uncertainties become.

Testing lines of pipe

(a) *The air test*

There are two recognized methods of testing sewers other than those which will operate under pressure: these are the water test and the air test. Both have their advantages and disadvantages and preferred circumstances in which they should be used. The air test is the preferred method of testing lines of clay pipes, concrete pipes, and asbestos-cement pipes after they have been laid in trench but before they have been covered with earth. After the pipes have been covered with earth the air test may have no value whatsoever, because of the possibility of subsoil water having risen above the level of the pipe preventing any leakage of air. It is then that the water test should be used. Accordingly, some engineers call for an initial air test, followed by a water test after the trench has been filled.

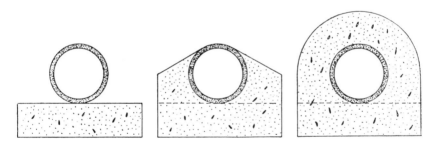

Figure 24   Methods of concreting pipes

In the air test both ends of the length of pipe to be tested, and any lateral connections, are sealed with expanding plugs or inflated bags. Air is blown into the sewer via a nozzle on the expanding plug at one end and recorded on a U-tube at the other end. The reason for blowing in air at one end and recording at the other end is to ensure that no secret intermediate plugs have been inserted. In British practice the pressure in the pipe is raised to 4 in. (100 mm) of water gauge, after which pumping is stopped. If the pressure on the water gauge drops more than the amount permitted in the specification any faults have to be repaired and the test repeated until the length of pipe has been proved sound. Recommended drops of water pressure that may be permitted when salt-glazed ware pipes are tested are given in Table 26, but the amount of drop of water-gauge given in the table may be doubled in the case of slightly porous material such as concrete or unglazed vitrified clay pipes if tested when thoroughly dry.

Table 26   Air and water tests for sewers and drains

Internal diameter of pipe (mm)	Air test Duration (min)	Maximum water drop (mm)	Duration of water test for 50 mm max. drop (min)							
			2½	5	7½	10	15	20	25	30
			Minimum length of pipe under test (m)							
75	1½	25	60	30	20	15	10	7½	6	5
100	2	25	80	40	27	20	13	10	8	6½
150	3	25	120	60	40	30	20	15	12	10
225	4½	25	—	90	60	45	30	22½	18	15
300	5	20	—	120	80	60	40	30	24	20
375	6	20	—	—	100	75	50	37½	30	25
450	5½	15	—	—	120	90	60	45	36	30
525	4½	10	—	—	—	105	70	52½	42	35
600	5	10	—	—	—	120	80	60	48	40
675	5½	10	—	—	—	—	90	67½	54	45
750	6	10	—	—	—	—	100	75	60	50

Clay, concrete, or asbestos-cement sewer pipes should not be tested under high air pressure as this can lead to serious accidents.

One distinct advantage of the air test is that the length of pipe being tested has no effect on the severity of the test. This is not true of the water test, for in the latter the shorter the length of pipe the less severe the test becomes unless allowance is made for this. The water test has the additional disadvantage in that it must be made on short lengths only where a sewer is laid on a steep gradient, as otherwise the pressure during the test may be more than the sewer may be expected to withstand. Still another disadvantage is that the water used in the test has to be disposed of afterwards.

(b) *The water test*

In the water test all outlets are plugged except at the top end where a temporary bend and upstand pipe is jointed to the end of the pipe and so arranged that a minimum water pressure of 4 ft (1.2 m) can be applied over the crown of the sewer at that end. This upstand must be of the same diameter as the sewer under test, otherwise it will give a false impression as to the amount of leakage. The length under test should not be more than that which would result in a maximum head of 20 ft (6 m) over the lower end of the sewer under test. Table 26 gives recommended durations of water tests according to the length of pipe under test.

Cast-iron, steel, and asbestos-cement pressure-pipe rising mains are tested under hydraulic pressure in the following manner. First, all bends and junctions are prevented from movement by concrete support provided up to above the centre line of the pipe and carried to the side of the trench. Each straight pipe is kept in position by earth being filled over its centre, but the joints are left·

exposed for inspection. All junctions are sealed and supported as necessary, or valves closed. Pressure is then applied by hydraulic pump to twice the maximum working pressure but to not more than the maximum pressure appropriate to the class of pipe used, and this pressure must be maintained without further pumping and without any detectable fall on the gauge for a period of at least half an hour, or as specified.

Brickwork and concrete construction

At one time, when practically all sewers were made large enough for men to enter them for cleansing purposes, the majority were constructed of brickwork. Many of these sewers consisted of one or two half-brick rings bedded on the excavation that had been cut to shape. Later practice included concrete backing to brickwork, giving considerably added strength at a moderate increase of cost.

At the present time sewers sufficiently large for men to enter are seldom constructed except when the flow is sufficient to justify their size.

When large brick sewers are constructed in trench they now usually consist of a brick lining backed by concrete. The brickwork makes an excellent lining because good-quality drainage bricks laid with close joints resist scour, and brickwork is more resistant to the corrosive action of hydrogen sulphide than unprotected concrete. To form a lining, a single ring of brickwork is sufficient provided that there is concrete backing to give structural strength. Even when corrosion of cement in the joints takes place as a result of hydrogen sulphide, the bricks should stay in position for a considerable time if they are well laid. The concrete backing to brick rings should be filled to the face of the excavation if work is constructed in trench. It can then take all thrusts and varied loads from above.

The bricks for lining sewers should be smooth vitrified bricks that absorb not more than 4% of their weight when soaked in water.

When brick sewers are constructed in heading, the form of the sewer will depend upon the geological formation. In rock it is possible to excavate to a true circle and a circular barrel may be constructed in the excavation with little back-filling or packing over the crown. This is true also of work in tunnel with the aid of a shield. In formations where a square timbered heading is necessary, the form of construction is altered to suit the circumstances.

In cases where scour or corrosion is not to be expected, and where large but inexpensive sewers are required, unprotected mass concrete may be used. Mass concrete, carefully worked to the face of the shuttering and rubbed down after the formwork has been struck, is a very economic form of construction. For foul sewers, or where there is any possibility of scour or corrosion, a hard brick invert is desirable. In many instances mass concrete, when used underground and filled to the excavation, makes a very sound form of construction and shows a marked economy on reinforced concrete work. All mass concrete should be vibrated.

Reinforced concrete may be used with advantage when a form of sewer that varies considerably from the circle has to be used, i.e. when it is found advantageous to adopt a design involving bending moments in the sides, crown, or invert. It also serves for the construction of sewers intended to function under pressure, such as inverted siphons of large size or large low-pressure rising mains. For other purposes it should not be used indiscriminately or without making comparative estimates with works properly designed and constructed in other materials. Reinforcement of concrete usually denotes both high-quality concrete and external shuttering which often, in underground work, involves costs out of proportion to the advantages gained.

In the design of sewers and sewage tanks, as in all watertight works, the stress in the reinforcement should not exceed 4,000 kg/cm^2.

Cast-iron segments

Large-diameter cast-iron pipes may be obtained on special order, and when the weights of normal-length pipes become too great for convenient handling pipes may be made in short lengths. But once a diameter has been reached at which it is possible for construction to be done from the inside, cast-iron segments are to be preferred for work in tunnel. These are particularly useful for the construction of large-diameter sewers, excavated in deep tunnel in clay with the aid of a shield, and where tunnelling has to be executed under compressed air.

Cast-iron segment sewers consist of a number of curved rectangular segments with flanges on the inside, the longitudinal faces of which are radial to the circle, except for two faces at the crown. At the crown is a small wedge-shaped key segment which has to be inserted from inside to complete the circle and therefore cannot have radial flanges, and this naturally applies to two of the flanges in direct contact with it. These segments are bolted together to form rings which in turn are bolted together with circumferential flanges.

If construction is in dry ground the circumferential joints are caulked with creosoted deal packing, but where it is expected that moisture will have to be excluded during construction of the work, the joints are first caulked with lead wire and red lead and afterwards pointed with rust cement. When rust cement is used it is necessary for the metal to be thoroughly clean, the surface to which the rust cement is to be applied being completely freed from the coating of bituminous solution, otherwise the rust cement will not adhere to the metal. As this cleaning adds to the cost of the work, rust cement is not used in conjunction with deal fillets, but the joints are pointed with ordinary cement mortar which, incidentally, some engineers definitely prefer to rust cement in all circumstances.

The rust cement used by some authorities consists of cast-iron borings mixed with powdered sal-ammoniac in the proportions of 400 to 1. The ingredients are first mixed dry, moistened, and mixed again. Other authorities have used mixtures incorporating flowers of sulphur, but in view of the injurious effect of sulphates on Portland cement and concrete the advisability of using such a mixture can be questioned.

The longitudinal joints have machined faces which may be jointed with red lead and lead wire, and bolted up; but as it is difficult when working in heading to be certain of securing a watertight joint by this means alone, it is usual to cast a groove 6 mm wide and 19 mm deep on the edge of the flange, in order that lead wire or lead wool caulking may be applied where necessary. This groove is finally pointed with cement mortar, or with rust cement if it has been left with a clean machined surface.

In straight work the circumferential joints are bolted up, adjustment being made as necessary to maintain the straight line. In slightly curved work slight curvature may be obtained by opening out the joints on one side and proportionately increasing the thickness of the deal fillets on that side. When more than slight curves are required special segments are cast to radius.

In erecting it is usual to stagger the key segments so as to break joint throughout. This is continued round curves of large radius, but when special segments are used on curves it ceases to be possible and the key segments occupy the centre at the top of the arch.

The bolts should have suitable washers and should be tightened down on to surfaces of the casting that are truly parallel. Leakage of subsoil water through the bolt holes is prevented by treated grummets.

After the cast-iron segments have been erected the space between the outer face of the iron and the face of the excavation is grouted under pressure with cement or lime grout through holes in the casting formed for the purpose.

Junctions of pipes with cast-iron segmented sewers, except those made at manholes or bellmouths, require special castings made to bolt to the curved faces of the segments, in which openings are cut or burnt for the purpose.

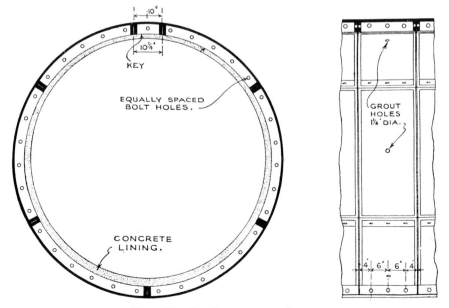

Figure 25   Cast iron segment of sewer

The completion of a cast-iron segmental sewer consists of lining the interior in order that the flanges will not interfere with the flow or cause solids to hold up in the invert. The lining usually consists of mass concrete formed to a circle which is not concentric with the circle of the cast-iron work. The invert should be faced with brickwork for at least one-eighth of the circumference.

Table 27, which is based on the proportions of cast-iron segments and accessories used in the construction of various sewers, should serve as a guide to designers.

As an alternative to cast-iron segments for work in tunnel, various proprietary pre-cast concrete segments have been devised. Where external water pressure exceeds 10 m head it can be difficult to make watertight joints with some types of concrete segments and cast iron is usually preferable.

Table 27   Dimensions of cast-iron segmental sewers (mm)

Internal diameter of concrete lining	1525	1600	1675	1830	1980	2135	2285	2440	2590	2745	2895	3050
Internal diameter of iron	1651	1727	1803	1956	2108	2261	2413	2565	2718	2870	3023	3175
External diameter of iron	1855	1931	2007	2160	2324	2477	2629	2793	2946	3112	3265	3417
Thickness of iron	19	19	19	19	19	19	19	19	19	19	19	19
Depth of flanges	83	83	83	83	89	89	89	95	95	102	102	102
Thickness of flanges at base	25	25	25	29	29	29	29	29	29	29	29	29
Thickness of flanges at edge	22	22	22	25	25	25	25	25	25	25	25	25
Diameter of bolt circle	1734	1810	1886	2039	2197	2350	2450	2660	2813	2972	3125	3277
Number of bolts	19	23	23	23	23	27	29	29	34	36	39	39
Size of bolts, diameter	19	19	19	22	22	22	22	22	25	25	25	25
Size of bolts, length	100	100	100	100	108	108	108	108	108	108	108	108
Number of ordinary segments one key and two taper segments) per ring	3	4	4	4	4	4	4	4	5	5	5	5

Sewer renewal and renovation

Recent interest in this subject has led to intensive investigation into the possibilities of avoiding expensive trench excavation and road reinstatement. Pipe-jacking from a series of pits, later to be utilised for manhole construction, has been an accepted technique for some years, usually employing some form of reinforced concrete pipe or segment. In its latest form, what is known as the "earthworm" technique is being applied. However, work of this type almost inevitably involves the selection of an alternative route for the renewed sewer. For renovation, as such, internal chemical grouting, slip lining with some form of thermoplastic tubing, pushed or pulled through the sewer, or in the case of man-entry sewers, the insertion of standard units to give a lightweight though rigid finished lining, are finding their applications, as described by Gale in the Water Research Council's recent Report TR87A of November 1981.

Shapes of sewers

For pipe sewers and culverts in which there is no free-flowing surface in contact with the air, the circular section is theoretically the most perfect. In nature, tubes that accommodate hydraulic flow in plants and animals are most often circular in cross-section, and in engineering practice nearly all small pipes—and a good proportion of large pipes, tubes, and barrels—are constructed to the circular form. But as radial work can be expensive and excavations in heading or in open cut are often trapezoidal or rectangular, some of the larger-size sewers and culverts are shaped so as to approximate only to a circle; the sides, crown, and invert sometimes being tangential thereto.

In addition to the considerations of hydraulic flow and simplicity of construction, the designer has to bear in mind that sewers must be self-cleansing and free from deposit other than that which may remain only for a matter of a few hours, and that the larger sizes of sewer have to be entered occasionally for repair purposes.

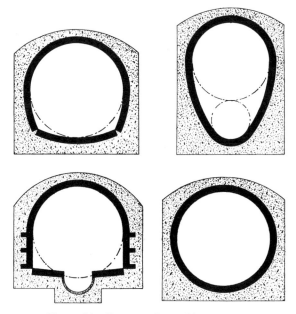

Figure 26   Cross-sections of large sewers

When sewers are so large that they fill the trench or heading in which they are constructed, and it is not practicable to excavate to the circular form, it pays to vary the shape of the sewer from the circle and approximate to a square or trapezium, provided that a flat bottom will not lead to silting. Sewers of truly rectangular form have been constructed but more often crowns are arched or flat, and sides are either curved or vertical.

U-shaped sewers have been found distinctly economical for construction in open cut. In these the invert is semicircular, or flat and swept up at sharp radius to join the sides, the sides are vertical from above the centre of the invert radius and the crown is of flat reinforced concrete.

The economic shape of a sewer varies with the type of construction that can be used in the circumstances. With modern steel shutters that can be used over and over again, shapes which were once considered too expensive no longer present difficulties and are economical provided that the lengths involved are sufficient to pay for the special shutters.

Figure 26 shows some well-known sewer shapes. Top left is the horseshoe-shaped sewer as once used in America and Britain. The crown is a semicircle the radius of which is equal to that of the sewer; the radii of the sides and invert are equal to the diameter of the sewer; depth equals width. Top right is a 'standard' or 'old' egg-shaped sewer which at one time was largely used because it was considered to give a higher velocity at slack flows. The crown is a semicircle; the radius of the side is three times the radius of the crown; the radius of the invert is half that of the crown, and the overall depth from crown to invert is three times the radius of the crown. The fallacy of the theory was that the extra depth of the sewer over that of a circular sewer of equivalent discharge reduced the available gradient and, as a result, the slack-flow velocity usually proved to be less than that in a circular sewer of equivalent capacity. Bottom left is a design that has been used for tank-sewers intended for storage of sewage during times when discharge to the sea is impracticable or undesirable. Bottom right is a typical section of a circular sewer lined with brick or other hard material backed with concrete.

Proportions of arches and sides of sewers

Sewers can fail by being distorted out of shape by earth pressures from above or by spot loads from vehicles or foundations. They can also fail when subjected to very great water pressures either from subsoil water from the outside or sewage pressure on the inside. The Marston theory (which, in most instances, reduces vertical earth pressure to less than the weight of earth which lies above the pipe) can be used in many instances; in others, theory of the design of arches can be used. It is not, however, always appreciated that internal or external water pressures can greatly exceed earth pressures and that generally, if a sewer is constructed to withstand the equivalent of a water pressure from ground level, it is very unlikely to fail under any other load. Table 28 gives thicknesses of barrel sufficient to withstand such water pressures.

Table 28 Maximum depth in metres of brick-barrel
sewers (ground level to centre of barrel)

Internal diameter (nominal mm)	Thickness of barrel (nominal mm)			
	113	225	340	450
750	15	—	—	—
825	12	—	—	—
900	10	32	—	—
975	8	30	—	—
1050	6	27	—	—
1125	4	24	—	—
1200	—	22	—	—
1275	—	20	—	—
1350	—	18	32	—
1425	—	16	30	—
1500	—	15	29	—
1650	—	12	25	—
1800	—	10	22	32
1950	—	8	19	30
2100	—	6	17	27
2250	—	4	15	24
2400	—	—	13	22
2550	—	—	11	20
2700	—	—	10	18
2850	—	—	8	16

Pipes in embankments or other filled ground

There have been several instances of failures of cast-iron pipes, asbestos-cement pressure pipes, and reinforced concrete beams supporting pipes, where these have been constructed to span between structures or to be supported by piles in embankments or other filled ground, and these cases do not appear to be adequately covered by the Marston theory.

In one instance a large-diameter cast-iron pipeline was laid above ground level supported on piles behind each socket and covered by an embankment filled to a foot or so above the crown: the earth load was sufficient for the pile heads to be driven into the invert of the pipes. This shows that it is necessary to provide large pile heads to spread the load, not merely to calculate bending moments in cases of this kind.

In another instance a cast-iron pipe was carried on piles and reinforced concrete beams in fill which extended down below the pipe to natural ground level about an equal distance to that at which the pipe was below the new formation level: both pipes and reinforced concrete beams were shattered. More frequent are failures of pipes between tanks and manholes in the embankments of sewage-works structures.

Because of such failures the author arranged for model experiments to be made to determine the loads on pipes laid about halfway down a depth of fill. These showed that when the fill was moved down relative to the pipe in the manner in which it would subside in course of time, the pipe had to support the whole weight of a triangular prism of earth, the apex of the inverted triangle resting on the pipe for its full width and the sides sloping outwards at an angle that varied from time to time although there was no observable difference in the conditions of the several experiments. On the basis of these experiments it was decided to design beams carrying pipes on the assumption that the beam had to carry the whole weight of a wedge of earth with sides sloping out from the sides of the beam to finished ground level at an angle of 30° to the vertical, and that the stress in the steel should not exceed 30,000 lb/in² (2100 kg/cm²) under the load caused thereby. No further failures were reported.

Sewers above ground level

Sewers constructed above the level of the ground are usually considered even less desirable than siphons because of their unsightly appearance, and for this reason they are very seldom employed except when of short length only. They do not involve complicated flow calculations and they are not difficult to construct.

Sewers above ground level may be constructed of cast-iron pipes or welded steel tubes, or steel tubes with special joints. When spigot-and-socket cast-iron pipes are used they are supported under each socket. If the height above ground level is moderate a separate pier can be built to support each pipe.

Welded steel tubes are very suitable for above-ground work. If exposed to the weather they should be coated with glass- or asbestos-fibre and bitumen. They may be carried over wide spans without intermediate support and some makers provide details of permissible bending moments. Expansion joints are necessary at calculated intervals to allow for contraction and expansion. A gap should be left between the butt ends of the pipes at the expansion joint, and it should be varied according to temperature of pipes at time of laying. A maximum figure of 12 mm expansion in 13 m of pipe is sometimes adopted for temperate regions when pipes are laid in cold weather. When special expansion joints are used at intervals on welded pipes, each length of pipe between expansion joints should be anchored half-way to prevent permanent creep. Angles in the line should be heavily anchored.

Pipes raised above ground level for more than a matter of a few feet should be provided at the ends with collars of spiked iron railings to prevent children from walking or climbing over them.

When sewers have to be carried at a considerable height above ground level, it becomes necessary for an aqueduct to be constructed. This may consist of brickwork, masonry, fabricated steel work, reinforced concrete, etc., according to circumstances. Such aqueducts should be constructed in straight lines, particularly if built with materials that make no allowance for expansion or if the direction of the aqueduct is from east to west, otherwise heat expansion may cause lateral movement resulting in fracture.

Inverted siphons

When it is desired to drain from one point to another which lies at a lower level, but a valley which cannot be circumvented lies between the two points, the designer has the alternatives of constructing an inverted siphon or carrying his sewer above ground level.

Inverted siphons are pipes which connect from the lower end of a gravitating sewer, pass under a valley or obstacle, and rise again to the top end of another gravitating sewer at a level sufficiently below the first to give an adequate hydraulic gradient. Inverted siphons, being under pressure, must be constructed of materials capable of withstanding tension; for this reason cast-iron or steel pipes are most frequently used. Other materials are reinforced concrete, pre-stressed concrete pressure pipes, and asbestos-cement pressure pipes.

The principal disadvantages of the inverted siphon are:

(1) because the gradient of the invert is not continuously downhill, large particles sliding along the invert will collect unless the velocity is periodically high;

(2) because the siphon is below the hydraulic gradient it is always full of sewage and therefore at times of slack flow the velocity may be low and silting occur unless precautions are taken;

(3) unless the inlet end of the siphon is carefully designed, floating solids may be separated and accumulate in the manhole or sewer upstream;

(4) side connections cannot be made to inverted siphons.

The first two factors, working together, make inverted siphons very liable to become silted. The remedy is correct design and this depends on the circumstances. A few examples are given to illustrate approach to design.

Example 1. In a simple case where the siphon is on a sewer which has to pass a peak daily rate of flow of $2-3 \times$ DWF and a maximum rate of flow of $6 \times$ DWF, and where there is no difficulty in obtaining the fall required to give the siphon a satisfactory hydraulic gradient, the siphon may be constructed to such a diameter that the velocity is self-cleansing at least once a day. For example, it could have a velocity of 2 m/s at the maximum rate of flow of $6 \times$ DWF, and $\frac{2}{3}-1$ m/s at the peak daily flow of $2-3 \times$ DWF. If the fall available were sufficient, it might be reasonable to allow a maximum velocity of 4 m/s at the maximum rate of flow, which would ensure a very good velocity at average DWF, in which case silting would be very unlikely. At high velocities the loss of head at entry to the pipe should also be taken into account in design.

Example 2. If there is a reasonable amount of fall available but the maximum rate of flow is very high compared with the peak daily flow, self-cleansing conditions may be obtained by connecting the lower leg of a flushing siphon to the top end of the inverted siphon. In this case a flushing tank would have to be constructed into which the sewer would discharge, the tank having a capacity of not less than the storage capacity of the inverted siphon. At times of slack flow the flushing tank would fill with sewage, and when full it would discharge through the flushing siphon at the maximum rate of flow, giving a self-cleansing flush to the inverted siphon. At times of maximum rate of flow the flushing siphon would function continuously. An overflow weir should be provided to pass the flow should the flushing siphon develop a stoppage or the flow exceed the siphon's capacity. The flushing siphon should be of a type suitable for use with crude sewage.

The calculations are comparatively simple. The manufacturers of the siphon should be able to supply information regarding the loss of head through it: the remaining head must be sufficient to produce the maximum rate of flow required to pass through the inverted siphon. The practical difficulties are greater, for the flushing tank needs to have a reasonable capacity and if there is little head available this capacity must be provided in a shallow tank with large surface area. The result is that sedimentation is liable to take place and self-cleansing conditions in the tank are difficult to ensure. Thus, the use of flushing siphons in conjunction with inverted siphons is limited to a minority of cases.

Example 3. In cases where there is very little fall available and the maximum rate of flow is comparatively high compared with the peak daily flow, inverted siphon design may be difficult. In such instances two or more siphons have been constructed, one to deal with DWF and the others to accommodate peak or storm flows. Such arranagements must be carefully designed if they are not to give trouble.

The normal case is the one in which there are two siphons, one to carry the peak DWF and the other to accommodate the difference between the peak DWF and the maximum rate of flow. To find the diameter of the dry-weather siphon, it is necessary first to find the actual top water level in the sewers at the upper and lower end of the siphon (by reference to Table 2) because if there is little head to play with the actual heads, dependent on the levels when the sewers are flowing partly full, must be known. It is assumed, of course, that the manholes at each end of the siphon are designed so that entry and exit losses are almost nil.

The siphon to take the storm flow should be so arranged that no flow enters it from either upper or lower end except when flow exceeds the peak daily flow of, say, 3 × DWF. To ensure this a weir should be provided at the top end which will come into action in the manner of a storm overflow weir at the moment that the flow exceeds 3 × DWF. At the lower end the invert of the siphon should be above the water level in the outgoing sewer at 3 × DWF, or else a weir must be provided to produce the same effect, otherwise there may be a tendency for silt to enter the lower end of the siphon when it is not in use.

In calculating the size of the storm-water siphon, the actual depth of water in the sewers at the ends must again be found (by reference to Table 3, p. 59), for it is most probable that the hydraulic gradient at storm time will be different from that during dry weather.

Accumulation of scum in the manhole or sewer above a siphon occurs if the design is such that the top water level in the upstream chamber does not fall to or below the crown level of the siphon pipe and the velocity of flow into the siphon is not sufficient to drag in the solids and carry them down. This is a matter of design which is not difficult, but should not be overlooked.

Inverted siphons should always be constructed so that they can be inspected and cleansed and, if practicable, a wash-out should be placed at the lowest point. If the lowest point can be close to the point of discharge, cleansing of the siphon is facilitated. The wash-out should discharge to a manhole from which the contents of the siphon may be gravitated or pumped away whenever maintenance is necessary. Hatch boxes should be inserted on the siphon, if possible, at intervals of not more than 360 ft (100 m) to facilitate rodding.

Siphons should be provided with sluice valves or penstocks at each end in order that they may be isolated, and arrangements for flushing should be installed. The manholes at both head and foot of siphons should be provided with ventilating columns or other adequate means of ventilation, for the siphon intercepts the natural flow of air in the sewerage system.

References

1. Whitman, N. D. Large concrete pressure pipes. *Civ. Engnr.*, (Sept. 1935).
2. *Surge Problems and UPVC Pipes*, Instn. Public Hlth. Engnrs. Symp., Exeter (Jan. 1978).
3. Clarke, N. W. B. and Young, O. C. Some structural aspects of the design of concrete pipelines. *Proc. Instn. Civ. Engnrs.* **14** (Sept. 1959).
4. Clarke, N. W. B. and Young, O. C. Loads on underground pipes caused by vehicle wheels. *Proc. Instn. Civ. Engnrs.* **21** (Jan. 1962).
5. Bldg. Res. Station. National Building Studies Special Report No. 32. HMSO (Apr. 1962).
6. Building Research Station–Ministry of Public Buildings and Works. *Simplified Tables of External Loads on Buried Pipelines.* HMSO (1970).
7. Spangler, M. G. Underground Conduits—An Appraisal of modern research. Paper No. 2337, *Am. Soc. of CE Proc.* (June 1947).
8. Clarke, N. W. B. The development and use of modern flexible joints for underground piplines. *J. Instn. Public Hlth. Engnrs.* **LXIII** Part 2 (Apr. 1964).
9. P. B. Lindley. Rubber O-rings as rolling seals—design considerations. *J. Instn. Public Hlth. Engnrs.* **LXVII**, Pt. 3 (July 1968).
10. Walton, J. H. The up-to-date position in regard to the theory of the strength of buried piplines. *J. IPHE*, **LXIX**, Part 4 (Oct. 1970).
11. Report of Dept. of the Environment Working Party on Sewers and Watermains. HMSO, 1975.
12. Carter, R. C. The structural requirements of plastic gravity sewers. Plastics Pipes Symp. Southampton Univ. (Sept. 1970).
13. Carter, R. C., A basis for the structural design of PVC gravity sewers. *J. Instn. Public Hlth. Engnrs.* **LXX**, Pt. 3 (July 1971).
14. Bland, C. E. G. and Picken, R. N. The strength of vitrified clay pipes on minimum bedding. *J. Instn. Public Hlth. Engnrs.* (Pt. 5) (Sept. 1973).

CHAPTER 7
Sewer Appurtenances

Working in sewers is hazardous, and more so than most industrial occupations. In the American publication *Occupational Hazards in the Operation of Sewage Works,*[1] it was pointed out that the number of accidental deaths per thousand at sewage-treatment works exceeded those for machine shops by more than 200%. The dangers were similarly appreciated in Great Britain, and the Ministry of Health issued in 1934 the publication *Accidents in Sewers: Report on the Precautions Necessary for the Safety of Persons entering Sewers and Sewage Tanks.*[2] There have been other publications, including those prepared for internal circulation by large authorities. The Institution of Civil Engineers has made recommendations regarding breathing apparatus for use in sewers and published, in 1967, *Safety in Sewers and at Sewage Works.*[3]

The dangers to men working in sewers and sewage tanks include drowning, asphyxiation, other harm due to poisonous and explosive gases, and accidents such as falling from ladders, platforms, or benchings. While the last are most frequent, and sometimes prove fatal, they do not account for the majority of fatalities. There is also the danger of infection by Weil's disease which is generally rare but to which sewer men are particularly exposed.

The danger in flooding arises from the possibility of a sudden rush of storm water or the release of a large quantity of sewage or waste. The only precaution that can be taken against storm water is for a man to stand at the surface and give warning if he considers the weather justifies such action. The avoidance of accidents by release of large quantities of water from industrial premises or the emptying of swimming baths, etc., calls for coordination between persons responsible for the discharge of large quantities of liquid to the sewers and the officers in charge of men working in sewers.

The gases liable to be found in sewers include hydrogen sulphide, which is highly poisonous and—even in a concentration of 0.2%—can cause death in a few minutes, and petrol vapour which, while not so poisonous as hydrogen sulphide, can rapidly cause death when in a concentration of 2.4% or over. Being heavy the latter is liable to accumulate in low places; it is also explosive. Carbon dioxide, produced by decomposition of the organic content of sewage, cannot be endured for more than a few minutes when in a concentration of about 10%. Carbon monoxide produced by incomplete combustion, or due to leakage of coal gas, produces unconsciousness in about half an hour when in concentration of 0.2%. Methane is not poisonous, but when mixed with air in quite a small proportion is explosive. If in a high concentration it can cause asphyxiation by oxygen depletion. It is produced by decomposition of sewage and therefore is often found associated with carbon dioxide. It is also a constituent of coal gas and sometimes natural gas from the ground.

The precautions to be taken by men working in sewers or sewage tanks include the provision of proper equipment and its use. Every gang or group of men should have at least two lifelines of adequate length, and the harness to which they can be fixed; at least one breathing apparatus, which must be regularly examined and maintained in working condition; a special lamp which will detect oxygen deficiency and the presence of explosive gases; and lead acetate paper which, when moistened in a solution of glycerol and distilled water, changes colour in the presence of hydrogen sulphide.

A recommended procedure before entering sewers is that not only the manhole to be entered, but also those immediately upstream and downstream thereof, should be opened to ventilate the sewer for at least half an hour before entry. Next, the special lamp and acetate paper should be lowered and no-one should enter the sewer until it is ascertained that no dangerous gases are present and there is no depletion of oxygen. The first man to enter the sewer should be secured by lifeline. Throughout the whole of the period of work there should be not fewer than two men at the surface prepared for rescue work. No unprotected lights or smoking should be permitted in the sewer or within 3 m of the entrance.

Because of the dangers of Weil's disease, sewer men should be warned not to neglect even the smallest abrasion. Suitable first aid kit should be carried and maintained in efficient condition, and used at once when necessary. At the depots lavatories should have pedal-operated taps in order that the men cannot foul their hands again after washing.

Apart from precautions to be taken by men working in sewers or sewage tanks and those officers in charge of them, all design of sewers and sewer appurtenances should be from the point of view not only of efficiency and reasonable cost, but also the safety of the workers. This chapter has been framed from that point of view.

The appurtenances of sewers include manholes, flushing tanks, overflow chambers, soakaways, arrangements for ventilation, and other special structures as may be required. Of these the most important, being most frequently used, are the manholes that give access to small sewers for rodding or cleansing by other means, and those which provide entry for men to work in large sewers.

Manholes

Manholes in various countries are broadly similar, but there are some noticeable differences in local practice, built up over the years. In America manholes, constructed of brickwork, mass concrete and other materials, are often circular or sometimes elliptical on plan and, from the size of the opening below the manhole cover, taper outwards to the larger diameter of the chamber in which the men may have to work. This tapering can be achieved by corbelling of brickwork or suitable shuttering of concrete. The manholes on small sewers lead directly down to the benchings at the sides of the channel inverts. Those on large sewers give direct access to the inverts.

In Britain manholes constructed of pre-cast concrete pipes, together with taper pieces to increase diameter from that of the manhole access shaft to that of the chamber, are circular on plan and often used in rural work. Most British manholes for urban work are rectangular on plan and do not taper except for a very short distance underneath the manhole cover. Instead, when the depth is greater than required for the construction of a chamber in which a man can stand on the benchings without stooping, an access shaft with vertical sides is constructed to lead from the manhole cover to a chamber which has an arched, or reinforced-concrete slab, roof. The manholes on large sewers do not lead down into the invert or flow of sewage but to a platform above the level of the dry-weather flow and arranged to one side.

The description of manholes herein is not, however, based on any manuals or codes of practice published on either side of the Atlantic. All the designs illustrated are by the author, and the principles of design described are those which he has evolved as a result of experience.

Manholes can be classified in several ways, but most commonly they are described according to size and depth. The smallest manholes, sometimes known as inspection chambers, are those on drains in private premises, where the invert is 450 to 600 mm below ground level. These are in a class by themselves, for work in them is done from ground level, the workmen standing on the ground or, if necessary, kneeling or lying prone so as to reach to the invert. Such manholes can be quite small. Generally they need only be large enough to permit sewer rods of normal flexibility to be used without difficulty. The benchings can be formed so as to keep clean but without regard for the need for a man to stand on them. Step-irons are not required.

As soon as a manhole becomes more than about 600 mm deep there is difficulty in doing work in it from ground level. If the depth is only 1000 or 1300 mm work in the manhole is particularly awkward, for not only is it impossible to reach to the invert without entering the manhole but also, if a man gets into the manhole, he cannot duck down and see what he is doing because the upper half of his body is confined by the opening in the manhole cover. Such manholes occur in large numbers on building sites and occasionally in the highway, but few designers feel that they are justified in making these awkward shallow manholes more expensive by providing an extra-large manhole cover.

The average depth of manholes on housing estates and rural sewerage, and generally on small sewers, is in the region of 2 m from ground level to invert. This depth of manhole is still not sufficient to permit the construction of a chamber high enough for the operative to stand upright on the benching; but such manholes can have chambers amply large in lateral directions for men to work in a squatting position.

A depth of 2.5 m, or a little more, from ground to invert makes it possible for the chamber to be deep enough for men to stand upright. This is the typical manhole of moderate depth, with working chamber and an access shaft leading from ground level. Still deeper manholes can be classed 'deep' and 'special deep', and may have one or more intermediate platforms between ground level and the invert in order that there shall not be too long a climb without a rest.

Manholes can be classified again according to the size of sewer to which they connect. First, there are the manholes on drains and small sewers of diameters from 100 mm to 750 mm, which are considered as the means of access to the ends of the pipes for rodding or other methods of cleansing. The second class connect to larger sewers which can be entered for maintenance purposes, and are therefore considered as means of entry to the sewers. In some instances these are described as access shafts, for there is no chamber other than the sewer itself or a small bay to the side of the sewer. In this class comes the side-entrance manhole which has a platform to the side of the sewer where men can stand away from the flow or a lateral gallery leading from the footway (sidewalk).

Further classifications are according to type. Backdrop manholes incorporate backdrops or tumbling bays, where sewage is allowed to fall from one level to another.

Each part of a manhole and its equipment should be designed or chosen bearing in mind that an efficient manhole is one which is structurally sound but not extravagant, is efficient in permitting flow of sewage and not becoming fouled by the solids of the sewage, is easy of access and, above all, safe. From these points of view the various components will be discussed, commencing from the surface.

The access shaft of an ordinary manhole should be sufficiently large for a man to be able to go down in comfort and yet small enough for the nearness to the walls to give a sense of security. Generally a width of 650 mm is sufficient in the direction of the man's shoulders and 800 mm as measured from the face of the wall to which the ladder or step-irons are fixed. Anything much less than 800 mm is cramping, but if this dimension is greater a man cannot stop on the ladder and rest his back against the back wall should he feel the need to do so. Circular shafts of pre-cast manholes are usually 650 mm in diameter, which is somewhat too small.

For the sake of economy the horizontal dimensions of a manhole chamber must not be excessively large, for an over-large chamber can add considerably to cost not only of the chamber itself but of the excavation, refilling, and making good of the surface above it. The desirable dimensions of the chamber of a manhole on a small-diameter pipe are 1370 mm long in the direction of flow and 900 mm wide. Slightly smaller manholes are sometimes used on private sewers or drains; larger chambers are required for special purposes.

Side-entrance manholes on large sewers often have quite small chambers

180

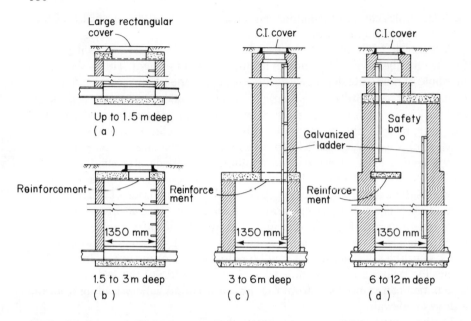

Large rectangular cover

Up to 1.5 m deep
(a)

Reinforcement

Reinforcement

1350 mm

1.5 to 3 m deep
(b)

C.I. cover

Reinforcement

1350 mm

3 to 6 m deep
(c)

C.I. cover

Safety bar

Galvanized ladder

Reinforcement

1350 mm

6 to 12 m deep
(d)

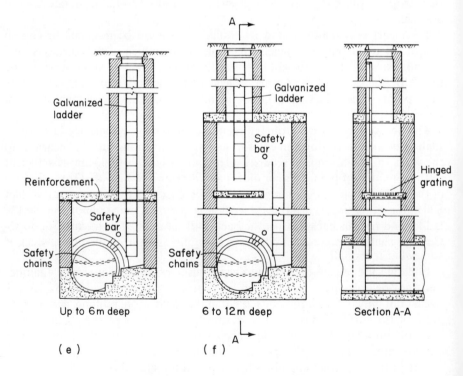

Galvanized ladder

Reinforcement

Safety bar

Safety chains

Up to 6 m deep

(e)

A

Galvanized ladder

Safety bar

Safety chains

6 to 12 m deep

(f)

A

Hinged grating

Section A-A

Figure 27   Typical brick manholes

900 mm wide in the direction of flow, and of length sufficient to accommodate the width of the sewer plus a standing platform at least 675 mm wide to the side of the sewer. In all cases (where practicable) chambers should be sufficiently high for men to stand upright, for it considerably adds to the difficulty of a man's work if he cannot straighten his back. A vertical height of 2 m over platform or benching is desirable.

Where the invert of a sewer is more than about 6 m deep below road level it is desirable to break the climb by ladder into two stages and place a platform half-way, or a number of platforms at not more than 6 m intervals may be provided in very deep manholes. These platforms are arranged in chambers which should not be less than 900 mm wide by 1350 mm, or slightly longer. The ladder from road level comes down on to the platform and to one side, the next ladder starts at least 900 mm above the platform level. A safety bar or chain is fitted at the edge of the platform.

An alternative arrangement is to carry one ladder vertically downwards and to provide platforms at intervals, the ladder passing through comparatively small openings in the platforms which can be fitted with hinged gratings.

There is a point to bear in mind when specifying gratings. These should be of wrought iron (where obtainable) or mild steel of about 50 mm by 9.5 mm section, the bars of the grating being tenoned into the metal that forms the frame. In no case should gratings made by welding be permitted, for too often welding is not sufficiently well done and bars have been known to come out, more than one at a time, under the weight of sewer men standing on the grating. Gratings, ladders, etc., should always be heavily galvanized after complete manufacture.

The arrangement of ladders and platforms in manholes is liable to interfere with direct hoisting from sewer to the surface. This matter should be looked into by the designer, and should he find that ladders, etc., prevent a direct vertical hoist, he should make an additional clearway by widening the shaft and providing openings fenced to prevent accidents. The reasons that possibility of direct hoisting to the surface is desirable are that, first, a straight haul is practically essential for rescue operations; secondly, a direct haul reduces costs in the handling of materials during repairs and maintenance.

Inverts and benchings of small manholes on pipe sewers should be neatly formed. The socket ends of pipes should be cut off and not project into the manhole; the invert should be curved to the radius of the invert of the pipe, carried up in flat vertical faces from the centre of the pipe to exactly crown level, then turned over sharply to join the faces of the benchings which should be plane surfaces sloping gently towards the channel. The angle between the benchings and the vertical side of the channel should be struck to a sharp curve of not more than 25 mm radius. The benchings should slope at about 1 in 6. At this slope it is safe to stand on one benching without danger of the feet sliding towards the channel, or stand astride the channel without danger of the feet sliding apart on a slippery surface.

In the author's view, the rounding of the benchings in cushion-like curves of large radius starting from half pipe diameter above invert and finishing at the chamber walls should not be allowed. Such benchings are very dangerous as there is no level surface on which to stand: when a man steps off the bottom of the ladder there is every possibility of his stepping on to a steeply sloping portion and slipping into the channel! For the benchings to be struck level or slightly sloping from too low a level, e.g. half diameter of pipe above invert, thereby becoming flooded at high rates of flow and heavily fouled with sludge, is also bad practice.

The inverts of manholes generally should be constructed to the cross-sections, levels, and gradients of their respective sewers, and of not less durable material than is used for the sewers.

Drop manholes

There is no universally accepted terminology for the various devices used when sewage has to be dropped from a high-level to a low-level sewer. The following use of terms, which at present are applied indiscriminately, might be suggested as a means of differentiating between the various types of drop:

(1) ramp: a pipe or channel sloping steeply, e.g. at 45°;
(2) vertical drop: a vertical pipe erected inside a manhole;
(3) backdrop: a vertical drop in the form of a pipe constructed outside the wall of a manhole;
(4) cascade or tumbling bay: a small drop over a sill, or a series of such drops in a flight of sills or steps.

These structures should be used according to circumstances, for each design has its advantages and limitations. Copying from previous designs, and adherence to the established practice or the firm of authority, has led to frequent misapplication of types and also the use of drops where they are not necessary.

All drops are, to some extent, undesirable and generally should be used only when they are needed to save capital or appreciable maintenance costs. The most frequent legitimate use for drops is when lateral sewers, laid at shallow depths, connect to deep sewers. In such cases the only alternative to drops would be the deepening of the lateral sewers, and a very little extra depth, such as 300 mm at the lower end of a 90 m length of sewer, often costs more than a backdrop.

There was a time when some engineers provided ramps or backdrops at manholes so as to reduce the velocities of the incoming sewers because they were afraid of the effect of scour. Now that it is known that increase of velocity does not increase scour in the straight pipe, but that it does increase scour in the bend of the bottom of the backdrop itself, this practice is discontinued.

The ramp is a somewhat out-of-date structure and of very limited application. It consists of a 45° junction pipe and a pipe sloping downwards from the underside of the incoming sewer at 45° then, with a 45° bend, turning into the bottom of the manhole. The disadvantage of this design is that the whole of the space between the 45° sloping ramp and the invert of the incoming sewer has to be filled with concrete, otherwise the sewer would collapse. This requires a large amount of concrete if the drop is more than a small distance.

The vertical drop is an erection, inside a manhole, of cast-iron soil pipe as used for sanitation of buildings, with caulked lead joints. At the top is a radial junction which, if it has to be jointed to clay pipe, is provided with a cast-iron connector, or else a length of straight cast-iron pipe is used and jointed to the clay pipe with a collar. The junction is open at the top end to permit rodding from above, while a bolted-on access plate is provided in order that the horizontal drain may be rodded from the manhole. Where there are several connections these can be accommodated either one above the other on the same vertical drop, or two or more drop-pipes may be erected in the manhole. The bottom of the drop-pipe terminates in a cast-iron bend fitted to discharge in the direction of flow in the channel. The whole of the materials, and means of jointing and fixing vertical drops in manholes, are the same as those used for the cast-iron soil stacks of buildings.

The term backdrop means a drop at the back of—i.e. outside the wall of—a manhole; therefore it is perhaps best restricted to the type of drop most generally useful in sewerage practice, which consists of a vertical pipe set in concrete outside a manhole and discharging through the wall into the invert. Backdrops range from simple structures designed to rule-of-thumb, to large and/or deep drops, the design of which can involve hydraulic calculations.

A simple backdrop consists of a vertical pipe of cast iron, stoneware, or concrete, to which the incoming sewer is connected at a cross-pipe. The sewer continues on past the drop and into the chamber or shaft of the manhole, where it terminates with a light tide-flap which permits rodding. The vertical pipe is carried up to ground level for rodding purposes and is covered by a lamphole cover. The bottom of the drop should always terminate in a bend of cast iron, for this is where the most severe wear occurs. Whether or not a duckfoot bend is used is a matter of opinion. The whole is surrounded with concrete, usually specified as being not less than 150 mm in thickness and filled to the face of excavation (see Figure 28).

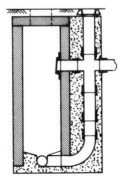

Figure 28   Typical backdrop

Such simple backdrops constructed on small-diameter pipe sewers are made of the same diameters as the sewers which they serve. The vertical pipes are, of course, more than ample in diameter to take the full sewer discharge.

The bottom bend of a simple backdrop should be turned toward the opening of the outgoing sewer and so arranged that there is no danger of sewage splashing on to the benching.

A still simpler variety of vertical backdrop is similar to that described above but without the rodding eye carried to ground level. Instead, a special stoneware radial 90° junction, having a spigot on the junction instead of a socket, is jointed in the line of sewer and connected to the drop.

When backdrops are constructed on large sewers, careful individual design becomes necessary if the construction is to be economical and the design sound.

In the first place the diameter of the drop-pipe can be reduced to less than that of the sewer and should be so reduced when an appreciable economy would result therefrom. Care, however, should be taken that the backdrop is not made too small to take the flow, and that its top end is large enough, or so bellmouthed, to be able to take the flow that has to go down the drop.

If the vertical pipe is of the same diameter as the incoming sewer, the head is usually of sufficient size to provide the velocity head required for the flow to enter the top end of the pipe.

Some engineers are nervous of constructing deep backdrops in the form of a single drop-pipe because of the theoretically high velocity that should develop. In practice, however, it has been found that when water falls down a small pipe which is larger than necessary to take the flow, very high velocities do not occur, because the water either clings to the sides of the pipe or falls as drops. Thus, very deep backdrops on small-diameter sewers can, like the soil pipes of high buildings, be safely constructed in one drop. At the bottom of such a backdrop the bend retards the flow and energy is dissipated in a standing wave. Such backdrops should, of course, be constructed of cast iron throughout.

The energy of a jet of water can be dissipated if the jet is directed into a mass of water which is at rest. This is the principle of the water cushion, which is a tank placed at the foot of a vertical drop. Water cushions are particularly useful in connection with waterworks, but they are not too easily applied to sewerage because they tend to retain solids. For this reason they are liable to become dirty unless well designed. Water cushions for sewage should have small outlets at the bottom to permit sediment to wash out.

Water cushions will dissipate energy, preventing scour and splash, if properly designed, but they can be quite ineffective. The capacity of the cushion must be appreciable and the inlet must be so arranged that there is no danger of the contents of the water cushion being driven bodily out by the incoming flow: also the depth of water must be sufficient effectively to reduce velocity. Preferably the jet should be brought vertically down into the centre of the tank and the walls should be vertical and leave the floor at sharp angles. Rounded corners assist the jet in sweeping the water out of the tank.

Cascades, which consist of a series of small steps in the invert, are used for accommodating very large flows. They are particularly applicable to moderate drops on sewers so large that they can be entered by men, for they then serve as flights of steps leading from one level to another, as well as for dissipating energy. For the steps to be useful to sewer men, they should not be more than 225 mm high, and preferably lower. Where this consideration does not arise, the depth of step can be increased to about 600 mm without producing a velocity liable to cause serious scour.

Special manholes

Manholes constructed on large sewers are intended to be used as means of access to the sewers themselves; they therefore differ in form and proportions from manholes on small sewers. When the sewers are sufficiently large for men to stand upright in them no manhole chambers are needed, the manholes being little more than shafts. But direct access to the invert of a sewer from above is not conducive to safety, and side-entrance manholes, in which the shaft is a little off the line of the sewer and terminates on a landing above the level which the sewage normally reaches in dry weather, are to be preferred.

When manholes are constructed on sewers which have been laid under streets where traffic is unusually heavy it is preferable, if the sewers are sufficiently deep, for access galleries to lead from the footway to the manhole shaft in order that men may enter the manholes without interfering with the flow of traffic. In such instances the vertical shaft and manhole cover are still required to facilitate the handling of building materials for reconstruction purposes, etc.

When sewers of very large diameter meet it is usual for the one of smaller size to sweep into the larger in such a manner that the latter is tangential to the radius of the curve of the smaller sewer. Junctions of this kind are known as bellmouth chambers. In them the inverts are formed so as to give the minimum resistance to hydraulic flow and, as far as practicable, the crown of the arched chamber is tapered so as not to give too rapid change of section. Side-entrance access is provided; preferably in the fork between the two sewers.

Manholes on cast-iron rising mains of large size are best constructed in the form of cast-iron hatch boxes; or cast-iron flanged T-junctions with blank flanges may be inserted in the main and chambers built round the blank flanges to serve as access manholes. This form of access may also be used for large-diameter inverted siphons.

On comparatively rare occasions pressure manholes are required on existing sewers because of surcharge which it is considered may be permitted. Such manholes are provided with water-tight doors built into the access shafts. When this is done the doors have to be tied down by bolts passing down the shaft for a sufficient distance for the weight of the shaft to counterbalance the upward thrust of water pressure.

It should be appreciated that when watertight doors are provided, the pressure exerted is not only that due to the head of sewage above the door; it may be the pressure due to the head of sewage above the level to which the sewage may rise after compressing air in the manhole. It should also be remembered that if air is compressed in the manhole during times of surcharge it would be dangerous at such a time to open the watertight door; therefore the door should preferably open downwards.

Construction of manholes

The chambers and shafts of manholes may be constructed of brickwork, mass concrete, pre-cast concrete tubes, reinforced concrete and, rarely, cast iron.

Brickwork may be used for manholes in both town and country and it is one of the best materials, although not the cheapest available for the purpose. First-quality drainage bricks only should be used.

Mass concrete is not infrequently used in sewer-manhole construction, although it does not always show a very great saving on brickwork. It has the disadvantage of requiring high-quality workmanship and supervision if it is to compare with brickwork on completion. The concrete should be of well-proportioned fine aggregate, carefully worked to the face of the shuttering during construction. Vibrating is an advantage. Table 29 indicates wall thicknesses and maximum depths appropriate for brick or concrete manholes with various lengths of wall face.

Table 29　Thickness of manhole walls. Approximate maximum depth in metres for internal or external water pressure on manhole walls

Length of internal face of longest wall (m)	Thickness of wall (mm)						
	225	338	450	563	675	788	900
1	4.4	10.0	17.6	—	—	—	—
$1\frac{1}{4}$	2.8	6.4	11.3	17.6	—	—	—
$1\frac{1}{2}$	2.0	4.5	7.8	12.2	17.6	—	—
$1\frac{3}{4}$	1.4	3.3	5.8	9.0	13.0	17.6	—
2	—	2.5	4.4	6.9	10.0	13.5	17.6
$2\frac{1}{4}$	—	2.0	3.5	5.4	7.9	10.6	14.0
$2\frac{1}{2}$	—	1.6	2.8	4.4	6.4	8.6	11.3
$2\frac{3}{4}$	—	—	1.5	3.6	5.3	7.2	9.3

Pre-cast concrete-pipe manholes are excellent, particularly for rural work. They cost less than other forms of construction and at the same time, unless badly supervised, the work appears well on completion. Very little skilled labour is required and the minimum of time is necessary for construction. Simple pre-cast concrete manholes consist of a chamber ring or a number of chamber rings supported on the benchings, a straight-back taper piece, a series of shaft rings, a making-up piece by which the correct level of the top is adjusted, and a heavy cover slab on which the manhole cover rests. Step-irons are built into the chamber, taper, and shaft rings during manufacture. When the manhole has been properly assembled the step-irons should form a regular vertical range. On no account should they be used for hoisting purposes.

The first pre-cast concrete manholes consisted of pipes with ogee joints, but the segments may now be obtained with spigot and socket joints, and these should be used in all cases where the manhole is likely to be below ground-water level, for weeping joints are unsightly and difficult to cure.

The aggregates of which pre-cast concrete manhole segments are made vary considerably, and the engineer should ask for the submission of samples of pieces of broken tube before the contractor places orders.

The procedure in the construction of a pre-cast concrete manhole is first to lay a 150 mm concrete slab below the invert. Pre-cast concrete inverts are obtainable, but they are heavy and some designers prefer not to use them, but to form the inverts and benchings in mass concrete. Before the latter are made, a circular mould of diameter equal to the internal diameter of the chamber ring should be placed in position and a ring of mass concrete formed up to the level of the bottom of the chamber ring. The chamber ring may then be placed in position on a layer of cement mortar and the manhole constructed, each ring being bedded in cement mortar upon that below it, and the joints carefully pointed. Mass concrete is poured round the outside of the chamber ring and continued to just above the top of the taper piece, but it need not be carried any higher unless the manhole is to be constructed where it is liable to be disturbed by traffic or excavations. When the manhole chamber and shaft have been completed the inverts and benches may be formed and the cast-iron manhole cover grouted in position on two rings of brickwork to allow for future changes of road level.

Figure 29 (on page 188) illustrates a typical concrete manhole shown in cross-section in view (a), with information as to available sizes of chamber and shaft rings, tapers, base units and other appurtenances, at (b).

Pre-cast concrete manholes are not only applicable to small-diameter sewers. They may be used with advantage for most purposes, particularly the shafts to deep manholes and special chambers.

Reinforced concrete manholes are often required in conjunction with other reinforced concrete work, but for general purposes, i.e. the construction of manholes on lines of small-diameter sewers, reinforcement is unnecessary except in roof slabs. A sewer manhole is an inherently strong structure: it does not need to have very thick walls, and the economy in wall thickness effected by addition of reinforcement is usually exceeded by the extra cost of steel, fine aggregate, and the difficulty of placing reinforcement in a confined space.

Manholes of cast-iron tubing are manufactured for use in unusual circumstances when it is considered that normal forms of construction would not be either practicable or sufficiently watertight. They are jointed in a manner similar to that used for cast-iron segmental sewers.

Ironwork

The cover of a manhole needs to have an opening sufficiently large for a man to be able to enter without difficulty, and also for a casualty to be lifted out with

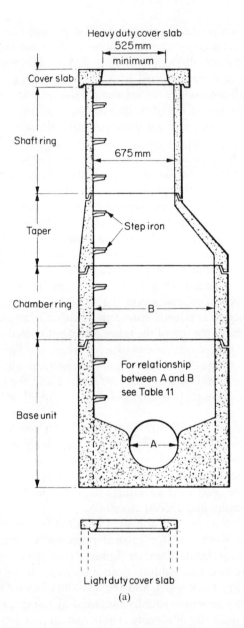

Heavy duty cover slab

525 mm minimum

Cover slab

Shaft ring

675 mm

Taper

Step iron

Chamber ring

B

For relationship between A and B see Table 11

Base unit

A

Light duty cover slab

(a)

rescue harness. Apart from exceptions, e.g. where an extra-large opening is required for a special purpose, the opening should not be unduly large, for it is more difficult to enter a manhole via a large opening than via one of normal proportions. Usually a man places his hands on the ground, one at each side of the opening, to take his weight until he can get hold of the ladder or step-iron, and he cannot do this if the opening is too wide. The difficulty is overcome, of course, if sewer gangs are accompanied by a vehicle which has a handhold to

Relationship between diameter of outlet and diameter of chamber	

Diameter of outlet Pipe (A) (mm)	Diameter of chamber (B) (mm)
150 to 375	900, 1050, 1200, 1350, 1500, 1800
150 to 525	1050, 1200, 1350, 1500, 1800
150 to 600	1200, 1350, 1500, 1800
150 to 675	1350, 1500, 1800
150 to 990	1500, 1800
Over 900	1800

Description

Cylindrical Manhole units with ogee joint, with or without base unit.

Manholes are usually classified as 'Shallow', up to 2.1 m, or 'Deep', over 2.1 m in depth.

Shaft and chamber cover slabs

Heavy Duty Slab, for use under main traffic routes and roads likely to be used for the temporary diversion of heavy traffic. It is reinforced to withstand one or two wheel loads, according to its size, of 112 kN on the slab when supported on its outer rim.

Light Duty Slab, for use under roads carrying light traffic only, or in fields, gardens or access tracks. It is reinforced to withstand a single wheel load of 35 kN on the slab when supported on its outer rim.

Shaft rings

Shaft rings are manufactured in lengths from 0.15 m to 0.90 m, in increments of 0.15 m.

Chamber rings

Chamber rings are manufactured in lengths from 0.30 m to 0.90 m, in increments of 0.15 m.

Tapers

Taper units (straight back) are the best means of reducing from the manhole chamber diameter to the access shaft diameter.

Reducing slabs

Where insufficient depth is available to include a taper, a reducing slab can be used.

Landing slabs

In deep manholes a reducing slab can be used as a landing or rest slab.

Base units

Precast base sections can be supplied with or without benching and/or full or half holes to suit customer requirements.

Standard diameters for precast base units are the same as for chamber rings.

Step irons

Galvanised malleable cast iron step irons to B.S. 1247 can be supplied pre-fixed to the units. These are staggered 150 mm each side of a centre line and spaced 300 mm vertically.

Ladders

For manholes over 4.5 m in depth step irons are usually replaced by ladders. Holes to accommodate the ladder fixings can be provided.

Handling

Lifting holes are provided in all units.

(b)

Figure 29   Typical concrete manhole (Courtesy Mono Concrete Co.)

assist entering manholes, or if a special dome-shaped cage is used to cover the opening while the manhole is open. Where ventilating covers are used it is common practice to place the cover on the ground near the open manhole and to use the ventilating openings as handholds.

Taking these factors into account, the usual diameter of a manhole cover opening is 600 mm in America and 560 mm in Britain. Covers with 500 mm openings are obtainable, and there are some covers, including the British

Standard designs for heavy road covers,[4] which have still smaller openings! For the reasons given, the last are not suitable for general use. Larger openings have been found necessary for manholes on large sewers that can be entered by men for maintenance purposes, and where building materials may have to be lowered, or detritus or debris lifted in skips.

Manhole covers have to be designed according to the loads which will bear upon them. They must be sufficiently strong to take the live load of the heaviest vehicle likely to pass over. In this connection it should be mentioned that the breakage of heavy cast-iron covers in some localities is not a negligible maintenance item. The covers should not rock when initially placed in position, or develop a rock with wear.

The design of light manhole covers for inspection chambers in back yards, or medium-weight covers for carriage drives, etc., does not involve a problem. It is when the covers have to be sufficiently strong to take heavy loads that efficient design presents some difficulty. They must not be too expensive and must not be too heavy for two men to lift out with comparative ease. On the other hand, it is generally considered desirable that manhole covers for use on sewers in public places should always be so heavy that one man cannot lift the cover alone. The reasons for this are that a cover too light may be lifted by mischievous boys, and where a cover can be opened by one man he may be tempted to break safety rules and go down a manhole without leaving someone at the surface to watch over his safety.

Manhole covers for use in heavy traffic should be capable of withstanding the loading tests recommended in the countries for which they are required. At the present time a manhole cover is usually strong enough if it will withstand a load of 35 tonnes applied via a 300 mm diameter hardwood block.

It can be said with justice that many commercial designs of manhole covers achieve the required test strength by virtue of the weight of the material in them; and generally there is not much difficulty in designing a 560 mm diameter circular cast-iron cover which will take the aforementioned load without involving excessive stresses in the low-grade cast iron which is often used for cover manufacture. When, however, a cover of larger diameter opening is required an engineering problem arises, for the bending moment becomes so great that a cast-iron cover capable of taking the load needs to be of excessive weight, i.e. more than 140 kg. This means that cast steel must be used, and then fresh problems arise for cast steel is a very expensive material compared with low-grade cast iron and, moreover, it has a low yield-point relative to the breaking load. A steel cover must therefore be designed not only to stand the test load without fracture, but it must stand a severe load or impact without developing a permanent set.

The most common type of cast-iron manhole cover is that which has a plate surface with ribs on the underside. Such covers are easy to make, but are inefficient in design because, as the ribs are on the underside, there is the least metal where it is wanted to take tension, whereas the compressive load, which cast iron can easily stand, is at the surface where there is an excess of metal.

The type of cast-iron cover which is therefore stronger than the ordinary type is that in which the plate is at the lower surface, the ribs on the top, and the spaces between the ribs filled with hardwood. These covers are excellent for use in roads taking very heavy traffic, although the hardwood wears and rots out in time, and has to be replaced. Of late years a satisfactory, and lighter, design in steel has been widely used.

Cast steel is particularly applicable to large-diameter covers of the ventilating type. In these circumstances the cover can consist of a network of ribs with no top or bottom plates, and the steel is used very economically. Nevertheless these covers are very costly.

Rocking manhole covers are generally annoying, and are a particular cause of complaint by nearby property-owners. The rocking of a cover also wears the seating to an extent that has to be seen to be believed! To overcome rocking the manhole cover must be:

(1) fixed down with screws;
(2) bedded on a machined seating which is cleaned every time the cover is replaced, for a perfect seating will wear irregularly if rocking is set up as a result of dirt preventing perfect bedding;
(3) supported at three points at the corners of a triangle.

Covers of the first type are manufactured for special purposes, but are not generally popular because of the extra work involved in opening and closing, and the small parts that can wear and need replacement.

Covers of the second type are generally to be preferred, but it must be admitted that the machining of the seating costs money and becomes ineffective if the sewer gangs are not trained in the habit of cleaning the seating and making sure the cover is properly bedded down and unable to rock every time it is put into position.

The triangular cover supported at three corners appears superficially the theoretically perfect solution to the problem and it is, in fact, possible to make such a cover which cannot be rocked with the feet. The practical design of a triangular cover which will not rock under heavy traffic is, however, a very different matter, for the bearing surface at the corners must be large enough to take the applied load without the compressive stress being exceeded or causing undue wear, and at the same time the area under compression must be sufficiently small to ensure that the applied load does not get outside the centre of compression by a sufficient amount for the moment produced by the superincumbent load to exceed the moment of resistance of the weight of the manhole cover: this is theoretically impossible and has not been achieved in practice. Generally the designs of triangular covers allow for a large bearing surface, and for this reason severe rocking has been reported in many instances.

Other disadvantages of the triangular cover are that some designs would be excessively large if made to allow a hole of size adequate for easy access, and are therefore made with unduly small holes. This makes them dangerous because men wearing safety harness cannot be quickly rescued through them.

Ladders and step-irons

Ladders, not step-irons, should be used in all deep manholes and in all manholes that are frequently entered, for they are much safer and better in many respects, although more costly. It is far safer and easier to go down a ladder when carrying tools or equipment than to use step-irons, and with a much greater sense of security. It is also possible to hold on to the stringer of a ladder, avoiding touching the rungs which may be infected with Weil's disease by the wet footprints of the last man who went up. Step-irons used in places where there is much work done have been known to develop sharp cutting edges where worn by treading. They are not always securely fixed in position. If of common cast iron they snap off very easily; for this reason only malleable, not common, cast-iron step-irons should be used and every step-iron newly put in should be tested by being given a smart blow with a 2 kg hammer.

Step-irons are often badly placed by the designer, or more often by the man who puts them in. It is of importance that the step-irons should be equally spaced vertically and, if staggered, regularly staggered about a vertical line; deviations to avoid openings or pipework should on no account be permitted because only too easily can such irregularities cause an accident. Where narrow-width staggered step-irons are used a very common fault is to place them too far apart. The centres should not be more than 190 mm apart otherwise a sense of insecurity and discomfort develops. Double-width step-irons which do not need to be staggered are better. Ladders or step-irons should start not more than 450 mm below road level and be continued to within the distance between two rungs on step-irons of floor, platform or benching.

Ladders, gratings, etc., being constantly in a damp atmosphere, are very prone to corrosion and in time will become dangerous unless replaced as necessary. Where, however, they are regularly submerged in sewage, they receive a protective coating and will last for a long time.

Sewer flushing

Sewers with adequate flows, laid to proper gradients, and with properly finished inverts, should not need to be flushed. Nevertheless, many old sewerage systems are not perfect, and in most new schemes arrangements for flushing have to be made because of difficulty in obtaining satisfactory gradients. In such cases flushing is the remedy for the silting that would otherwise occur.

It may suffice for manholes at the top ends of lateral sewers to be provided with disc penstocks in order that they may be filled with water from nearby hydrants, and the water suddenly released by opening the penstock. But when flushing needs to be of regular occurrence as, for example, in the case of a main sewer of inadequate gradient, an automatic flushing tank should be provided. A number of devices have been used in the past, but at present the most common arrangement is a flushing tank which consists of a chamber of suitable capacity which discharges water into the sewer through an automatic siphon.

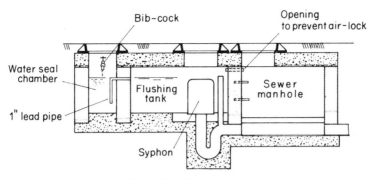

Figure 30   Flushing tank

These tanks are normally filled from the water mains and, in Britain, a usual requirement of water authorities is that the water main shall not discharge directly into the flushing tank but shall fall freely from a bib-cock into a chamber separated from the flushing tank by a trap. A meter is provided in a separate chamber (see Figure 30). The flushing tank fills slowly at a rate found by experiment, the bib-cock being adjusted accordingly. When the tank is full the siphon automatically discharges at a rate sufficient to ensure that the sewer runs at least half full. A suitable size of flushing tank is one-tenth of the cubic capacity of the total length of sewer to be flushed.

It is essential that flushing tanks shall have an air inlet, as otherwise the siphon may fail to function. When the tank is built in conjunction with a sewer manhole, as in Figure 30, it is sufficient for an opening to be formed above top-water level between the manhole and the tank.

To economize in the cost of using company's water, natural sources should be used for flushing as far as is practicable. If a small natural spring can be diverted to discharge to a flushing tank this should serve admirably, but the rate of flow should be controlled. In some circumstances it has been justified for a well and pump to be provided, so as to supply clean but untreated water for flushing a large sewer.

Ventilation of sewers

In places in America where a main- or house-trap is required between the plumbing and the house drains or, in Britain, where the local authority (usually without any power so to do) requires an intercepting trap between the house drains and the public sewers, the sewers do not receive any ventilation via the soil and ventilating pipes of the building as they would if there were no such trap. In this case special ventilation for the sewers has to be provided to prevent the accumulation of dangerous gases or those, such as hydrogen sulphide, which would attack the structure.

It might be argued that, if the sewers had been properly constructed, no such gases would accumulate. In spite of all precautions, however, they do; unless ventilation is provided. In some places such as in the Near East, there are greater risks because even if sewers are laid to what are elsewhere considered adequate gradients, there may be very large amounts of detritus entering the sewers at a greater rate than can be washed away. This can hold up organic material which decomposes, producing objectionable gases in sufficient quantity to damage the sewers.

In Britain it has been the practice to provide ventilating columns which are commonly erected and connected to the manholes at the top ends of all lateral public sewers. The practice, however, is not so common as it was because opinion has changed and few local authorities now require intercepting traps; therefore, the ventilating pipes of the houses provide all, or nearly all, the ventilation required for the sewers. The ventilating columns connecting to small lateral sewers were normally of 150 mm internal diameter. They were carried to above the eaves of the highest building nearby. Where columns are required, in residential areas a standard height of 10 m may be adopted.

Ventilating columns are still provided in places where pressure relief may be necessary on main sewers, or where it is considered additional fresh air is desirable. These are of larger diameter than those at the ends of the laterals but not necessarily of great size, always being considerably smaller in cross-sectional area than the sewers which they serve.

A ventilating column consists of a heavy cast-iron base into which a length of steel tube is inserted. A small cast-iron door is formed in the base to give access to a chained rust box. The base should be securely set in about 1 m³ of concrete, which should not be brought to ground level, as this would interfere with replacement of paving. To prevent birds from nesting in the top of the column a copper-wire ball should be fixed in position: galvanized iron-wire balls are obtainable but they should not be used because they do not last long. Spun-concrete and plastic columns are also available.

The type of ventilating column to be used should be one which fits in well with existing lamp or other standards, for ventilating columns can never be considered ornamental and should not be made too obvious. They become less noticeable if they serve a double purpose; e.g. ventilation and supporting street lamps. In some circumstances their presence is not detected by the general public. In carrying out a drainage scheme it is worth while erecting ventilating columns before commencement of other works, as by this stratagem complaints about imaginary foul odours may be easily refuted.

Ventilating manhole covers are used in places where ventilation is possible without causing nuisance, e.g. in streets well away from buildings. They have the disadvantage of permitting the entry of road grit.

House connections

When new sewers are laid, or existing sewers reconstructed, it is necessary to make provision for house and/or gully connections. Junction pipes are jointed in the line of sewer as the work proceeds, each junction canting to the side of the road from which the appropriate lateral connection will be made.

The term house connection or lateral is usually applied to the length of pipe which extends from the boundary of private property to the sewer.

In the normal case where sewers are of moderate depth, the junctions are set to slope upwards at an angle of about 45° to the horizontal. A bend of appropriate radius is connected to the junction and thence a line of pipe is laid at an even gradient, and by the shortest possible route, to the last chamber on the system of drains which the junction serves. The junction pipes are usually 45° junctions, facing upstream, and these are generally preferred. The line of pipe connecting to the sewer, however, is normally at right angles to the sewer in order that the line shall be no longer than necessary.

When sewers are very deep, direct connection from lateral drains is not easy, and the local authority has to consider the alternatives of permitting vertical connections, or else of providing an additional shallow sewer connecting to the main sewer at a backdrop manhole. Such provision of an additional sewer is preferable, but some authorities permit connection to deep sewers in the following manner. The lateral connection is brought out from the boundary of the property at not less than the minimum permissible gradient according to the diameter, until it is almost above the sewer to which connection is to be made. At this point the lateral connects to a T-junction set vertically, and a vertical line of pipe which terminates at the bottom end with a right angle bend and junction connecting to the sewer. The upper end of the vertical shaft is brought to about 300 mm below ground level and provided with a removable cover. The whole of the shaft, bend, and junction are surrounded with concrete at least 150 mm thick.

The rules for concrete protection of sewers apply equally to house connections which should be of the same materials as the sewers to which they connect, clay pipe drains connecting to clay pipe sewers, and cast-iron pipes to cast-iron sewers.

Practice regarding the extent to which house connections are constructed when new sewers are laid varies considerably. Perhaps the most common procedure is to lay the sewer, at the same time to insert the junction pipes, and afterwards to permit the individual householders to make their connections under the supervision of the local authority inspector. When this is done, it is necessary for a junction book to be kept in which are accurately recorded the positions of all junctions and the directions in which the branches are turned.

In laying a sewer and making junctions at the same time, either special junction pipes may be used, or ordinary junction pipes inserted and plugged with stoneware stoppers sealed in position with a mastic preparation of lime mortar. Some authorities prefer not to use junction pipes, except when connections are to be made at once, on the grounds that too often junctions are inserted in positions that are found not to be suitable when the future connections have to be made, and that the number of unnecessary junctions involves needless expense. The alternative is to insert junctions only when they are required for immediate use.

When further connections are needed the sewer is cut with hammer and chisel (in the case of clay or concrete pipe) and a stoneware saddle connection cemented in position. This work requires care if the sewer is not to be damaged. Saddles for connecting to clay pipe sewers form either right-angle junctions or obtuse junctions, the oblique angle being usually more obtuse than 45°, owing to ease of manufacture of more obtuse angles. Very large numbers of saddles are used; nevertheless, the article as commonly manufactured is very unsatisfactory in design and generally does not permit a clean neat junction between the saddle and the cut sewer.

Construction of soakaways

Simple, small soakaways may be constructed by filling pits with large boulders or rubble, or by laying drains with open joints in trenches filled with rubble. However, these methods, which are useful for taking the moderate run-offs from individual roofs or for soaking away the comparatively steady flows of sewage effluent, are too expensive when a large storage capacity is required to absorb the run-off due to intense storms of short duration. When excavations are filled with rubble, the rubble itself occupies much more than 50% of the volume excavated, and good rubble is not a very cheap material. A lined cavity is therefore usually more economical, even allowing for the higher quality of the materials used.

The more usual form of soakaway see (Figure 31) is a lined cavity or chamber which consists of a circular steining, and is somewhat similar to a large-diameter shallow well. It is, in fact, a well in which the action is reversed, i.e. instead of water being drawn from it out of the ground, water is discharged into it to soak away.

The lower part of the chamber may consist of 225 mm circular brickwork with openings constructed at intervals. Another method which has been found satisfactory is to commence at the bottom with four courses of brickwork in

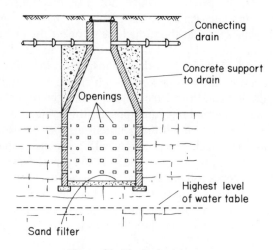

Figure 31   Typical soakaway

cement mortar, and above these to lay a further four courses of dry brickwork. In this manner dry and cemented rings are laid alternately until the level is reached at which it is not intended that water shall pass through the walls. At this point corbelling commences and the soakaway reduces, in bottle-neck form, to the diameter of the manhole cover at the top.

In most instances the bottoms of soakaways are left open, but in certain places where large quantities of leaves or other organic material are found to collect, floors of concrete are provided to make cleansing easy—the whole of the water soaking away through the perforated walls. Sometimes a layer of fine sand is deposited at the bottom of the soakaway to serve as a filter and to assist the removal of debris. Filtration of this kind has been provided (hopefully) in chalk areas where surface water from roads could be a cause of pollution of water supplies, and up to 0.6 m depth of sand has been used in such localities, but this can have no effect on the water which passes through the walls, and 150 mm thickness is sufficient to assist cleansing.

The surface-water drains connecting to soakaways usually connect to the shafts and therefore have to be supported by concrete walls where the ground has been excavated. In many instances water is allowed to fall down from the drain level to the bottom of the soakaway, but it is better to install backdrops or water cushions.

As soakaways are satisfactory only where the geological conditions are favourable, it may be expected that local geological variations will influence design. For this reason engineers experienced in the work mention in their specifications that variation in the design of chambers shall be made as the work proceeds, according to the nature of the ground as found.

References

1. *Occupational Hazards in the Operation of Sewage Works*, Fed. of Sewage Wks Assoc., Manual of Practice No. 1, USA (1944).
2. *Accidents in Sewers: Report on the Precautions Necessary for Safety of Persons entering Sewers and Sewage Tanks*, Ministry of Health (1934).
3. *Safety in Sewers and at Sewage Works*. ICE (1967).
4. British Standards for Manhole Covers. (B.S. 497—1967 and Amendments) British Standards Institution, 2 Park St, London W1A 2BS.

CHAPTER 8
Sea Outfalls

The discharge of sewage into the sea will always tend to arouse controversy. Much of this could probably have been avoided if the theory of dispersion rather than dilution had been better understood. Standards for the discharge of effluents are often related to the dilution factor available in the receiving stream, and in this sense, the discharge of crude sewage into a body of reasonably clean moving water whose equivalent volume of flow is 500 times that of the discharge should easily have been capable of absorbing the impurity load, provided satisfactory mixing of the discharge and the receiving water could be achieved. Increasing interest in environmental conditions, and factors such as the siting of atomic power stations in isolated coastal areas, coupled with the failure of many existing outfalls to discharge sewage without visible repercussions on crowded bathing beaches, has stimulated a good deal of work on the risks involved in discharging sewage to the sea. It is now accepted, for instance, that many industrial wastes, particularly those containing heavy metal salts, should be treated to prescribed standards before being accepted into any sewerage system, particularly one providing secondary treatment. It is also becoming mandatory that some form of disintegration of solids should be applied before discharge to sea, but this does not necessarily imply solids separation.

Informed opinion now is that rates of actual dispersion are, in fact, generally much lower than the 500 to 1 implied by the old conception, but, at the same time, given the right conditions a rate of say 100 to 1 dispersion of crude domestic sewage in sea water is more effective in removing pollution than secondary or even tertiary treatment, as provided by a sewage works. By disintegration and return of solids in a finely divided state—and in quantities roughly corresponding to the amount of solids in the original sewage—for discharge through properly designed jets arranged along an outfall pipe of adequate length the most effective, as well as the most economical, method of disposal may well be achieved.[1] Bacterial activity and ultraviolet light, coupled with the oxygen content of the sea water, should, in the right circumstances, provide a complete means of treatment and disposal without any detriment to the environment.

Much work has also been done on the effect on discharges of tidal movement, currents, wind effects, etc. It is becoming increasingly clear that limitations to the body of water such as would apply in a lake, or even an estuary or a bay, at once introduces many complications that do not apply to a straightforward sea discharge. Mostly these problems are associated with the movement, or lack of movement, of the receiving water, which in turn greatly affects dispersion. Environmental problems also tend to become more severe in enclosed waters and estuaries, and it is pointed out later that, for coastal towns, of the three alternatives available, full inland treatment and discharge of effluent say to the head of an estuary, or partial treatment with discharge further down that estuary or to a bay, may both compare unfavourably, in terms of cost and environmental effect, with disposal of disintegrated sewage through a well-designed outfall to the open sea.

The seas around Great Britain are all subject to considerable tidal range. This in its turn induces strong currents running parallel to the shore and is, in theory, capable of providing very suitable conditions for disposing of any discharge, particularly if positioned clearly beyond low water of spring tides. However, the currents usually reverse according to whether the tide is rising or falling, and other factors such as wind strength and direction, and even eutrophication where strong seaweed growth is taking place, render careful preliminary survey essential before any clear picture of area affected by the discharge can be built up. A report and survey of the discharges of foul sewage to the coastal waters of England and Wales is contained in a publication of the Department of the Environment and the Welsh Office.[2] This document also indicates that at least 45 schemes for new outfall installations have been carried out since 1960. The basic criteria applicable to discharges of domestic sewage and industrial effluent in Great Britain are that solids, identifiable as of sewage origin, should not reach areas used by the public for recreation; that discharges should be either sufficiently diffused or so remote that general offence is not caused to the public; and that the effects on local fishing interests should be minimized—such measures are considered adequate for the protection of public health. As is made clear later, this definition falls short of the requirements issued by the European Common Market,[3,4] to which all future outfalls will have to conform. Wood,[5] following a general study of conditions in the North Sea, concludes that the observed effects of sewage outfalls into the North Sea are less than might be expected because of the tidal regime and the free exchange of water with the Atlantic Ocean. Where tidal currents are vigorous—as on the British, French, and Belgian coasts—dispersion of wastes is rapid. On the western side, where currents are relatively weak, the mixing and dispersal of wastes is less effective. Wood regards it as unlikely that conditions generally will deteriorate in the long run, but warns that care is needed to ensure that the capacity of coastal waters to receive sewage wastes is not exceeded. Aubert and Breittmeyer[6] report that bacterial contamination is present nearly everywhere in the Mediterranean coastal regions, but not *very* high in relation to the size of the virtually tideless Mediterranean. Such discharges near major urban areas are responsible for very substantial enteric contamination in some localities, and this is increased by industrial wastes which can change the natural biological balance—as in the north-west and north Adriatic Basin. However, disposal techniques which take into account the hydrology of the area, and make full use of dispersion, show that existing conditions could be greatly improved. The Baltic has sometimes been assumed to be the most polluted sea area in the world, but as Svensson[7] explains the Baltic is like a large estuary with a very long residence time, or a fjord with a shallow sill and relatively deep parts, where stagnation is a regular phenomenon. Recently a convention for Baltic pollution put a ban on discharge of PCB, DDT, many other pesticides, and heavy metals.

Berg,[8] in a survey of *Regional Problems with Sea Outfall Disposal of Sewage on the Coasts of the United States*, points out that half the states are contiguous with the sea, and within them are some of the nation's largest population centres and most difficult sewage disposal problems. Most of the liquid waste from these large urban centres is discharged through outfalls to the oceans. Tonnes of solids, even from inland cities, are barged out to sea. Any problems associated with these discharges have been of major concern only in circumscribed areas. However, government policy as defined in the Federal Water Pollution Control Act Amendments,[9] which became law in 1972, set 1985 as the year when discharges of pollutants into the nation's waters must cease. The official version gives a rather different interpretation in that 'the administrator [of the Act] shall issue regulations [by July 1983] for effluent limits to require the elimination of discharges of all pollutants if he finds that this elimination is technically and economically achievable'. It may be significant that for the Fiscal Years 1972–74, the biggest spenders on sewage treatment facilities were California, Michigan, New Jersey, and New York—all with over 1000 million dollar expenditures. These references form an interesting comparison with those of Warren and Rawn, and Weston in *Modern Sewage Disposal*, in 1938.[10,11]

Tidal experiments for location of outfalls

A sea outfall may discharge into a tidal estuary either well upstream or near the sea, or it may lead directly into the open sea. In the case of an estuary the flow of the tide in and out makes it difficult for one to ascertain where the sewage will go. When the tide is flowing out any float put into the water will usually tend to be carried out to sea: on the return of the tide there is a time of comparatively slack water when large eddies are formed and the floats may be carried ashore. There would therefore appear to be a suitable time after high water at which stored sewage or effluent could be discharged.

Figure 32 shows a typical plotting of the tracks of floats put into the water in a large tidal estuary (in which the sea is to the right). The numbers against the tracks indicate the number of hours after high water at which the floats were released at the point of outfall. It will be seen that in this case the best time for discharge would be during the first 3 h after high water, but that discharge from then onwards would be undesirable, for at various times between 4 and 9 h after high water sewage would be expected to wash ashore or go upstream.

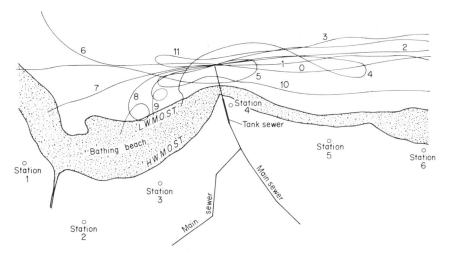

Figure 32   Typical result of tidal experiments

A complication not always appreciated—and which the behaviour of floats does not bring to light—is that while fresh water in the tidal reaches of a river and estuary flows out to sea at the surface, salt water, being heavier than fresh water, flows from the sea upstream at the bottom, eventually mingling with the fresh water. This will carry sewage upstream from an outfall that discharges near the bottom of the stream.

The behaviour of floats indicates reasonably the speed and direction of the water near the surface, apart from their being influenced to a small extent by the wind. They do not, however, give any indication of what else is happening. When crude or partially treated sewage is discharged it is possible that a sludge bank will form at the end of the outfall unless there is a scouring current that carries the sludge away. Observation from a high point will show a greasy track from the outfall which indicates the surface flow: this shows not only the direction which floats might be expected to take at the same state of the tide, but also the spread of the sewage flow from a point to a wide stream.

When it is considered that discharge of crude or partially separate sewage is reasonable, the engineer has to study tides and currents so as to determine the most suitable place of outfall and the times at which discharge may take place. This involves float experiments by which the direction and speed of currents may be determined, and investigation of the limits of tital ranges and the form of the tidal curve.

Ordinary neap and spring tides and maximum and minimum recorded tides may be obtained from tide tables; e.g. the Admiralty Tide Tables available in Britain. In the absence of records of the kind required it is necessary for an automatic tide recorder to be installed in a position where its float will not be unduly disturbed by wave action and where its readings will be truly representative. When records have been taken for a sufficiently long period, say from the spring equinox to the end of June, the results may be studied and average curves drawn for ordinary spring and neap tides.

The rise and fall of the surface of the sea is known as the tide, and correctly the term should be applied only to the rise and fall. The lateral flow that occurs at the same time is the tidal stream or current. The terms flood and ebb refer to motion of water towards and from the land, and in and out of harbours, and the time when there is no horizontal motion is known as slack. Set and drift are terms denoting the direction and velocity of tidal streams. At all times of new and full moons spring tides occur; at first and last quarters of the moon neap tides occur.

It is usual to base sea outfall calculations on ordinary spring tides, i.e. the tide that has a range from the low water mark to the high water mark as shown on maps, but extraordinarily high tides have to be kept in mind because of the possibility of their preventing discharge of sewers: therefore highest recorded tides should be shown on drawings, together with ordinary spring high and low water levels.

Tides are considerably affected by wind; therefore when records have to be taken they should preferably be taken over a long period, together with notes of the wind and barometric pressures, preferably obtained from a local observatory.

In Britain the heights of tide are usually referred to the sill level of a local dock, the particular dock being stated. If the level of the sill above or below Newlyn datum is not mentioned, this will have to be ascertained in order that the position of Newlyn datum relative to the tidal range may be shown on the drawings. This will be somewhere near the centre of the tidal range, but mean sea level is not generally the same as Newlyn datum and may be considerably different.

The direction and speed of currents are determined by plotting the movements of floats. The floats for tidal experiments are of two kinds: those that float deep showing the general trend of the currents a short distance below the surface, and those that float on the surface indicating the direction likely to be taken by floating particles under the influence of both wind and tide. The floats need not be elaborate. They are satisfactory if they float deep without sinking and are easily visible. For deep floats, billets of wood about 2.5 m long and painted bright orange are weighted at one end so that they float upright with about 450 mm out of the water. For the shallow floats, billets about 600 mm long may be used. Spare floats should be made in case any become waterlogged and sink.

Float experiments may be carried out with the aid of both deep and shallow floats, one of each being put in the water at the same time in order that the influence of the wind may be estimated. More frequently only deep floats are used.

It is necessary for the direction of the currents at all states of the tide to be known in order that the direction of the flow of sewage discharged at any time after high water may be determined. A full investigation into the directions of flow that sewage may take if released at all states of the tide requires considerable time for its completion. Floats should be placed in the water at regular intervals after high tide and the conditions should be known for both spring and neap tides, and intermediate tides in all weathers. Practical considerations frequently reduce the scope of tidal experiments to work that can be carried out at reasonable cost in a period of a few weeks.

A procedure somewhat on the following lines may be adopted. One or more floats are placed in the water at the proposed point of outfall, at high tide or an exact period in hours after high tide, and followed in a boat, their positions being recorded every quarter of an hour until they wash ashore, or until they are obviously drifting out to sea and practically certain not to return shorewards. Opinion varies as to the period of time that the floats should remain in the water if they have not washed ashore, or the distance they should have travelled before their progress is no longer of interest. The importance of distance travelled depends very much on direction relative to beaches, etc., the treatment proposed, and the relative importance of protecting beaches.

Having followed the floats for the prescribed time and distance, or until they have gone ashore or out to sea, the surveyor takes them out of the water and returns to the proposed point of outfall. He puts the floats back into the sea again at an exact period in hours after high water. This work is continued day after day until the behaviour of the floats when put into the water at hourly intervals at all times after high tide has been recorded. More frequently $\frac{1}{2}$h and $\frac{1}{4}$h intervals are used. Work of the last class takes about a lunar month to complete in good weather.

The course of the floats may be plotted by theodolite readings taken simultaneously from two points on the shore, but more often the surveyor approaches the float in a boat and takes nautical-sextant readings on three points of known position on the shore as follows:

Referring to Figure 33, S is the position of the surveyor's boat; P1, P2, and P3 are three points of known position ashore. The surveyor reads with the

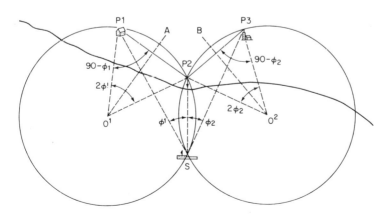

Figure 33   Construction of three-point problem

Table 30   Float experiments log book

Date	Time	Min after high water	Float no.	Stations Left	Stations Right	φ^1 (left)	Stations Left	Stations Right	φ^2 (right)	Remarks (weather, wind, etc.)
5 April 1973	2.31 p.m.	60	2	P1	P2	32°31′45″	P2	P3	28°4′30″	no wind
						etc. etc.				

nautical sextant the angles φ^1 and φ^2 and books them against time, stating also the points on which they were read. The booking would be as in Table 30.

When the survey is completed in the field, it is plotted on a map on which have been indicated the various shore stations, P1, P2, P3, etc. The various positions of the floats may be plotted with the aid of a station finder. In the absence of this instrument the two angles φ^1 and φ^2 may be drawn on a piece of tracing paper which may then be used as a station finder. It will be found that by moving the station finder on tracing paper over the map on which the stations are plotted there is only one position in which the three lines drawn on the tracing paper will touch the three points on the plan, except when the two circles referred to in the three-point problem coincide. In that case it is not possible by any means to determine the point S (but such coincidence is extremely rare.)

This method is sufficiently accurate for most purposes, but if desired, the positions may be calculated by the three-point problem in the following manner:

Let P1, P2, and P3 be three points on the land observed from the surveyor's boat, and let S stand for the surveyor's boat. The angles observed are P1, S, P2, and P2, S, P3, φ^1 and φ^2 respectively. Bisect P1, P2 and P2, P3 by perpendiculars A and B. These will pass through the centres of circles as shown on the diagram. At P1 or P2 by means of protractor draw angle of 90° minus φ^1, and the line will pass through the centre of the circle which cuts P1, P2, S because angle P1, φ^1, P2 is equal to twice φ^1. Similarly find the centre of the circle which passes through P2, P3, S. The intersection of the two circles occurs at P2 and S, thereby the position of S is found. A number of different cases occur, of which one is shown in Figure 33.

When the positions of all the floats at all times have been plotted, and the times marked against them on the map, lines may be drawn showing the courses taken by each float and each line may be distinctly marked with the time after high tide at which the float was placed in the water. It is then easy to see at what states of the tide sewage is likely to wash ashore or, alternatively, be carried out to sea, and the engineer may decide from the results of the experiments the times at which sewage may be discharged.

Soundings

Although in many cases useful information will be available from navigational charts, these will almost certainly have to be supplemented by systematic soundings close inshore, wherever an outfall pipe is likely to be located. This information is usually essential, so as to obtain some check on sand and shingle movement, and may well have to be repeated to check the effects of winter conditions, or even particularly severe storms on the area in question. Success in the design and operation of the new outfall will be just as dependent on this information as it will be on the vagaries of the tides and currents as determined by float tests.

The equipment available for such work becomes ever more sophisticated, but the combination of echo sounding and location by fixing systems such as the Decca navigational system can be operated in boats down to the size of small fishing smacks. Information on the practical use of such equipment is available in the *Admiralty Manual of Hydrographic Surveying.*[12] As pointed out, these sounders measure time, not depth, and their accuracy depends on the velocity of sound in water. Thus pulse frequency, duration, and shape are important variables. Different pulse transmission is required for deep-water soundings, but a damped pulse, in which the oscillations ascend and then descend, is most generally used. The undamped pulse becomes very necessary at depths over 2500 fathoms. In all cases, reflectors are used to concentrate the echo beam. Although it is possible to time the echo pulses by stopwatch, the instrument normally includes its own recorder, producing a trace of the sea bed according to the movement of a stylus, activated by the electronic circuits, and leaving a trace on sensitized paper. The movement of the stylus, driven by an accurately governed electric motor, depends on the relation between the speed of revolution of the dial to which the stylus is attached, and this in turn depends for accuracy on the actual speed of sound through the water. Adjustment up to a maximum of about 10 m is available in most instruments. Likewise, if radar is used for fixing, normal errors in radar ranges obtained by a moving vessel may be as much as 10 or 15 m, and this should be taken into account when deciding the scale on which the survey is to be plotted. The speed error also should always be found and allowed for. The performance of the set should be checked each day before starting work by comparing with fixes obtained by sextant. The use of radar reflectors, such as the corner reflector, the Luneberg lens, or more recently by the use of transponders, enables radar fixing to be applied to soundings. The index error of a transponder, which may be as much as 100 m, should be checked from time to time to make sure it remains constant. Another possible hazard is the skywave effect, which can only be avoided by limiting sounding times to 'Decca daylight', which is free of any morning and evening skywave affects. Experience has also shown that the presence of drying banks or a large expanse of tidal foreshore in a transmission path can cause irregularities in the readings.

As a variant to hyperbolic and two-range Decca, the Air Fix system can be used, with light portable equipment and masts not above 10 m in height. There is also Hydrodist, a development of the tellurometer which fixes by means of two time measurements automatically converted into spheroidal distances.

Sewerage systems for coastal towns

The sewerage systems selected for coastal towns depend on the number and location of outfalls and the necessity for treatment of sanitary sewage. In a case where all sewage may be discharged into the sea through an outfall pipe without any treatment other than screening and disintegration followed by storage or sedimentation, and the whole of the town may be easily drained to one outfall, the combined system is indicated. When a town straggles along the coast and there is only one suitable place for discharge of foul sewage, or where more complete treatment is advisable, the separate system may be preferable. In the latter case the main sanitary sewer would probably follow a line more or less parallel with the coast, connecting all parts of the town to the treatment works or point of outfall, while the surface water would be discharged at many points on the foreshore, into streams or, where possible, to soakaways. Surface-water sewers can in many cases discharge direct on to the foreshore, or through shorter outfall pipes.

In most cases some point may be found on the coast to which the sewage may gravitate, but this will not always coincide with the point at which discharge is satisfactory, and sometimes the natural fall of the land is not towards the coast. In the last case, inland disposal and full treatment may be preferable to a scheme involving an excessively expensive outfall together with the cost of pumping.

In those parts of the country where the land adjoining the sea is low-lying and flat, pumping schemes, sometimes involving a number of small stations, become necessary. The point and method of discharge may be determined not by the fall of the land but by other characteristics of the site, with the result that a number of alternative locations of outfall may be possible.

Discharge of sea outfalls

In cases where it is possible for sewage to be discharged continuously at all states of the tide the sea outfall has to be of sufficient size to pass the maximum rate of flow at the time of highest tide. An outfall is often a surcharged pipe and, in nearly all cases, flow calculations are based on the hydraulic gradient, not on the gradient at which the outfall is laid. In those calculations allowance should be made for the difference of specific gravity of sewage and sea water and the hydraulic gradient expressed

$$\text{Length divided by fall} = \frac{L}{H - (h/36)} \qquad (24)$$

where: H = height of maximum sewage level in tide flap chamber above highest sea level, in metres;

h = height in metres of highest tide level above centre of bottom end of outfall pipe;

L = length of outfall in metres.

Tank sewers and storage tanks

When it is only permissible to discharge sewage at certain states of the tide, for example in cases in which sewage is liable to wash ashore except for a few hours after high tide, storage must be provided. Storage is also necessary when high tide rises above the crown of the main outfall sewer and causes tide lock. In such cases either a tank sewer, or else a storage tank somewhat similar in form to a reservoir, is constructed. In the case of tide lock an alternative is to pump the flow.

The discharge of tank sewers and storage tanks involves mathematical problems of varying complexity. The simplest example is that in which a rectangular tank is emptied by a pump discharging at a practically constant rate. The most complex case is that of a tank discharging into the sea, the rate of discharge being influenced by the rise and fall of the tide and by the flow of sewage into the tank at the same time. A mathematical solution to this case has never been found.

Where storm water has to be pumped against continuous tide lock the problem is simple, and formulae (15), (16) or (17), see pp. 91–2, may be used as appropriate. But where a calculation has to be made to find the time to empty a tank by gravity when the flow of sewage into the tank varies according to storm flow or dry-weather flow, altering the head in the tank, and the level of the sea is rising and falling according to the tide and, therefore, also altering the hydraulic gradient from time to time, a straight mathematical solution is not practicable. What the engineer then has to do is to find the hydraulic gradient when the tank is full and the tide is at the highest; calculate the flow through the outfall at this hydraulic gradient; and then, by deducting from the rate of outflow the rate of inflow of sewage, find the new top-water level in the tank after 10 min of flow. With this new level in the tank and new tide level after 10 min, a new hydraulic gradient is found and, on this, the calculation is repeated to find the top-water level in the tank after the next 10 min. This calculation is repeated again and again until the tank is theoretically empty. The use of computers will, of course, make these operations much simpler and quicker.

In the case of a proposed new outfall it is necessary, in the first place, to guess at the size of outfall required and the calculation is then made on the basis of this guess. If the outfall proves to be either too large or too small it must be changed, and the whole calculation has to be repeated as many times as necessary to determine the required diameter of the outfall pipe.

The curve of the tide is theoretically a sine curve but actual tide curves vary considerably from this. The rise and fall of the tide in open ocean is very little;

merely some 600–900 mm. But in shallow seas the affect of diminishing velocity is to increase head and cause a build-up of very high tide. For example, at Bristol, England, the tidal range is about 15 m. Double tides occur at Southampton. Tides building up in large river estuaries cause tidal wave or bore. As has been mentioned, there are places without tides of sufficient magnitude to need consideration, including Lake Michigan, which has a tidal range of about 50 mm, and the Mediterranean and the Baltic, which have tides in the range of 300–600 mm.

Construction of sea outfalls

Sea outfalls for sanitary sewage are expensive to construct because they usually have to be carried to at least low water level ordinary spring tide and therefore have to be constructed in or on the foreshore and below high tide level. Work has to be carried out intermittently, in coffer-dam, or with the aid of divers, often at the risk of damage and loss.

Sea outfalls, being exposed to wave action, and having to be laid over sand and pebble beaches or in trench in rock, require to be constructed of the best available materials and be given adequate support. Where the foreshore consists of rock and it is possible for the outfall to be laid below ground level, the outfall may consist of pipes laid in trench cut in the rock, the trench being completely filled with concrete and surfaced with spalls of local material. If it is necessary for the outfall to be above ground level for part of its length, the trench should be continued and the pipe surrounded with concrete extending above the ground and also filling the trench (see Figures 34 and 35). If the rock is of a sufficiently sound nature a trench need not be cut but the pipe can be secured to the rock by means of straps and long rag bolts grouted in, and the whole covered with 300 mm or more of concrete.

Where the foreshore consists of loose material and the outfall is not completely buried, it may be necessary for cast-iron pipes to be used and securely held in position by timber or steel piles unless the outfall is constructed in a sheltered position (see Figure 36). For this class of work heavy cast-iron pipes with turned

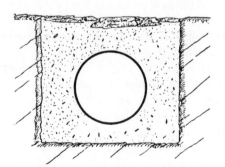

Figure 34   Sea outfall laid in concrete
on rock

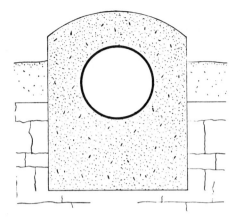

Figure 35   Sea outfall laid in concrete on rock

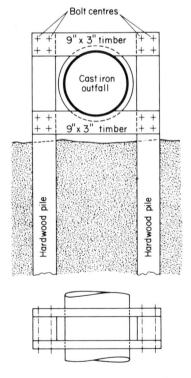

Figure 36   Sea outfall carried on piles

210

and bored joints are suitable. To aid laying cast-iron pipes, and to keep them in position after they have been laid, bolts in addition to the normal joints are advisable. Sometimes pipes with both flanges and turned and bored joints are used, but more commonly turned and bored pipes with three heavy lugs near spigot and socket are bolted together with three bolts, one at the top and one at each side. The side bolts should be high enough above invert to be accessible to a spanner (see Figure 37). Whenever an outfall is laid above foreshore level the possible groyne effect on littoral drift should be studied.

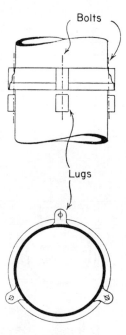

Figure 37 Joint for
cast iron outfall

Tidal flaps are provided at the top ends of sea outfalls to prevent the sea from entering the sewers. The end of an outfall has to be marked by a buoy or beacon, on which is fixed a sign made in accordance with the requirements of Trinity House (Britain) or other controlling authority.

While heavy works are necessary where outfalls are constructed in exposed positions in which they are liable to be damaged by storms or undermined by littoral drift, much more simple construction is possible in sheltered waters such as inside estuaries or natural harbours. In such cases it may be satisfactory merely to lay unprotected pipes on the bed of the sea or river. In some instances outfalls of steel or suitable plastics may be used. Steel outfall pipes may be jointed by welding and then the whole length rolled sideways into the sea, the ends being plugged in order that the outfall will float. The outlet end of the outfall is then towed out to the required position, when the plugs are removed and the outfall permitted to sink to the bottom. Plastic pipes may be laid at right angles to the coast and winched or towed out.

According to Snook[1] the ideal submarine pipeline will be heavy, strong, and flexible so as to resist drag or uplift forces when not buried, or prior to burial, to withstand heavy stresses both permanent and during construction, or due to accidental damage; also to take up bending due to permanent undulation of the sea bed and supports or buoyancy aids during construction. Torsional stresses may also be set up in construction, hoop stress by pressure testing and surge, or by external hydrostatic pressure during construction when empty under-water. Tensile stresses may also result from the foregoing causes.

As for materials of construction, Snook regards steel with internal and external protection as the most popular medium. Both internal and external coatings can be epoxy-based, the former with high anti-abrasive and anti-corrosive characteristics, the latter either an epoxy or bituminous enamel reinforced with fibreglass. In addition a reinforced concrete weight coating is required both for added protection and to resist the forces of drag and up-lift. Other metallic materials successfully used have been ductile iron, aluminium, and stainless steel. He also states that in certain circumstances a whole range of plastic materials have also been used vary successfully, particularly UPVC and HdP. The paper by Vestbo[13] describing the installation of flexible steel armoured polyethylene pipes, as long sea outfalls in the Bay of Naples and in Capri; although subject to size limitations and therefore restricted in capacity, may indicate 'the shape of things to come', in terms of construction and sub-sequent repair.

Several outfalls incorporating diffusers recently constructed in Great Britain, have involved tunnelling. One of these is at Brighton and Hove and is described by Osorio.[14] Figure 38 shows a longitudinal section of the outfall tunnel which

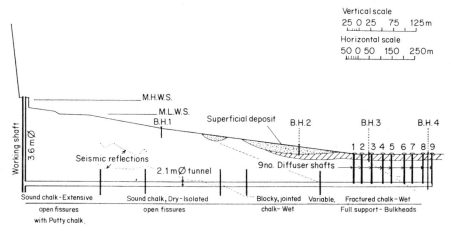

Figure 38   Brighton and Hove sea outfall—longitudinal section (*J. Instn. Public Hlth Engnrs.*)

212

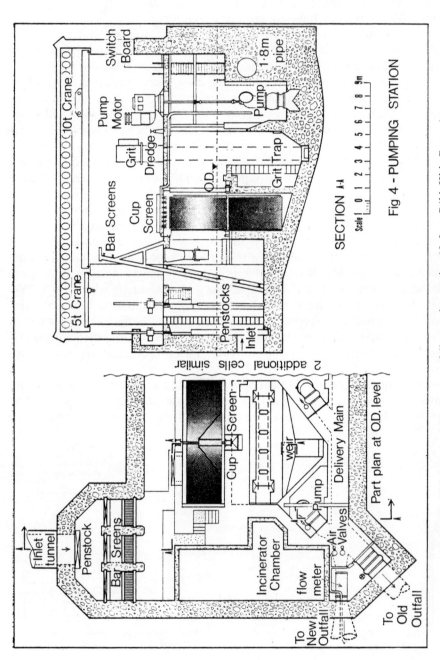

Figure 39 Brighton and Hove sea outfall pumping station (*J. Inst. Publ. Hlth. Engnrs.*)

is in fissured chalk and mostly lined with smooth-bore concrete segments. The diffusers terminate in octagonal concrete blocks above the sea bed each containing four 250 mm diameter ports connected to a mortar-lined 700 mm diameter steel riser by a swept double PVC tee wrapped in glass-reinforced polyester resin.

The pumping station, of which a part-plan and section are shown in Figure 39, also involved a large excavation in the chalk cliff. As with the tunnel construction cement grout was used to seal fissures and stabilize the ground. The pumping and screening plant was designed for a flow of 4000 l/s and the outfall and diffusers for 6800 l/s, or six times the anticipated average dry-weather flow. A novel feature of the design is the arrangement to de-water and incinerate screenings. Grit is removed by chain-and-bucket scrapers. The pumps are mixed-flow bowl pumps with variable-speed drives to ensure maximum dilution of effluent in the tidal current.

Outfall length

Although modern laying techniques have to some extent favoured the tendency of designers to regard length as the criterion for a satisfactory outfall, rapidly increasing costs have intensified the interest in preliminary investigations, particularly as regards dispersion of discharge, and observations must now cover considerably more than the evidence to be derived from surface floats. In Great Britain the decisions taken with regard to the location of atomic power stations, where discharge of large quantities of cooling water (with the ever-present risk of minor contamination with radioactive waste) made a carefully located outfall to the open sea one of the essentials, and thereafter the setting up of the eleven large water authorities in England and Wales, with powers to promote sewage treatment and disposal over wide areas, has led to a number of investigations. Some of these were directly sponsored by the Government, others by the Atomic Energy Authority, and others again by the water authorities, all concerned with the location of suitable points for discharge of waste waters to sea.

Techniques aimed at shorter outfalls as described by Crofts[15] are being employed by the Welsh Water Authority. Following the determination of tidal advection by drogue (or float) tracking, one or more moored, self-recording current meters are deployed to obtain information on tidal velocities, slack water times, variation of residual current with meteorological effects and, if the necessary sensors are fitted measuring temporal variation of salinity and temperature. Economy dictates the use of only one such meter, and this is usually located at the intended point of discharge. It should be deployed sufficiently below surface to minimize wave action at low water, but sufficiently above the bottom so that boundary effects are also minimized. The data so collected are analysed by computer to produce velocity analogue plots, percentage exceedance diagrams, progressive vector diagrams, tidal residual plots, scatter diagrams, etc.

In addition, vertical profiles of velocity, salinity, and temperature are obtained at various points throughout the survey area for varying durations, to provide detailed information on the general current structure, stratification, and extent of vertical or horizontal shear.

For diffusion effects a good deal of work has been done with fluorescein dye, although Smart and Laidlaw[16] have recently proved it to be poor as a quantitative tracer. Bacterial cultures[17] have also been used as tracers, as have short half-life radioactive isotopes. Although both these alternatives may be more sensitive than fluorescein, the first require sample analysis to indicate results, and the second is so hedged about with restrictions as to its use, that its application often becomes unnecessarily complex.

Jet discharge

The numerous instances of complaints arising from the discharge of outfalls, many of them of long standing, originally located without the evidence now obtained from the various forms of survey just described, has led to a particular study of the form and character of discharge from a pipe projecting into water, and the effect of its position, whether this be in a river, a tidal estuary, or the open sea. Apart from any possible undesirable scouring effects, discharge to rivers can be controlled by known dilution factors and prescribed standards of effluent. Estuarial discharge introduces the complication of ebb and flow, and for this and other reasons estuaries in general offer nothing like as good a dispersion factor as does the open sea. Professor Pearson[18] rates the dilution for treated municipal and industrial waste waters, discharged around the periphery of a bay or estuary, as near zero at the head of a natural estuary (advective flow $\simeq 0$), to 20–30 for mid-bay or estuary discharge, and 30–50 for a well-designed outfall in a large bay. This is because the actual dilution of waste waters in a given restricted body of water depends on the waste flow rate, the advective flow, and effective tidal/turbulent exchange. However, outfalls located along open coasts can be designed for any magnitude of discharge, to achieve immediate initial dilutions of the order of 150–200. As conventional secondary waste treatment systems achieve removals of 90–95% (equivalent to dilutions of 10–20), whilst primary treatment achieves a maximum of perhaps 65% (equivalent to a dilution of 3), thus a terminal dilution (removal plus terminal dilution) of 450–600 would be available from primary treatment plus open sea discharge whilst the estuarial discharge (secondary treatment plus estuarial discharge) would have to obtain 30–40 dilution to equal it!

As the conditions referred to for river discharges only apply in locations well above the tidal influence, it is clear that it behoves any town near the coast to weigh the cost of initial treatment against sea outfall and as, for various reasons, at least primary treatment will be given, discharge to sea probably presents the best ecological alternative, avoiding as it does the estuarine region, which is likely to be a critical area, bound to be affected by the residence time of pollutants. Moreover, if pollutants of unknown character occur, for which treatment may or may not be adequate, then sea discharge, which may well be in an area of non-critical or lesser ecological significance, affects substantial reduction in pollution concentration by dilution.

A great deal of literature has built up on jet discharge, since some of the original work was done on this subject by Rawn, Bowerman, and Brookes,[19,20] in the US. In Great Britain, Agg and Wakeford[21] have described field studies of jet dilution, and Agg and White[22] have considered devices ranging from the type illustrated in Figure 40 to multi-jet arrangements.

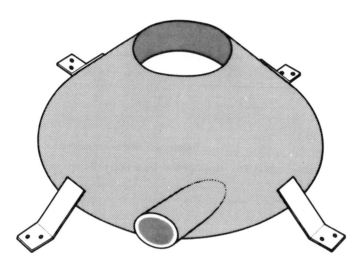

Figure 40   Pre-dilution device (*J. Instn. Public Hlth. Engnrs.*)

The device shown in Figure 40 was found to be the most efficient form of circular outlet capable of increasing initial dilution at water surface by a factor of about 2, for typical UK tidal situations. Dilution within the cone, with an inlet densimetric Froude No. of unity, is around 10. The outlet diameter is $0.9\sqrt{10} = 2.84$ m, and the emission of the mixed fluid is at a densimetric Froude No. of $1 \times 10^{0.25} = 1.78$. According to Brookes,[20] for low-density Froude numbers dispersion plume width in still water, just below the surface, is $w = 0.3y$, where y is the depth of submergence. Multi-jets together with improved methods of interpretation of preliminary tests, hold the prospect of much improved dilution with shorter length of outfall. Workers in this field include the Fisheries and Marine Service, Pacific Environment Institute, Canada, and Chalmers University of Technology, Goteburg, Sweden.

More recently, Snook[1] has presented a paper based on a long practical experience of sea outfall design and construction, under the title 'Submarine pipeline and diffuser design'. He refers to the EEC Directives concerning the standard of bathing waters to be achieved along the coasts of member countries, and points out that although some extension of time may be allowed for compliance so far as existing discharges are concerned, there seems no way in which a new outfall can do other than comply, in receiving waters where bathing is not prohibited, and is traditionally practiced. These standards are based on an *Escherichia coli* count of not more than 2000 per 100 ml in 95 % of samples taken, presumably in the bathing season only. Starting with raw sewage with *E. coli* 10^8/100 ml, this means reduction by a factor of 50,000, or 99.998 %. Full biological treatment should be capable of reducing *E. coli* by 90%, leaving 10^7/100 ml, to be reduced by a further 99.98 %, or a factor of 5000. Residual chlorination might be regarded as a possible solution to the problem, but there is evidence that chlorinated sewage discharged into the sea produces organo-halogens, which are prohibited from discharge into sea or rivers by a further EEC Directive![23,24] Snook is confident that a properly designed sea outfall and diffuser system is capable of meeting the requirements.

The discharge of finely disintegrated sewage into the sea is capable of providing dilution factors of several hundred on the surface—IDF (Initial Dilution Factor). As sewage/sea water mixture spreads out and away from the diffuser further dilution is achieved by eddy diffusion and density spread. The consensus of opinion is that the main factor contributing to *E. coli* mortality is the effect of ultraviolet light; thus the greater the IDF, the less turbid is the sewage/sea water mix, and the greater the attenuation of ultraviolet. A properly designed system will enable the reduction in *E. coli* to be assessed.

A jet of sewage rises to the surface by virtue of its kinetic and buoyant energy (see Figure 41). The turbulence and shear plane between the ambient sea water and the jet causes mixing processes—'jet diffusion'—as long as the column continues to rise it becomes increasingly diluted—the degree of dilution at the

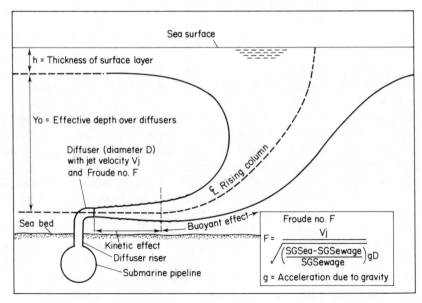

Figure 41   Jet issuing from a horizontal port (Courtesy W. G. Snook)

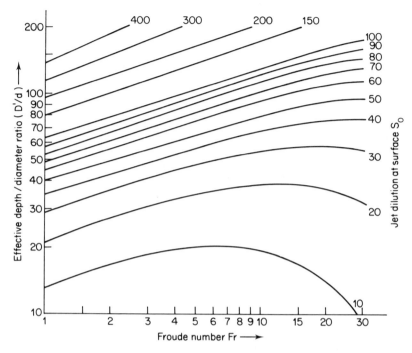

Figure 42   Jet dilution for horizontal discharge in still water (W. G. Snook)

head of a rising column is referred to as IDF. The degree of dilution is dependent on the Froude No. (F), relative to the discharge characteristics and the ratio Y_0/D. Many researchers have devised mathematical methods of calculating the IDF in still-water conditions. The resultant calculations can be expressed in graphical form as Figures 42 and 43 (after Abrams),[25] for horizontal and vertical discharge respectively. Figure 44, prepared by Snook and based on his own observations allied to the work of Cederwall,[26] and Hansen,[27] together with Figures 42 and 43, clearly show the relationship between IDF (S_0), effective depth (Y_0), diameter of jet (D), and the Froude No. (F), which, in turn is a function of the jet velocity (Vj), the relative specific gravity (SG) of the sea and sewage, and again, the jet diameter (D).

A diffuser system consists of a number of jets, so that for any given total flow (Q), the velocity is influenced by the number of ports (n)—Figure 44 shows influence lines for variation of the parameters Y_0, D, V, Q and n. Increasing Q only increases Vj and consequently F, and the total head required. Increasing Y_0 usually increases the length of the submarine pipeline between diffuser and coast. Decreasing D increases the factor Y_0/D and F. Combining an increase of n and a decrease of D usually produces the most cost-effective benefit insofar as increase in S_0 is concerned, and can be coupled with adjustments of the submerged pipeline, and consequently Y_0.

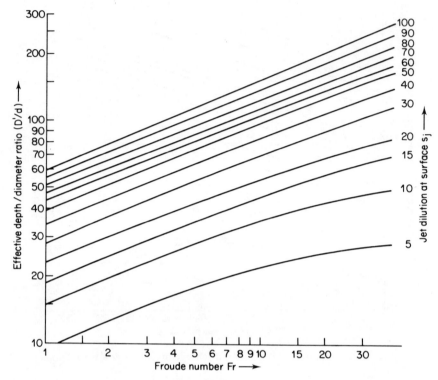

Figure 43   Jet dilution for vertical discharge in still water (W. G. Snook)

The grease content of the sewage must be taken into account in selecting the velocities, both in the pipeline and the jets, so that they are self-cleansing and also capable of re-agitating and suspending material already settled out. Velocity should also be sufficient to prevent build-up of grease, moreover the initial dilution factor (S_0), with grease present, should be greater than 50–70 to prevent slick formation.

Selection of minimum jet velocity is crucial—it should never be less than 1.5 m/s. Increase in velocity head is not great, varying from 0.11 m for $Vj = 1.5$ m to 1.83 m for $Vj = 6$ m. The extra cost of pumping may well be compensated for by reduction in number of diffuser ports necessary. Figure 45 shows a plot of jet diameters against Froude Nos. for various jet velocities. Port diameters usually range from 75 to 200 mm, with a tendency to 100–125 mm. For a minimum velocity, say, 2 m/s the Froude No. will vary between 9 and 14, and for 4 m/s between 18 and 30—often higher velocities and Froude Nos. are more economical and can give greater dilution factors, due to the increase in Y_0/D, as can be seen in Figures 46–51.

Much research has been conducted to determine the decay constant, and it has been shown that a 90% reduction in bacterial numbers can usually be achieved in 2–6 h. This period is known as T_{90} time. The T_{90} values, patterns

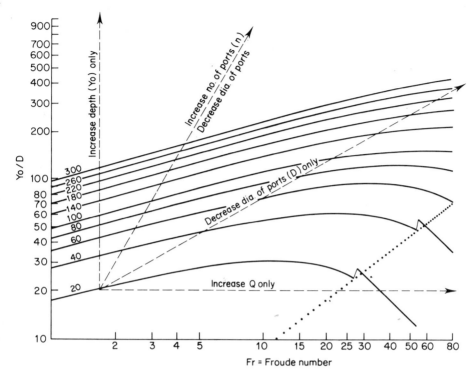

Figure 44   Jet dilution for horizontal discharge of three-dimensional jet in still water
(Courtesy W. G. Snook)

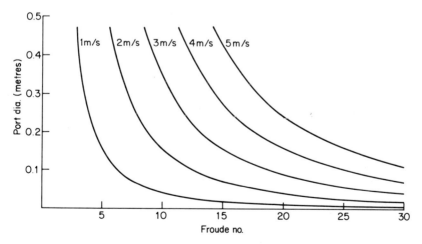

Figure 45   Graph showing relationship of port diameter to Froude No. for
given velocity (Courtesy W. G. Snook)

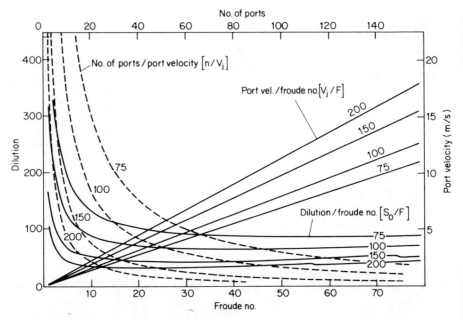

Figure 46   Graph showing relationship of dilution and port velocity to Froude No. and number of ports to port velocity for given port diameters (Courtesy W. G. Snook)

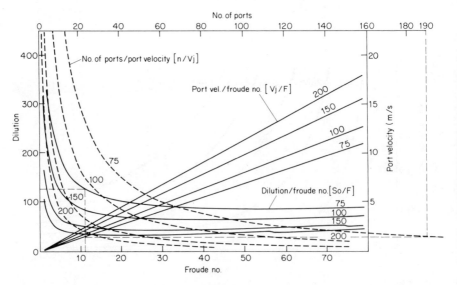

Figure 47   Graph showing relationship of dilution and port velocity to Froude No. and number of ports to port velocity for given port diameters (Courtesy W. G. Snook)

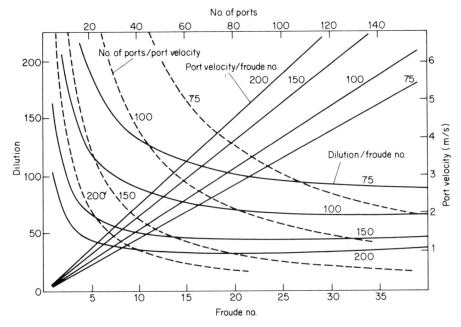

Figure 48 Graph showing relationship of dilution and port velocity to Froude No. and number of ports to port velocity for given port diameters (Courtesy W. G. Snook)

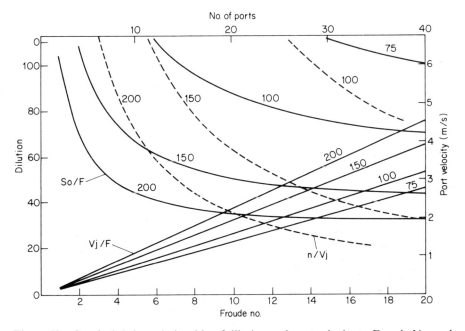

Figure 49 Graph showing relationship of dilution and port velocity to Froude No. and number of ports to port velocity for given port diameters (Courtesy W. G. Snook)

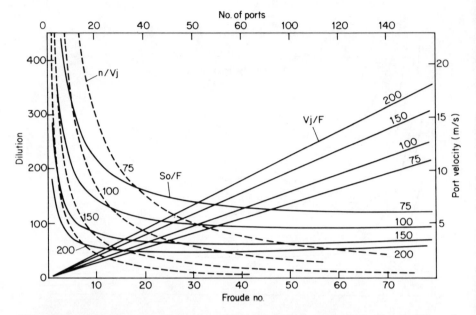

Figure 50   Graph showing relationship of dilution and port velocity to Froude No. and number of ports to port velocity for given port diameters (Courtesy W. G. Snook)

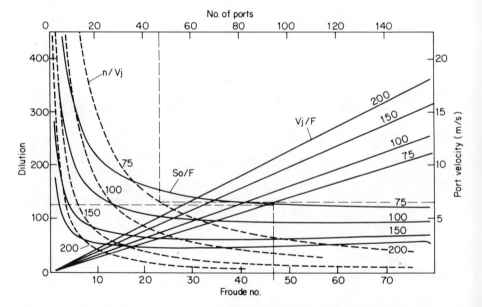

Figure 51   Graph showing relationship of dilution and port velocity to Froude No. and number of ports to port velocity for given port diameters (Courtesy W. G. Snook)

and slope of the sea bed, specific gravity gradients, and wind patterns, all affect the selection of initial dilution factor. This is usually calculated for low water at spring tides, when least Y_0 gives least S_0, and the flood tide, usually with a tendency to on-shore drift, is about to commence. Often dilution factors in excess of 100, as compared with only 70, for slick prevention, have to be achieved to satisfy other requirements.

Diffuser design, to achieve both initial dilution factors and port sizing and spacing, is very much a repetitive 'trial-and-error' process and calls for the services of the computer. Graphical aids, prepared by computer, assist in selection and final selection is made by computer. Examples of these graphical aids are shown in Figures 46–51, for a flow of 1.2 cumecs and a depth of 10–14 m, with a density gradient of 1.0262–1.0263. Curves of S_0, Vj and n are plotted against S_0, F, V_j and n for varying D. Each graph is relative to a fixed Y_0 and density gradient. Guidelines superimposed on Figures 45 and 49 for $S_0 = 125$, $D = 75$ mm, show that an extra 4 m of depth (Y_0) reduces the number of ports from 190 to 45, but increases V_j from 1.6 to 6.6 m/s. The corresponding velocity head increases from 0.13 m to 2.22 m.

Having determined the length of the diffuser, the effective depth from diffuser port to underside of surface layer can be determined. Knowing the variations in current speeds, the volume of sea water passing over the diffuser can be calculated, and when divided by total flow should exceed the dilution factor required. Snook contends that by downward inclination and baffling, the 'ideal' jet can be approached and dilutions in excess of 500 achieved.

Dilution and dispersal of solids

The proportion of suspended solids in raw domestic sewage averages 250 ppm (1/4000). Snook contends that provided these are extracted from the main flow, and very finely disintegrated, before being passed back into the discharge in the same proportion as they arrived in the in-flow, they will be subjected to the same diluting processes as the liquid portion—a diagrammatic layout of a typical pre-treatment arrangement is shown on Figure 52. Assuming an S_0 of 100, the suspended solids are reduced to a level of approximately 2.5 ppm immediately over the diffuser. Fine dispersion of faecal matter releases *E. coli* colonies but, in turn, subjects the additional *E. coli* to the batericidal effect of ultraviolet light, thus killing them off more rapidly than if they had been contained in undis-integrated solid matter.

He further contends that intermittent discharge with the longest possible intervals between discharges produces 'puffs' of sewage/sea water with complete peripheries, thereby greatly enhancing eddy diffusion, as compared with continuous discharge from a line diffuser. This implies more balancing capacity, as shown in Figure 52.

A highly diluted sea water/sewage mixture rising with a distinct velocity gradient from bottom to surface will cease to rise and become 'trapped', when the density of the mixture becomes more than that of the sea. The submerged

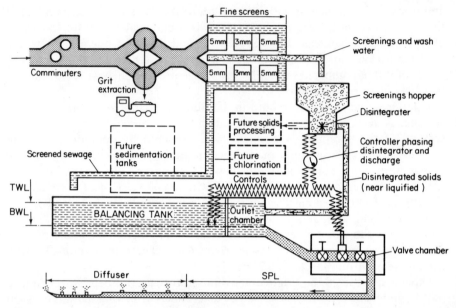

Figure 52   Diagrammatic layout of typical pre-treatment arrangements (Courtesy W. G. Snook)

fields so created are more common in warmer climates, but can occur in the UK, particularly during the holiday months, or due to reduced salinity caused by river water. It is essential to avoid the welling-up of a submerged field near a bathing beach.

The effect of 'swell' on low set velocity/low Froude No. discharges from a diffuser can cause 'pumping', and ingress of sea water, due to the differential head between peak and trough of the swell. This process will occur during periods of no discharge, but provided the main bore velocity and V_j have been properly selected any sea water, or material entering the system, would be rapidly evacuated without serious effect on performance. With high jet velocities and Froude Nos. a single jet diameter may well be adopted for all ports, without detriment to the hydraulic and dilution characteristics of the system. The densimetric head differential for diffusers laid on an inclined sea bed has little effect provided the slope is not too severe.

Snook recommends that the seaward length of diffuser main bore should be not less than 200 mm diameter, having been reduced progressively so as to maintain a selected velocity, to prevent settlement and provide re-agitation of material that has settled during the period between discharges. Termination should be in a hinged flange containing the final port, to facilitate flushing, which is achieved by blanking-off certain of the diffuser ports and opening the hinged flange (see Figure 53). A diffuser port should precede each reduction in main bore diameter and should be attached to an inspection hatch, to be used in the unlikely event of blockage. Main bore reducers should be 'level-invert', and diffuser port

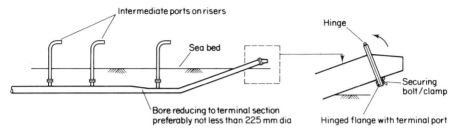

Figure 53 Diagrammatic representation of a typical hinged terminal port (Courtesy W. G. Snook)

fittings flanged for ease in removal, and temporary insertion of blank flanges. In the diffusers themselves Snook prefers not to use nozzles to keep up velocity. By terminating the risers in 90° bends, not only is horizontal discharge achieved —thereby securing maximum spread—but the outlets can be turned so as to discharge in a direction normal to the prevailing current direction. Jet spacings should be based on calculations of discharge 'spread' so that adjacent columns do not interfere or merge even on the highest tides. For protection of jets and risers against possible damage by trawl nets and anchors, domed pre-cast concrete 'igloos'—supported on pre-cast concrete walls—may have to be used, but where such risks are slight simpler devices in the form of reinforced concrete, or coated steel 'bedsteads' are much more economical.

Much attention has been paid to improving discharge factors by means of various devices located in the discharge ports.[28] Snook considers that a 45° wedge plate situated two jet diameters from the orifice will approximately double the dilution factor. Such devices are likely to be seriously affected by the presence of suspended or moving debris (seaweed) but if incorporated inside the 'door' of an 'igloo' this problem is overcome.

Construction of storage tanks and tank sewers

Tanks for storage of sewage over periods during which discharge is not permissible may be in the form of tank sewers or rectangular reservoirs.

Tank sewers combine the purpose of storage with that of conveying sewage to the point of discharge. They are merely sewers of large diameter so constructed that they will store the quantity calculated and at the same time remain free from deposit when emptied. The latter requirement is satisfied by sloping the invert steeply to a central channel or grip, which must have a gradient towards the outlet sufficient to ensure a self-cleansing velocity.

Figure 54 illustrates the details of a tank sewer. At the top end of the tank sewer the incoming sewer enters with its crown level with that of the tank sewer, but the sewage is carried down to the invert by means of a cast-iron backdrop. Access is given to the tank sewer by side-entrance manholes constructed at intervals not exceeding 110 m. At the bottom end the tank sewer discharges to the sea through one or more outfall pipes.

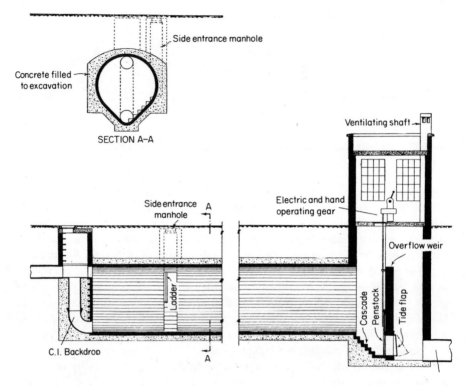

Figure 54   Tank sewer

At the top end of the outfall a chamber is constructed in which a tidal flap is installed, the purpose of which is to prevent backflow from the sea into the sewerage system or the partial filling of the tank sewer with sea water, reducing its storage capacity for sewage. The tidal flap is usually of heavy cast iron, hung on gun-metal or other non-corrodable hinges, having machined gun-metal faces and, if of large size, being divided into segments. Tidal flaps are usually placed at the top end of the outfall, very rarely at the bottom end. If tidal flaps are placed at the bottom end the outfall may become empty of sewage in certain circumstances and, as already mentioned, on the rising of the tide it may float and break up unless securely anchored down. Also, a tidal flap in such an exposed position would be very liable to damage or to become buried and in such a position it would be very difficult to repair.

Between the tidal flap and the lower end of the tank sewer is a penstock by which the outlet end of the tank sewer is closed, preventing discharge to the sea at certain states of the tide. This penstock is sometimes manually operated but electrical operation, in addition to manual operation, can effect economy in labour, if the motors are clock- or float-controlled. In the case of clock control, an electrically wound or other clock is arranged so that it will start the motors that open the penstock at regular intervals of 12 h 26 min and similarly close the penstock at the correct time after opening. With this arrangement it would only be necessary for men regularly to test, adjust, and inspect the clock and operating gear so many times a week, instead of labouring on hand-operated gear four times a day, often at inconvenient hours. Float control by means of the motion of the tide itself is possible, but presents more difficulties than clock control; for this reason the latter system is preferred. Auxiliary hand operation should always be provided in case of breakdown.

Between the tank sewer and the tidal flap chamber there should be a weir provided to permit overflow in case of the flow exceeding the storage capacity of the tank sewer or in case of failure of the penstock to open. This weir should be at or below crown level of the incoming sewer but above highest recorded tide.

Tank sewers should always be adequately ventilated, but vent columns on the sea coast are not attractive in appearance and therefore the means of ventilation should be disguised for example, as shown in Figure 54, or by fitting a topmast and yard to a steel ventilating column, giving it the appearance of a flagstaff.

Rectangular storage tanks, like tank sewers, must be self-cleansing unless they are to serve as sedimentation tanks for the removal of sludge. One of the best ways of ensuring self-cleansing conditions is to construct the rectangular tank as if it were a tank sewer folded backwards and forwards against itself. In this case there would be a number of parallel compartments; the flow would enter one and, passing down it, discharge into the next, return back towards the inlet end to discharge into the next compartment and so on until finally discharge would take place through the outfall. Tidal flap, penstock, and ventilating arrangements should be provided as for a tank sewer.

As far as possible storage tanks should be sufficiently inland to be protected from the action of the sea. In places where coast erosion is taking place they should be well behind the sea wall. In some circumstances it is unavoidable for the outer wall of the storage tank to serve as sea wall. In such a case the construction should be no less heavy than the heaviest construction used for sea walls in the area concerned. The engineer should ascertain local practice and study the design of existing works, because variation in the details of works required for coast defence depends largely on differences in local conditions. Advice of specialist consultants may be desirable.

Sea walls are often constructed of very heavy masonry consisting of the largest stones available bedded with close cement joints in order that the sea may not loosen and break the wall down piece by piece. Stone-faced concrete, mass, and reinforced concrete are also used, also less expensive materials where the action of the sea is not severe. Economy in initial cost is not good policy where seas are heavy: it pays in the long run to construct the strongest works and those which least often require replacement.

Sea walls are battered, curved, or stepped in order that they will not receive the full force of the sea. At the top they may turn outwards or be provided with a cornice to throw back the waves and spray. The foundations of the walls should be very deep and provided with protective aprons or sheet piles so that they do not become undermined by scour.

The height of waves can be estimated by the formula where:

$$H = 0.32\sqrt{L}; \tag{25}$$
H = height in metres;
L = width of sea in kilometres (i.e. distance coast to coast).

The construction of sea walls may add to scour, for which reason groynes may also be required. In the matter of construction of groynes, again local knowledge is of the utmost importance, for they will function satisfactorily only when there are sufficient suspended materials travelling along the coast for the groynes to cause accretion.

Surface-water sewerage for coastal towns

Some or all of the surface water at a coastal town may be discharged direct to the sea, surface-water sewers terminating with tidal flaps at the face of the sea wall or some other suitable points and discharging on to the foreshore. Where the subsoil is suitable for the use of soakaways, soakaway drainage is common. In those circumstances where sanitary sewage that has not been fully treated is to be discharged to the sea at certain states of the tide only, combined sewerage adds to the problem, for if the storage tanks are not to spill over at any state of the tide during storm time, they need to be of large capacity.

References

1. Snook, W. G. G. Submarine pipeline and diffuser design. IPHE. Wales District Centre, Symposium on Long Sea Outfalls for Sewage Disposal, Cardiff (1979), Instn. Public Hlth. Engnrs.
2. Dep. of the Environmt. and Welsh Office, *Report of a Survey of the Discharge of Foul Sewage to the Coastal Waters of England and Wales.* HMSO, London (1973).
3. Comm. of the Eur. Communities. Coun. Directive of 8 Dec. 1975, concerning the quality of bathing water. Official Journal No. L31 (5 Feb. 1976).
4. Commn. of the European Communities. Council Directive of 4 May 1976, on pollution caused by certain dangerous substances discharged into the aquatic environment of the community. Official J. No. L129 (18 May 1976).
5. Wood, P. C. The discharge of sewage from sea outfalls into the North Sea. Int. Symp., *Discharge of Sewage from Sea Outfalls* (Londn, Aug.–Sept. 1974), Pergamon Press.
6. Aubert, J. and Breittmeyer, J.-Ph. The Mediterranean Int. Symp., *Discharge of Sewage from Sea Outfalls* (Londn Aug.–Sept. 1974), Pergamon Press.
7. Svensson, A. Some problems in the Baltic. Int. Symp., *Discharge of Sewage from Sea Outfalls* (Londn, Aug.–Sept. 1974), Pergamon Press.
8. Berg, G. Regional problems with sea outfalls, disposal of sewage on the coasts of the US. Int. Symp., *Discharge of Sewage from Sea Outfalls* (Londn, Aug.–Sept. 1974), Pergamon Press.
9. Birgman, J. I. and Beeland, G. The Water Pollution Control Act (as amended in 1972, P.L. 92-500). *Handb. of Wat. Resources and Pollution Control*, chap. 19. Van Nostrand Reinhold Co., N.Y., 1976.

10. Warren, A. K. and Rawn, A. M. Disposal of sewage into the Pacific Ocean. *Modern Sewage Disposal* (1938).
11. Weston, A. D. Disposal of sewage into the Atlantic Ocean. *Modern Sewage Disposal* (1938).
12. *Admiralty Manual of Hydrographic Surveying.* The Hydrographer of the Navy, Taunton, Somerset, England (1969).
13. Vestbo, H. Flexible steel armoured polyethylene pipes used as submarine pipes. Symposium, *Long Sea Outfalls for Sewage Disposal*, Cardiff (1979). Instn. Public Hlth. Engnrs.
14. Osorio, J. D. C. The design and construction of a screening and pumping station and an outfall for the disposal of sewage from Brighton and Hove. *The Public Health Engineer*, **4** (5) (Sept. 1976).
15. Crofts, I. P. Practical guidelines for the evaluation of outfall lengths. Instn. of Public Hlth. Engnrs. (Wales District Centre). Symp., *Long Sea Outfalls for Sewage Disposal*, Univ. of Wales Inst. of Sci. and Technol., Cardiff (Mar. 1979).
16. Smart, P. L. and Laidlaw, I. An evaluation of some fluorescent dyes for water tracing, *Water Resources Res.*, **13**, 1 (Feb. 1977).
17. Pike, E. B., Bufton, A. J. W. and Gould, D. J. The use of *Serratia indica* and *Bacillus subtilis* var. *niger* spores for tracing sewage dispersion in the sea. *J. Appl. Bact.* **32**, 206–216 (1969).
18. Pearson, E. A. Conceptual design of marine waste disposal systems. Int. Symp., *Discharge of Sewage from Sea Outfalls* (Aug.–Sept. 1974), Pergamon Press.
19. Rawn, A. M., Bowerman, F. R. and Brookes, N. H. Diffusers for disposal of sewage in sea water. *J. Sanit. Engg. Div. Am. Soc. Civ. Engrs.* **86** (SA2), 65 (1960).
20. Brookes, N. H. Diffusion of sewage effluent in an ocean current, Envir. Proc. Ist. Int. Conf. Waste Disposal in the Marine Envir. Univ. Calif., Pergamon Press, 1960, p. 246.
21. Agg, A. R. and Wakeford, A. C. Field studies of jet dilution of sewage outfalls. J. Instn. Public Hlth. Engnrs. **71**, 126 (1972).
22. White, W. R. and Agg, A. R. 'Outlet design'. Paper No. 27, Int. Symp., *Discharge of Sewage from Sea Outfalls* (Aug.–Sept. 1974), Pergamon Press.
23. Commn. of the Eur. Communities, Coun. Directive of 8 Dec 1975, Concerning the quality of bathing water. Official J. No. L31 (5 Feb. 1976).
24. Commn. of the Eur. Communities Coun. Directive of 4 May 1976, Pollution caused by certain dangerous substances discharged into the aquatic environment of the community. Official J. No. L129 (18 May 1976).
25. Abrams, G. *Jet Diffusion in Stagnant Ambient Fluid*, Delft Hyd. Lab. Pub. No. 29 (July 1963).
26. Cederwall, K. 'The initial mixing of jet disposal into a recipient', Lab. investigations, Chalmers Inst. of Technology, 1963.
27. Hanson, J. and Schroder, II. *Horizontal Jet Dilution Studies by Use of Radio-active Isotopes, Acta Polytech. Scandinavica*, Copenhagen (1968).
28. Agg, A. R. and White, W. R. Devices for the predilution of sewage at sea outfalls. *Proc. Instn. Civ. Engnrs.* Part 2 **57**, 1 (1974).

CHAPTER 9
The Nature of Sewage

Sewage is a dilute mixture of domestic waste, trade waste, infiltration from the subsoil, and in greater or lesser degree, runoff of surface water. While the organic content is very small, being usually less than one part in a thousand, it is sufficient to case serious pollution of any watercourse to which it is discharged unless the latter is of very great size in proportion to the flow of sewage. This is because the organic content becomes decomposed by micro-organisms and the oxygen taken up in this process depletes the dissolved oxygen content of the water and may cause offensive conditions.

According to the Fifth Report of the Royal Commission on Sewage Disposal,[1] crude sewage may be discharged into a stream of sufficient size to dilute it by 500 volumes, subject to such conditions as to the provision of screens or detritus tanks that might appear to be necessary. In Great Britain generally, rivers and streams are not of sizes sufficient to give this degree of dilution, and all sewage discharged into inland waters needs to be given treatment so as to produce an effluent up to at least Royal Commission standard, (see p. 232).

The 1970s saw developments which put a new complexion on sewage treatment, in the sense that in England and Wales the responsibilities so long resting on local authorities to construct works so as to avoid discharge into a watercourse of any effluent below a standard acceptable to the local River Board, or Authority—whose inspectors performed the role of vigilantes to ensure compliance—passed to eleven large Water Authorities who are responsible both for treating the sewage and maintaining the standards of purity in rivers and streams. They inherited from the local authorities hundreds of treatment works of all sizes and conditions, most of them originally intended to produce effluents of Royal Commission standard, but many, for one reason or another failing to do so, and some, on the other hand, producing effluents well above that standard. With the economic climate unfavourable, there has been little evidence so far of the construction of large-scale treatment works to replace the majority of those for which they assumed responsibility.

A few quite large and modern works have been built, but the general strategy has been towards making the best of what is available, with attention, only to major shortcomings in the short-term, to be followed by major alterations and improvements in the long term. In Scotland, although reorganization of local authorities has taken place, the river purification boards have managed to retain their independance.

In America, where conditions in many places could be considered more favourable in terms of the dilution factor available in rivers, etc., there has been local abuse of such conditions on a major scale in some places, and these, coupled with the growing interest in environmental pollution, produced the Amended Water Pollution Control Act (PL 92-500) of 1972, and the setting up of the Environmental Protection Agency (EPA), which has the job of implementing its provisions. Contrary to the uncertain financial position in the United Kingdom, PL 92-500 promises 75 % State aid in bringing sewerage and sewage disposal to the standards required, and to up-grade all sewage treatment to what is known there as 'Secondary Standard' by 1983. The progress being made, according to official reports of the EPA, is illustrated in Table 31. It will be seen that, undoubtedly due to the demands outstripping the availability of funds, progress has not been quite as rapid as intended. To keep check on the position detailed 'Needs Surveys' are being carried out every two years, and copies can be obtain from the Office of Water Program Operations (WH-595) Wash. D.C.

Table 31   Anticipated progress of sewage treatment in the USA

| | Population Receiving treatment 1000s | | Percentage of total population |
	Resident	Non-resident	
1975	144,368	14,205	66.7
1990	232,290	26,121	89.7

On the Continent of Europe, the setting up of the Common Market, although not in any sense an ecological measure, has had a similar effect, in that it generated a common interest in pollution prevention and control and led to the promulgation of 'Requirements', governing discharge to streams, and even to the sea, as well as many other public health matters. These Requirements are to become mandatory on all new treatment or disposal works and are already causing concern in Great Britain, let alone in Italy, where national control has not been strong or consistent enough to bring the municipalities and manufacturers into line. No doubt a good deal will be done with the help of Common Market Grants, to which some British Water Authorities have already been beholden.

Although in Canada, Australia, South Africa, and Japan, British and American standards are largely observed, and in some cases even bettered, the so-called

Third World presents a confused picture, in that although some of the developing countries can afford any expenditure they wish on water supply and sanitation, others are so poor that the likelihood of their ever achieving what we regard as Western standards, seems very remote.[2] Though some large cities have gone in for sophisticated treatment works it appears that generally it has to be accepted that water-borne sewage disposal is not only out of reach, but, in any case, would not be accepted by the people, without a great change in their present standards of living. Some form of privy, and the collection and possible utilization of night-soil for agricultural purposes, is the only practical target, and subject to control of disease may be the only permanent solution.

Standards of sewage effluents

Sewage-treatment works are designed to reduce the strength of sewage to a figure which may be expected to ensure complete avoidance of nuisance in the circumstances in which the sewage is discharged. In Britain the Royal Commission on Sewage Disposal in the early 1900s laid down as a general standard for sewage effluents that an effluent should be considered satisfactory if it contained not more than 30 mg/l of suspended solids and did not absorb more than 20 mg/l of dissolved oxygen in 5 days. This standard, which is generally applied to effluents discharged into rivers, was intended for application to rivers which diluted the effluent by at least eight volumes of river water to one of effluent. Where this degree of dilution was not available, a higher degree of purification would be advisable.[3]

The Royal Commission standard has been used in Britain for many years as a general yardstick, regardless of the degree of dilution of effluent by river water, but low stream flows and ever-increasing interest in cleaner rivers led to demands from the former river authorities for improvement on these standards in many places. For works designed largely on the principles laid down by the Royal Commission such demands really meant the introduction of a third stage of treatment, and gave rise to many arguments concerned chiefly with the extra cost involved, as related to the subsequent benefit to the community. The doubt which often arose as to whether a wide enough view was being taken in any particular cases no longer applies, since the coming of the Water Authorities who have the power to establish their own effluent standards, but have recently become legally responsible for their efficacy. There is little doubt that proper management and maintenance should be capable of maintaining or improving standards of effluent at many works, but to bring the general standard of effluent produced significantly above 'Royal Commission' will involve a great deal of expenditure.

Allusion has already been made to the American scene. The so-called 'secondary treatment' standard there involves production of an effluent containing 30 mg/l both of BOD_5 and suspended solids, and with the comparatively weaker sewage prevailing, should be well within the capabilities of any works affording primary settlement followed by some form of biological treatment. Again a careful appraisal will be necessary before any substantially higher standard is achieved. The direction in which engineers are expected to look for improvement is laid down in the US EPA publication *Process Design Manual for Upgrading Existing Wastewater Treatment Plants.*[4] It will be interesting to see whether the emphasis in this manual on tried methods and 'cost-effectiveness' will affect the hitherto progressive approach attributed to some of the leading American cities and states.

Methods of sewage treatment

The usual methods of treating sewage first aim to remove solids by screening and sedimentation. These processes together can easily remove the greater part of the solids content of the sewage and, in so doing, appreciably reduce the biochemical oxygen demand also (the sedimentation being mostly responsible for this). In America it is usual to expect suspended solids reductions of 40–70% and biochemical oxygen demand reductions of 25–40%. In Britain, where suspensions are heavy and practice is to use larger sedimentation tanks than in America, a reduction of suspended solids of not less than 70% is expected, and a reduction of biochemical oxygen demand of not less than 42% where the sedimentation tanks have a detention period of about 6 h dry-weather flow. The screening and settlement reduce the load on the secondary treatment, which is oxidation of the remaining organic content with the aid of bacteria.

Secondary treatment was once effected by irrigation on to impervious land (broad irrigation) or filtration through porous land from ditches to underdrains (land filtration); but these methods are little used now except in an improved variety of land filtration, known as intermittent sand filtration, which is largely used in America but unknown in Britain other than for tertiary treatment. The aeration method most commonly used is biological treatment on trickling filters (percolating filters) while at a smaller number of works, usually those of the larger size, the activated-sludge methods are applied. In these the sewage is aerated in tanks and oxidized by organisms that are freely in suspension.

The primary aim of sewage treatment is to retain the sewage in circumstances in which it is in contact with the air and acted upon by aerobic organisms for a sufficient time to oxidize the organic contents to a sufficient degree for it to be safely passed to natural waters without any fear of causing a nuisance.

Most systems of aeration resemble one another in that oxidation takes place in water which has a large surface exposed to free air and the oxidation results from, or is chiefly due to, the natural activities of organisms living in water.

The decomposition of sewage, like the decomposition of humus in the soil, passes through stages, for the product of one type of organism is acted upon by another until finally stable compounds are produced. Briefly, after the initial oxidation or lowering of biochemical oxygen demand comes nitrification with the production of nitrates.

In activated-sludge works micro-organisms alone form the flora. On the other hand, trickling filters also contain moulds, fungi, and algal growths (for example, the blue-green layer often seen on the surface of the filter) and worms and the larvae of insects feed on these.

The flora and fauna of trickling filters can vary considerably according to location and to the strength and type of the sewage. There are also the well-recognized seasonal variations: the excessive growth of algae in the winter when the activity of larvae is reduced to a minimum is followed by the spring off-loading when the rise in temperature permits a sudden increase in the activities of larvae resulting in the rapid destruction of vegetation which is broken down and washed out of the beds.

The type of sewage-treatment works to be adopted in any particular instance depends on:

(1)   the system which will involve the minimum of running costs and annual repayments on cost of construction;
(2)   the extent to which minor nuisance due to flies or sewage works odour matter in locations is concerned;
(3)   the area and cost of land available, and the suitability of the land for each of the methods of treatment;
(4)   the degree of treatment required;
(5)   the available fall from incoming sewer to point of outfall.

These together form a common-sense economic problem to which there should be only one correct answer in any particular instance. In fact, however, decisions as to type of works too often depend on the preferences of individual engineers, and on what may be considered fashionable at the time.

Land treatment can be the least expensive form of treatment where land is cheap; nevertheless it should be remembered that the proper preparation of land for treatment purposes involves appreciable capital expenditure. Both level and falling land can be used, and the method can mean very little loss of head through the works. Land treatment also permits the growth of some useful crops. The objections to land treatment are:

(1)   much larger areas of ground are required for this method than for either trickling filters or activated-sludge works;
(2)   some odour is unavoidable.

The trickling-filter method has been the most widely used. It requires much less land than land treatment, and costs less in capital expenditure in those localities where land is expensive or, owing to its unsuitability, would require large areas to be used for the purpose. Broadly, modern trickling-filter schemes involve less overall cost than the activated sludge method in its traditional form, although for small schemes, the Passvere ditch is now in the lead. The filters produce large quantities of flies (particularly Psychodae) which, however, can be controlled to some extent. There is a moderate degree of sewage works odour which permits small works to be located near habitations without being noticeable from this cause, provided the maintenance of the other parts of the works is satisfactory.

The activated-sludge systems occupy the smallest areas of land and frequently involve noticeably lower capital costs compared with percolating-filter schemes. No flies are produced at activated-sludge works, and the slight tarry odour of the aerations tanks should not be considered offensive. Most activated-sludge systems involve very little loss of head through the works and, therefore, in some circumstances they obviate pumping.

On the other hand activated-sludge processes are not so robust as percolating filters, being more easily upset by difficult sewages and irregular loads. They require the continuous use of power in appreciable quantity (and this alone can in many instances rule them out on account of running costs) and some types of plant used involve heavy maintenance costs owing to rusting of iron-work in conditions ideal for the production of rust. They require skilful operation.

In 1963 The British Institute of Sewage Purification (now the Institute of Water Pollution Control) published a *Directory of Sewage Works* which was the result of 310 replies from 550 local authorities to whom a questionnaire had been circulated in Great Britain and South Africa. This included particulars of sewage works serving populations of more than 10,000, smaller installations being exluded so as to reduce the statistics to manageable proportions. This investigation showed that the activated-sludge process had increased in popularity, there being about half as many activated-sludge works as trickling-filter installations in Britain.[5] The number of works having land treatment was fewer than 10% of the total. In South Africa, out of 47 treatment works in all, 44 used trickling filters, one had diffused-air plant and 6 retained some degree of land treatment.

In the United States, a 1957 inventory of sewage works[6] gave the information shown in Table 32, and subsequent 'Needs Surveys' (see p. 231) provide a great deal more information.

Table 32   1957 Inventory of municipal and industrial waste facilities

Degree of treatment	No. of plants	Population served
Minor	41	1,860,330
Primary	2730	25,666,745
Intermediate	100	5,590,952
Secondary	4647	43,325,704

Siting of sewage works

As the majority of sewerage systems are gravitational, sewage-treatment works are most commonly located at the lowest part of the system and below the developed areas served. They must also be above the natural waters into which effluent is discharged. If either of these conditions is not practicable pumping becomes necessary and, in some instances, sewage has to be pumped to the treatment works or lifted at the works so as to provide the necessary fall through the components of the plant or, more rarely, the effluent has to be pumped from the works to the stream or river into which it is to be discharged. The general aim of the designer of the system is to locate the works so that pumping is avoided while, in other respects, the location of the works is as desired. The following are the main factors concerned:

(1) location of works so as to avoid pumping;
(2) adequate area of land available;
(3) location of site away from inhabited areas;
(4) suitability of land as regards fall of the surface, freedom from flooding, and quality of the soil and subsoil.

The position of the works, as mentioned before, needs to be one which places the treatment plant below the lowest part of the sewerage system and above the natural watercourse. This can be enlarged on. A trickling-filter scheme requires a depth in the region of 4 m from the invert of the incoming sewer to the highest water level of the river, although this figure may be reduced to 2.5 m in some circumstances. An activated-sludge system, on the other hand, provided there is nothing in the way of measuring weirs, or grit channels, needs only about 0.5 to 0.6 m fall through the works, and this is one reason why it may in some cases be chosen, in order to do away with the necessity for pumping. Land treatment is intermediate in the minimum loss of head which it involves.

It is always desirable for an adequate area of land to be purchased or reserved not only for present requirements but also for the maximum estimated future extension of works, plus a goodly margin to allow for unexpected future possibilities. Cases have occurred where sewage works have been built on sites adequate at the time, but later, when extensions far beyond those originally thought of became necessary, lack of adjacent land for the purpose of extension has led to major difficulties.

Examination of all twenty sewage-treatment works in Greater London, serving populations varying from 2,966,856 to 60 persons, showed that an adequare area for works of any type (other than land treatment) could be found by the formula

$$A = \frac{P^{2/3}}{74} \tag{26}$$

where: A = hectares;
P = head population.

(The average area occupied by those works was two-thirds of this figure, but included some which had no sludge-drying arrangements.)

On the matter of location of site away from inhabited areas, it should be understood that properly kept modern sewage works need not be particularly offensive or dangerous to health. But in the more highly developed countries sewage works, however ornate, are not popularly considered an amenity and thereby common sense requires that even an activated-sludge plant should be located, as far as practicable, away from the public eye.

It is not possible to state any uncontroversial opinion as to what distance houses or work places should be from sewage works, but the figures in Table 33 give some indication as to what has been found acceptable for conventional works at several existing sites. Filter flies are another reason for keeping all except activated-sludge works well away from developed areas, particularly residential estates.

Table 33   Depth of isolation zone for sewage works

Population served	Minimum depth of isolation zone (m)
1,800,000	375
750,000	275
225,000	175
95,000	150
28,000	100
12,000	75
3,500	50

Choice of land is important from several points of view, the first consideration being that the land selected should not be liable to floods which could put the works out of action. Next to this comes the consideration of ease and cost of construction. The land selected for sewage works should, if possible, consist of sound ground on which works and buildings can be constructed without expensive foundations. Unfortunately, the best sites from other points of view are sometimes alluvial areas on the banks of rivers where the subsoil is not sound and where special precautions have to be taken.

For a trickling-filter scheme, land sloping moderately at a fall of about 1 in 50, according to the size of the scheme, is best because this permits the location of the various components of the works comfortably on or in the ground, with an easy fall from each component to the next, without either excessive expenditure on excavation or stilting up and embankment.

Moderately sloping land is usually suitable for surface irrigation, although very level ground is best for land filtration, and for any form of treatment requiring oxidation ponds or ditches.

An activated-sludge plant is most easily constructed on land which is level, or almost level; if the land is not level it has to be levelled at the cost of excavation.

The quality of the soil and subsoil matters most when land treatment has been chosen—for light, friable soils, pervious to water, are by far the best for this purpose, and a low level of subsoil water is essential.

Chemistry of sewage

Sewage-treatment works are designed on the quantity of sewage to be treated and the strength of the sewage, i.e. the degree of organic pollution. Were sewages generally more constant in consistency than they are, it would be quite possible to design accurately in terms of head of population to be served, but sewages vary considerably according to the amount of water used per head of population, the degree of infiltration of subsoil water, and the quantity and nature of trade wastes. In the case of entirely, or almost entirely, domestic sewage from a normal town, where water closets are generally installed, the size of the aeration unit of the sewage works (land, trickling filter or aeration tank) can be, and sometimes is, determined according to the population to be served. In most other cases the aeration unit is sized according to the product of strength of the sewage and the average flow.

There is a tendency, in recent years (probably due to the work of Imhoff and perhaps more so to his disciple, the late Gordon M. Fair) to evolve equations to cover all types of processes, separational or biological. In the hands of the highly trained and experienced sewage chemist these equations may be useful tools, but to anyone else, including the engineer trying to arrive at a quick answer to some major design problem, they must surely be applied with the utmost caution as any mis-judgment in evaluating the terms of a particular equation may well produce an erroneous answer. As most of these terms must, of necessity, represent accurately the characteristics of the sewage, or process-stage effluent in question, the ever-present risk of unforeseen variations which led to the commonly expressed opinion that all sewages are different, surely makes this approach questionable, except perhaps to give a lead to the causes for process failure, or partial failure, at an existing works. If for any reason reliance cannot be placed on conventional treatment methods, or it is hoped to achieve economy either in construction or maintenance by introducing new processes, surely the pilot-plant method carried out on the actual sewage to be treated is more acceptable than the answers to any number of equations? It is possible, of course, that in Britain, at any rate, the experimental work now being done at Coleshill,[7] and Davyhulme, aided by the use of computers, may enable a wide range of variables to be considered and so provide answers in many cases without the need for further experimental study. The same may be said of the EPA in America.

Chemical analyses

Various tests have been devised for ascertaining the strength of sewage, but none is absolutely satisfactory. The recommended tests, and the means of carrying them out, are set out in the American Public Health Association publication *Standard Methods of Water and Sewage Analysis* (latest edition) and in the Ministry of Health publication *Methods of Chemical Analyses as Applied to Sewage and Sewage Effluents*, as revised by the Ministry of Housing and Local Government.

The following are the usual tests made on sewages and sewage effluents. Following American practice, the results of analyses are expressed in parts per million or, more recently, milligrammes per litre.

(1) *Suspended solids*

The suspended solids consist of organic material varying from large particles to non-aqueous fluids, and inorganic material varying from gravel to colloidal minerals. The test for suspended solids when applied to average daily flow (not dry-weather flow) of sewage is an indication of the amount of sludge to be expected. In the case of an effluent the test is made for determining whether or not the effluent is sufficiently free from suspended solids to comply with the required standard.

A test for what are known as settlable solids is made by letting sewage settle in either a graduated glass cylinder or in an Imhoff cone for a period of 1 h or such other time as may be decided. The volume of sludge settled is considered an indication of the settlability of the solids. An alternative method is to draw off the settled sludge and find the weight of suspended solids after drying. This test, while useful to the operator of sewage works, should not be used for comparison of results obtained with different sedimentation tanks because it does not give a true indication of either the suspended solids in the sewage or the amount of solids likely to be settled in a good design of sedimentation tank. Efficiencies of sedimentation tanks expressed in terms of percentages of settlable solids should be viewed with distrust, particularly if stated to be more than 100%!

(2) *Dissolved solids*

Sedimentation removes the greater part of suspended solids in sewage. The remaining organic content requiring treatment after sedimentation is mainly in the form of dissolved solids.

(3) *Chlorine*

The chlorine content of sewage is mostly due to the presence of common salt, which is a normal constituent of urine. Thus the chlorine content is more or less constant, proportionate to the population, except when influenced by trade wastes. As the treatment of sewage has no effect on the chlorine content, the test for chlorine is useful as a check when comparing samples of sewage and effluent, for if the chlorine content in the effluent does not correspond to that in the sewage, one of the samples must have been more diluted than the other, as would be the case for sewage arriving at the treatment works at night as compared with daytime sewage.

(4) *Albuminoid ammonia*

Nitrogen being a constituent of protein is an indication of the amount of organic pollution. The albuminoid ammonia test determines the amount of undecomposed nitrogenous material in the sewage.

(5) *Free ammonia*

Free ammonia is ammonia produced by decomposition of nitrogenous material. This ammonia is capable of further decomposition.

(6) *Nitrites and nitrates*

Nitrites and nitrates are the final products of decomposition and when they, particularly nitrates, are in large proportion comparatively, they indicate a nitrified effluent that is well purified and stable.

(7) *Oxygen absorbed from permanganate in 4 h at 27°C*

This test is made as a means of ascertaining organic pollution in terms of oxidizable material. It is a short-period test and does not give more than an indication of the amount of oxygen that can be absorbed in a long period.

(8) *Oxygen absorbed from permanganate in 3 min*

This test is similar to the above except that all solutions must be at 27°C before the sample and permanganate solution are mixed.

(9) *Chemical oxygen demand (COD)*

This is a test for oxygen consumed from potassium dichromate in acid solution. It is of value as an estimate of the strength of sewages and industrial wastes in which biochemical oxygen demand (BOD_5) cannot be determined because of the presence of substances that are toxic to the organisms that have to act during the biochemical oxygen demand test. This test, using a sealed digestion method, is now regularly used by Water Authorities in this country.

(10) *Dissolved oxygen absorbed in 5 days*

This test, which is applied to sewage effluents to determine whether they conform to approved standards, is also a recognized test for sewages as an alternative to strength McGowan (which will be described). The test determines the amount of oxygen absorbed by the bacteria, etc., in the sewage or effluent, the sample being incubated for 5 days at a constant temperature. The result is known as biochemical oxygen demand and is most commonly described as 5-day BOD or BOD_5.

(11) *Acidity and alkalinity*

Domestic sewage is usually alkaline but can be rendered acid by septicity or by acid trade wastes. The test commonly applied is the test for pH value, which is a measure of the hydrogen-ion concentration, and hence of the acidity or alkalinity of a solution, expressed on a scale of numbers ranging from zero for a solution containing 1 gramme-ion of hydrogen ions per litre (corresponding to extreme acidity), to 14 for a solution containing 1 gramme-ion of hydroxyl ions per litre. A neutral sewage has a pH value of 7. The scale is logarithmic, so that to take the average of a number of readings for several days does not give the average pH for the period. The procedure should be to find the anti-logarithm of each figure, average these, and find the logarithm.

Strength (McGowan)

While the BOD_5 is now used in most countries as a measure of the strength of sewage, partially treated sewage, and fully treated effluent, the McGowan formula still has occasional use in Britain. Formulae (McGowan's formulae) were once given for sewages and septic tank liquors and for precipitation liquors. The latest publication of *Methods of Chemical Analysis as Applied to Sewage and Sewage Effluents* gives that for sewage only as follows:

Strength of Sewage = 4.5 × (ammoniacal nitrogen + organic nitrogen) in parts per 100,000 + 6.5 × permanganate value using N/8 permanganate in parts per 100,000.

The factor of 4.5 represents the amount of oxygen required to convert the nitrogen to nitrate and is correct provided that all the nitrogen is originally present as ammonia or one of its substitution products. The factor 6.5, however, is empirical and depends upon the proportions of the oxidisable carbonaceous matter which reacts with permanganate under stated conditions, and this proportion varies very widely with different trade wastes. The McGowan formula suffers from the disadvantage already mentioned, that it is based solely on oxygen requirements, although it has proved useful in the absence of appreciable concentration of trade wastes.

It should be noted that the McGowan strength is calculated from the permanganate value using N/8 permanganate and not N/80 permanganate. . . . The relation between the two is no doubt variable. A figure of 1.6 has been given as the general ratio between the N/8 test carried out at 25.7°C and the N/80 test carried out at 18.3°C. The ratio must be smaller when the N/80 test is carried out at 27°C . . . and probably does not exceed 1.2. This figure could be used to obtain an approximate figure for the McGowan strength, but analysts still using it are recommended to adhere to the original method of determination.

There is no direct correlation of BOD and McGowan strength, but on the average a strength McGowan of 100 could be the equivalent of a BOD_5 in the region of 350, or perhaps 375 mg/l. This is considered to be about the average strength of British sewage. After sedimentation for about 6 h British sewage of average strength has a BOD_5 of about 200 mg/l The average BOD_5 of American crude sewage would appear to be about 200 mg/l.

Some indication of the average analyses of a British sewage are the figures in Table 34 which are taken from the Fifth Report of the Royal Commission on Sewage Disposal. At the time of the analysis the dry-weather flow was 162 litres per head per day.

Kilogrammes BOD$_5$ per head of population

The load on sewage works in kg BOD$_5$ per head of population is often used in connection with design. In the case of mainly domestic sewages the author has recommended for Britain 0.055 kg BOD$_5$ per head per day. The Water Pollution Research Laboratory found at Stevenage New Town that the load from a purely residential area was 0.045 kg. The author found that for large industrial areas in Britain sewered on the combined system the figure of 0.077 kg was representative. American figures given by Metcalf and Eddy[8] were 0.052 kg for separate sanitary sewers without industrial wastes and 0.063 kg for combined sewers without industrial wastes. Other figures that have been commonly used in the United States are 0.065 for cities with separate sewers and 0.075 for cities with combined sewers. Recognized design figures in the United States and Canada are 0.077 kg BOD$_5$ per head per day and 0.10 kg suspended solids per head per day.

Table 34   Analysis of an average British sewage

	mg/l
Albuminoid nitrogen	9.1
Ammoniacal nitrogen	35.3
Total organic nitrogen	22.5
Oxidized nitrogen	Trace
Total nitrogen	58.5
Oxygen absorbed at 27 °C, in 4 h	112.7
Chlorine	91.6
Solids in suspension	294.0
The strength (McGowan) was approximately 100 (p.p. 100,000)	

Night soil content of sewage

The following information was gleaned from various sources. Its uses are mostly in relation to conservancy sanitation. The average amount of faecal matter has been stated to be 135 g per person per day at 75 % moisture content. The quantity of urine has been stated as being, for the average man, 1500 g of water per day, containing 72 g of dissolved solid matter. According to some experiments of limited extent, it was found that closets were used for defaecation one and a half times per day by each person in an establishment and four times as often for urination only.

Hourly variation of rate of flow and strength

Not only does the rate of flow vary throughout the day (being in Great Britain at a maximum from 11 a.m., in the case of a small drainage area to as late as 4 p.m. in a very large area) but the strength of the sewage varies also and in the same manner, as is shown in Figure 55. The organic load on the sewage works is the product of the rate of flow and the strength, and it will be seen that, because

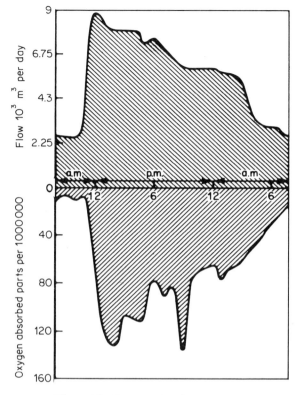

Figure 55   Dry-weather flow diagram

of this, the greater part of the load on the works is confined to quite a small period of the day. This fact is an important consideration when sewage samples are taken. A spot sample taken at the peak rate of flow could give an impression of a sewage having a strength twice that of the true average, and vice-versa. It is for this reason that sampling of sewage, settled sewage, and effluents must be taken throughout the day and mixed in such a manner as to give a true average daily sample, as will be described.

Sampling

The samples most often taken at sewage works are those of the final effluent, and the most important analyses are for suspended solids content and bio-chemical oxygen demand, for these indicate the degree of treatment and whether or not the effluent is satisfactory. But more complete analyses are made of sewage effluents as a routine at large, or even occasionally at small works. In addition, there are other tests including those for sewage, settled sewage, mixed liquor, and returned activated sludge at activated-sludge works. Where sludge is digested it is usual to analyse the raw sludge and partially and fully digested sludge for total suspended solids, solids, and oil content and the sludge gas, particularly for CO_2 content.

Sewage-works analyses, in addition to providing information on the performance of the plant, are helpful for determining qualities, such as strength and difficulty of treatment, which would be useful in calculating the sizes of any future extensions of works. The analyses of the crude sewage also give the quantity of sludge to be expected, and those of the sludge and sludge gas the number of thermal units likely to be available for power and heating purposes.

It is difficult to take truly representative individual samples of sewage-works liquids, in particular of crude sewage. If samples are taken near the surface· they may contain excessive quantities of light matter such as oils, and if near the bottom, an undue proportion of sludge. Even samples taken at an identical point in the flow of sewage and simultaneously have been observed to vary considerably, and an individual sample shaken up and divided into two parts to be analysed by different laboratories has been found not to give the same analytical results. Thus, whether samples are collected by hand or by machine, great care should be taken in choosing the point and manner of hand-sampling, or the location and type of the automatic mechanisms.

Broadly, it can be said that no individual sample or analysis of a sample can be taken as a truly quantitative representation of the content of a sewage or waste. It is only when many samples have been taken, and methods used which obviate any persistent bias one way or another, that the average sample can be considered representative.

All samples should be taken gently in such a manner as not to cause aeration of the liquid, yet they should be taken from turbulent flow and from about mid-depth in a straight pipe or channel in order that disproportionate quantities of heavy and light materials are not included. Even with these precautions the samples may not be representative, for except when turbulence is considerable there is a measurable difference of analysis between liquid near the top and near the bottom, and sometimes between liquids on the inside and outside of a bend in the channel.

In Britain samples are usually placed in $2\frac{1}{2}$ litre bottles filled to the top and stoppered with glass stoppers, so as to exclude any air bubbles, and kept in ice. If they are to travel not in ice, an air space must be left for expansion, but then the sample will be influenced by the oxygen present and continuous biological action. Strictly, they should be analysed at once for the determination of several of the tests. Light should be excluded, for it has an effect on the chemical content. Once removed to the laboratory, and most of the tests put in hand, that portion of sewage which is to be examined for nitrogen compounds may, at the discretion of the chemist, be aged so as to permit the hydrolysis of urea to facilitate the test.

As both flow and strength of sewage vary throughout the day, the strength tending to increase at the same time as the flow, any individual snap sample taken at one time during the day is unlikely to be representative, and constant-size samples taken at regular intervals throughout the 24 h will almost certainly give misleading values. If samples are taken at regular intervals it is necessary for the size of each individual sample to vary in proportion to the rate of flow at the time if the aggregate sample for the day is to be representative. Without this precaution the sample may have as little as two-thirds the strength and suspended solids content as the crude sewage.

When hand-sampling the method is for the sampler to take each individual sample as already described, read the rate of flow from such weir or instrument as may be provided, measure out the appropriate quantity from the sample taken into a measuring glass, and tip this quantity into the vessel which is to hold the 24 h sample. Samples so taken are known as weighted samples.

Automatic sampling

In the design of an automatic sampler the aims should be to ensure representative sampling and an inexpensive reliable mechanism. Sewage samples should be weighted as already mentioned. The weighting can be achieved not only by varying the size of samples taken at regular intervals, but also by taking samples of constant size at intervals of time which vary inversely to rate of flow, a fact which is a great help in sampler design.

Care should be taken that the sampler does not in any way segregate the components of the sewage before sampling. For example, if crude sewage is sucked up a pipe by a pump or similar mechanism, gas will be given off and this will carry grease and solids to the surface. If a sample is then taken from the first of the flow to arrive at the machine, it will contain a disproportionate quantity of grease and solids and vary considerably from the sewage in these and other respects.

The liquid stored in the sampler should be kept in such a manner that sludge does not settle out or light particles rise to the surface in any fixed vessel, so that when the sample is drawn off and taken to the laboratory, sludge and scum may be left behind or come off spasmodically from day to day, giving irregular results. It is, in fact, essential that the liquid should gravitate directly into a removable sample vessel that can be taken to the laboratory, emptied under expert surveillance, and thoroughly cleaned and sterilized before being returned to use.

The sampler storage vessel or vessels should be so designed and handled that organisms cannot develop in them and cause organic or chemical changes in the sample during its storage in the machine. The writer is of the opinion that the vessel should be enclosed in a refrigerator if its contents are stored for many hours (on the average for more than 12 h per day), giving time enough for considerable changes to take place in hot weather. It is obvious that a sample stored without temperature control on a hot day must give a considerably different analysis from one taken from the same sewage and kept on ice.

The effect of the neglect of the above-mentioned precautions is illustrated by the following. The author had been studying week after week the crude-sewage suspended-solids values and the settled-sewage suspended-solids values of a newly constructed plant. There had been slight variations in the crude sewage reflected in variations of similar order in the settled sewage giving an almost constant efficiency of sedimentation. Then a date occurred beyond which the figures for suspended solids in the crude sewage fluctuated from sample to sample in an extraordinary degree which was not reflected in the settled sewage sample. The author got in touch with the officer in charge of the sewage works asking him no more than what out of the usual had happened on the date when the change started. He was informed that nothing had occurred except that the crude-sewage sampler had been removed and replaced by one of a different type. The new sampler had all the faults mentioned above!

There are several types of sampling machine, both of proprietary design and manufacture and specially made for, or by, drainage authorities. They include machines that suck, pump, or scoop up the liquid; those which give weighted or unweighted samples into one vessel; and those which take unweighted samples into a number of small vessels each filled at regular intervals so as to determine the change of the qualities of the sewage or effluent throughout the day. These machines vary considerably in cost and efficiency, the latter characteristic having little bearing on cost but more on competence of design.

A sampler which the author designed should be of interest to any drainage authority considering installing their own machines. This is one of a number of types that the London County Council (Greater London Council) has constructed and tested at its sewage-treatment works. The principle of this machine is similar to that of the old-fashioned leap weir. Liquid is pumped from the sewage or effluent channel to waste so as to clear the pump, pipework, and sampler before sampling. Then, when the pump stops, the reduced velocity of discharge permits the measured sample to fall into a vessel. Weighting is achieved by arranging for the samples to be taken at varied intervals of time controlled by the sewage works flow integrator.

The site must be provided with a flow recorder and integrator to measure the rate of flow to the works. Whenever the integrator clocks up, say, one-eighteenth of a day's dry-weather flow, it must send out an electrical impulse of a few minutes duration to actuate all sewage, settled sewage, and final effluent samplers on the works.

The sampler unit consists of an electrically driven pump with its starting gear, and the sampler, which is constructed almost entirely of screwed galvanized-iron barrel and standard fittings, without any moving parts (see Figure 56), and a removable bottle to receive the sample. According to circumstances, the electric pump can be placed immediately by the sampler, or it may be some considerable distance away; it may be self-priming or placed at a low level so as to be primed via its suction.

When an impulse is sent out by the integrator, it holds in a relay which sets and keeps in motion the electric motor for the duration of the impulse. The sewage effluent is then drawn from the channel or culvert, delivered through the rising main (the right-hand vertical pipe of the figure), and through the sampling pipe which terminates in the carefully made gun-metal nozzle arranged

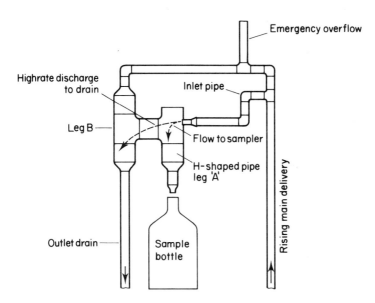

Figure 56   Automatic sampler

to discharge into the H-shaped large-diameter pipe of the sampler. When the velocity is high during pumping, the liquid is thrown over the near leg of the H (which connects to the sampler bottle) and falls into the far leg which connects to the drain. Any excess of pumping rate, over that which the nozzle will pass, passes through the overflow pipe at the top and to waste. A period of time is allowed sufficient for the suction pipe and rising main to be thoroughly washed out before the sample is taken.

On cessation of pumping the head in the vertical pipe on the inlet side falls and, as it does so, the jet of water curves downwards until finally the contents of the sampling pipe drop into the sample bottle. It will be noticed that there is a small dam welded into the cross-pipe of the H-piece to give a clear cut-off, so that there is no dribble or drain-back. This ensures samples of sufficiently even size being taken at all times. There is no tendency to dribble: it was found on test that, provided the gunmetal nozzle is well-made and clean, the liquid on start of pumping immediately leaps straight across to the waste pipe. The following are the considerations in the design of this sampler:

(1)   the rising main needs to be of a suitable diameter to be free from chokage with the kind of liquid concerned, e.g. 30 mm galvanized barrel for final effluent or settled sewage;

(2)   the pump must have a delivery that, at maximum and minimum working heads, will ensure a self-cleansing velocity in the pipework;

(3)   the nozzle must have a diameter that, under the maximum head permitted by the overflow pipe (achieved at the minimum pumping rate) will throw the liquid clear over the dam into the waste pipe;

(4) the sampling pipe must have a capacity equal to the required individual sample between the nozzle and the point on its vertical component at which the surface of the water will stand when the trajectory of the jet throws the liquid just inside the dam and flow starts to fall into the sample bottle;

(5) the sample bottle must be sufficient in capacity to receive the maximum number of component samples when the maximum rate of flow passes through the works all day.

It will be obvious that the size of this vessel has some bearing on the design of the sampler, in particular the size of the individual sample. If the vessel is unduly large it will be too heavy to handle when filled on a wet day. On the other hand, on the day of minimum flow the size of the component sample must be large enough for analytical purposes. If the individual sample is 0.4 litre and the maximum number of samples in a day during wet weather 43, the largest aggregate sample will be 18 litres, while with a minimum flow on a dry day giving at least 12 component samples, the sample will be little more than 4.5 litres.

Instrumentation and computer control at sewage works

At one time very little use was made of recording and other instruments at sewage works. There were, of course, the normal instruments used in connection with pumping plant, e.g. ammeters and pressure gauges. Later it became usual to record the total flow arriving at the works and, in the case of activated-sludge plants, records were necessarily kept of quantities of returned or recirculated sludge and air delivered to diffusers.

At the present time practice has changed very much. Not only are numerous records of flow taken, but consideration can also be given to the introduction of automatic control. The questions therefore arise, to what extent are instrumentation and automatic control desirable or practicable, and how can they best be implemented?

Many industrial processes are now largely operated with the aid of instruments centred in control rooms and automatic machines, which themselves make the necessary adjustments from time to time, maintaining the required conditions of the process and reducing the amount of intelligent operation. At sewage works, possibilities of this kind tend to be limited. However, the coming of the computer has greatly increased the possibilities in this direction, and it is worth quoting Dunbar and Steven[9] writing of their experiences in improving the Carbarns Sewage Treatment works in Scotland:

Computer control of sewage treatment works offers definite advantages provided the decision is made at an early stage in the planning of the new works to install an automated control system. The benefits include a reduction in power costs, improved knowledge of plant behaviour and plant potential, improved performance of the unit processes, and correct use of established staff with consequent increased efficiency and staff satisfaction.

They further recommend that where complex processes are involved there should be a clear operational programme defined at the outset, so that only slight, if any, amendments will be necessary when the system is commissioned. Switchgear and instrumentation should be standardized where possible, and compatible with the electrical and control system, so as to reduce the stock of spares required, and guarantees must be sought from the manufacturers as to their availability over at least the repayment period of the equipment. The main electrical contractor should be responsible for the complete electrical installation, including the control system, associated telemetry, and power cabling. Moreover the preparation of contract documents for mechanical plant, electrical installation and control system should be the responsibility of one individual, so that complete integration of the separate contracts may be obtained and consistency of specification achieved.

Jones, McVie, and Cotton,[10] speaking of progress at Norwich, England on commissioning of the computer-controlled Whittingham sewage-treatment plant, recommend the concept of stage-by-stage introduction of process controls so that the staff feel in command of the central processor unit, rather than vice-versa. A dilemma has to be faced by the designers, as to whether to put off the purchase of a computer system till the last minute to allow for incorporation of the latest developments, at the risk of severe cabling and termination problems at a later date.

At certain works, difficulties have been experienced with sludge density meters and an answer has been found in closed-circuit television. At Beckton[11] these cameras are mounted on the movable part of telescopic bellmouths, so that viewing distance remains constant. Screen washers and wipers are provided, and artificial light is always used. At this works all 16 tanks can be desludged in turn by one man in the monitor building.

Short of automatic control, use of instruments can greatly assist the sewage-works operator and, if their provision is not carried to excess, reduce costs. On the other hand, if too many instruments are installed their maintenance becomes a considerable charge on the works, and there is always the danger that instruments, if too numerous, may be neglected and become inoperative.

At very large sewage works several flows have to be measured, including total flow of sewage arriving at the works, storm water, and flow separated and delivered for full treatment. Alternatively flow to treatment and storm flow may be measured separately, and totalled to give total flow to works. In large activated-sludge works, it is usually advantageous to split the plant into a number of completely independent units; then the flow to each unit should be measured separately and also the separate flow of returned sludge. In the case of a diffused-air plant, the quantities of air delivered to each unit must be recorded.

It is not, however, altogether good policy to install flow meters as a means of distributing flow to individual sedimentation tanks, aeration tanks, etc., if this can be achieved by other and simpler means; also records should not be taken and filed unless a useful purpose can be served by them.

Total, but not individual, flows of raw sludge should be measured, total flow of surplus activated sludge, and total flow of liquor from digestion tanks, etc., in order that a day-to-day check may be kept on the way that the works are operated. Quantity and analysis of sludge gas from digestion tanks should be measured.

Other useful measurements are water levels, pressures, temperatures, and electric loads.

At pumping stations it is usually desirable to install a recording ammeter on the total electricity demand, and revolution counters on the individual pumps. Records of water levels in pump sumps are sometimes necessary or desirable. The use of float or electrode control of pump starting gear is a form of automation that has been standard practice for as long as can be remembered. This is sometimes combined with time-switch control.

In screen houses automatic starting and stopping of screens by difference of water level upstream and downstream has been not uncommon for many years, and the automatic starting of screenings disintegrators is usual except at very small works. The use of mechanically raked bar screens in which the raking is intermittent and often occupying, in the aggregate, little more than an hour a day, contrasts with such methods as comminution, where, by continual operation, control—automatic or manual—is virtually done away with.

Most of the modern methods of detritus separation are largely automatic. This applies to installations like the Door Detritor with separate grit washer, and the constant-velocity detritus channel, from which the grit can be removed sufficiently clean not to need washing if it is intended to be used about the works for any other purpose than making cement mortar or concrete. The constant-velocity detritus channel can be completely automatic in everything except grit removal, without the use of any machines, if it is of the flume or special weir-plate controlled type. But channels in which the constant velocity is maintained at all rates of flow by the opening and closing of penstocks must be either manually operated (a not-too-happy arrangement) or electrically controlled by floats or electrodes which completely or partially open or close penstocks according to a prearranged scheme.

Separation of storm water can be by discharge over separating weirs; pumping to storm tanks at all flows in excess of the maximum rate of flow to be treated, the pumping being electrode controlled; or automatic electrically controlled penstocks. Which method is preferable depends on circumstances. If sewage is pumped to the works, separation of storm water by pumping is usually most economical and convenient. At small works where sewage is not pumped, separating weirs are simplest and generally to be preferred. Automatic penstocks are sometimes the most economical arrangement at very large works where sewage is not pumped.

Sludge digestion is a process in which instrumentation is required. Where there are several tanks the temperatures should be taken of the sludge in each tank, of the total raw ingoing sludge, and the total outgoing sludge. Records are required of the quantities of hot water used for heating sludge and ingoing and outgoing temperatures, so that total heat demand can be determined. In addition to quantity of gas delivered from digestion tanks to utilization, it may be desirable to measure the amount discharged to waste. In addition, the chemist's regular determination of gas analysis and calorific value, an automatic recording of CO_2 content can be taken. Quantity of decanted sludge liquor should be measured and integrated, the records again being used in conjunction with the chemist's analysis of samples.

In sludge treatment plant there are sometimes dangers—in particular, of producing a partial vacuum and drawing air into the tanks with the danger of producing an explosive mixture. Automatic audible and visual warnings of the approach of such dangers are desirable.

The final questions that arise are; where shall the various readings of instruments be recorded and how shall the plant be controlled; are simple local records best, or should there be one or more control and instrument centres? The answers to these questions depend very much on circumstances. At small works, where there are few records to be taken or instruments inspected, local instruments are best. At large works the converse is true. There should be at least one control centre in which are located the instrument boards and such manual remote controls as exist.

These instrument boards are best arranged in such a manner that the relations of the various instruments to the various processes of the works can be seen at a glance. In one arrangement the instruments can be set out on a conventionalized plan of the works, each in their correct position. On the same boards on which the instruments are mounted, coloured lights can show where motors are running or where emergency conditions have arisen. Another arrangement is to place the instruments on the board in a convenient compact manner, but paint on the board in various colours the several flows of sewage, sludge, liquor, etc., so that the purpose of each instrument is immediately obvious.

Flow gauging and recording

Weirs are often used for recording sewage flows, but as settlement of solids cannot easily be avoided on the upstream side of a weir, either a venturi flume or a standing-wave flume is generally preferred. In some circumstances a standing-wave flume may be used in conjunction with a weir with satisfactory results; for example, at a sewage disposal works the dry-weather flow may pass through a standing-wave flume and be recorded thereby, while the storm flow may pass over storm weirs, the depth over which may again be recorded. Solids may be retained at the foot of the storm weirs, but in time they will pass away through the standing-wave flume if the design is satisfactory.

There are several types of weirs and flumes, all of which have their particular applications. The choice of type of weir or flume used in any particular circumstance depends on:

(1)   the degree of accuracy required;
(2)   convenience of construction;

(3) the necessity for avoiding settlement or interference with self-cleansing conditions;
(4) the volume of flow.

For sewage the standing-wave flume has the distinct advantage of offering no resistance to the movement of solids; nevertheless thin-plate weirs are not infrequently used.

The plate of a thin-plate weir should be brass-edged, the edge 1.6 mm thick, sharp on the upstream side and chamfered at 45° on the downstream side. There should be a free fall of 75 mm from the sill level of weir to the top-water level downstream and the invert downstream should be not less than 150 mm below sill level.

In the case of rectangular weirs the sides of the channel upstream of the weir should be not nearer than a distance equal to four times the maximum head from the sides of the weir, and the upstream invert should be three times the maximum head below the sill level.

In the case of V-notch weirs the invert and sides of the upstream channel should be not less than one and a half times the maximum head from the nearest part of the notch. Unless the channel is straight for some distance above the weir plate baffles should be placed in the channel upstream, the nearest being ten times the maximum head upstream.

The head above sill or lowest part of notch should be measured at a distance of six times the maximum head upstream of the weir. The velocity of approach should be a maximum of 150 mm/s, otherwise conditions are complicated.

Proportions of standing-wave flumes

In order that a standing-wave flume may record accurately, the proportions of the flume should be carefully designed. The details of this work may be best left to the manufacturer of the recording mechanism, but for determining the amount of space required and for drawing purposes. Figure 57 is included.

The quantity of flow is measured by the values of H and B, according to the formula

$$Q = 1.706BH^{1.5} \tag{27}$$

where: Q = cubic metres per second;
B = width of the contraction of the standing-wave flume in metres;
H = depth in metres from invert of flume to upstream top-water level.

The depth H is measured from the level of the top of the hump or, in the absence of a hump, from the invert level of the throat, to the top-water level, at a distance of at least $3H$ upstream of the hump or contraction. The length of the throat should be $2H$. Except where humps only are used, the value of B is preferably approximately equal to the maximum value of H. The radius of curvature of the side contractions upstream of the throat should not be less than $2B$. When there is a hump the radius R should be adjusted so that the curved length is

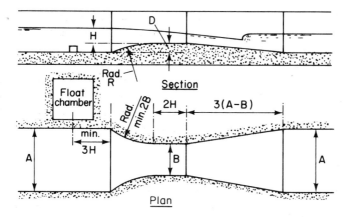

Figure 57   Proportions of standing-wave flumes

equal to the length of the curve of the sides as shown in the figure, but in the absence of side contractions R should not be less than $2H$. Downstream of the contraction the flume should increase in width, each side sloping outward at about 1 in 6, and the floor should fall away to such an extent that at minimum rate of flow the water level in the channel below the flume is at least 25 mm below the invert of the throat.

In order that no disturbance shall occur in the flume there should be a length of straight channel above the throat equal to at least eight times the width of the incoming channel.

The float chamber should be connected to the main channel by a small opening in the position shown on the drawing, measuring about 150 mm square. Arrangements should be made for flushing the float chamber with sewage or otherwise to remove any silt that may collect therein. The float chamber should measure about 750 mm square and its invert should be sufficiently below the invert at the throat to allow for the submergence of the float.

In the United States the Parshall flume is often used to control the rate of flow in grit channels of semi-parabolic cross-section, and the Sutro weir for rectangular channels, though the latter must have free discharge, and is therefore wasteful of head. A Sutro weir is so shaped that its width is inversely proportional to the square root of the height of that point above the crest which will have a theoretical discharge exactly proportional to the head.

Flows of sewage or sewage effluent in rising mains or any permanently surcharged pipes can be measured by meter. When sewage flows are measured, the venturi meters must be so constructed that all openings and pipes communicating between the throat, or upstream of the throat and the columns or other places where the differential pressures are recorded, can be easily cleared of solids. This can be done by means of probing rods or high-pressure water. Magnetic flow motors can also be used for most applications.

Flows of primary sludge, digested sludge, and activated sludge are sometimes measured by venturi meter, but such measurements cannot be considered accurate in all circumstances because the nature of sludge varies from time to time, even at the same works, and consequently the reading of the meter cannot always be taken as being a true indication of the rate of flow. If meters are used for sludge measurement they should, on installation, be calibrated after testing with the actual sludge normally produced: thereafter they may be considered as giving reasonable indications, not necessarily actual recordings of flow. There is a magnetic-flow meter (Foxboro-Yoxall Ltd) which will measure flow of sludge without being affected by solids content.

References

1. Fifth Report of the Royal Commission on Sewage Disposal HMSO (1915).
2. Pickford, J. Sanitation for buildings in hot climates. J. Instn. Public Hlth. Engnrs. **7** Pt. (4) (Oct. 1979).
3. *Technical Problems of River Authorities and Sewage Disposal Authorities in Laying Down and Complying With Limits of Quality for Effluents more Restrictive than those of the Royal Commission.* Minist. of Local Govt. and Housing. HMSO, London (1966).
4. *Process Design Manual for Upgrading Existing Water Treatment Plants.* U.S. Environmental Prot. Agency (Oct. 1974).
5. *Directory of Sewage Works.* Inst. of Sewage Purification (1963).
6. Thomas, J. R. and Jenkins, K. H. *Statistical Summary of Sewage Works in U.S.* Public Hlth Serv. Publ. No. 609 (1957).
7. Clough, G. F. G. and Maskell, A. D. Coleshill advanced wastewater treatment project. *Proc. Instn. Civil Engnrs*, **60**, Pt. 1 (Aug. 1976).
8. Metcalf, L. and Eddy, H. P. *American Sewerage Practice* (3 vols), McGraw-Hill, N.Y. (1928).
9. Dunbar, J. H. and Steven, W., Computer control of sewage works; progress at Motherwell and Wishaw. *J. Instn. Public Hlth. Engnrs.* (14) (Mar. 1975).
10. Jones, C. E., McVie, A. and Cotton P. Computer control of sewage works, progress at Norwich, *J. Instn. Public Hlth. Engnrs*, **3** (14) (Mar. 1975).
11. Dainty, S. H., Drske, R. A. B., James J. E. and Shephard, T. D. Design of extensions to the Beckton sewage treatment works of the Greater London Council. *Proc. Instn. Civil Engnrs*, **51** (Apr. 1972).

CHAPTER 10
Site Planning

Because sewage-treatment works are usually situated close to a stream into which the effluent is to be discharged, they are liable to be built on alluvium such as peat or silt; materials which are unable to bear loads. There is also a danger of tanks, when empty, floating upwards in waterlogged ground and breaking their backs in the process. The protections against these dangers are the support of structures on piles, or the carrying of heavy foundations down to bedrock. It is therefore necessary before design that a subsoil survey shall be made, consisting of borings and/or trial holes. At the same time samples of subsoil and subsoil water should be taken to ascertain whether or not sulphate-resisting materials may have to be used in the work.

These investigations not only give information on natural conditions, but may bring to light the existence of tipped refuse or other made ground: but there are also other things to be done. Records regarding the site should be searched for information on pipelines, electric cables, etc., which may have been forgotten. Inspection of maps should show which places are liable to floods, and there is almost certainly some local knowledge of highest recorded flood level.

After the engineer has learnt as much as he can about the site he is in a position to plan.

The designer of a sewage-treatment plant should aim at a scheme which does not cost more than the average for works of equivalent size but without being below standard in any respect; that permits some elasticity of operation according to the practice or opinion of the works manager; and that has a good appearance, making it look well designed.

As in all design, efficiency of works in proportion to economy of cost forms the primary consideration. In the arrangement of sewage works economies can be made in the use of land, labour, and materials. It is the first of these that is most commonly neglected. Many works occupy as much as three times the area that is necessary for their convenient construction and yet are so arranged that future extensions are difficult.

While a design permitting several methods of operation may be popular with a competent operator, it also introduces risk of improper use. Having the

256

benefit of long experience, a designer may do his best to achieve an efficient and reasonably fool-proof scheme but have it wrecked by its being used in a way that had never been intended. This does not mean merely the production of a bad effluent but actual damage to plant, sometimes on a large scale. Such calamities are very hard to anticipate and even when foreseen cannot be wholly excluded.

Finally, as regards appearance, sewage works are not popularly considered as works of art but they can be very attractive and even more so than many

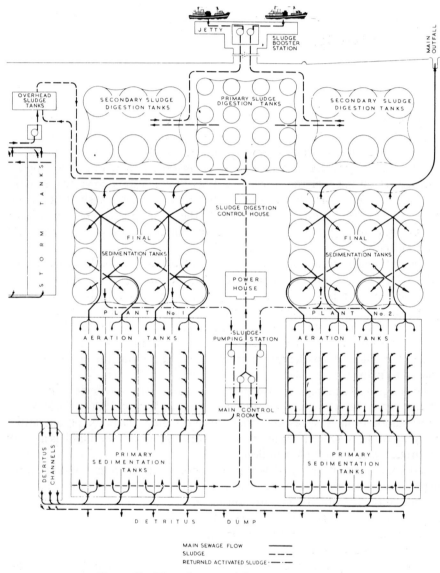

Figure 58   Flow diagram of Crossness sewage works

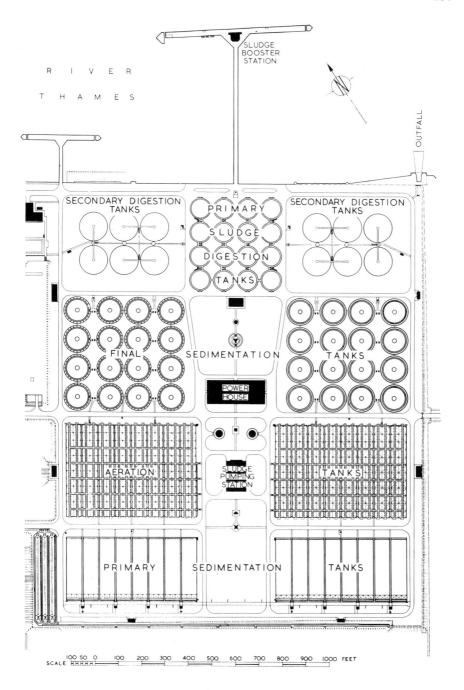

Figure 59　Layout of Crossness sewage works

public parks. It is always desirable that the general plan of a sewage-treatment plant should be symmetrical and pleasing in appearance. Regularity of outline and ordered grouping of the components often tend to aid efficiency of working, economy of construction, and ease of future extension. In the spacing and layout of units, particularly of very large works, consideration must be given to how the contractor will use cranes, lorries, and other plant during construction.

Figure 58 shows a logical flow diagram for a very large activated-sludge scheme and Figure 59 is the general plan based on this. It will be observed that this plan is orderly, balanced, and of attractive appearance on paper. With the park work—including lawns, tree plantations, and flower beds—it is, in fact, pleasing and impressive to visitors.

Whilst the points concerning fool-proof operation are still relevant in most parts of the world, it is probable that the setting-up of the Water Authorities in Great Britain, and the emphasis being put on the training of works operators—both there and in the United States, Canada, New Zealand etc.—will in future leave the consulting or design engineer more scope for introducing alternative flow-paths, and other refinements aimed at possible improved performance. Up-to-date design practice involves a much closer link between representatives of the design and maintenance side. Some have gone as far as to say that the 'works manager' is a thing of the past. If this is so it is regrettable, as nothing is more effective, particularly in emergency, than a thorough understanding of how a works operates and, under normal circumstances, how the best results can consistently be obtained.

Procedure in preparing a layout plan

Assuming that an adequate area of suitable land has been found, the first part of the procedure is to find the highest point on this land to which the sewage can be delivered and there, assuming the works to be of conventional design, locate either the screen chamber or detritus-settlement unit as high as practicable on the ground. The rest of the works must follow in correct order below the highest unit, except in cases where pumping of tank effluent or treated effluent is essential. The highest part of the works is placed near a corner or boundary of the land but allowing sufficient space round the unit for contractor's plant during construction. Tentative plans showing the different positions of the units are then made and, at the same time, flow-through sections drawn to show the top-water levels of the various tanks or top-medium levels of trickling filters and sludge beds, and to ensure that the lowest units are above flood level. By this means the practical positions of the units, and those involving the least excavation, are determined. When the general limits of location of the various units have been found the final plan, giving their positions in orderly and economical arrangement, can be drawn. It is at this stage that the aesthetic considerations come in, and the plan is made to look balanced and orderly.

The largest item of excavation in a trickling-filter scheme is usually that for the trickling filters. For this reason the positions and levels of the trickling filters tend to control the plan. On a level site the undersides of the floors of the filters are preferably set at just below top-soil level. On a sloping site, the undersides of the floors may be slightly above ground level on the downhill side of the filters, a small amount of 1:5:10 concrete being used under the floors to prevent the necessity for the much larger amount of excavation that would be necessary were the filters arranged with floor levels generally below ground level.

The level of the trickling filters is, of course, not the only factor to be kept in mind: except when pumping is unavoidable the levels of the components of the works are always limited by the invert level of incoming sewer and flood level of the stream to which effluent is discharged.

The sedimentation tanks must have their top-water levels 600 mm or more above the top of the medium in the trickling filters to allow for dosing siphons and pipework losses. This means that either the filters must be a fair distance downhill of the sedimentation tanks or the sedimentation tanks must be stooled up above ground level.

The sludge-drying beds (assuming that there is no digestion) are usually at least 1200 or 1500 mm below top-water level of sedimentation tanks, and lower by an amount due to the head loss in the sludge main leading from the tanks to the beds. These should be arranged so as to be just in the ground on their downhill side, where the access road should be located. Excessive excavation should be avoided. For the sake of economizing in length of sludge main and often of access road, the drying beds should not be too far from the sedimentation tanks. Quite frequently, difficulty with inlet and effluent discharge levels will leave more room at the top of the site than elsewhere, and if it is intended to employ drying beds, the best layout will be achieved if the separated sludge is piped to a low area on the site and pumped up to drying beds arranged on the higher ground. This arrangement is even more appropriate if any form of sludge digestion is incorporated in the design. Incidentally, the use of the old type of drying bed is no longer finding favour in Great Britain, where interest has turned to vacuum filtration, followed by sludge pressing, thus reviving interest in a process long practised in the industrial towns of the 'North country'.

Humus tanks are often unavoidably deep in the ground. Generally, they should not be set too deep as this makes operation difficult, and they should, of course, have their top-water levels above highest flood level. Failing this, it may be necessary to arrange to pump the effluent to the river in times of flood.

Provision for future extension

Rarely is an extension of works carried out as originally envisaged. Therefore arrangements of works, and space left for future extension, should be such that they permit the best and most elastic use being made of the site in the future. The extension of sewage works is not necessarily an identical duplication of the complete works, although it is often advisable to make allowance for a

100% extension. In many cases it is found that after entirely new works have been in use for some time, the proportions of the units are not quite right and that, while one component is amply large, another begins to exhibit deficiency as rate of flow or strength of sewage increases. Thus, the extension of works may develop piecemeal. For this reason each unit should be so arranged in the original scheme that an extension of 100% or 200% can be made immediately adjacent to it without undue interference with the existing pipework. Nevertheless, the first layout should be complete in itself, forming a tidy rectangle with no gaps, irregularities, or unexplained spaces.

All this sounds very difficult but, in fact, it is very seldom that a reasonable degree of thought will not produce an economical and tidy arrangement, with possibility of an equally economical and tidy future extension.

Cut and fill

Designers should keep in mind the balance of cut and fill, or make sure that either there will be sufficient excavated earth to form the embankments, or sufficient embankment shown on the drawings to absorb virtually the whole of the surplus excavated material.

It is always desirable to avoid the expense of disposal to a distant tip, and too often the alternative chosen is the objectionable procedure of spreading earth on the site of works. There are several reasons why surplus excavated material should not be spread on the site. In the first place this buries good top-soil which could be used for agricultural purposes, grassland, or ornamental gardens. Secondly, earth tips encourage the growth of undesirable types of weeds. Thirdly, should there be any further construction of tanks, buildings, etc., on the site, the presence of recently-made ground adds to the cost of foundations.

Designing a scheme so as to balance cut and fill and to use the whole of the surplus excavated earth, apart from a small margin, in the formation of embankments, involves a little extra calculation of quantities of excavation and fill during the design of the scheme, but this is justified for it saves appreciable cost.

Whilst for very large works the rectangular plan may be almost essential, it is often possible to improve the aesthetic qualities of the small to medium-sized works by relating the layout of the largest units to the swing of the contours on the site.

The form that embankments take has much to do with the appearance of the works on completion. It is, of course, necessary for the earthworks to be well executed and for the embankments to be properly trimmed to profile, with level berms, even slopes, and sharp arrises; but for the contractor to be able to achieve a good effect the design should be good and the drawings technically correct.

Generally, the rule to work to is that embankments should be simple and square, with as few valleys or irregularities as possible. They should be shown on the drawings as the designer intends them to appear in the finished work. For this last reason the designer should go to the trouble of making sure how the true intersections of the various horizontal surfaces, sloped plane surfaces, and the curving surfaces of the natural ground will appear on the ground. If he guesses instead of working out the true intersections, or makes his drawings vague, he cannot expect the contractor to carry out the work satisfactorily.

Feed pipes and channels

In the earlier designs of sewage works it was quite common practice to lay from the detritus tanks an open channel from which the sedimentation tanks were fed in turn, and to arrange a similar long line of feed pipes to supply the trickling filters, each of which obtained its individual supply by connecting laterally from the main at varying distances from the source. This method of feeding the units has not been found to be very satisfactory. Where sedimentation tanks are fed in this manner, whether their inlets are simple openings or weirs, the tanks nearest to the source of supply take the larger proportion of the flow and deposit the greater quantities of sludge. In a similar manner trickling filters so fed are liable to receive badly proportioned flow, although the sluice valves which control such filters usually permit better control than do the penstocks on the inlet side of sedimentation tanks.

It is now becoming usual practice to design feed channels and pipes in such a manner that the distribution of flow is automatically regulated and thereby the work of the sewage-works operatives is simplified. The normal arrangement is as follows. Where there is a number of units to be supplied, the feed pipes are bifurcated from time to time, each leg of the bifurcation being of equal proportions until the total number of connections has been obtained. This method may require a little more piping than would be required by a simple feed, but it adds to efficiency and is never difficult to design. In carrying out a design of this type, adequate provision should be made for future extension.

This procedure, adequate as it is for the small installation, will not do for very large works serving several hundreds of thousands or a million or more head of population without modification, for it involves excessive lengths of large-size channel, culvert, or pipe with many bends and, consequently, considerable loss of head. Thus, other means of distribution have to be found. One of these is the distribution of flows to the various units with the aid of measuring devices and accurate control valves. Another is for the works to be divided into parallel sections each complete with their own primary sedimentation tanks, aeration channels, and final sedimentation tanks. The flow to each unit is measured and controlled by meter and valve, then divided to each of the primary sedimentation tanks by progressive bifurcation of channel, culvert, or pipe. From then on there would be no further measurement of flow.

The pipework, valves, and other ironwork of sewage works are very seldom discussed in technical literature, but this does not mean that they are not important. There is a correct material or article for use in each circumstance, and the designer who wishes to produce the most effective schemes at the lowest cost should study the available materials and their uses (see Chapter 5).

The pipework at sewage-treatment works includes gravitating mains, inverted siphons, and rising mains, which carry liquids of various solids content. The flow in the majority of these pipes needs to be at sufficiently high velocities to ensure self-cleansing conditions, but the same velocities do not apply in all instances because of the differences in the liquids. Crude sewage entering the works, discharge of wash-outs of sedimentation tanks, and all liquids containing large solids should be delivered by gravity at not less than 0.75 m/s or pumped at about 0.85 m/s.

Feed pipes leading from sedimentation tanks to aeration units, pipes discharging settled effluent from storm tanks to outfall or returning decanted top water from storm tanks to treatment works, wash-outs of the filter-feed system, and pipes receiving drainings from sludge drying beds carry liquids which are comparatively free from heavy solids (but the solids content of which is sufficient to necessitate self-cleansing conditions of flow), all require to be suitably graded. It may be surprising that, after prolonged sedimentation, sewage still contains suspended solids capable of settling in outlet channels if the rate of flow in the channels is not sufficiently fast to prevent settlement, but it has been observed that this is the case. The flow in all the pipes and channels mentioned above should be at a velocity of not less than 0.6 m/s at the peak daily rate of flow.

Clarified effluent from humus tanks does not call for high velocities of flow and no precautions need be taken.

Aerated channels for activated sludge should remain clean regardless of velocity of flow because of the turbulence due to aeration. These consequently are often constructed to give low velocities so as to economize in head. As the capacities of such channels can be included as aeration capacity in the design calculations, it is not extravagant to make them large. However, in activated-sludge schemes which do not include air diffusion, recirculation channels or culverts should be of such proportions that, at the minimum rate of recirculation, the velocity is not lower than 0.46 m/s. As present-day British practice is to allow for recirculation at rates from one third dry-weather flow to dry-weather flow, this would mean a maximum velocity of 1.37 m/s.

Flow of sludge

Primary, semi-digested, and digested sludges have high critical velocities. Above the critical velocity flow is turbulent and (as the author has determined by experiment) head losses are the same as those for either clean water or sewage in the same pipes and at the same temperatures. Also, according to all the information so far available, the critical velocities are always high enough to ensure self-cleansing conditions.

Below the critical velocities flow becomes laminar and head losses do not correspond to those calculated on the basis of any turbulent-flow formula.

Table 35 Minimum gradients for sludge mains

Diameter (mm)	Gradient (length/fall)	Discharge (m³/min)	Critical velocity (m/s)
100	18	0.91	1.86
125	28	1.31	1.72
150	41	1.74	1.59
175	56	2.24	1.50
200	72	2.80	1.43
225	90	3.62	1.38
250	105	4.16	1.37
300	145	5.72	1.31
350	185	7.59	1.27
375	205	8.64	1.26
400	225	9.77	1.25
450	270	12.15	1.23

Moreover, if the velocity is reduced to the region of 300 mm/s, accumulations of silt in the invert and greasy material in the crown of the pipe are to be expected.

For the foregoing reasons it is safe practice to design all primary, digested, and semi-digested sludge mains to flow at all times at not less than the critical velocity appropriate to the diameter of pipe. Table 35 gives critical velocities which are slightly more than the average of those determined by the author and other experimenters.

Recommended reading

Accurate interpretation of the information made available by trial holes or borings is of the utmost importance, both in the interests of economy in structural design, and to rule out the possibility of future structural movement. Most textbooks covering the subject, in addition to theory, contain a section or chapter on site exploration, sampling, and testing, from which the engineer will obtain useful practical guidance. Three such volumes are:

(a) *Soil Mechanics in Engineering Practice*, by Terzaghi and Peck (N. Y.: John Wiley & Sons. Inc.; Lond.: Chapman & Hall, Ltd.) 1948.
(b) *The Mechanics of Engineering Soils*, by Capper and Cassie. Spons Civ. Engnrg. Series. (Lond.: E. & F. N. Spon) 2nd ed. (1953).
(c) *Practical Problems in Soil Mechanics*, by Reynolds and Protopapedakis. (Lond.: Crosby Lockwood & Son).

CHAPTER 11
Preliminary Treatment

Preliminary treatment processes are either for the protection of subsequent units at sewage-treatment works or as a form of crude partial treatment for sewages that are discharged with no other treatment, except perhaps sedimentation. Included in preliminary treatment are detritus settlement for the removal of heavy solids such as road grit, sand, etc., and screening to remove large suspended or floating solids including faeces, paper, contraceptives, rag kitchen waste, and various materials which may have been improperly put into the sewers. The order in which detritus settlement and screening come is not always the same. Where screenings are broken up with the aid of disintegrators, or where the screenings are removed and broken up simultaneously with comminutors, the placing of the screens after detritus settlement units reduces the wear on the plant in question. On the other hand, if the screens are upstream of the detritus-settlement units this arrangement gives protection to the pumps that handle the detritus when it is removed from the tanks or channels. Thus, the decision as to the relative positions of the detritus-settlement and screening units has to be made at the same time as the choice of equipment.

Screens

Screens are classified as manually operated, which means that they are raked by hand, or 'mechanical', having power-operated mechanism for removal of screenings, often automatic. They are also classified as fine, medium, and coarse and there appears to be reasonable agreement on the meaning of these definitions in both America and Britain. By the term fine screens is meant perforated plates with round holes or short slots having diameters or widths of (say 1–6 mm). These are in the form of rotating drums or rotating circular plates; there are also some more unusual devices. Medium screens are usually bar screens or racks having clear spaces of say 6–38 mm. These, of suitable grade, are the most common equipment at sewage-treatment works and for the protection at sewage-pumping stations of pumps of a kind which cannot handle unbroken solids. Coarse screens are bar screens or racks with clear spaces of more than

38 mm. They can remove large floating solids that might cause damage, but are otherwise of limited value.

When sewage is discharged to the sea without other treatment than screening, the purpose of the screening is to take out those solids the nature of which is so obvious that they would be objectionable if encountered by the public. The screens therefore remove the most undesirable elements from the discharge but, while they do to some extent reduce the biochemical oxygen demand of the effluent (by 5–10% in the case of fine screens) they are doing little more than rendering the discharge less offensive. In such a case the use of fine screens has the most effect: but they involve machinery often much more expensive than other installations and require comparatively heavy maintenance. The required net area of the perforations is 1 m^2/9 m^3 per minute at the peak rate of flow.

On the new outfall at Brighton (England)[1] cup screens were specified to have a perforation size of 5 mm, and a total area of perforation in each screen below water level, 12 m^2, to give maximum velocity through the holes of 0.15 m/s. (The screens are shown in Figure 39) p. 212 with a speed of rotation of 7.2 m/min. It was anticipated that blinding of the screen would be limited to about 50%. Initial operating experiences showed that the mesh invariably emerged from the liquid completely blinded, causing an increase in head loss from the anticipated 12 mm to 300 mm. The basic reason for the difficulties experienced was that almost all the screenable solids were found to arrive at the station between noon and 3 pm! The use of expensive screening plant is surely justified on sea outfall schemes, as it is, in many cases, the only treatment process, and its presence affords a practical safeguard against trouble from possible variations in the level of diffusion of the discharge to sea. There is of course an accompanying problem of screenings disposal, and to help with this a twin-screw press, originally developed for the sugarbeet industry, is used at Brighton, reducing moisture content from 90% or above, to 50–55%.

Bar screens of medium grade serve the purpose at sewage-treatment works of removing solids which could damage pumps, choke small-diameter pipelines, or be unsightly should they settle on the weirs of tanks. They can also remove those materials such as contraceptives and corks which accumulate in digestion tanks because they are lighter than water and virtually indestructible; but for the last purpose the screens must have narrow end spaces. There is considerable difference of opinion as to the clear spaces of bar screens. The writer's experience is that to intercept corks and a fair proportion of contraceptives, the clear spaces between bars should not be more than 19 mm and preferably not more than 16 mm.

Hand-raked screens

Hand-raked screens with close spaces, e.g. 12 mm, are usually empoyed at small works. These should have submerged areas of not less than 0.14 m^2 per 1000 head of population, on the assumption that they are raked three times a day. If the screens are raked only once a day the area should be increased threefold.

They should be set at an angle which is easy for raking, and at the same time ensures a good area of submergence.

One of the problems involved in the design of screen chambers is provision for detritus removal. Screens necessitate some increase in the cross-sectional area of flow of sewage, thereby reducing the velocity and causing settlement of the heavier solids. This applies particularly to the chambers containing hand-raked screens which are unavoidably of comparatively large proportions, and it is for this reason that hand-raked screens are usually placed in detritus tanks (see Figure 60).

At small works treating flows from separate soil sewers, settlement of detritus is not always necessary, because the amount of grit is small and the screen chamber can be arranged with a hopper bottom, from the apex of which the outlet carries away all solids to gravitate or be pumped to the sedimentation tanks. But heavy detritus can settle and pack so as to make sludging difficult under hydrostatic head. For this reason small tanks incorporating screens and

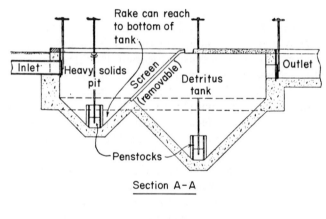

Section A–A

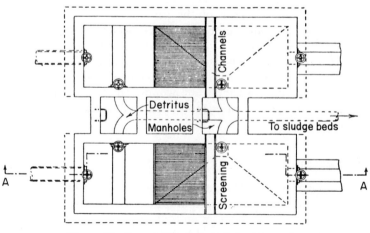

Figure 60   Screen chambers and detritus tanks

arranged for settling detritus should have sludge outlets that are very easy to clear.

In fact, of late years there has been a tendency for the 'separate' sewage to be pumped to the works, to dispense with any preliminary treatment at all, and apportion the flow to the settlement tanks direct from the pump delivery chamber. This is done on the grounds that the pumping process will break down the solids, and whilst this is to some extent true, only a 'stereophagus' type of pump will really achieve this; in other cases a watch must be kept for rags, etc., which may render the 'sludging' of the settlement tank more difficult.

Mechanically-raked screens

There are several types of bar screen available, and the type selected must depend on circumstances and the price justified. For deep to very deep sewers, vertical or nearly vertical screens are usual with tines that carry the screenings to the surface, sometimes against a back-plate which continues all the way from the top of the screen to floor level of screen house. The tines are usually moved on a continuous circuit carried by chains or steel cables. Where the sewer is not at a great depth below ground level, other types of screen, such as the design manu-factured by Dorr-Oliver,[2] can be used (see Figure 61). Screenings are removed

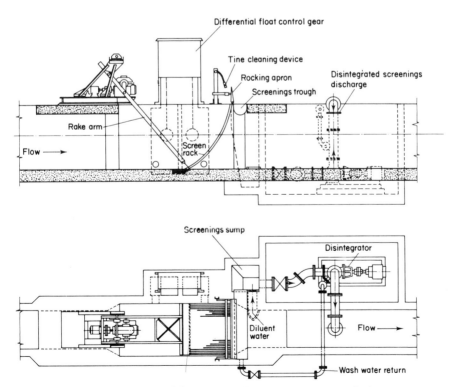

Figure 61   Dorr-Oliver screen (Courtesy Dorr-Oliver Co.)

from tines by sparge, brush, or rotating comb and carried away by water-carriage or belt conveyor.

When screens are mechanically raked they can be set in operation by hand when required, provided with a time switch so as to be raked at regular intervals, or provided with differential-float controls so that they are automatically raked whenever chokage with screenings causes a difference of water level on the upstream and downstream side, or they may be left running continuously.

Continuous raking is uneconomical as a rule, and in the majority of instances the use of differential float control is most satisfactory, for it minimizes the size or number of screens required and does away with unnecessary wear and tear and use of power. For this method to be used it is necessary to calculate the differences of head across the screens when they are clean and when they are fouled, to the extent of reducing the clear spaces between the bars of the submerged part of the screen by one half. The difference of head between the two conditions should, satisfactorily to work the differential-float mechanism (see Figure 61) be at least 150 mm. The necessary differential head must be calculated to make sure that the screen-raking gear comes into operation when the screens are partly choked with screenings, and the gear should be adjustable so that the rakes are not in operation when the screens are clean. The loss of head across the screens can be calculated near enough by the orifice formula

$$Q = mA\sqrt{2gH} \tag{28}$$

where: Q = cubic metres per second (cumecs);
H = Differential head, in metres, of water above and below screen;
A = Submerged area of openings in screen, calculated by multiplying the sum of the width of the openings in metres by the vertical depth less one-third of the head in metres (assuming that the upstream water level is not above the top of the screen);
g = acceleration under gravity, 9.80665 metres per second per second;
m = a constant for the type of orifice, which in the case of clean bar screens is about 0.6.

It is necessary to maintain self-cleansing conditions in the chamber that houses a bar screen at all rates of flow; otherwise detritus will settle and call for means for removal. The flow velocity through the chamber should never drop below 0.6 m/s for ideal conditions, but this means that the velocity between the bars of the screen will be considerably higher because the bars themselves occupy no small part of the width of the chamber, and usually 150 mm of the width is occupied on each side by the side frames of the mechanism. Thus, suppose a screen chamber is 1.5 m wide, the space available for the screen after deducting the side frames will be 1.22 m and, if the bars are 1.3 mm thick and have clear spaces of 19 mm, the total width of flow through the screen will be 0.73 m. Thus, when the velocity of flow through the chamber upstream of the screen itself is 0.76 m/s the velocity through the spaces of the bars must be no less than 1.6 m/s, calling for a differential head of about 0.33 m.

According to Fair[3] the maximum head loss through bar screens is normally limited to 0.76 m. Babbitt[4] considered that the screens should be capable of bearing a differential water pressure of 0.6–0.9 m. The author has observed a mechanically raked screen working satifactorily with a differential head of 1.1 m.

Screens should be isolated upstream and downstream by penstocks, and the chambers arranged so that they can be drained by gravity or pumping and in order that men may enter the chambers for maintenance purposes.

Disposal of screenings

At very small works screenings are disposed of by burial in trenches or composting with other materials. Burial is not altogether satisfactory, and is quite unsuitable for large works because of rat infestation. In America, incineration of screenings is considered satisfactory, the material first being dewatered by pressure then burnt in gas, oil, or coal furnaces at high temperature involving the consumption of 900–1400 kW/kg, or about 3700–5800 kJ/kg of screenings destroyed on the basis of weight before dewatering.

Incineration is not popular in Great Britain for two reasons. First, it is costly in fuel; secondly, in view of the high efficiency of British primary sedimentation, there is no reason for the screenings to be kept out of the flow of sewage after they have been disintegrated. In Britain screens are used to protect the works, not to reduce the biochemical oxygen demand.

By far the most popular method of dealing with screenings in Europe is to break them up with the aid of disintegrators such as those manufactured by Sulzer Bros. Ltd. The screenings are first diluted with a flow of water in the ratio of 100 parts of water to one of screenings and gravitated through the disintegrators which break them into small pieces. The flow is normally discharged back to the flow of sewage upstream of the screens so as to intercept any screenings which have become matted together again.

It has been remarked that once the screenings have been removed from the sewage they should not be returned because this is putting an unnecessary load on the treatment plant. But the amount of screenings taken off bar screens is quite small compared with the quantity of similar material which passes between the bars without being intercepted and is insignificant as compared with the total organic load on the works. Moreover, the whole purpose of disintegration is to avoid the unpleasant duties of handling screenings or having to dispose of this very objectionable material.

The particulars of Sulzer disintegrators as given in the manufacturers' catalogues are shown in Table 36. The dilution water into which the screenings are mixed before they are disintegrated is usually drawn from downstream of the screens, and the pipe used should be of the same diameter as the suction branch of the disintegrator. The soffit of this pipe must be below the lowest possible water level and a screw penstock or valve should be fitted to adjust the flow. The soffit of the disintegrator suction should be at the level of the invert of the dilution-water pipe.

Table 36   Recommended working conditions for sulzer disintegrators

Size	Suction diameter (actual) (mm)	Delivery diameter (actual) (mm)	Delivery (m³/min)	Output of screenings (m³/d)	Total head (m)
D8	203	203	1.82	5.7	1.83
D10	254	203	2.27	7.1	1.58
D12	305	254	4.55	14.2	2.44
D15	381	305	7.73	25.5	2.28

(Published by courtesy of Sulzers)

Hammer mills, also known as impact grinders, have been used for breaking up wet screenings but, whereas these are very useful for grinding dry solids, their horse-power demand is uneconomical when applied to materials with moisture contents of 80% upwards.

The quantity of screenings collected varies according to the size of the sewerage system and the spacing of the screens. Large sewerage systems give smaller quantities per head of population, because the larger solids become broken up in transit through the sewers. Generally the quantities tend to vary between 0.01 to 0.03 m³ per 1,000 head of population per day. Screenings weigh from 700 to 900 kg per m³ and have a moisture content of from 80% to 85%. About 85% of the dry matter is volatile.

Screen houses should be rendered fly-proof by having windows protected with coarse copper-wire gauze. It is also advantageous if air can be blown into the building by gauze-protected fans and allowed to escape via the sewer.

As an alternative to the disintegrator, and capable of a wider range application, a device known as the Mono-Muncher, manufactured by Mono Pumps Ltd., Muncher Division, of Audenshaw, Manchester, has recently become available. This is a low-speed machine which nips all solids between intermeshing, contra-rotating toothed cutters, operating at different speeds, thus setting up a tearing action, capable of shredding nylon, timber, plastic, metal cans, rags and small particulate matter. It is capable of passing much larger volumes of flow than a disintegrator, and can be installed in a normal inlet channel, in a similar way to comminutors, as described under the following heading. It renders screens unnecessary, is low in power consumption and by its more positive action is known to avoid the problem sometimes associated with comminutors of the 'balling-up' of fibrous matter after passing through these machines. The Muncher will operate in pipes of various diameters, as well as in channels.

Comminutors

Comminutors are largely used on both sides of the Atlantic.[5] They serve as both screen and disintegrator in one, stopping large solids and breaking them up until they can pass through with the flow of sewage. The comminutor has an electrically-rotated drum with horizontal slots which form a screen. The flow approaches the drum from the outside and passes through the slots to the inside, from which the screened sewage passes downwards. Fixed to the drum is a series of shear bars and cutters which bring any solids too large to pass the slots in contact with a fixed comb, against which the particles are sheared. The horse-powers involved are very low.

Comminutors are limited in their application in that the head must not be too small or the screenings are not held against the cutters by the flow; also the head must not be too great. The manufacturers recommend a minimum head of 50 mm for the smallest and 100 mm for the largest, and a maximum head of 175 mm and 375 mm respectively, the variation of head being according to the rate of flow, the comminutor behaving as a submerged orifice. In some circumstances control of water level downstream by means of weir or standing-wave flume is necessary. Generally, the comminutor should be set at such a level that the head loss at maximum rate of flow does not exceed the permitted maximum and is not less than the minimum figure at $1\frac{1}{2}$ times the average daily flow.

The slots in the comminutor vary in width from 6 mm to 9 mm according to the size of the machine: thus they give good screening. Comminutors are suitable for both large and small works.

Comminutors should be placed downstream of detritus tanks or channels so as to reduce the wear on the cutting parts. There is, however, no need to protect Comminutors by screens, for they are quite capable of dealing with most of the floating solids likely to be found in sewage and are less liable to damage than are many mechanically raked screens.

Where any form of flow separation takes place some economy can be achieved by locating the comminutors (or screens) downstream of the separating weirs, or other device. If a standing-wave flume is being used to control the level of separation a graph can be prepared combining the head loss through the flume with that through the comminutor, to determine the relative separation level. On small or medium-sized works where flows are within the range of a single comminutor, still further economy is possible by providing a manual screen instead of duplicating the comminutor to cater for periods of maintenance or repair.

Detritus settlement

The design of detritus tanks or channels has passed through stages of evolution and there is now a range of types, which may be classified as follows.

Small detritus tanks or grit chambers

These are small tanks or chambers usually provided in duplicate, and frequently housing the screens at small sewage works. While in the past in Britain these tanks were made so as to have a total capacity of one-fiftieth of the dry-weather flow, a much smaller capacity is desirable. Generally, where the simple detritus chamber is unavoidable it should be made no larger than necessary to house the screens (see Figure 60).

Large detritus tanks or grit chambers

The large detritus tanks or grit chambers, which at one time were installed at comparatively large works, were never used to house the screens, being too big for that purpose, and generally came before the screens. They were in the form of long deep tanks with the bottoms sloping towards central channels, and were sludged by means of grabs, dredgers, or moving grit pumps. The detritus so removed contained a large proportion of organic matter.

Mechanically-swept detritus tanks

These are provided with raking gear similar to that employed in flat-bottomed circular sedimentation tanks. One of the best known is the Dorr Detritor which comprises a shallow tank, either circular or square with round corners, in which the sewage flows from one side to another. The grit which settles in this tank is swept by a rotating raking gear to a pocket on one side of the tank, from which it is evacuated by means of a reciprocating rake mechanism, the action of which is to wash the grit free from entrained organic matter. Grit can also be evacuated from the pocket by means of a pump to a separate cleaning mechanism if desired (see Figure 62).

The size of the grit-collecting tank is based on surface area according to Hazen's theory (see Chapter 13). The area chosen for any particular installation is that required to settle out the smallest desired particle at the maximum rate of flow. Under low-flow conditions much finer material and considerable organics will be settled in a tank of given area, but the subsequent washing process classifies out the fines and the organic matter, returning them to the sewage flow. The falling velocities of small particles of siliceous sand, as determined by experiment, are given in Table 37. This shows that if the tank were to be 1 m deep, and it were desired to settle with reasonable efficiency a particle of 0.02 cm diameter, the detention period of the tank would have to be not less than $1 \times 55 = 55$ s. Some engineers design on the net figure as calculated above but, because turbulence in the tank militates against settlement and reduces the actual falling velocity, a factor of safety of, say, 2 is advisable.

The Dorr–Oliver grit washer consists of an inclined channel with a ladder-like mechanism to which a reciprocating motion is induced so as to push the grit up the slope against a flow of wash water. Eventually the comparatively clean detritus is discharged into a tip-wagon while the organic material gravitates back to the flow of sewage upstream of the detritus tank. Trouble has been experienced with tea leaves, which readily settle in the detritus tank, and are washed out of the detritus by the grit washer and therefore tend to accumulate in the circuit, though more recent installations do not seem to have suffered from this problem.

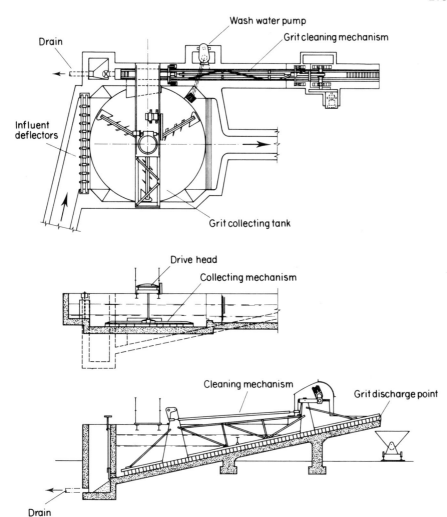

Figure 62 Dorr–Oliver Detritor and grit-washing mechanism (Courtesy Dorr–Oliver Co)

Table 37 Falling velocities of siliceous sand

Diameter (cm)	Falling speed (cm/s)	Seconds per metre of fall
0.013	1.0	100
0.02	1.8	55
0.05	5.0	20
0.1	10.0	10
0.4	25.0	4
1.0	43.0	2

The Dorr–Oliver grit washer or classifier is probably the most widely used machine for this purpose. It is, however, being challenged by a method involving the cyclonic principle. In this the grit-laden water is pumped into a cyclone tangentially, causing it to rotate at a high speed and thereby centrifuge the heavy solids to the outside while the water containing organic material is drawn from the centre of the eddy.

Detritus channels

Under the head of detritus channels come several designs varying from long detritus tanks, in which the velocity of flow is comparatively slow, to constant-velocity detritus channels. A design much favoured some years ago incorporated a number of channels which were brought into use one at a time in accordance with the rate of flow, so that the velocity of flow was kept in the region of 300 mm/s at all times of the day. This was sometimes accomplished by automatic electric penstocks, float controlled. The design is applicable to large works only.

Flume- and weir-controlled constant-velocity detritus channels

Constant-velocity detritus channels are usually two or more in number, each being controlled by means of an orifice, weir, or flume downstream. The cross-sectional area of the channel is so proportioned to the orifice, weir, or flume that, whatever the rate of flow, the velocity in the channel will remain constant. There are two main classes of channel that come under this head:

(1) the channel of parabolic (or approximate equivalent) cross-section, controlled by a rectangular standing-wave flume;
(2) the channel of approximately rectangular cross-section, controlled by a specially shaped weir plate.

In the design of constant-velocity detritus channels, a moderate variation of velocity above or below 300 mm/s is permissible, it being held that, provided the velocity does not fall below 225 or more than slightly exceed 300 mm/s, the channels will perform reasonably. This allows a little latitude in the design of the controlling weir or of the channel, making possible some economies in cost of construction. However, any drop of velocity below 200 mm/s will permit faecal matter to settle, and any increase above 375 mm, particularly at low rates of flow, will cause some detritus to be swept away. Generally in design, because of the difficulty of doing otherwise, velocity is permitted to drop at very low rates of flow only, but otherwise to remain close to 300 mm/s.

Constant-velocity detritus channels are sometimes used so many at a time, the flow being controlled by automatic-electric penstocks in addition to the downstream weir or penstock, but generally this is an unnecessary elaboration.

Parabolic detritus channel controlled by rectangular standing-wave flume

The discharge of a rectangular standing-wave flume is according to formula (27), p. 252. The cross-section of a channel which increases in area directly as the rate of flow through a rectangular flume increases is, parabolic, the sharpest part of the curve forming the invert. The width of the channel at any point above the invert to give a velocity of 300 mm/s is

$$X = 4.92Q/H \qquad (29)$$

where: X = width of channel at water level in metres;
H = depth of flow in metres;
Q = rate of flow at that depth in cubic metres per second.

The first figure to be determined is the depth of the channel. Where fall is not important this may be varied considerably, according to the designer's preference. But where there is not much fall it may be found advisable for the value of H to be equal to the depth from crown to invert of the incoming sewer, plus the friction loss through the channel so that, when the sewer is flowing full at the maximum rate of flow, the flume and channel are also flowing full. With the depth of flume fixed, the necessary top width of flume can be found. In many instances it may be found that the channel, the width of which is so calculated, is too wide and flat for mechanical cleansing, or is liable to encourage stranding of faeces etc., at the edges. If this is so, all that is necessary is to arrange for two or more channels, the total widths of which are equal to the calculated width, to discharge to one flume. Then, for obvious reasons, the sides of the channel will be steeper and more liable to prevent stranding of organic solids and to make the grit gravitate towards the centre of the invert.

When there are no difficulties as to fall, the designer may commence by choosing a width of single channel equal to about the depth, and in any case not more than twice the depth, thereby securing steep sides, and he can determine the proportions of flume accordingly.

Theoretically, the length of channel must be sufficient to permit the smallest particle to be settled to fall from top-water level to invert of channel during the detention period, in spite of the unavoidable turbulence which tends to prevent settlement.

Detritus which settles at the rate of 300 mm in 16 s (and, more recently, 300 mm in 10 s) has been allowed for in the design of some detritus tanks or channels. Obviously, such fine detritus would not all be settled in a channel just sufficiently long for a particle at the surface at the inlet end to fall to the invert before it reached the outlet. Lacking a better rule, the writer has allowed channels of twice the length that would be required to settle a particle of the predetermined size in the absence of turbulence. This has meant that the length of each detritus channel has had to be not less than 32 times the maximum depth in the case of particles that fall 300 mm in 16 s. There is usually no settlement near the inlet because of inlet turbulence.

The working of constant-velocity detritus channels depends on whether the channels are manually or mechanically cleansed. When there are two channels only, one takes the whole of the flow (including storm flow), the other remaining idle; when there are more than two channels, a number together are able to take the total maximum rate of flow, one or more remaining out of commission.

In the case of manually cleansed channels, as the accumulation of detritus spreads towards the outlet ends the channels are put out of commission by closing the penstocks at the inlet ends and the flow diverted to other channels. The sewage then runs out through the standing-wave flumes, the inverts of which should be well above the downstream top-water level in order that no backflow from other channels can take place. The detritus is then dug out. This arrangement is applicable to medium-sized and comparatively small sewage works, not sufficiently large to justify mechanical equipment. The labour involved is not great.

At large sewage works detritus is removed, either by mechanical dredger or by moving grit pump, while the channels are in commission. A dredger is preferable if the grit is to be removed comparatively free from water and dumped for drying; grit pumps are preferred if the detritus, after removal, is to be passed through a grit washer so as to render it clean and either usable on site or saleable.

At very large sewage works, where rates of flow vary considerably, it may be found advantageous to increase the number of channels in accordance with the rate of flow—automatic electric penstocks, float-operated, being installed for this purpose. But in most instances an installation consisting of two channels only, one of which takes the total flow, is to be preferred at works treating the flow from separate sewerage systems, because of its simplicity.

The two-channel installation can be applied to very large flows without difficulty. For example, a parabolic channel in which the width at top-water level is about twice the maximum depth, is convenient for all sizes of works, provided the maximum rate of flow is not too great compared with the average rate of flow. This channel can be applied even to works receiving so small a flow flow as 500 l/min. In the case of such small detritus channels, a simple form of construction is a half-round stoneware or pre-cast concrete channel set in concrete. A 300 mm half-round channel will accommodate such a flow. It can be cleansed with the aid of a purpose-made spade or scoop.

An increased number of channels, and automatic electric control, become desirable when there is a great difference between dry-weather flow and maximum rate of flow, owing to the inclusion of appreciable quantities of storm water.

The figures in Table 38 give the results produced by some constant-velocity detritus channels equipped with an Ames Crosta detritus-removing mechanism.

Rectangular channels controlled by proportional-flow weir

The Rettger proportional-flow weir is an opening in a thin plate like an inverted T with curved sides (see Figure 64). The formula for the discharge of this type of

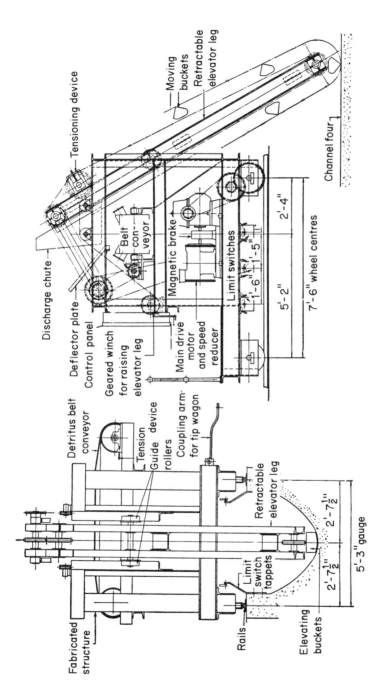

Figure 63   Dredger for parabolic constant-velocity grit channel (W. E. Farrer)

weir has been given as

$$Q = 5.38C\sqrt{2g}\,b\pi h \qquad (30)$$

where: Q = discharge in cumecs;
$\quad\quad h$ = depth of flow in metres over theoretical crest;
$\quad\quad b$ = weir constant;
$\quad\quad C$ = coefficient for thin plate weir, usually taken as 0.62.
$\quad\quad g$ = acceleration due to gravity = 9.80665 metres per second per second

From this formula the value of b is calculated. Then the curves of the sides of the plate are plotted by the formula

$$z = \frac{b}{0.33\sqrt{x}} \qquad (31)$$

where: x = the vertical distance in metres measured from the theoretical crest of the weir to the curve of the plate;
$\quad\quad z$ = the distance in metres from the centre line of the opening to the curve of the plate.

It will be noted that the two arms of the T run off to infinity; therefore in practice they are shortened and an additional area added to the orifice by lowering the actual crest below the theoretical crest sufficiently to compensate for the lost area.

Table 38   Analysis of detritus taken from channel (percentages)

Content	At inlet	At middle	At outlet
Moisture	30	30	32
Organic	5	6	5
Inorganic	65	64	63
Mechanical analysis of inorganic content			
Above 16 mesh	25	23	10
Above 30 mesh	13	13	16
Above 100 mesh	50	50	57
Silt	12	14	17

In English practice it is usual to arrange this type of weir, not in the form of the theoretical Rettger plate or Sutro weir, but to an approximation which in most instances amounts to the use of a standing-wave flume or a trapezoidal notch in conjunction with an orifice. If, for example, a rectangular plate weir or standing-wave flume, with invert level to the invert of the detritus channel and of such a width as to accommodate about one-half the flow in the channel, is arranged in parallel with a wide low orifice set at the invert of the channel and proportioned so as to take the remainder of the flow, it will be found that the combined flow through weir and orifice varies almost directly as the head in

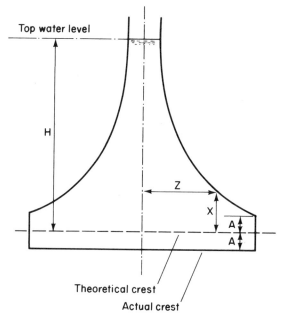

Figure 64   Proportional weir for rectangular constant-
velocity grit channel

the channel except at very low rates of flow, with the result that a practically constant velocity is achieved, except that velocity falls off when the channel is running at very shallow depths. By use of methods of this kind it is possible to devise a suitable weir to serve a channel of almost any cross-section.

Although for years a feature of many works, both large and small, particularly in Great Britain, the constant-velocity grit channel is giving way to the Dorr Detritor, or the cyclone separator, both of which are more compact on large works, and which with their attendant cleaning devices provide reasonably satisfactory means of grit disposal without the problem of lifting it from long channels, which, despite the development of a number of mechanical aids, has sometimes proved difficult, and without such aids, unduly time-consuming. Fischer,[6] writing in 1976 of U.S. practice, although mentioning velocity control by weirs and flumes, refers only to grit collector tanks, and uses the Dorr Detritor as his illustration.

Quantity of detritus

The quantity of detritus is very variable and, owing to modern methods of road construction, is probably less than early in the century. It is more in some countries than in others. It is least in domestic soil sewage from separate systems and greater in combined sewage where flows are taken from heavily gritted roads. In old sewerage systems with badly graded sewers large quantities may be washed down to the works in storm times. In one instance the amount of detritus was found to average 0.007 m³ per head of population per annum: about two-thirds of this figure has often been adopted for the basis of design. Dry detritus weighs in the region of 1200 kg/m³; nearly all the material will pass an 8 mm sieve.

Grease removal

In America grease is removed from sewage prior to sedimentation; in Great Britain generally, this has been found unnecessary. There does appear to be some difference in the grease content of the sewages, which may be due to different culinary practices. During the World War of 1939–45 the author had to deal with American military camp grease troubles which were due to canteen practices and which he did not find elsewhere.

One of the most effective means of removing grease from sewage is passing the flow through a channel having a detention period of about 3 min at the maximum rate of flow, and applying diffused air as in the diffused-air activated-sludge process. By this means the grease is absorbed by the bubbles and forms a scum which may be removed from the surface.[7]

Grease can be removed by aerating sewage and then placing the sewage in a partial vacuum to cause air and other gas to be driven off, the bubbles carrying the grease to the surface whence it can be removed. The arrangement is a circular tank closed at the top and with entrance and exit at the bottom submerged in the flow of liquid. Suction is applied by vacuum pump, to raise the level of the sewage in the tank to near the roof, and for vacuum to be maintained. Equipment for this type of plant is obtainable from Dorr-Oliver.

Application of air in mechanical flocculation

Although the advantages of aeration before settlement are widely appreciated in the United States, the practice has so far found little popularity in Great Britain, where the possible advantages have either not been appreciated, or for some reason do not seem to have materialized. The practice in the United States[8-11] is to provide a 30 min aeration period with an air supply from 0.1–0.4 m³/min per metre of tank or channel. The number and size of diffusers required is found by dividing the estimated air requirements by the optimum difference per unit, commonly 0.1–0.2 m³/m² of diffuser area. Water depths are usually 2.5–4.0 m. For rectangular tanks the ratio of flocculating compartment length to width may be as high as 6 to 1.

Standard US practice[8] recommends 30 min of mechanical flocculation before sedimentation, to obtain 10–30% higher removal of suspended solids in a sewage of normal strength, thereby achieving 35% greater effectiveness of settling tank capacity.[12] Air agitation as previously described is claimed to be equally effective. Tanks may be combined units with mechanical sludge collection in both flocculation and sedimentation compartments. In longitudinal tanks flocculation and settling may be achieved in a single straight-line tank, comprising two compartments divided by baffles, with sludge-collector flights traversing the entire length in the direction of flow. Square or rectangular tanks may be used with mechanical paddles inside. With two radial-flow impellers in a rectangular basin, length should be double the breadth, and water depth 2–4 m. Circular tanks may have flocculators either in a central compartment, or in a peripheral inlet channel, tangential to the incoming flow. Mixing in a central compartment can be by means of rotary paddles on vertical axes, or by air-diffusers, tubes, or nozzles. The direction of flow is downward, and then outward to an annular settling compartment. The 30 min detention time should be followed by 1.5–2.0 h settlement at an overflow rate of 40 m^3/day per m^2. Flocculation periods may be as low as 15 min, followed by say $2\frac{1}{4}$ h settlement, if followed by activated-sludge treatment; with trickling filters flocculation may be followed by 1 h settlement with an overflow rate up to 50 m^3/day per m^2. The number of installations on these lines has increased rapidly since 1945.

Recent developments 'Conditioning' with oxygen

This chapter would not be complete without a reference to true preliminary treatment such as has been introduced in recent years by the application of oxygen in rising mains, and sewers. This technique has resulted not only in the eradication of septicity in the sewage reaching the treatment works, but in actual reduction in BOD_5, enabling economies to be made in the capacity of the treatment units.

A good deal of interest has developed, at any rate in Great Britain, in the possibilities for the 'conditioning' of sewage, in either sewers or rising mains[13] prior to arrival at the treatment works, by the injection of oxygen. Whilst originally aimed at avoiding septicity, interest has grown in the possibilities of reducing the load on the treatment works, both in terms of BOD_5 and suspended solids, thus postponing the necessity for capital expenditure on works extensions.

Commenting on these possibilities Boon[14] remarks that sewage rarely becomes septic in gravity sewers, because turbulence in the flow ensures adequate aeration. The potential for sulphide formation is greatest in rising mains, where sewage is pumped out of contact with air. Hydrogen sulphide formed in the sewage does not usually result in corrosion of a rising main and will only affect the sewerage system if liberated to the atmosphere when it can corrode concrete and steel, and produce odour nuisance at any nearby point of ventilation.

Skellett[15,16], describes how it was possible to inject oxygen at an hourly rate of 14 mg of oxygen per litre of sewage contained in a rising main and at 700 mg O_2/m^2 of pipe surface. The demand depends on many factors: 'age' of sewage, concentration of substances inhibitory to biochemical oxidation, pH value, and temperature. Uptake varies widely. A 'young' domestic sewage may have a demand of only 2–3 mg O_2/l per hour, but this can increase to values greater than 20 mg O_2/l per hour, as the sewage ages, but finally declines as the

282

concentration of organic matter in the sewage decreases. The use of oxygen rather than air is justified largely because of the high concentration (perhaps 200 mg/l) that would be required. Given sufficient oxygen, the BOD_5 of the sewage will be significantly reduced during passage through the main. It should be possible to construct a rising main of sufficient length and internal surface area to ensure that, when it becomes coated with slime, sewage flowing through it would be substantially treated by the time it was discharged and would require little more than settlement to remove accumulated solids before being discharged to a watercourse. It has been shown that on a trickling filter works, known to be overloaded by 40%, treatment of crude sewage in the rising main enabled it to produce a consistently high-quality effluent (31 mg/l suspended solids, 19 mg/l BOD_5, and 75% reduction in ammoniacal nitrogen) during the summer months, when flow in the river provided little dilution, avoiding an increase in the pumping power requirements, which averaged 50% higher during oxygen injection, for the same volume of sewage. This was due to gas-locking at high points on the main, and emphasizes the need for careful study of conditions in any rising main, where oxygen injection is to be applied. Although alternative methods of preventing septicity, such as injection of hydrogen peroxide or chlorine were tried, and could possibly have been less costly, the overall saving due to the benefits obtained from oxygen injection showed a clear and economic advantage in its use.

Study of many existing sewerage systems may well reveal conditions where septic sewage from rising mains is adversely affecting the condition of a much larger quantity of sewage in the lower reaches of a gravity system. Application of quite small quantities of oxygen, in such cases, might materially lighten the load on the treatment works. Oxygen injection may conveniently be applied at a pumping station. A schematic diagram of the injection system at Twerton pumping station, Bath, is shown in Figure 65. Liquid oxygen stored in an in-

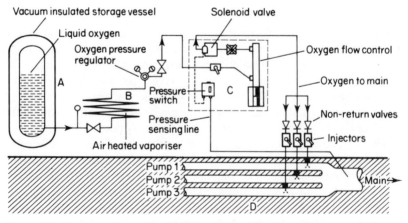

Figure 65   Twerton Pumping Station—Schematic Arrangement of Oxygen Injection System (*J. Instn. Public Hlth. Engnrs.*)

sulated vessel (A) from which oxygen is vapourised (B) and piped into the pumping station, to a small control box (C) containing a gas-flow meter, and a pressure switch to provide automatic shut-down of oxygen supply when pumping stops. Oxygen is then fed via a fine nozzle injector into the inlet pipe (D), from each pump. Injection points are chosen to provide maximum pressure and good mixing. The station is fully automated and the only attention required to the oxygen equipment is a daily check of injection rate and tank pressure.

References

1. Osorio, J. D. C. The design and construction of a screening and pumping station and an outfall for the disposal of sewage from Brighton and Hove. *J. Instn. Public Hlth. Engnrs* **5**, pt. 4 (1976).
2. Dorr-Oliver Inc., 77 Havenoyer Lane, Stamford, Connecticut, USA, and in England at Norfolk House, Wellesley Road, Croydon, London.
3. Imhoff, K. and Fair, G. M. *Sewage Treatment*, 2nd edition, John Wiley & Sons N.Y. (1956)
4. Babbit, H. E. *Sewerage and Sewage Treatment*, 7th edition. John Wiley & Sons, N.Y. (1953).
5. Campbell, L. J. Preliminary treatment processes in sewage purification. *The Surveyor, Lond.*, **111**, 55 (1952).
6. Hays, T. T. Air flotation studies of sanitary sewage. *Sewage and Ind. Wastes*, **28** pt. (1), 100 (Jan. 1956).
7. Imhoff, K., Müller, W. J. and Thistlethwayte, D. K. B. *Disposal of Sewage and other Waterborne Wastes.* Butterworth, London (1956).
8. *Sewage Treatment Plant Design.* Jt. Comm., Am. Soc. of Civ. Engnrs. and Fed. of Sewage and Ind. Waste Assocs. (1959).
9. Fischer, A. J. Sanitary sewage treatment. *Handbook of Water Resources and Pollution Control* (edited by Gehm, H. W. and Bregman, J. I.). Van Nostrand-Reinhold, N. Y. (1976).
10. Wisely, W. H. Trend towards pre-aeration. *Sew. Works J.*, **21**, pt. (4) (July 1949).
11. Roc, F. C. Pre-aeration and air flocculation. *Sew. and Ind. Wastes*, **23**, pt. (2) (Feb. 1951).
12. Camp, T. R. Studies of sedimentation basic design. *Sew. and Ind. Wastes*, **25**, pt. (1) (Jan. 1953).
13. Pomeroy, R. D. and Parkhurst, J. D. Self-purification in sewers. *Proc. 6th Int. Conf., Wat. Pollution Res.*, Pergamon Press, Oxford (1972).
14. Boon, A. G. Technical review of the use of oxygen in waste-water treatment. Symp., *The Use of Oxygen in Public Health Engineering*. Cranfield Inst. of Technol., Cranfield, Bedford. Instn. Public Hlth. Engnrs. (Apr. 1978).
15. Skellett, C. F. Oxygen injection into rising mains and its effect on treatment. Symp., *The Use of Oxygen in Public Health Engineering*. Cranfield Inst. of Technol., Beford. Instn. Public Hlth. Engnrs. (Apr. 1978).
16. Boon, A. G., Skellett, C. F., Newcombe, S., Jones, J. G. and Forester, C. F. The use of oxygen to treat sewage in a rising main. *Wat. Pollution Control*, **76**, 98–112 (1977).

CHAPTER 12
Separation and Treatment of Storm Water

When sewerage is on the separate system storm water usually is discharged without treatment, while the soil sewage from the sanitary sewers has all to be fully treated. The pipework, channels, etc., of the treatment works have to be able to accept the maximum flows that the sanitary sewers can discharge, otherwise flooding will occur; and the units must be able to treat these flows without the processes being upset. If, for example, the main sanitary sewers can pass six times dry-weather flow (DWF), the treatment works should be designed to take this flow and produce an effluent of acceptable quality.

If the sewerage is on the combined system, or even partially separate, the storm flows will often be far too great to be treated as they come: therefore they must be retarded and later fed through the works as and when they can be taken. It is only of late years that it has been appreciated that short peak storm flows equivalent to about $40 \times \mathrm{DWF}$ are not uncommon, and that even a storm of several hours duration may well be equal in intensity to $12 \times \mathrm{DWF}$. The total quantity of storm water in the year, and the amount of pollution it could cause, is also far from negligible.

The usual solution of this problem is separation of storm water and the provision of stand-by tanks.

Separation of storm water

The whole of the flow of sanitary sewage combined with storm water that arrives at the sewage works is commonly passed through the screens and usually through the detritus-settlement units, before separation takes place. As much as can be given full treatment in the primary sedimentation tanks, aeration unit, and final sedimentation tanks is accurately measured and allowed to pass through, while the remainder of the flow is separated and passed to the storm tanks. At small works it is sufficient to provide an adjustable orifice to limit the flow to be treated, and overflow weirs for the storm water, as illustrated in Figure 66; but more elaborate means are required at large works. The flow to be treated may there be controlled by one of the proprietary unchokable modules that are

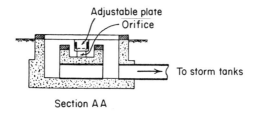

Section A A

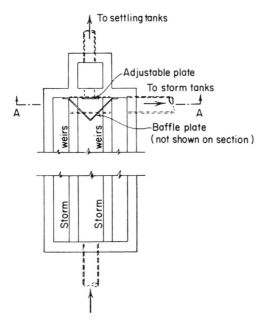

Figure 66   Storm water separating chamber

available, or by float-controlled automatic electric penstocks. Where sewage pumping is involved the best arrangement is to have a separate installation of pumps which will deliver the desired quantity of sewage to be treated, when the water level in the suction well is at its maximum, and other pumps to lift the storm water to the storm tanks.

Storm tanks

The purposes of storm tanks are:

(1)   to store excess of storm flow over the maximum rate of flow to be treated, and retain it for return to the sewage works after the storm;

(2)   to partially treat by settlement any excess of storm water over the quantity that the treatment plant can accept and the storm tanks can store.

The capacity of the storm tanks has to be calculated on the basis of the estimated maximum storm flow likely to occur in, say, one year, the rate of flow that can be put through the treatment works, and the amount of pollution that can be permitted on the rare occasions when the outlet weirs of the storm tanks are compelled to spill over.

The marked difference formerly to be observed between American practice— where combined systems were often provided with storm overflows, thus reducing the rate of outflow to the treatment works, and rendering storm water tanks unnecessary and British, has possibly diminished as a result of the Report by the Joint Committee, in 1969,[1] where it is stated that combined systems of sewers are not considered good modern practice. The report recommends that intercepting sewers be designed to take 2–4 × average flow, although McKee[2] has shown that a capacity of 80–160 × peak DWF may be required to prevent overflow in combined systems. Palmer[3] showed that, at Detroit, diminution of pollution in storm water run-off does not occur after the 'first flush' as successive peaks were found to arrive later. McKee, however, showed that at Boston only about 1.8% of the year's sanitary waste water was lost through overflows set at 3 × DWF, and this was reduced to 0.6% by provision of interceptor sewers having 10 × DWF capacity. At Detroit overflow was found to occur 85 times per year with interceptor sewers rated at 1.5 × DWF capacity. This reduced to 65 times per year with interceptors at 6 × DWF. However, speaking more generally, it is admitted that 60% of sanitary waste water may be lost at a rainfall intensity of 1.5 mm/h, and more than 90% at 7.5 mm/h, and that no satisfactory reduction in pollution from storm overflows can be accomplished by increased interceptor capacity.[4]

In England and Wales, as mentioned in Chapter 4, the provision of storm overflows is at variance with Sections 30 and 31 of the Public Health Act, 1936. More recently, the Technical Committee on Storm Overflows and Storm Sewage have given more precise definitions,[4] also referred to in Chapter 4. It should not be forgotten that every since the Act just mentioned, action can be taken under common law to abate nuisance due to storm-water discharge.

The Royal Commission on Sewage Disposal[5] laid down recommendations as to the provision and design of storm tanks and for several decades these suggestions were worked to, until they were challenged as unsatisfactory. Briefly, the recommendations of the Royal Commission were:

(1) as a general rule not fewer than two stand-by tanks should be provided, and that in most cases would probably suffice for their total capacity to be 6 h DWF;
(2) the sewage-treatment works should be capable of giving full treatment to 3 × average DWF.

The second of these recommendations, to the effect that the rate of flow to be given full treatment should be limited to 3 × average DWF, can be criticized on the grounds that if works serving separate sanitary sewers must be able to accept flows in the region of 6 × DWF and, in fact, often do accept such flows,

Table 39 Greater London area—details of sewerage districts

District	Area sq. miles (km²)	Pop. millions	Average sewage flow mgd (m³/s)	Flow per head gpd (litres/day)	Imp. area sq. miles (km²)	Stormwater from 'Standard Storm' million galls (1000 m³) to river	to treatment	Total million galls (1000 m³)	Estimated tonnes EOL* from 'Standard Storm'
Central	200 (518)	4.01	339 (17.8)	84 (382)	60 (155)	603 (2741)	327 (1487)	930 (4228)	294
North West	170 (440)	1.36	104 (5.47)	77 (350)	27.2 (70.4)	235 (1068)	187 (850)	422 (1918)	53
North	96 (249)	0.70	42 (2.21)	60 (273)	14.4 (37.3)	142 (646)	88 (400)	230 (1046)	32
North East	60.5 (157)	0.37	22 (1.16)	59 (268)	8.15 (21.1)	100 (455)	32 (145)	132 (600)	23
South East	212 (549)	0.82	40 (2.10)	49 (223)	21.2 (54.9)	306 (1391)	22 (100)	328 (1491)	70
South West	107.5 (278)	0.73	45 (2.37)	62 (282)	15.6 (40.4)	155 (705)	93 (423)	248 (1127)	35

* EOL = Effective Oxygen Load. (*J. Instn. Public Hlth. Engnrs.*)

the works serving combined sewers should also be able to treat flows of this order. As will be discussed in Chapter 14, trickling filters should well be able to treat such flows during wet weather even without any reduction in the standard of effluent, and this should apply also to good-quality land treatment. The circumstances could, however, be different in the case of the activated-sludge processes; but in most cases increasing the rate of flow to be given full treatment is accompanied by dilution and, therefore, reduction of strength. This will result in much less river pollution than discharge of untreated or merely screened and settled sewage.

It has frequently been commented that at the beginning of storms following dry periods dirt from road surfaces and settled material from ill-graded sewers cause an initial flush of sewage containing more than the average amount of suspended solids. This is true but, contrary to some opinion, it does not last a few minutes only. This is because the initial flush of that part of the catchment system that is near the point of outgo occurs at the start of the storm, whereas the initial flush of that part of the catchment that is farthest from the point of outgo arrives at the point of outgo later, by a time equal to the time of concentration. Thus, the period of extra-heavy pollution lasts for a little longer than the time of concentration of the sewerage system: this was shown to be a fact by automatic sampling of storm flows made by the Water Pollution Research Laboratory at the Oxhey Estate of the London County Council.

After this period the solids content of the sewage reduces and eventually the sewage will be much weaker than the normal sanitary sewage by virtue of its dilution by rain water. It is therefore difficult to say just what the pollution value of a storm-water discharge will be at any particular time. However, it can be said that, volume for volume, the storm-water discharge contains about the same amount of solids as the sewage that is passed to full treatment at the same time.

In the case of an average fully-combined urban sewerage system it was found that the amount of pollution due to the spill-over of a once-a-year storm from storm tanks having a capacity of 6 h DWF would be greater than the amount of stream pollution caused by a Royal Commission effluent in a week. Further, if tertiary treatment to polish effluent so as to halve the pollution caused by a Royal Commission effluent were provided, the once-a-year storm would offset the advantage of these works for over a fortnight. Also the whole of the storm water spilt over during a year could at least offset the advantage of installing tertiary treatment, and possibly be as much as twice that caused by a Royal Commision effluent in that time. More recent observations on London's storm water problems are given in a paper by Horner, Wood and Wroe[6] from which Table 39 is taken.

Storm-tank capacity

The storm-tank capacity is a matter of calculation. It has no relation to DWF but, as already mentioned, should be determined on storm flow, rate of flow to be treated, and degree of pollution permissible on rare occasions. The first approach can be made with the aid of formulae (15), (16), or (17) as appropriate: as this is most particularly a British problem, formula (17) has been selected for the following discussion, see pp 91–2.

Suppose:

(1) a combined sewerage catchment serves a population of 10,000 persons;
(2) the DWF is 250 l per head of population per day;
(3) the ultimate impermeable area is 39 m² per head of population;
(4) it is decided to fully treat 6 × DWF;

then P (the quantity to be stored) in formula (17) will be:

$$\frac{(6 - 1) \times 10,000 \times 0.25}{24 \times 60} = 8.68 \text{ m}^3$$

and the impermeable area will be:

$$\frac{10,000 \times 39}{10,000} = 39 \text{ ha}$$

if it is decided to calculate on the storm liable to occur once a year, so that $N = 1$, and to neglect time of concentration, the required capacity according to formula (17) is

$$C = \frac{88 \times 39^{1.4}}{8.68^{0.4}} = 6295 \text{ m}^3$$

The approximation to Bilham's formula which was used in formula (17) always gives a factor of safety that varies according to the case. The amount of this factor of safety can be found by the method shown in Table 40 which is constructed as follows: the quantity of rainfall run-off that occurs during 1 h, as caused by a once-a-year storm according to Bilham's formula, is inserted in the second column of the table. In the third column is inserted the maximum quantity that can be passed to full treatment during 1 h, less 1 h average DWF. The fourth column is the difference between the second and third columns. Next the storm-water run-off for 2 h during the once-a-year storm of 2 h duration is inserted in the second column, the maximum possible flow to full treatment during 2 h, less 2 h average DWF is inserted in the third column, and the difference in the fourth column. This procedure is continued for increasing periods of rainfall until the maximum value of storm-water storage is found. The table shows that the maximum value is reached at about 4 h, the quantity being 5069 m³ as against the 6300 m³ calculated by formula (17). By reference to tables of Bilham's formula it will be found that the figure calculated by formula (17) is in fact the amount of storage required to accommodate the once-in-2-years storm: this may be accepted for design purposes or, alternatively, the required capacity reduced to the maximum figure as calculated by the method illustrated in Table 40, as the engineer may consider desirable.

Table 40   Example calculation of storm-tank capacity

Hours rainfall	Run-off during rainfall (m³)	5 × DWF during rainfall (m³)	Storage in storm tank (m³)
1	4565	519	4046
2	5756	1038	4718
3	6550	1557	4993
5	7146	2076	5069
5	7642	2595	5046

Another method adopted in several major schemes, where new trunk sewers were constructed to receive the flow from existing systems, was to specify a maximum flow figure into the new system usually of the order of 6 × the normal DWF, and to guarantee full treatment up to the limit, by using storm-water tanks to store surplus flow, and return it for treatment when the storm had passed. The limit on rate of flow into the new trunk sewers meant that a series of storm overflows had to be devised at or near the junction between the existing and new sewers. At one of the more recent of such undertakings—the extensions to the Davyhulme works at Manchester—the primary settlement tank capacity of some 45,460 m³ was supplemented with over 227,300 m³ storm-water tank capacity. It is interesting to note that the Americans are turning to flow equalization by means of balancing as a positive aid in upgrading existing works. It is pointed out, however, that though equilization may embrace storm flow, it is in this case aimed at equalising the rate of DWF over the 24 h.

Storm-tank details

Early in the century storm tanks were almost identical in construction to sedimentation tanks. They were rectangular and flat-bottomed, about five times as long as they were wide, and the depth varied from 5 to 10 ft (1.5–3.25 m). Inlets were at one end and the outlets, in the form of weirs protected by scumboards, were at the other. They were arranged for sludging by manual sweeping. The differences from the primary sedimentation tanks of the time were:

(1)   each tank had a weir on the inlet channel and these weirs were set at different levels so that a storm of small magnitude would not cause more tanks to come into operation than necessary;
(2)   the scumboards at the outlet ends were floating, not fixed, so as to be effective while the tanks were filling;
(3)   each tank had a floating arm, protected by a floating scumboard, for decanting the contents back to the flow of sewage to be fully treated after the storm.

Many storm tanks of later construction retained the general rectangular shape and proportions, but were provided with mechanical sludging mechanisms of the kind used for rectangular primary sedimentation tanks. There is, however, no reason why storm tanks should not be like the modern sedimentation-tank design adopted and be circular, so as to simplify the sludging mechanisms; but they should still maintain the special arrangements mentioned above. The inlets of the tanks should be controlled by weirs so that each tank fills in turn: but the outlet weirs should all be at the same level so that, if there should be any spillover, all tanks would discharge simultaneously. The contents of the tanks should be returned to full treatment via floating arms or floating weirs.

The operation of storm tanks can be completely automatic. They should fill by gravity or pumping when the rate of flow from the sewers exceeds the maximum rate of flow to be given full treatment, and they should start to empty immediately the flow from the sewers reduces below the rate that the treatment works can accept. Sludging of each individual tank can be started at any time after the tank has become partly full, or it can be arranged not to start until decanting of top-water is completed. All this routine can be operated by float or electrode control.

Although on large works some economy can be achieved by utilizing a travelling scraper to serve a battery of four or more tanks, any such machinery becomes expensive on smaller works. Self-cleansing floors can be arranged by breaking the floor up into a number of V-shaped channels arranged just below decanting level, so as not to interfere with the decanting arms. The channels end just short of the outlet end of the tank where a common sludge channel can be provided, served by a single low-level outlet valve or penstock. Rather than automating all decanting valves the first tank, having received the bulk of the first flush of storm water, would be equipped with an automatic decanting device, and the others left for manual operation when the attendants are available. In the rare event of another storm coming before the attendants could decant the other tanks, the first tank would still be available to take a good deal of the 'first flush', but overflow of storm water would then take place immediately. As on smaller works it is commonly the practice to discharge all surplus flow, partially treated or not, onto grass plots, it is unlikely that the arrangement described would be of any serious detriment.

The extent to which pumping is required depends on circumstances. If there is adequate fall, the flow from the sewers can gravitate to the storm tanks and later be decanted to the treatment works by gravity. If there is not adequate fall for gravitation throughout, it is usually best to arrange for emptying the storm tanks by pumping, as this requires a much smaller pumping installation than would be needed for delivering storm flows to the storm tanks. Where, however, circumstances require that all flows must be pumped, it is best to deliver the storm flow to the storm tanks and arrange the tanks at a suitable level for them to empty by gravity to the works after the storm.

Storm tanks must normally be kept empty, as otherwise their capacity will not be available when wanted.

Storm tanks other than at the treatment works

The vast majority of storm tanks have been constructed at the sewage-treatment works but they can be used elsewhere in suitable circumstances. Shortly before the 1939–45 war the Ministry of Health approved the use of storm tanks on a discharge of storm water to a tidal estuary in lieu of an overflow for crude sewage. In proposals which the author has made he has suggested that, in cases where flows were to be diverted from small existing sewage works to other and larger works some considerable distance away, the existing sedimentation tanks should be converted to storm tanks which, by automatic operation, would discharge their contents of sewage and sludge to the sewers after the storms were over.

Balancing flows from pumping stations

When sewage is delivered to the treatment works by pumping, the rates at which the pumps operate may at times exceed the maximum rate that the works should receive although the actual rate of flow to the pumping station is, at the time, less than this. In such a case, unless precautions are taken sewage will spill over, or the treatment works may be periodically overloaded, with resulting inefficiency, These troubles can be obviated by the provision of balancing tanks which receive and store the pump discharge and pass an even rate of flow to the works.

A balancing tank should be provided with a module or constant draw-off floating arm which discharges exactly the same amount at all depths of sewage in the tank. The capacity of the balancing tank need be no more than the total capacity of the suction well or wells of the pumping station or stations discharging to it.

Balancing tanks should be designed so that they may be readily sludged: hopper bottoms with sides sloping at 60° to the horizontal are desirable. Their walls can serve as separating weirs for storm water for delivery to storm tanks.

In some instances balancing is arranged in the upper portions of old-fashioned detritus tanks or screen chambers. Balancing has also been provided in sedimentation tanks, but the latter practice is not advisable, because unless accompanied by some form of control on the rate of outflow, requirements of a balancing tank are so different from those of efficient sedimentation that tanks incorporating both purposes unless carefully designed would be comparatively inefficient for sedimentation.

On small or medium-sized works serving a system of separate sewers it is common practice to limit the top pumping rate to 6 × DWF. As balancing tanks are probably more of a nuisance at smaller works where labour is at a premium, it is quite practical to accept all flow into the settlement tanks, but control the outflow in stages to a maximum of say $4\frac{1}{2}$ × times DWF, which would be the top rate accepted by the secondary process, to have an emergency overflow weir in the outlet balancing chamber to keep peak flow levels under control, and to arrange for any surplus flow that may arise due to prolonged operation of the largest pumps to be piped away to grass plots. This arrangement involves effluent draw-off by means of slotted weirs at 0.6 m or more below the emergency top-water level. These weirs will, of course, become drowned as the extra tank capacity is taken up, but this fact will not affect the rate of outflow which will still be under the control of the floating arm outlets.

References

1. Jt. Comm. of the Am. Soc. of Civ. Engnrs., and the Wat. Pollution Control Fed., *Design and Construction of Sanitary and Storm Sewers* (1969).
2. McKee, J. E. Loss of sanitary sewage through storm water overflows. *J. Boston Soc. of Civ. Engnrs.*, **34**, 55 (1947).
3. Palmer, C. L. The pollution effects of storm water overflows from combined sewers. *Sewage and Ind. Wastes*, **22**, 154 (1950).
4. Technical Committee on Storm Overflows and the Disposal of Storm Sewage. *Final Report*. HMSO, London (1970).
5. Royal Commission on Sewage Disposal. *Final Report*. HMSO, London (1915).
6. Horner, R. W., Wood, L. B. and Wroe, L. R. London's storm water problem. *J. Instn. Public Hlth. Engnrs.* **5**, pt. (6) (Nov. 1977).

CHAPTER 13
Sedimentation

In sewage-treatment works of conventional design sedimentation removes more suspended solids than any other process and, in so doing, reduces the biochemical oxygen demand considerably; yet the cost of the primary sedimentation tanks may well be less than one-sixteenth of the total capital expenditure on treatment works. Because of these advantages sedimentation, preceded by screening, is used for partial treatment as well as a component of full treatment; but alone it cannot effect complete treatment. While the removal of suspended solids makes a considerable reduction in the biochemical oxygen demand, there is little change in the dissolved-solids content which remains to be reduced or rendered harmless by other means. For these reasons most sewage-treatment works incorporate primary sedimentation tanks to come before the aeration units, and final sedimentation tanks (known as humus tanks in trickling-filter schemes) to remove as much as practicable of the suspended solids produced by the aeration process together with those which the primary sedimentation unit have allowed to pass through.

There are two theories of sedimentation-tank design, the first of which stresses the importance of surface area and the other, capacity or detention period. The former theory is that the smallest particle to be settled must be given time to fall from the surface of the water to the invert of the tank, or it will be swept out with the flow. If the tank is made deeper the particle has a longer distance to fall and therefore the capacity of the tank is increased in direct proportion to depth, while the surface area remains the same: this is Hazen's theory.[1] On the basis of this line of thought it is argued that the surface area of a sedimentation tank must be of such magnitude that, where the whole of the flow is to be upwards, the velocity at the maximum rate of flow must be less than the falling velcoty of the smallest particle to be settled, it being assumed that there is no turbulence.

However, in all continuous-flow sedimentation tanks or channels there is turbulence in the form of eddies which produce velocities in all directions including upwards and downwards. Wherever there is an upward flow sewage containing a greater than average proportion of solids is lifted towards the surface, and where there is a downward flow, sewage from an upper layer containing a lesser concentration of solids moves towards the bottom of the tank. This means that turbulence must militate against sedimentation and, consequently, some allowance for this has to be made when tanks are designed on the continuous-flow principle.

The surface-area method is often used in the design of detritus tanks as described in Chapter 11. In the design of other sedimentation tanks, while it is appreciated that a ridiculously small surface area would seriously reduce or prevent settlement, there are other factors of more importance, and this brings us to the second theory of design.

When sewage enters a sedimentation tank the sudden reduction of velocity produces a state of turbulence which continues throughout the whole of the detention period except in a minority of circumstances. In quiescent sedimentation tanks (which will be described) turbulence dies out at the beginning of the settling period, and in large septic tanks receiving small intermittent flows there must be some periods of quiescence, but the majority of continuous-flow tanks are in a state of turbulence throughout.

The experiments of Clifford and Windridge[2] showed that a measurable reduction of efficiency, as compared with that of most other forms of inlets, was caused by bringing the sewage into the tank so as to fall even as little as 25 mm over an inlet weir set at a higher level than the outlet weir. The author had to investigate a sewage-treatment plant which was producing an unsatisfactory effluent because of very poor primary sedimentation. The trouble was found to be due to the inlet which was 225 mm above top-water level, and the energy dissipated by this fall was sufficient to ruin the performance of the tank in spite of a large capacity and very large surface area. Correcting the fault made the works satisfactory.

The above observations show that the design of the inlet of a continuous-flow sedimentation tank is of great importance. If the kinetic energy of the sewage entering the tank can be made as small as possible in the first instance, the least harm will be done to sedimentation. Nevertheless the remaining energy has to be dissipated, and this necessitates a large body of water on which the turbulence can have little disturbing effect. If a tank has a negligibly small capacity there will be no settlement at all, just as there will be none in a tank of very small surface area. If the tank is infinitely large the energy of the incoming sewage will be negligible and the condition may be considered quiescent.

In between these two extremes the efficiency of settlement depends, in the main, on two factors: the initial concentration of suspended solids in the influent and the length of the detention period. On the basis of published data from various sources the author derived the formula

$$S_2 = \frac{S_1}{C_1 \log_{10} S_1 \, D^n} \tag{32}$$

where: S_2 = suspended solids in tank effluent in parts per million;

S_1 = suspended solids in crude sewage in parts per million;

C_1 = a constant (approximately 0.837 for crude sewage);

C_2 = a constant, which in the main experimental data was approximately 10;

D = detention period in hours;

$n = (\log_{10} S_1)/C_2$ (for crude sewage n is approximately 0.25).

The constants are influenced by the design of the tank, the nature of the fluid and suspension, and other factors not determined. The detention period, in hours, is 24 times the capacity of the tanks, divided by the daily flow.

Practice of continuous-flow sedimentation

Rules both as regards minimum surface area in relation to flow and to minimum detention period are used in America and Britain, but standards are not quite alike. In America it has been suggested that the surface area should not be less than $0.1 \, \text{m}^2$ per $3.5 \, \text{m}^3$ of sewage per day and preferably $0.1 \, \text{m}^2$ for every $2.25 \, \text{m}^3$ per day at the mean rate of flow for primary sedimentation and $0.1 \, \text{m}^2$ for every $5.5 \, \text{m}^3$ per day for humus tanks and for final sedimentation tanks of activated-sludge works. Detention periods of 1–3 h, and a general average of 2 h, have been recommended for primary sedimentation tanks, 1–2 h for humus tanks and $1\frac{1}{2}$–$2\frac{1}{2}$ h for the final tanks that come after activated-sludge treatment, sometimes with an allowance of one-half of the rate of recirculation of activated sludge.

In Britain $0.1 \, \text{m}^2$ has been allowed for every 0.75–$1.33 \, \text{m}^3$ of sewage per day at the maximum rate of flow for primary sedimentation tanks, 1.75–$2.0 \, \text{m}^3/\text{day}$ for humus tanks and 1.25–$1.5 \, \text{m}^3/\text{day}$ for the final tanks of activated-sludge works. However, these figures are often ignored and only detention periods considered. Detention periods of 6 h dry-weather flow (DWF) for primary tanks, 4 h DWF for humus tanks and 4–6 h DWF for either primary or final tanks of activated-sludge works have been allowed.

In America the comparatively low suspended-solids content of the crude sewage, and short detention periods allowed, make it unsafe to rely on efficiencies of primary sedimentation tanks of much more than 56% reduction of suspended solids and 34% reduction of biochemical oxygen demand (BOD5), whereas in Britain designers expect not less than 70% reduction of suspended solids and not less than 42% reduction of BOD if the sewage is of average strength and the detention period is about 6 h DWF (see Tables 41 and 41a). This comparison of figures might suggest that practice in Britain is better than in America, but theoretical examination shows that the American figures are

Table 41　Approximate suspended-solids content of settled sewage after primary sedimentation (mg/l)

Crude sewage	Hours detention							
	1	2	3	4	5	6	8	10
100	60	50	46	42	40	38	36	34
150	83	70	63	58	55	53	49	47
200	104	98	79	74	70	67	62	59
250	125	105	95	89	84	80	74	70
300	145	122	110	103	97	93	86	82
350	165	138	125	116	110	105	98	93
400	194	155	140	130	123	117	109	103
500	222	187	167	157	148	142	132	125

Table 41a　Approximate biochemical oxygen demand of settled sewage after primary sedimentation (mg/l)

Crude sewage	Hours detention							
	1	2	3	4	5	6	8	10
100	76	70	67	65	64	63	61	60
150	110	102	98	95	93	92	90	88
200	142	133	128	124	122	120	117	115
250	175	163	157	153	150	148	145	142
300	207	193	186	181	178	176	172	169
350	239	223	215	210	206	203	199	196
400	270	253	244	238	234	231	226	222
500	333	312	301	294	289	285	279	275

the more economically sound, particularly where activated-sludge works are concerned. The economical detention period is about 1 h at the maximum rate of flow preceding diffused-air treatment, and $2\frac{1}{2}$ h preceding trickling-filter treatment. Calculations made by the writer showed that any detention period exceeding $1\frac{1}{2}$ h DWF would, on the basis of capital cost only, be uneconomical for tanks preceding surface-aeration treatment, and that the economical size of tank prior to trickling-filter treatment was by no means critical, there being very little difference in overall capital costs of schemes having tanks of from 1 to 10 h detention period, the filters being appropriately adjusted in size.

As Tables 41 and 41a show, unless a sewage is very strong the improvement in removal of solids and BOD5 after the first hour's detention is very gradual. The well-known curves prepared by Sierp[3] emphasize this even more clearly. This being so, and in the face of the ever-increasing realization that over-prolonged storage encourages septicity, British practice has become closer to American, and maximum rates of flow up to 3 $m^3/0.1$ m^2 are no longer uncommon.

Flocculation

Solids suspended in sewage include very small particles which, in unfavourable conditions, will not settle but remain suspended indefinitely. Nevertheless they tend to flocculate; that is, when they come in contact one with another they cling together producing floccules or woolly masses which, being of larger dimensions than the initial particles, settle easily.

Flocculation can be stimulated in a number of ways. In the first place, it has been found that if rods are moved slowly through the liquid they push the particles together and cause them to flocculate. The motion of the rods must be slow so as not to break up the floccules mechanically or disturb the liquid and inhibit settlement; there appears to be an optimum velocity for this purpose.

Flocculation is greatly assisted by chemical precipitants such as aluminium sulphate. The combined effects of chemical precipitation and mechanical flocculation are very efficient, and spectacular results can be obtained with laboratory working models. The pH value of the liquid has a marked effect on flocculation, in particular chemical precipitation, and control of pH value is necessary to ensure the best results with some precipitants.

The upward-flow sedimentation tank is designed to effect flocculation. The principle involved is that the sewage flowing upwards from a low-level inlet to surface weirs carries with it the smaller suspended particles which fall at a lower velocity than the upward-flow velocity. These particles encounter falling floccules and, combining with them, are carried to the bottom of the tank to form sludge.

Methods of sedimentation

Sewage can be settled in either quiescent or continuous-flow sedimentation tanks, in both cases with or without chemical precipitants. Of the methods, quiescent chemical precipitation effects the most settlement in proportion to detention period; next comes continuous-flow with chemical precipitation, followed by quiescent settlement and continuous-flow settlement. However, continuous-flow sedimentation is by far the most usual method. This is because chemicals are expensive and difficult to obtain, for which reasons they are used only in special circumstances, and quiescent sedimentation, while it reduces the net detention period, does not necessarily reduce the gross capacity of tankage that has to be provided. It also involves more labour than continuous-flow sedimentation. Moreover, the design of continuous-flow sedimentation tanks has greatly improved of recent years and, with increased efficiencies, capacities have been cut down. Quiescent tanks necessitate either pumping plant or sufficient head to decant the tanks by gravity.

Quiescent sedimentation tanks are virtually obsolete except in connection with special treatment, e.g. of trade wastes. They should not be too deep, but there is no practical reason why tanks should be long and narrow where purely quiescent sedimentation is concerned. However, such tanks were usually designed so as to function as continuous-flow sedimentation tanks should the operator neglect to stop the flow into a tank on it becoming full. Continuous-flow sedimentation tanks of the rectangular flat-bottomed type are preferably long, because length, if sufficiently great, reduces short-circuiting, the major eddies failing to reach the outlet ends.

Quiescent tanks are often filled through crude inlets, i.e. penstocks discharging directly from channels. The sewage is allowed to flow into each tank in turn. When a tank is full the penstock is closed and the sewage allowed to stand for a sufficient length of time for settlement to take place; say, 2 h. At the end of this period the settled sewage is decanted from the surface with the aid of a floating arm surrounded by a floating scumboard.

It is important in all forms of sedimentation that the effluent should be drawn from as near the surface as possible. After a period of settlement it is found that the suspended solids content increases from the surface downwards (not by any means constantly, for there is a marked distinction between settled sludge and unsettled sewage, but there is a variation throughout the depth of the tank).

In order that overflow may take place and the tanks serve as continuous-flow sedimentation tanks, weirs are provided at the ends of the tanks farthest from the inlets and protected by scumboards which are sometimes arranged to float, kept in position by chases in the sides of the tank. They should be set well back from the weirs in order that they do not induce an upward velocity towards the weirs.

It is interesting to note that in the 1972 revision of CP302,[4] on small sewage works, it is recommended that the scumboards at the outlet end of these tanks should have a deflector board or plate set at an angle of 45°, projecting from the end, or outside, wall to the same distance as the scumboard is offset from it. This would have the effect of deflecting rising particles *inside* the scumboard. Apart from its possible interference with any mechanism provided for scraping the walls of circular tanks, such a device might be usefully employed in larger tanks, either rectangular or circular.

In continuous-flow sedimentation the tanks are kept full and sewage continuously flows into them and passes through them; finally to depart over the outlet weirs. To take advantage of this method, many old quiescent tanks were converted to the continuous-flow principle. This made available four times the capacity in use in the quiescent process and simplified operation. With the further provision of mechanical sludging mechanism, the tanks could be left full indefinitely. The present-day tendency, however, is to use circular, not rectangular, tanks and to have central inlets and peripheral weirs, although there have been some large rectangular tanks of recent construction where such design suited the circumstances.

Sedimentation should always be in one stage: double-stage units are less efficient; also the use of baffles is detrimental, although, as explained on p. 316, batteries of tube settlers have recently given improved results.

Sedimentation-tank details

The shape, proportions, and details of tanks depend on their specific purposes. Primary tanks have to settle a great variety of solids, from large organic particles to very fine suspensions which are not easily separated. They also have to intercept floating material referred to as scum, and this means that they need scumboards and means of scum removal. Humus tanks have to deal with a type of sludge which is easily settled but, if neglected, will gas and come to the surface. They also need scumboards. Final tanks for activated-sludge works settle out a heavy suspension of activated sludge but there is virtually nothing in the flow that is capable of floating unless the treatment process has been upset.

The sewage entering primary tanks is usually warmer than the contents of the tank which has cooled during the detention period and, therefore, in spite of its solids content, it may tend to flow over the surface. Mixed liquor entering final tanks of activated-sludge works is always heavier than settled liquor and therefore tends to fall to the bottom.

Differences of specific gravity, and the direction in which flow is turned on entering the tank, set up rotation of the main body of the contents. Much has been said to the effect that these eddies must be stopped, and suggestions have been made as to the use of baffles. It has been found by experiment, however, that under practical conditions it is impossible to prevent rotation of the main body of sewage in a sedimentation tank. The sewage *must* rotate in some direction, and no matter in what direction the sewage is turned at the inlet, a straight-line flow from inlet to outlet cannot be effected. Only if rectangular tanks are of very great length in proportion to depth, e.g. twenty to twenty-five times as long as they are deep, does the vertical eddy die out without reaching the outlet end of the tank: results with such very long tanks have in some instances been very good, perhaps because of this feature.

As in most instances the vertical eddies cannot be prevented, it is best to recognize them and design for their control. The easiest condition to maintain is the swirl in which the sewage falls on entering the tank, passes along the bottom to the outlet, and rises towards the surface; that proportion not passing over the weirs returns towards the inlet. This principle can be applied with effect to the design of final sedimentation tanks for activated sludge, and optimum results can be obtained when sewage is arranged to fall at the inlet. In the case of rectangular tanks the outlet weirs are not set at the far end of the tank but are in the form of a bridge at a distance from the outlet end of not less than 0.7 of the tank depth. The advantage of this arrangement is that sludge, which normally is lifted from the bottom of the tank towards the outlet weirs, is able to resettle while it flows back towards the inlet end. The principle is also applicable to central-inlet tanks, in which case the weirs are suspended away from the peripheral walls. The method, which was originated in America by Anderson,[5] has been adopted in Great Britain for a number of large circular tanks and is proving satisfactory.

In the recently published volume *Process Design Manual for Upgrading Existing Wastewater Treatment Plants*,[6] the US Environmental Protection Agency (EPA) illustrate three 'typical clarifier configurations', all circular, one with centre feed and peripheral outlet weir Figure 67(a), and two with peripheral inlets from behind deep annular baffles and suspended channel

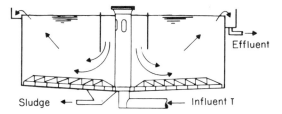

(a) Circular centre-feed clarifier with a
scraper sludge removal system

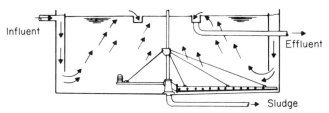

(b) Circular rim-feed. centre take-off clarifier with a
hydraulic suction sludge removal system

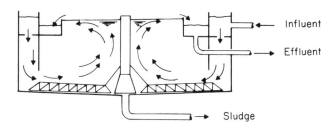

(c) Circular rim-feed, rim take-off clarifier

Figure 67   Typical clarifier configurations (a) Circular centre-feed clarifier
with a scraper sludge-removal system; (b) circular rim-feed, centre take-off
clarifier with an hydraulic suction sludge-removal system; (c) a circular
rim-feed take-off clarifier (Courtesy US, Environmental Prot. Ag.)

outlets, both annular, one suspended only about one-third of the diameter
from the centre of the tank Figure 67(b), the other inside the inlet baffle and of
generous proportions Figure 67(c)—depths of 3–5 m are recommended. A
successful modification of an existing square clarifier is also illustrated (Figure
68). There are inlet pipes on two sides, extending to about two-thirds of the
tank depth. An annular effluent channel is provided, and flow circulates across
the opposite halves of the tank and back to the outlet weirs.

The Agency has done a lot of work on this subject, the most comprehensive
study to date being that published in May 1979[7] wherein all available information is considered, whilst carefully designed new works will almost certainly

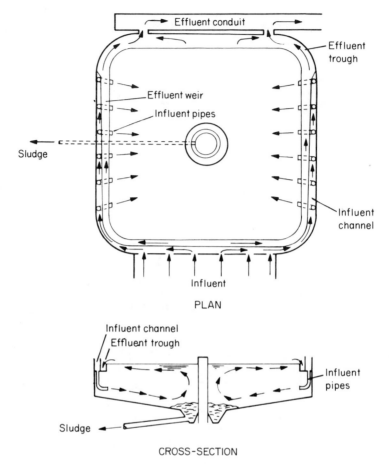

Figure 68   Clarifier modifications at the Greater Peoria Sanitary District
sewage plant (Courtesy US, Environmental Prot. Ag.)

bring advantages, the many problems associated with existing tank design and
fluctuating impurity load, as distinct from actual flow variation have led to
disappointing results in some of the cases analysed.

Although efficiency of sedimentation is important, the designer must not
consider this factor only, to the exclusion of all others. From the practical point
of view, convenience of working must be taken into account. Included under
this heading are the means of sludge removal (and, in the case of pyramidal
tanks and mechanically swept primary tanks, the storage capacity for sludge),
the prevention of sludge from hanging up and becoming septic, the means for
removal of scum and the means of preventing it from passing on to the trickling
filters, the proper proportioning of inlet and outlet channels in order that they
shall be adequate but not so large as not to be self-cleansing, and the arrangement
of feed pipes in such a manner that they distribute flow equally to all tanks and
not an undue proportion to any one tank.

The proportions of sedimentation tanks are primarily determined by the effects of velocity of flow in the tanks. This is easily demonstrated by imagining extreme cases. Suppose a cross-flow tank is so shallow or so narrow that the velocity is such that settlement cannot take place and settled sludge will be scoured; the tank will obviously have an efficiency of nil. Suppose in an upward-flow tank the surface area is so small that the upward velocity is at all times greater than the falling velocity of the largest particle to be settled; the efficiency of the tank is again obviously nil.

In the case of horizontal-flow tanks it is usual practice to make the tanks long in proportion to their width (say, five to one) and wide in proportion to their depth, the purpose of the length being to confine the reduction of velocity and dissipation of eddies to one part of the tank, leaving the remainder to act as sedimentation capacity.

The minimum depth of horizontal flow might be determined by the velocity at which settlement is prevented by scour. A number of observations have been made on the effects on tank efficiency of rate of cross-flow, but so far no velocity has been found that corresponds with any particular velocity of upward flow as regards the efficiency of sedimentation. There have been opinions expressed which relate to this matter but, in practice, it is usual to make flat-bottomed tanks 1.5–3.0 m deep and about five times as long as they are wide.

Taking both considerations, upward velocity and cross velocity together, there would appear to be optimum proportions and sizes for different types of sedimentation tank, but experiments have suggested that the shape of a tank is not important provided that it is reasonable and that its capacity and inlet arrangement are adequate.

Inlets and outlets

The design of inlets is deceptive, and the designer of them is very liable to go astray unless he keeps in mind the principle that inertia is largely responsible for the direction of currents and that, broadly speaking, currents entering a tank continue in a straight line until they are deflected by an obstacle. The inlets to a horizontal tank should be so designed that the velocity at entry is as low as practicable without causing sedimentation within the inlet device itself. In a circular tank with central inlet the flow is usually directed to the surface and a baffle is placed round the inlet to prevent the flow from passing directly to the outlet weir. A very successful inlet for all purposes is the well-known Clifford inlet, in which the inlet pipe discharges into the open end of an 'eddy bucket' out of which the sewage emerges slowly and comparatively free from turbulence. The design of sedimentation-tank outlets is simpler. It is fixed by two criteria:

(1) because the least concentration of suspended solids is near the surface, the highest efficiency of sedimentation is obtained when flow is drawn off from the surface in as thin a film as possible: therefore the outlet should always be a weir;

(2)   in order that unsettled sewage or partially-settled sewage does not pass prematurely over the outlet weir, the outlet weit should be throughout its length, as far as is practically possible, equidistant from the inlet.

The first requirement is of importance but it is easily achieved by designing the weirs so that, at the average rate of flow, the sewage passes over them in as thin a film as is possible without the flow breaking up into patches under the influence of surface tension. With some of the materials used for weirs, surface tension breaks up the flow into streams when the depth of flow over the weir is less than 2.5 mm. There is no accurate information on the amount of flow that passes over 1 m of weir at this depth, but using the formula for flow over flat-crested weirs, an approximate estimate can be made.

With pyramidal tanks, particularly those at smaller works (say below 500 population), the continuous weir is virtually impossible to achieve, and resort has been made to various expedients, such as three castellated weir openings in each wall, or possibly a saw-tooth type of weir plate. The former type can be designed quite readily, preferably to achieve full coverage at less than average DWF. The saw-tooth, or V-notch, type of opening clogs very easily, and if clogging is neglected the flow pattern suffers. For the reasons given previously, suspended channel outlets for circular tanks will function better with one side only acting as a weir. Another device, used with some success with suspended channels, is the provision of numerous perforations, about 25 mm diameter, in the floor. With steel channels it should be remembered that such perforations be cut *before* and not *after* the main rust-protection coating is applied, so as to avoid the possibility of localized deterioration. With holes in the floor this risk is minimized as the metal surfaces are seldom exposed to air.

Sludge removal

The method of sludge removal decides the type of tank to be used. If mechanical sludging is employed the tanks may be rectangular, square, or circular. For hand-cleaning by sweeping after decanting sewage, flat-bottomed rectangular tanks used to be constructed, but this type of tank is now virtually obsolete because of the labour involved and, for removal of sludge under hydrostatic head, tanks are now designed square or circular on plan with pyramidal or conical bottoms.

At large works mechanical sludging is very commonly employed. The sludging is carried out in continuous-flow tanks while the tanks are full, and with the aid of power-operated gear; but it may be either continuous or intermittent.

Flat-bottomed rectangular tanks, similar except in detail to old-fashioned hand-swept rectangular tanks, can be arranged for intermittent sweeping by travelling scrapers. This machinery is applicable not only to new installations, but is particularly useful when existing quiescent or continuous-flow rectangular hand-swept tanks are modernized. The only alterations necessary, apart from improvements in detail, are the provision of sludge hoppers at the inlet ends of the tanks to hold at least 1 day's sludge deposit.

The mechanical sweeper runs on rails and pushes the sludge towards the inlet ends of the tanks where it falls into a series of pyramidal hoppers, from which it can be withdrawn under hydrostatic head in the same manner as from a hopper-bottomed tank (as will be described). To facilitate sludge removal in this manner, the tank should slope slightly in the direction in which the sludge will be swept. The sweeping speed is usually in the region of 1.2 m/min (peripheral speed in the case of circular tanks).

The sweeping mechanism also removes scum. In some designs the same blade that sweeps the sludge is brought to the surface and, moving in the reverse direction to that employed when sludging, pushes the scum towards the scum-board against the outlet weir. From this position the scum is removed by opening a penstock at the side of the tank and, as the sewage runs out, an operator pushes the scum towards the outlet. This method is not too satisfactory, for some of the scum is almost certain to pass under the scumboard and escape with the effluent. Also, while scum is being removed for admixture with the sludge, far too much sewage escapes with it, thereby increasing the moisture content to an undesirably high figure.

An alternative arrangement is to have a second blade which pushes the scum towards the inlet end of the tank at the same time as the sludge is being swept in that direction. Finally, the scum is pushed up a ramp and falls into a channel which serves also as the channel for withdrawing sludge. Thus the scum is carried away by the flow of sludge and there is no undesirable increase of moisture content. For this method to be satisfactory the sweeping mechanism must be of good design, otherwise some scum may be left on the ramp and have to be washed off by means of a hose.

In order that one sweeping mechanism may serve for a number of rectangular tanks placed side by side, a transporter carriage is provided on to which the scraper can run under its own power. The transporter carriage, also power-operated, moves along a track from tank to tank.

Unless there is provision for manual sweeping of tanks, the works should have at least two sweeping mechanisms so that one can act as stand-by. Apart from this, the number of mechanisms provided depends on the number of tanks and their operation. For example, if there were sixteen tanks each of which required to be swept once a day, and each mechanism was capable of comfortably sweeping four tanks during a day of 8 h allowing time for sludging the hoppers, a total of four mechanisms should be provided, each with its own transfer carriage.

Methods of rectangular tank cleansing largely used in North America provide for continuous sludging by scrapers attached to belts or chains which are power-driven. They are losing favour because of high maintenance costs.

There are many designs of scrapers for circular tanks, all of which are rotating mechanisms. Some are for use with flat-bottomed tanks, others for tanks having moderately sloping bottoms. One design consists of a single spiral blade which draws the sludge towards the centre. Most other mechanisms have a series of blades or sweeepers which move the sludge, theoretically stage by stage, either towards the centre or towards a hopper at the periphery. It has been said that the multi-blade type moves the sludge to the centre too slowly because of the time that the sludge rests between the successive pushes that it receives from the several blades. However, tests have proved that, where many blades are provided on two or more arms, these cause a general flow towards the central outlet and there is no excessive delay.

What has often been overlooked is that although the arms of rotating sweeping mechanisms are beams of some considerable depth, the reduction of their weight when the tanks are filled with water is sufficient to lift them from the bottom of the tank so that they do not sweep it clean. In one instance the floor of the tank was finished with cement screed by fitting wooden blades set fractionally below the permanent blades and by rotating the mechanism. This, it was thought, would ensure that when the tank was in use the arms would rotate only just clear of the tank floor. But when the tanks had been in use for some time it was found that a deposit of completely decomposed sludge up to 30 mm thick was left on the floor and, on calculation, this proved to be due to the upward deflection of the arms by semi-flotation.

One way of overcoming the above difficulty is to have the scrapers trailing on the floor under their own weight behind the arms that drag them. For final tanks for activated-sludge works, a not uncommon design is that of a tank with a sloping bottom and a chain which is dragged round the floor by a peripheral ring: this is all that is necessary if the slope of the floor is sufficient (see Figure 69).

Where primary sedimentation tanks are sludged mechanically, whether the tank is circular or rectangular, the hopper into which the sludge is pushed should have a capacity sufficient to store all the sludge swept into it between the times when the sludge is drawn off from the hopper under hydrostatic head. At large works where regular attention can be guaranteed the tanks may be sludged every day, but it is desirable to have a sludge-storage capacity in the hoppers equal to twice the maximum quantity likely to be deposited in 24 h. Incidentally, the practice that has sometimes been adopted of sludging twice a day results in a thin sludge, and is not recommended.

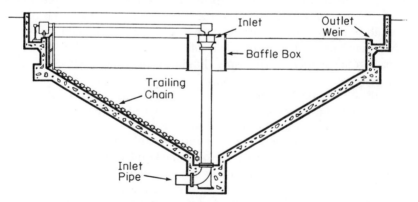

Figure 69   Final settlement tank for activated sludge with trailing chain to disturb sludge

In the majority of instances primary tanks, while mechanically sludged, have their hoppers emptied by manual control, the bleeding-off at intervals process being used to ensure the removal of a thick sludge without undue quantities of sewage. Attempts at constant draw-off by adjustable valve are not too satisfactory because the sludge tends to become thin before the operator realizes it. There is no need for very large sludge hoppers in the final sedimentation tank for activated sludge from which removal is constant.

The method known as tractor sludging is applicable to the conversion of very large sedimentation tanks originally designed for manual sweeping. This method has the advantages of exceptionally low capital expenditure and reasonable economy in labour. A vehicle fitted with a special blade either runs down a ramp or is lowered by crane into the tank, where it proceeds to sweep the sludge towards the sludge outlet. The advantages of mechanical scrapers are:

(1) reduction of cost of tank construction by avoidance of deep hoppers;
(2) the production of a dense sludge.

The disadvantages of mechanical cleansing are:

(1) increased quantity of ironwork and machinery requiring maintenance;
(2) the necessity for using power; and, in some cases
(3) reduction of sedimentation efficiency as a result of the disturbing effect of the sweeping.

Sludging under hydrostatic head

Continuous-flow sedimentation tanks are usually sludged under hydrostatic head, i.e. the sludge under the head of the sewage in the tank rises up a sludge pipe to a manhole from which it is discharged. The sludging arrangements depend very much on the purpose of the tank, and those applicable to primary tanks are not suitable for final activated-sludge tanks and vice-versa.

The sludge pipes for primary sedimentation tanks should be brought up in dead straight lines from the sludge hoppers and continued above top-water level to form rodding eyes. A short branch pipe should connect from each straight pipe and, controlled by a penstock or sluice valve, discharge to a sludge manhole.

The method of sludging primary tanks is to bleed off the sludge. At the start of sludging the valve is fully opened to start the flow and, if necessary, the sludge pipe is rodded. At first there may be no flow at all, or a slow creep of thick sludge. But soon the sludge will thin down and the rate of flow increase greatly. Then the penstock or valve must be closed sufficiently to reduce the flow to a small steady stream, otherwise sewage and not sludge will be drawn off, for the sludge tends to hang up on the sides of the hoppers and time must be given for it to slide down. When sewage only, and no more sludge, comes away, the valve should be closed; then after a period of time reopened to ascertain if any more sludge can be drawn off. In time the operator gets to know the most satisfactory routine.

Humus tanks for percolating filters are sludged in the same manner except where humus is drawn off for long periods and passed to the primary sedimentation tanks.

Final sedimentation tanks for activated-sludge works are sludged continuously, this being necessary to the process and made possible by the even consistency of the sludge. The hydrostatic head in this case is only a few centimetres and the flow is controlled by means of telescopic weirs or other adjustable weirs which are set by experiment and only altered from time to time according to the need to adjust the rate of return of activated sludge. The sludge pipes must be large to take the flow.

Whereas with humus tanks (or as the Americans prefer it, final clarifiers), it is probable that a suitably slow but continuous bleed-off of sludge is the best way of ensuring the satisfactory operation of the tank, to adopt such a practice with primary sludge would inevitably lead to an increase in water content and, at this stage, a very appreciable increase on the quantity being fed to the sludge digestors. Automatic valves have been tried but, again, are not proof against variations in the quantity of sludge collected. Another device uses partial air lift to control the rate of flow of the sludge,[8] and this appears to have had some success in reducing the amount drawn off, and at the same time lowering the water content. Many works still rely on the skilled operator to control the quantity and moisture content of the sludge and at the same time get the most efficiency from his tanks.

Hopper-bottomed tanks

At all small works, at most works where mechanical plant is not desirable and, by no means infrequently, at large works including those incorporating activated-sludge treatment, hopper-bottomed tanks are preferred because of their mechanical simplicity and low cost of maintenance. Hopper-bottomed tanks work on the principle that sludge will settle to the bottom of an inverted pyramid or cone and can be withdrawn slowly but efficiently under gravity, provided the sides of the pyramid or cone are sufficiently steep (see Figure 70).

Hopper-bottomed tanks, frequently called semi-Dortmund tanks, work more or less on the upward-flow principle. They are descended from the original Dortmund tank, which was deep and cylindrical in form, and in which the sewage flowed truly upwards from a very low inlet to a system of weirs arranged at the surface. The majority of modern hopper-bottomed tanks have central inlets placed a few feet below water level and peripheral weirs, and the flow is more outwards than upwards. Hopper-bottomed tanks can be highly efficient provided that they are well designed; but their design is full of pitfalls and deserves careful study.

While the depth occupied by sludge in a rectangular tank is usually only a few centimetres, a tank bottom in the form of a cone or pyramid makes the depth occupied by sludge a larger proportion of the total. For example, the amount of sludge that would collect in 1 week in a conical or pyramid tank (without any

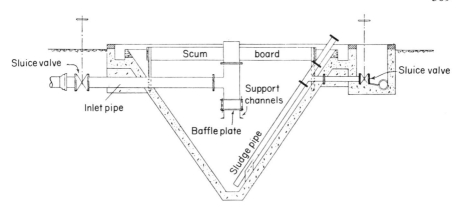

Figure 70   Pyramidal-bottomed sedimentation tank

vertical sides) would, in many cases, occupy as much as two-thirds of the total depth. At small works, sludging is sometimes as infrequent as once a week, and therefore the designer should be sure that his inlet is not so low that it will become submerged in sludge before the tank is sludged. He must, therefore, calculate sludge capacity and arrange the position of his inlet and the direction of its discharge accordingly. To be on the safe side he should make allowance for storing in the lower two-thirds of the depth of the tank about 10 litres of sludge per head of population.

In one instance the author had to investigate a pyramidal-bottomed sedimentation tank from which no sludge whatsoever could be withdrawn, the sludge pipe discharging nothing but crude sewage. On the other hand, scum collected on the surface and this had to be removed with a scoop. It had been suggested that the trouble was due to a large amount of disinfectant in the sewage. However, the contract drawings showed that the inlet was a small-diameter pipe carried down nearly to the apex of the pyramid. Thus the velocity of entry disturbed any sludge that ever had a chance to settle, bringing it back into suspension whenever there was appreciable flow. The author arranged for this pipe to be cut off above the theoretical sludge level and a baffle to be fitted, after which there were no further reports of trouble.

Having determined the proportions of his sedimentation tanks by finding the best length of weir, required surface area, space occupied by sludge, and adequate depth of flow from centre inlet to outlet weirs, the designer will find that he requires a certain number of tanks of decided proportions but unknown capacity. He may then find what this capacity is by cubing up and, provided that it is not unreasonably large or suspiciously small, he may accept the size.

In Britain a slope of 60° to the horizontal for the sides of pyramids or cones has become almost universal practice, probably because it is easy to draw such tanks with a 60° set-square. In America, steeper slopes such as 2 in 1 have been preferred. Conical tanks are theoretically better than pyramidal tanks, but they are more costly to construct.

The Imhoff tank

The Imhoff tank, invented early in this century by Dr Karl Imhoff of Germany and introduced into the United States about 1907, became popular throughout Europe and America but failed to achieve popularity in England. The tank consisted of a sedimentation tank, of a normal detention period according to American practice, in conjunction with a cold digestion tank, again usually of normal capacity for the purpose. The sedimentation chamber had sloping sides terminating with a narrow slot that admitted sludge to the digestion chamber below. The floor slabs overlapped at the slot to prevent gas from rising through it and disturbing settlement.

The largest Imhoff tank ever constructed was that at Chicago where it has to deal with a flow of 2,138,160 m^3/day. The method became less popular because more economical sludge digestion could be effected in separate thermophilic tanks. Shortly after World War II a very small Imhoff tank was constructed in England for research purposes with regard to small works for isolated premises. The author examined the results and came to the conclusion that, as far as the necessary data were available, the Imhoff tank installation showed no advantage over the other designs of septic tank or sedimentation tank tested.

Besides being widely used in Europe, tanks of this type have had a vogue in Australia. The settling compartment is kept to a minimum, usually $\frac{1}{2}$ h at $3 \times$ DWF, whilst the sludge compartment is rated at 60 l/head of population. Fairly careful maintenance is required, as successful operation depends on the narrow slot giving access to the digestion compartment remaining free from clogging and on the 'conditioning' of the digesting sludge by seeding with fresh material drawn in during the desludging periods. The digestion process is, of course, directly comparable with simple 'cold' digestion in a separate tank where the capacity provided would normally be considerably greater. Any attempt to apply heat for digestion purposes would upset the operation of the settling compartment.

Sludge pipes

Sludge pipes are usually of cast-iron: in American practice they are 200 mm in in diameter and English, 150, for nearly all sizes of pyramidal tanks (other than activated-sludge final tanks). For sludging under hydrostatic head the valve connections to the sludge pipes should be about 1 to $1\frac{1}{2}$ m below top-water level.

When cast-iron sludge pipes are provided inside the tank it is preferable that they should be in a straight line from the bottom of the tank to the rodding eye. The rodding eye can come outside the tank (which is convenient because it facilitates rodding without the risk of the operative falling into the tank). This may mean that the outlet sluice valves must be arranged with their axes parallel to the near side of the tank. Bellmouths at the bottoms of the sludge pipes are undesirable, because they encourage choking.

Vitrified clay sludge pipes set in concrete outside the tank are not so common as they used to be, although they have advantages in that they do not provide places for sludge to hang up. The reason for their having gone out of favour is not obvious, unless it is that if anything goes wrong with them they cannot easily be replaced with new. But in most instances, if a stoneware pipe outside the tank should be damaged or rendered unserviceable, which is unlikely, it could be replaced by a new cast-iron pipe inside the tank. The rodding eyes to sludge pipes are essential and should always be provided. Obviously, their ends must be brought above top-water level as otherwise water would escape out of them. Nevertheless, a common mistake on beginners' drawings is for them to be below top-water level and closed with blank flanges!

Scumboards

Primary sedimentation tanks and humus tanks, but not the final tanks of activated-sludge works serve a secondary purpose in the removal of floating solids or scum. To prevent these materials from passing over the outlet weirs, scumboards are provided near, but not too close to, the weirs. The reason that scumboards must not be too close to the weirs is that such positioning produces a local upward velocity that has the effect of reducing the efficiency of sedimentations.

For many years scumboards were invariably of wood—usually well-creosoted soft wood or suitable hardwood, preferably not elm, because although it can easily be obtained in large scantlings, it is not suitable for use in a position partly in and partly put of the water. In small works, particularly those in out-of-the-way places, there is no reason why wood should not continue to be employed, and when it is used the thickness of the boards should be 58 mm and the width (or depth) should be about 325 mm; submergence, say 300 mm. On larger works the trend has been to turn to steel, aluminium alloy, asbestos sheets and, inevitably, plastics. Unless stainless steel is used, at considerable extra expense, there will always be a risk of deterioration by rusting, although more effective rust-proofing methods are now available. With asbestos and plastics, frequency of supports has to be carefully watched to avoid the possibility of warping, although the purpose made compressed asbestos board with the top-edge turned over in a small quadrant, achieves considerable stiffness. In plastics, not only does frequency of supports need careful watching, but expansion and contraction due to temperature variation is, of course, quite significant and unless properly allowed for can result in small gaps or distortions which, apart from probably loss of efficiency, make for very poor appearance.

Steel has also been used, but rusts badly when between wind and water. Reinforced concrete or slate are durable materials but comparatively expensive. Timber boards are usually bolted to angle-irons which are again bolted to the concrete or brickwork of the tank. Gunmetal rag-bolts should be used, so as to avoid the need for cutting out the walls when replacements have to be made.

Wash-outs

It is always desirable for tanks to be capable of being completely emptied by gravity for maintenance purposes; but it is not always practicable. Some hopper-bottomed tanks are so deep that wash-out drains leading from their bottoms would involve unduly heavy costs for excavation and refilling of trenches with concrete where they lie under, or near, structures. When, however, it is possible to drain tanks by the provision of a wash-out pipe at reasonable cost, this should be done. Such drains usually discharge to the pumping station which deals with sludge liquor.

Channels

Inlet and outlet channels of tanks, etc., should be strictly proportioned to the flow and not designed by eye, even at the smallest works. A common fault in sewage-treatment works is that, where large sedimentation tanks and small humus tanks have been installed, the channels of the sedimentation tanks are large in proportion to the flow, while the channels of the humus tanks are small, although they have to carry the same flows. Another common fault is the over-sizing of channels, which results in silting.

   Channels should be designed in the same manner as sewers are designed, and laid to self-cleansing gradients and calculated sizes. In determining the gradients allowance should be made, either by direct calculation or rule of thumb, for the loss of head at sharp bends. For the inlet channels of sedimentation tanks, gradients ensuring a velocity of not less than 0.76 m/s at the maximum rate of flow are desirable. Half-round channels may be used, or channels of rectangular section, but in either case the depth of flow at the maximum rate of flow should not be less than one-half the width of the channel. Outlet channels of tanks may be laid to similar gradients or, in the absence of adequate fall, to gradients giving velocities of not less than 0.6 m/s at the daily peak rate of flow. Channels should have adequate but not excessive freeboard: as a rough guide, this may be taken as not less than the maximum velocity head although, if there are sharp corners, rather more might be advisable.

   Channels should be constructed with smooth inverts rendered flat to true falls or, alternatively, semi-circular stoneware may be used. As far as possible all angles should be in the form of sweeping curves, but where this is impractic-able—for example at the corners of tanks—the angles should be swept at the outside to prevent the hold-up of solids in corner eddies.

   On smaller works it may be worth remembering that the ordinary broom is the tool used for cleaning, and any width less than 200 mm, particularly if semi-circular, becomes difficult to get the broom into, and is, practically speaking, undesirable.

   When designing either pipework or open channels the designer should always keep in mind two things: the hydraulic gradient at the maximum rate of flow, and the gradient of the invert which will be the hydraulic gradient when flow is at a minimum.

Future trends

If the trend being recommended in America to equalize flow to the secondary processes by the introduction of an aerated balancing tank (Figure 71) catches on, such tanks may well supersede the old type of preliminary balancing tank, for receiving pumped flow, and also avoid the need for any control on the outlet side of the settlement tanks themselves. Apparently these tanks would also receive any returned flow from the pumping station dealing with dewatering of sludge, chamber washings, etc., which would benefit considerably from the aeration. The price of an efficient evening-out of the load may well be additional pumping; here again, the growing popularity of the low-lift screw pump may receive another boost.

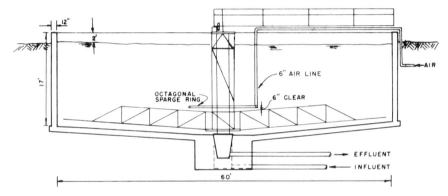

Figure 71   Cross-sectional view of walled lake/novi equalization basin

Note for aerating and mixing raw waste water

To maintain acrobic conditions	To prevent solids deposition
1 to 2 m³/min/10000| of storage	0.05 to 0.10 kW/10000| of storage
0.02 to 0.035 kW/10000| of storage*	0.0 to 0.08 kW/10000| of storage*

* Those figures apply to mixed liquor.

Whilst admitting that any attempt at flow equalization for a works serving a 'combined' system of sewers is likely to involve equalization at various points on that system, and will almost certainly involve pumping, in the case of a works serving a separate system normal operation of a treatment plant can be achieved under near-ideal load conditions. The result will be a reduction in the overflow rates of the primary settlement system, or the sizing of additional clarifiers to meet average, rather than peak, rates. It avoids disruption of settlement due, say, to the starting and stopping of pumps. As the design includes pre-aeration it can significantly improve primary settlement, although such benefit may be diminished if the equalization flow is pumped, as this may induce shearing of floc.

Design capacity is based on a hydrograph showing divergence of actual and average flow. Theoretical capacity is derived by drawing lines parallel to the average flow line, enclosing the deviating 'actual' flow line on either side and scaling the distance between them. Actual basin sizes require to be greater because operation of aeration equipment will not allow complete emptying of the tank, and some contingency must be allowed for possible changes in the flow pattern. Capacity requirements, however, should be in the range of 10–20% of average flow.

No elaborate construction is envisaged, but rather the use of available facilities, such as aeration tanks, clarifiers, digesters, or even sludge lagoons. Figure 72 illustrates an earth-embanked tank with concrete floor slab. A square form is preferred, as elongated tanks encourage 'plug' flow, and complete mixing is essential to optimize the damping effect. Inlets and outlets should be designed to prevent short-circuiting. A main objective, and a very desirable one, is that the aeration and mixing devices must prevent the deposition of solids. This implies that grit-removal facilities must precede equalization. The aeration requirement itself is sufficient to prevent septicity. By American standards, with suspended solids up to 200 mg/l, power required would be 4.5–8.0 kW 1000 m^3 storage, for aerobic conditioning at 1.00–1.5 m^3/min per 1000 m^3. The capacity for transfer of oxygen by mechanical aerators, operating in tap water in standard conditions, is 1.00–1.35 kg oxygen/kWh whilst minimum operating levels generally exceed 1.5 m, thus requiring low-water cut-off control to protect the tank. It is admitted that power requirements to prevent deposition of solids may greatly exceed those for oxygen transfer and blending, although this difficulty can be met either by mounting an aerator blade on the mixer, or by

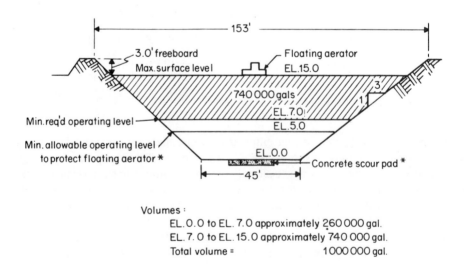

Volumes :
EL. 0. 0 to EL. 7. 0 approximately 260 000 gal.
EL. 7. 0 to EL. 15. 0 approximately 740 000 gal.
Total volume =   1 000 000 gal.

*These dimensions will vary with aerator design and horsepower

Figure 72   Earthen equalization basin (Courtesy US, EPA)

meeting the oxygen requirements through a diffuser system. Facilities are required for flushing solids and grease accumulations from the walls of the tank, as well as a high-level draw-off for withdrawing floating solids.

Although pre-aeration has not proved popular in Britain, provision of facilities in the form described, as an alternative to a balancing tank subject to heavy and objectionable sludge deposits, seems worthy of attention. With the whole, or part, of the works flow delivered by intermittent pumping, this tank could well replace the present inlet chamber, the balancing effect being achieved by means of a Kent regulator, or some other form of automatic flow control. The obvious advantages to be gained by utilizing the settlement tanks under optimum conditions as to rate of flow, and of aeration to counteract the effects of any tendency to septicity in pumped flow, appear to outweigh the added difficulties postulated by higher BOD and suspended solids which would probably apply.

Addition of chemicals

The so-called physiochemical process, although mainly concerned with the later stages of treatment, includes the concept of chemical dosage and a flocculation process as a preliminary to sedimentation. It is perhaps a little surprising that experiments should still continue on the addition of chemicals as an aid to sedimentation. Perhaps it is true to say that the interest displayed at the Coleshill experimental plant, in Great Britain,[9] and in a number of full-scale applications in the US, is aimed as much at conditioning the sewage, so as to ensure removal of phosphates and control of nitrates in later processes, as to aid the separation of solids, although there must inevitably be an increase in the amount of sludge to be removed from the settled sewage. At Coleshill, lime and ferric sulphate were the chemicals used, and comparisons were made with settlement tanks of different capacities, so that the effect of different up-flow rates and detention times could be studied without altering the rate of flow through the plant. Each tank had a mechanical scraper and air-activated sludge extraction equipment and Mono pumps for sludge handling. The sludge could pass to a thickening tank or be partly recirculated, either to the flash mixer or the inlet, or any settling tank, or be transferred to a dewatering plant. The sludge pumps were controlled by level probes and equipped with hand-set variable-speed drives.

Although a biological process is expected to form part of the complete treatment cycle, the Coleshill experiments allowed for re-carbonisation of settlement tank effluent, with carbon dioxide to reduce the pH to 7, when reactivated carbon adsorption was to follow. Use of a biological process instead of carbon adsorption could avoid the need for the reactivation process.

In the US, chemical clarification followed by carbon adsorption was evaluated as a process in the 1940s, and abandoned on the score of cost. However, more stringent effluent requirements, arising from the Clean Water Act, have revived interest in these methods.[10] Chemical dosing aimed at a desired level of suspended solids and/or organics, and/or phosphorus, has resulted in higher

Table 42   Achievements of chemical clarification

Plant	Chemical	Organic removal (%)	Suspended solids required (%)	Phosphorus removal (%)
Ewing Lawrence	170 mg/l FeCl₃	80	95	90
New Rochelle (2M)	Lime, pH 11.5	80	98	98
Westgate, Va	125 mg/l FeCl₃	70	—	—
Salt Lake City	80–100 mg/l FeCl₃	74	—	80
Blue Plains	Lime, pH 11.5	80	90	95

organic removal than would be expected from complete removal of suspended solids, as well as up to 50% 'soluble' organics, due possibly to chemical adsorption of colloidal and soluble organisms. After clarification the waste water passes to carbon adsorption and, in general, very good effluents have been achieved, as well as high carbon adsorption capacity, resulting from biological action on the carbon. It is emphasized that, used as a second stage in physiochemical treatment, the carbon is subject to a heavier load than is usual in water treatment, or in the tertiary stage of sewage treatment.

Some recent results of chemical treatment as shown in Table 42.

Tube settlers

As a means of improving their performance, and added protection against undue loss of solids caused by upsets in biological processes, or peak flow conditions, a vogue for applying 'tube settlers' has grown in the US[11,12]. These consist of banks of small-diameter tubes (say 50 mm) 600 mm long, suspended in

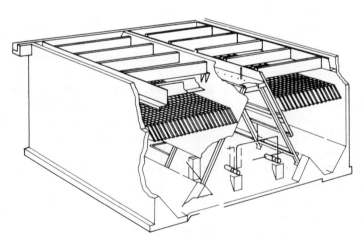

Figure 73   Typical tube settler installation in a conventional horizontal-type clarifier (Courtesy Permutit Ltd.)

the outlet zone of a clarifier. The tubes are suspended at about mid-depth of the tank and operate in conjunction with baffle-boards, inducing the flow to rise through them, to the outlet weir. In passing through the sloping tubes the light suspended solids tend to adhere to the walls, thereby achieving a material decrease in the suspended solids content of the effluent. To produce the required results the application of tube settlers requires careful evaluation. Provision for cleaning the tubes, either manually or automatically, possibly by means of air diffuser tubes located below them, must be provided. A typical tube settler installation in a horizontal type clarifier is shown in Figure 73.

References

1. Hazen, A. *Trans. Am. Soc. Civ. Engnrs.*, **53**, 45 (1904).
2. Clifford, W. and Windridge, M. E. Experiments with model tanks. *J. Inst. Sewage Purification*, Part 1 (1935).
3. Imhoff, Dr. K., Müller, W. J. and Thistlethwayte, D. K. B. *Disposal of Sewage and Other Water-Borne Wastes.* Butterworth, London (1956).
4. British Standards Institution, Code of Practice No. 302 (1972).
5. Anderson, N. E. Design of final settling tanks for activated sludge. *Sewage Works J.* (Jan. 1946).
6. *Process Design Manual for Upgrading Existing Wastewater Treatment Plants*, Environmental Protect. Ag., US.
7. The Hanover Park Study, Envir. Protection Ag., 1982.
8. 'Sewpas' desludging Equipment, W. E. Farrer Ltd, Birmingham.
9. Clough, G. F. G. and Maskell, A. D. Coleshill advanced water treatment project. *Proc. Instn. Civ. Engnrs.* **60**, pt. (1) (Aug. 1976).
10. Kugelman, I. J. Status of advanced waste treatment. Chapter 14, *Water Resources and Pollution Control.* Edited by Gehm, H. W. and Bregman, J. I. Van Nostrand-Reinhold Co., N.Y. (1976).
11. Slochta, A. F. and Conoley, W. R. Recent experiments in pilot-scale application of the settling tube concept. *J. Wat. Pollution Control Fed.*, **43**, p. 1724 (1971).
12. Hansen, S. P., Culp, G. L. and Stuckenberg, J. R. Practical application of idealised sedimentation theory in waste water treatment. *J. Wat. Pollution Control Fed.* **41**, 1421 (1969).

CHAPTER 14
Trickling Filters

Before the invention of trickling filters the only method of sewage aeration other than on land was the use of contact beds. These were beds of clinker, broken stone, or other suitable material which were filled with settled sewage to thoroughly wet the medium, then emptied and left for a period of aeration. The process was repeated, usually by hand operation of valves of penstocks. There were always two stages and sometimes more, for the effluent of a first stage was generally not good enough. Contact beds proved to be much less efficient volume for volume than trickling filters and, being dependent on meticulous operation, suffered from misuse. They rapidly fell into disuse during the first quarter of this century, their place being taken by trickling filters and activated-sludge works.

The first experiments on trickling filters are understood to have been made in America at the Lawrence Experimental Station in 1889. But practical development was mostly due to English engineers and chemists and, broadly, it can be said that the trickling filter is a little over 80 years old, or 20 years older than the activated-sludge process.

The terms percolating filters or continuous filters were used in Britain to distinguish the method from the former contact beds. In America the usual term is trickling filters and, in some quarters in England, the new term biological filters has been applied with the risk of confusion with biological processes in which real filtration is involved. Trickling filters are *not* filters: they are designed to let the sewage flow through and, in fact, if by misfortune or misuse they should act as filters they will quickly choke and become inoperative. As, however, the term trickling filter is known internationally it will be used in this text.

Trickling-filter treatment is by far the most frequently adopted process of sewage aeration. Those who have had experience of medium-size and small sewage works will know that trickling filters have almost invariably been adopted at these, whilst large works tend to adopt the activated sludge process.

In the United States, in 1973, there were more than 3500 trickling-filter plants serving over 28 million people, and approximately 3750 activated sludge plants serving 48 million people. Over there the trickling filter has been considered ideal for populations of 2500 to 10,000.

The advantages of trickling filters are:

(1)   reliability when subjected to very varied flows and/or difficult sewages;
(2)   low costs, including annual operating costs plus repayment of loan capital as compared with the activated-sludge processes;
(3)   ease of operation, often without specially skilled staff;
(4)   comparative economy of land as compared with land-treatment works.

The disadvantages are:

(1)   filter odour, which is similar to that of land treatment but more noticeable than that of activated-sludge works;
(2)   filter flies, in particular *Psychodae* which, however, do not fly long distances and are not known to be a danger to health;
(3)   required depth between incoming sewer and outfall, which is usually greater than needed for most other processes.

It should be mentioned, with regard to comparative costs, that the activated-sludge processes have frequently been used in cases where they would not be less expensive than trickling-filter works, and that it is probable that activated-sludge treatment effects marked economies only when the works are very large indeed. According to American standards of construction, trickling-filter schemes cost about 60% more to build than activated-sludge works of equivalent performance, regardless of size of works. At the same time the annual running costs of trickling filters average about one-third of the cost of running activated-sludge works. If the capital expenditure is expressed in terms of equal annual repayments of loan and interest and added to the running cost, activated-sludge and trickling-filter schemes can be directly compared on the basis of cost.

All small and many fairly large trickling-filter schemes are cheaper than conventional activated-sludge works when both capital and running costs are taken into account. But as capital costs per head of population reduce slightly for large populations, running costs reduce much more rapidly and, accordingly, very large activated-sludge works can be more economical than very large trickling-filter schemes. The point at which it becomes more economical to use the activated-sludge process depends on individual circumstances and particularly on the percentage interest payable on loans at the time.

Trickling filters are beds of material similar to that used in contact beds; but they are not flooded. The settled sewage is evenly distributed over the surface either by moving sparges or by spray nozzles so that it will trickle through the beds down to the floor. In course of time organic film, 2–3 mm thick, develops on the surfaces of the medium wherever it is wet, and it is the aerobic organisms which constitute this film that effect the oxidation of the organic content of the sewage. The water trickling over the organic film, having a large surface exposed to the air and being in motion, dissolves oxygen as rapidly as it is taken up by the oxidation of the organic content of the sewage. This means that the bed must be ventilated and, except in rare processes where air is blown through the medium, the ventilation is caused by convection. When the sewage is warmer than the outside air, air rises through the bed; when the air outside the bed is warmer than the bed the air flow is downwards. For these flows to be possible the spaces between the particles of the medium must be such as to permit free flow and there must be adequate ventilation to top and bottom of the bed.

It is the outer surface of the film that consists of the aerobic organisms. Beneath this is a layer of dead organisms undergoing anaerobic decomposition. The surface growth tends to fall away in time, being removed by the flow of sewage, and a new growth develops in its place. There are many other organisms in the filter. At the surface, as far as sunlight can penetrate, algae develop and would tend to choke the bed, but these are broken up by worms and the larvae of insects, except at such times when the weather is too cold and the macro-organisms descend to a lower level. Thus in temperate countries there is an offloading of vegetable material in the spring when the warmer weather permits the macro-organisms to come to the surface.

The amount of work that can be done by a trickling filter, or any other method involving healthy organic film coating solid surfaces and under water, is in almost direct proportion to the surface area of the film.

A good indication of what this area is in relation to organic load was given by tests on the Ames Crosta Babcock 'BioDisc'[1,5,6] process. The BioDisc unit consists of a tank, fabricated in steel or other suitable material, somewhat similar to an Imhoff tank (see end of Chapter 18), but which has a preliminary sedimentation compartment. For this compartment the sewage enters a chamber containing a series of vertical circular discs made of plastic mesh material, mounted on a horizontal shaft located above water level so that about 40% of the disc area is immersed. The shaft is rotated slowly so that the discs pass through the settled sewage and through air when oxygen is absorbed by the biological film which develops on the plastic mesh. With suitable loading of influent, an effluent of 'Royal Commission' quality is produced.

This mechanism permits a fairly accurate measurement of the film surface to be made and it has been found that about 75 m² of disc area is required for the removal of 1 kg of BOD_5 from the sewage per day.

This figure can be compared with maximum loads that can be applied to intermittent sand filters (as described later) which Imhoff and Fair[2] gave as 1 kg BOD_5 per day per 55–120 m² of filter, the figure of 55 m² being given also by Babbitt.[3] The recommendations of the Royal Commission on Sewage Disposal as to the maximum load on the best land treatment was that the area of land actually in use should not be less than 2.5–5.5 m² per person. If the BOD_5 per head of population per day were 0.05 kg (a figure recommended by Metcalf and Eddy[4] for separate sanitary sewage without industrial wastes) the area of film would be 104 and 52 m² respectively of film per kg BOD_5 per day, the average of which is 78 m² of film per kg BOD_5 per day. There is a good degree of correspondence between these figures from various sources, and therefore it can be said that about 80 m² of film in active condition is required to reduce 1 kg BOD_5 per day.

The net area of film in a trickling filter is, however, very uncertain and invariably only a small percentage of the total area of the particles of medium in the bed. The area of all the particles of medium of any particular shape in a cubic metre varies inversely as the average size of particle, and thus a medium consisting of small particles provides a larger surface on which film can grow than does a medium consisting of large particles. To some extent this is true but where the particles are in contact with one another there is a loss of surface area on which effective growth can take place because the surfaces of one particle and the next are within about 5 mm of each other, leaving no space for flow of air or water. Thus, for any particular shape of particle (say spherical), an optimum particle size could be calculated; but practice is to select a size of particle according to experience.

In addition to loss from this cause there is a still greater proportion of loss because the settled sewage, after it has been sprinkled over the surface of the bed, gravitates through the medium down to the floor, taking the shortest route. This means that, at any particular rate, part only of the medium is wetted by the flow while the rest remains dry and, as no film develops on dry surfaces, that amount of the capacity of the filter is lost.

It will be seen that a purely mathematical approach to the problem is not practicable and the performance of trickling filters must be determined on experience. This has led to different practices being adopted according to individual experience, local conditions, and local standards. It has, however, been found that the volume of medium necessary to treat sewage at a steady rate of flow is virtually in proportion to the load in kg BOD_5 per day.

The opinion of engineers as to the size of stone or clinker to be used in a trickling filter varies considerably and while fine media, as described by the Royal Commission, are no longer used except, perhaps, for special purposes, at the present time filters are constructed of media varying from 25 to 75 mm grade, and larger.

There is a very considerable difference between the surface area of the particles of a cubic metre of medium of 25 mm diameter and that of a cubic metre of 75 mm medium; therefore it would not be unreasonable to expect much better results from finer media than from coarser within this range of sizes. In a series of tests made by Levine[26] the effluent from a medium of 20 mm grade proved to be 14% better as regards BOD_5 removal than the effluent from a medium of 55 mm grade. Experiments made by the Water Pollution Research Laboratory on domestic sewage at Stevenage, using slag, clinker, rock, and rounded-gravel media of 25 mm nominal and 65 mm nominal size, showed that the smaller media gave the better results, and that the percentage removal of BOD_5 was related to the surface area of the particles of media in the bed. The best results were obtained with clinker and slag which exposed larger surfaces per unit particle than the other materials.

These experiments definitely showed the advantage of the smaller size of particle during a limited period of tests. There was, however, some tendency towards ponding of the beds containing the smaller media in cold weather and it could be argued that, over a period of several years, beds of small media might become choked whereas those with larger media would not, and the present trend in British practice is to use larger media than formerly, particularly for treating strong sewages or in high-rate filters, so as to avoid ponding or other chokage and thereby increase the life of the bed.

About 225 mm of the depth of the bed at the bottom is made up of very coarse material to assist drainage and ventilation. This, which consists of large stones or clinker 100 or 125 mm in diameter, is packed round the under-drains or rests on a false floor. Above the level of the coarse material the main body of medium is deposited. Generally this should be of even grade throughout the remainder of the depth of the bed. As ponding is most likely to occur at the surface of the filter, because light is necessary to the algae which are mostly responsible for ponding, some designers have allowed for about 300 mm depth of very coarse medium at the surface.

It is of importance that the size of particle in any grading of medium should not be too varied. If all the particles are of one size the maximum percentage of voids is ensured for the grading concerned. If, on the other hand, very varied sizes of particle are permitted, the smaller particles fill the spaces between the larger particles and the medium tends to pack. This not only encourages ponding, but militates against effective ventilation. Generally, the smallest particles in the bed should not have a diameter of less than two-thirds of the diameter of the largest particle in the same part of the bed.

The medium selected should consist of roundish or squarish particles, not flat or flaky. It should be durable, to stand the weather and to resist damage by being walked on by works operatives. Preferably it should have a rough

Table 43   Comparison of plastics media

	Surface area (m^2/m^3) (claimed)	Void space $(\%)$	Material
Tube media			
Cloisonyle	220	94	PVC
Sheet media			
ICI Flocor E	90	95	PVC
ICI Flocor M	135	95	PVC
Munters Plasdek B27060	100	95	PVC
Munters Plasdek B19060	140	95	PVC
Munters Plasdek B12060	230	95	PVC
Random media			
Norton Actifil 90E	101	95	Polypropylene
Norton Actifil 50E	124	92	Polypropylene
Norton Actifil 75	160	92	Polypropylene
MT Filterpak 1127	120	93	Polypropylene
MT Filterpak 1130	190	93	Polypropylene
MT No. 2 Mini Ring	118	93	Polypropylene
MT No. 3 Mini Ring	79	94	Polypropylene

Journal IWPC.

surface. The price should be reasonable, as filter medium cost amounts to no small part of a sewage-treatment plant. The weight and angle of repose of the medium should be ascertained prior to design of filter walls. All dust should be screened out on the site before the filters are filled.

In the United States and Germany the preference appears to be for filter media having particles in the region of 50–90 mm and 38–75 mm diameter respectively. The author has recommended, as a fair representation of British practice, that media graded from 38 to 50 mm diameter should be used in preference to somewhat smaller media, i.e. 25–38 mm, that many British engineers have been specifying. For this grade, he has recommended that for every kg BOD_5 per day in the crude sewage, 7.5 m^3 of medium should be provided or, on the assumption of 42 % reduction of BOD_5 by 6 h detention in sedimentation tanks, about 13 m^3 of medium should be provided for every kg BOD_5 in the settled sewage, as calculated on the average DWF. This figure makes allowance for peak flows up to 3 × DWF and increase of strength above the average more or less in proportion to rate of flow.

Types of plastics media

Plastics media are usually classified by shape[7]:

(a)   vertical tube packing;
(b)   vertical sheet packing;
(c)   random packing.

The first type is epitomized by 'Cloisonyle' made by Cupralex of Paris, for whom Air Products of Acrefair, near Wrexham, Denbighshire, are agents for the UK and Ireland. The medium consists of 80 mm outside diameter tubes, divided internally into a honeycomb structure of fourteen small tubes, each about 15 mm, square, usually of unplasticized PVC. These tubes are erected vertically, to heights of 4–6 m. Flow is evenly distributed over the top of the tubes. The surface area of the medium is 225 m^2/m^3, and it is particularly intended for reducing the BOD_5 of strong wastes prior to secondary treatment, or discharge to local authority sewers.

'Flocor' made by ICI Ltd, is probably the most widely used. 'Flocor' R was developed as a polishing medium and 80–90 % of its surface area can be utilized with good distribution. At a hydraulic load of 1.75 m^3/m^3 per day the retention period of Flocor R, and 50–100 mm slag, was 24 and 8 min respectively. Temperature loss has been measured as 2°C as against 4°C with stone medium, the hydraulic load being nearly four times greater, at the time. The makers consider its properties ideal for low-rate filtration, as well as high.

Random packings such as Actifil and Filtripak, both developed from Biopak and made of polypropylene, are more recent forms of plastics media. These are in the form of right cylinders with perforated walls and internal and external ribs. Actifil 90 is 90 mm long and 90 mm diameter; Actifil 50 is 50 mm long and 50 mm diameter.

Table 43 gives comparative figures for the various types of plastics media mentioned.

Comparative BOD₅ loadings

Table 41a on p. 297 gives a rough guide to the percentage of BOD_5 remaining in sewages of various strengths after various periods of settlement. Suppose that an American sewage of medium strength has a BOD_5 of 200 parts per million and the sedimentation tank detention period is 2 h, the BOD of the tank effluent could be about 133 parts per million, or a reduction by the factor 0.665. If it is desired to produce a Royal Commission effluent it will be safe to allow, according to the author's rule, 13 m^3 of medium for each kg BOD_5 per day in the settled sewage or (in this case) $13 \times 0.665 = 8.6$ m^3/kg BOD_5 per day in the crude sewage.

Although this standard can always be achieved, American practice is often aimed at lower standards. For example, Imhoff and Fair[2] on the basis of Standards for Sewage Works, Upper Mississippi River Board of Public Health Engineers and Great Lakes Board of Public Health Engineers, 1952, suggested a maximum loading for low-rate filters of 4.2 m^3 of medium per kg BOD_5 in the crude, sewage which, in the case in question, would be about 6.2 m^3/kg BOD_5 per day in the settled sewage. Babbitt[3] suggested 3.1–6.2 m^3 of medium per kg BOD_5 per day (presumably in the crude sewage).

The reasons that British practice and, accordingly, the author's rule, require more than the quantities of medium called for by the other rules quoted presumably are:

(1) British sewage works for more than half a century have been designed to accommodate $3 \times DWF$;
(2) designers aim at a quality of effluent that is well within the requirements of the Royal Commission, whereas in the United States a percentage of BOD reduction is often required, not a specific quality of final effluent.

In comparing figures, effect of sedimentation-tank capacity should be taken into account.[6]

Rate of flow per unit area of top-surface of bed

As the flow per unit of top-surface area of bed is increased a greater proportion of the surface area of the particles of medium becomes wetted, thus improving the performance of the bed until an optimum figure is reached beyond which there is no further advantage. This opinion is based on experiments into performance of trickling filters operated in series, on trickling filters through which effluent was recirculated, and on exceptionally deep filters which were enclosed and provided with forced ventilation.

Mills[8], in experiments at Birmingham, 1940–1944, stated:

From results of the continuous operation of the large-scale experimental plant at Minworth for four consecutive years, it has been shown that settled sewage can be treated by the process of alternating double filtration at four times the rate of treatment at which similar sewage can be treated by single filtration.

The same writer also stated[9]

it is concluded from these experiments that the recirculation process is comparable in efficiency with alternating double filtration, if effective action can be taken to counteract the effects of obstruction of the sub-surface layers of the filtering medium which occurred in the recirculation process in the winter.

Hunter and Cockburn[10] stated:

On an average over the four years 1940–43 inclusive, the enclosed filter, when compared with our open filters, has treated with an equal degree of purification twice the volume of tank liquor per cube yard.

There were some other experiments at Minworth, Birmingham made with the aid of a number of small filters, all 2 m deep which, operated as normal straight-run filters, were loaded to very high rates of flow. During the course of these experiments, which are described in *Water Pollution Research*, 1950, tests were made for the purpose of determining the time taken for liquid to pass through the filters at various rates of flow. The figures were as given in Table 44a. As the

Table 44   Period of contact in trickling filters

Rate of treatment of settled sewage litres/day/m^3	Mean period of contact in filter (min)	
	First determination	Second determination
594	110	105
1188	76	50
1783	71	55
2675	59	59
3567	33	38
4756	22	25
5946	20	27

Table 44a   Retention of water in trickling filters

Litres per day per m^3	Litres per day per m^2	Litres retained per m^3 of medium		
		First determination	Second determination	Average
594	1090	45.43	43.4	44.4
1188	2180	62.79	41.3	52.0
1783	3270	88.0	68.1	78.0
2675	4893	109.4	109.4	109.4
3567	6524	81.8	94.2	88.0
4756		72.5	82.6	77.5
5946		82.6	69.9	76.2

Filters say, 2 m deep.

depth of these filters was given as 2 m, it is possible to make the calculations set out in Table 44a which show that, as the rate of flow per m^2 of bed is increased, the quantity of liquid retained in the medium increases to an optimum, when the rate of discharge on to the bed is somewhere in the region of 6000 l/m^2, after which further increase of rate of flow causes some reduction of retention of water in the medium. Thus, if the trickling filter is considered as a tank holding water during a process of aeration the tank is, as it were, capable of growing larger and increasing its detention period until the optimum figure is reached. It is therefore not unreasonable to assume that, at the optimum rate of flow per m^2 the filter should be capable of doing more work than at other rates of flow.

This is what the Minworth experiments did in fact show. As the rate of flow of sewage was increased, the BOD_5 removed, in terms of kg/day per m^3 of filter medium, increased in direct proportion until after the rate of application exceeded 6000 l/day per m^2, or 5.4 m^3/day per m^2. On further increase of flow the percentage BOD_5 removal fell off progressively.

The foregoing suggests that, with any given quantity of filter medium, an optimum result can be obtained by so arranging the depth, or otherwise operating the filter, that an optimum rate of flow per superficial metre is maintained. There are three ways in which a high rate of flow per superficial metre may be produced. First, the filter can be made deep and the superficial area reduced accordingly. Secondly, the flow can be put through two or more stages of filtration in series. Thirdly, the effluent can be returned through the filter by recirculation at any desired rate.

As regards the first method, very deep normal filters are comparatively rare, and so far the writer has not been able to obtain information on results with them. But, for the sake of economy in roof construction, most enclosed filters were made 5 or 6 m deep and, as a consequence, they have been dosed at rates that should produce longer detention periods than usual in normal filters. It is possible, therefore, that the high efficiency of the enclosed filter is due not to the roof and mechanical air-blowing, but to the depth of the medium.

There are several sewage works in England where tank effluent has been passed through two successive stages of filtration, but little information is available on the results of this type of treatment. The method of alternating double filtration, however, requires the use of filters in series, with a consequently high rate of application per m^2.

The maximum consistent rate of flow of sewage treated at Minworth by alternating double filtration was 1430 l/m^3 (at a time when the normal filter could treat no more than about 360 l/m^3). The total depth of primary plus secondary filters was 4 m; therefore the rate of flow per m^2 was 5660 l/day, or practically the optimum figure.

In the recirculation experiments at Minworth, the rate of flow of sewage was 1430 l/m^3 and the average filter depth 2 m; but as the effluent was recirculated once, the dosage per m^2 was again 5650 l/day.

It is significant that the methods of recirculation, alternating double filtration, and enclosed filtration, in spite of their apparent differences, were all alike in that they produced virtually the same results when the rate of flow per m² of top surface of bed was in the region of 5500 l/day. It is reasonable to consider that the improvement of performance of the filters was due in all cases to the loading per unit of surface area at the optimum rate of flow.

For a time the spectacular results of the experiments on these methods persuaded engineers to allow flows of up to twice the loads adopted in previous practice in cases where recirculation or alternating double filtration had been provided for. But results obtained by testing filters fed at steady rates and not at fluctuating rates as at normal sewage works, should not be expected to give reliable indications as to what would occur in practice. Trickling filters at British sewage-treatment works which usually receive once a day about twice the average daily rate of flow, and at about twice the average strength, at the same time, could be expected to have a performance nearly as good as a filter provided with arrangements for recirculation so that the flow through the medium would be maintained throughout the day at twice the average rate of flow to the works, and the provision of recirculation in such a case could well prove disappointing. The advantage of recirculation is that it gives a means by which the rate of flow per unit of surface area can be adjusted so as to find and maintain the optimum conditions for the particular filter: it should not be relied upon to make possible spectacular reduction in filter size.

Further information on plastic media

Although more usually associated with high-rate filtration there has been considerable interest in the use of plastic media for low- or medium-rate filtration.

Hemming and Wheatley of ICI[11] started testing plastic media in 1961, and their conclusions are that the modular type of synthetic media rely on high irrigation rates to ensure wetting of available surface. This difficulty can be overcome by the use of random-packed media[12] to promote irregular flow, such as the pattern available in corrugated cylinders, tested at Brixham and used at Langley Mill, Birmingham, and Prestbury (Macclesfield).

Bruce and Merkens[13] define low-rate filtration as 'where organic loading does not exceed 0.6 kg BOD_5/m^3/day'. Anticipated stages of operation would include (a) coagulation, (b) oxidation to carbon dioxide, and (c) nitrification. A low-rate system would normally be expected to remove 85–90% of the organic load. For this purpose the medium should have a higher surface area than that required in high-rate filtration. This implies a random type medium with surface area in excess of 180 m² per m³.

Experiments with plastic media of the same dimensions as mineral media, but with three times the surface area, showed that three times the hydraulic loading could be applied to produce an effluent of the same quality. Plastics media with twice the dimensions of the normal media, but with 1.5 × the surface area, when loaded to three times the hydraulic loading did not produce a good-quality effluent. Small media are therefore required to ensure good flow and distribution.

Performance and efficiency increased with increased loading on the larger plastic media (Flocor R25) and this was attributed to improved distribution at higher loading. The same was found to be true of the smaller media (Flocor RC and R5). With these the best performance was achieved at 4.2 m^3/m^3 per day and 0.2 kg BOD/m^3 per day, when an effluent containing considerably less than 20 mg/l of BOD$_5$ was obtained. An increase in organic loading from 0.2 to 0.5 kg BOD$_5$/m^3 per day produced little deterioration in effluent quality. It was anticipated that loadings up to 0.55 kg BOD$_5$/m^3 per day could be treated with recirculation in the ratio of three of effluent for one of sewage, at the same time maintaining an effluent consistently at 20 mg/l of BOD$_5$.

It was noted that at higher than conventional loadings ammoniacal nitrogen was produced from deamination of organic nitrogen, thus reducing the efficiency of ammoniacal nitrogen removed, although this could probably be improved with recirculation, and/or double filtration.

High-rate filtration

Considerable use is made of high-rate filtration in the United States.[12] This is the passing of large flows of settled sewage through trickling filters so as to apply both a heavy organic load per m^3 of medium and a very large flow of water per unit top-surface area of bed. The load on the surface area is maintained between about 9 and 27 m^3/day per m^2 and, to accommodate the biochemical oxygen demand, from about 1 to 3 m^3 of medium per kg BOD$_5$ in the crude sewage are provided.

To obtain a high rate of flow per m^2 of top surface the filters can be made deep and of small surface area, or there may be two batteries of filters in series to reduce the surface area in proportion to flow, or the treated effluent may be recirculated through the beds. The last method has the advantages of being applicable where head is limited, and of being adjustable as the sewage works operator may find advisable.

Bruce and Merkens[14] have also made a study of high rate filtration with plastics media. Their main conclusion is that the efficiency of BOD removal is terminated usually by surface area per unit volume of medium. Results show that at the same rate of application the vertical sheet has a higher efficiency than the vertical tube. The latter type depends completely on the quality of initial liquid distribution. Randomly dumped media spreads the liquid in all directions, thus helping to achieve efficient utilization of surfaces. Sheet media, although superior to tube, tend to distribute the liquid in one direction.

Design by volume determines a volume and leaves the designer to determine overall dimensions.

Design by surface loading (or irrigation rate) method may well predict a small volume at high irrigation rates, implying less volume in a deeper bed. Kornegay[15] predicts 'filter volume necessary to reduce an influent concentration of 300 mg/l to 30 mg/l decreases from 2237 m^3 to 906 m^3 as hydraulic load is decreased from 49 to 29.2 m^3/m^2/day. Equations based on this method will give a specific filter height, whereas those based on volume leave this undetermined.

Pilot plant studies must still be used to decide the best method of treating any specific waste though theoretical studies help to check empirical equations.[16,17]

Increase in effectiveness at high irrigation rates is probably due to the increase in area of packing wetted by the liquid.

Porter and Smith,[7] comparing plastics and stone media, put the voidage in plastic media at about 95% as against 25% for stone, and the usual organic loading for plastics about 3 kg BOD$_5$/m^3 per day, or twenty times higher than for stone media. Bed depths with plastics media are usually from 3 to 7 m.

The removal of BOD$_5$ is achieved by the biomass which grows on the media. To achieve economic operation the entire surface area will be covered by a thick growth of biomass. This implies high organic loading but only partial treatment of the waste. To maintain a reasonable thickness of biomass, at the bottom of a filter, some BOD$_5$ must still remain. It is for this reason that high-rate filtration can be successfully achieved only at an efficiency of 50–80% in terms of BOD$_5$ removal, coming under general description of 'roughing filters'. It is possible to achieve similar rates of flow, when treating smaller weights of BOD$_5$, which can be done with a unit efficiency up to 90–95%, as 'finishing' or 'polishing' filters, following another biological process.

Roughing filters[18] are characterized by high BOD$_5$ and high hydraulic loads, with low retention periods and a large weight of heavy biomass, often up to 300 kg/m^3. The medium must support this weight and also possess sufficient void space to allow humus solids to pass freely with the flow. These conditions impose a limit of surface area to 140 m^2/m^3 on the medium, which must be of an ordered type, promoting flow without short-circuiting.

Polishing medium also depends for its performance on available surface area, but void space can be lower—up to, say, 250 m^2/m^3.

Recirculation for the purpose of obtaining the best results and giving ease of operation should be provided at all large sewage works incorporating trickling filters. The rate of recirculation should be capable of easy adjustment and the pump sizes should be sufficient to give at least the optimum rate of flow per m^2 of top surface of bed.

Recent trends

In recent years changes have come about both in the UK and the USA, where, in the first case, since 1974, all municipal sewage works have come into the hands of the big River Authorities, each now responsible for the operation of several hundred works, and in the other case the passing of the Clean Water Act (Amendment) in 1972, requiring all works to provide higher standards of treatment. The financial restrictions in the UK have largely limited activities to getting the best out of what was already there, and to accept this as the position until long-term improvements or alterations can be realized. In the USA the Environmental Protection Association (EPA) has taken upon itself virtually to specify the methods of improvement which should be adopted, and having approved them, to administer grants at the rate of 75% of the cost of the works!

In their role as advisers and administrators of the available funds the EPA are leading an ever-growing programme of laboratory, pilot-scale, and full-scale experimental work, co-relating the results and publishing them as a guide to all engaged on the works improvement programme.[19] One of their publications is the *Process Design Manual for Upgrading Existing Waste Water Treatment Plants* which may be said to supersede the joint ASCE and FSIWA manual *Sewage Treatment Plant Design* of 1959, 20–23. For trickling filters they explore the possibilities of the forms of recirculation already discussed, and also of high-rate, followed by low-rate, filters. Possible combinations of trickling filters, either preceded or followed by application of the activated sludge process, are also examined. Several conclusions reached are of interest. One is that the various types of synthetic media now available permit what is termed 'super-rate' performance, with hydraulic loading up to 170 m^3/m^2 including recirculation accommodated. Performance of such beds is considered to be as good as ordinary high-rate filters up to about 1.6 kg/m^3 of BOD_5 or hydraulic loading up to 60 litres/m^2. They conclude that recirculation rates greater than 4 × average flow do not materially increase plant efficiency—normal rates are only once or twice average flow. With synthetic medium recirculation must produce a rate of flow which will induce slime growth throughout the depth—this would be in the range 100 to 200 litres/min./m^2—any increase above this decreases BOD_5 removal. They find that recycling of filter effluent direct, is as effective as re-cycling as clarified effluent, and is certainly preferable when clarifier loading is high. On the question of the depth of beds, they quote 2–3 m for low rate, 1–2 m for high rate stone media beds and 5–10 m for synthetic medium, the extra depth being required to provide adequate contact time. On the interesting question of temperature they quote the formula $E_T = E_{20}\theta^{T-20}$, where $\theta = $ a constant 1.035 to 1.041, $E_T = $ filter efficiency at temperature T, $E_{20} = $ efficiency at 20°C, and $T = $ waste water temperature °C. In northern regions the effects of air and water temperatures are significant, especially with high rate filtration, due to the cooling effects of recirculation. Covering the filters does *not* substantially increase their performance, as it fails to affect waste water temperature.

These observations, based as they are on actual plant performance as well as experimental work, are worthy of great respect. The figures in Table 45, taken from the Manual, are interesting, perhaps, in their variety! Several of the works will undoubtedly have to bring about improvements in order to reach the standard called for in second degree treatment, of 30 ppm BOD_5 and suspended solids.

The present-day call for nitrification, which may well become more insistent as the quality of water in rivers supplying potable water is ever more closely monitored, puts an emphasis on trickling filters as one of the recognised types of nitrifying treatment and supports the continued use of either conventional

Table 45 Operating data for single-stage trickling filter plants at low, intermediate and high loading rates

Plant location	Influent flow (10^3/m^3)	Filter media depth (m)	Plant influent BOD (mg/l)	Recirculation ratio	Hydraulic loading (10^3m^{3}d/ha)	Organic loading (kg BOD/m^3/day)	Final effluent BOD (mg/l)	Filter and final clarifier BOD removal (%)
Low rate								
Aurora, Illinois	29.5	1.8	117	0	19.6	0.07	14	80
Dayton, Ohio	130.3	2.3	227	0	32.7	0.19	33	76
Durham, N. Carolina	6.8	2.1	375	0	17.8	0.21	68	74
Madison, Wisconsin	18.2	3.0	248	0	22.5	0.10	33	76
Richardson, Texta	5.7	2.0	166	0	36.4	0.21*	20	83
Intermediate rate								
Plainfield, New Jersey	16.3	1.8	629	0.6	27.5	0.40	13	83
Great Neck, New York	1.9	1.2	187	1.0	72.9	0.32	20	83
High rate								
Oklahoma City, Oklahoma	61.3	1.8	463	1.0	152.6	1.37	66	78
Freemont, Ohio	6.8	1.0	134	1.5	177.7	0.66	21	78
Storm Lake, Iowa	28.0	2.4	690	2.1	201.1	0.99	61	84
Richland, Washington	8.3	1.37	173	2.8	183.3	0.70	20	83
Alisal, California	2.3	1.0	293	3.1	194.6	0.85	24	87
Chapel Hill, N. Carolina	5.3	1.3	166	2.0	152.8	0.30	44	43

* Computed based on assumed 30% BOD removal in primary clarifiers.

low-rate or more highly dosed artificial medium beds, to follow either short-period activated-sludge treatment, or more probably high-rate primary filters in many parts of the world.

Although the stringent economic conditions in the UK have prevented the carrying out of large-scale works of later years, a good deal of attention has been paid to the improvement of existing works, usually in terms of accomodating higher loading without deterioration of effluent stands.[24,25] The use of tower-type trickling filters packed with artificial media, either as roughing filters preceding existing beds, or in some cases followed by activated sludge units has helped significantly in such circumstances. Replacement of stone or clinker media by plastic for the top third or quarter of the depth of existing beds has also been found beneficial. The ever-increasing cost of conventional media coupled with the relatively high cost of plastic media has caused many engineers to turn to an application of the activated-sludge process in an attempt to reduce capital costs. However, research is at present being carried out into the possibility of still further improvements in performance of trickling filter, and this may bring about a resurgence of interest.

Construction of trickling filters

The majority of trickling filters are circular on plan in order that the flow may be distributed over the surface by rotating sparges. Where it is necessary to economize in land rectangular beds are constructed and the flow distributed over the surface by sparges that travel from end to end of the bed, running on rails. These methods of distribution are liable to suffer from freezing in very cold weather, and therefore in colder climates distribution may be by spray nozzles. Circular beds are common in both America and Britain. Rectangular beds with travelling distributors are rare in the United States and fixed-spray nozzles are very rare in Britain.

When revolving sparges are used the number of filters to be built is determined by the maximum diameter that any sparge can cover on the one hand and, on the other, the minimum number of filters desirable for convenient working. For the sake of economy of construction there should not be too great a number of small filters. On the other hand, except at the very smallest works, there should always be more than one trickling filter so as to permit maintenance, and preferably more than two. Apart from this consideration, the larger the filter (within limits) the lower the cost.

In America circular beds up to 50 or sometimes 60 m diameter have been constructed. In Britain revolving sparges are obtainable from most manufacturers for filters of 30 m diameter but sizes suitable for beds of much larger diameter have been made to special order. Thus it will be seen that at very large works there must be a large mumber of filters, if of the circular type, because of the limit on the diameter which can be served by commercial designs of distributor; but at medium-size works it is not uncommon for filters to be of other than the largest available size, the size of filter being chosen to fit the plot of land.

There are several favoured designs for circular beds which are often used indiscriminately, but which should be applied according to the peculiarities of the site. When it is necessary for a trickling filter to be constructed mostly below ground level, and surrounded by a wall for the purpose of keeping out the earth, it is most economical to slope the floors towards the centre and construct a circular central manhole surrounding the distributor column to collect the effluent that drains from the bottom of the bed. From this manhole, which is usually about 2 m in diameter, the effluent drain is laid under the floor and away to the humus tanks. This design is economical because it does away with the excavation which would otherwise be necessary for the construction of collector channels at the periphery; but as ventilation is limited, air shafts or pipes should be arranged at the periphery, connecting with the underdrains or false floor. These should preferably be constructed outside the beds, not in the medium, to avoid the problem of direct discharge from the sparge-arms passing down them. (For arrangement of central manhole, see Figure 75.)

When, as is usually the case, the site is sloping and the beds are partly in the ground, but are out of the ground at one side, the cheapest design is that in which the floors slope in the direction of the fall of the land and the channels collecting the effluent are constructed on the lowest sides of the beds extending for about one-quarter of the circumference. These channels, being only just in the ground, are cheap to construct and the fall of the floors in the direction of the natural grade of land surface reduces excavation to a minimum (see Fig. 74).

A very satisfactory design which, however, is only applicable to flat sites, is that in which the filter is totally above ground level and surrounded by an effluent channel which gives access and provides ventilation to the underdrain system of the bed around the whole of the circumference.

Underdrainage

Small filters can be underdrained merely by constructing the floors of smooth-faced concrete laid to moderate falls of 1 in 100 or 1 in 200, and depositing very large material carefully on the surface so as to permit the free ventilation and drainage of effluent. It is usual, however, to lay parallel lines of agricultural tiles or field drains not more than 1 m apart, in addition to the large material. These assist ventilation and are not really intended to effect drainage. In most instances drainage takes place over the general surface of the floor, unless channels are provided in the floor. If such channels are provided, they are covered with half-round tiles.

Many trickling filters have been given complete false floors of special filter tiles as used for the gravity filters of waterworks; the reason for this choice of construction is that such false floors provide the most adequate ventilation. There is, however, no evidence that ventilation as provided by agricultural pipes open to the air at both ends is not adequate for any trickling filter and,

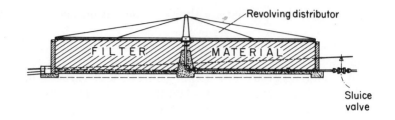

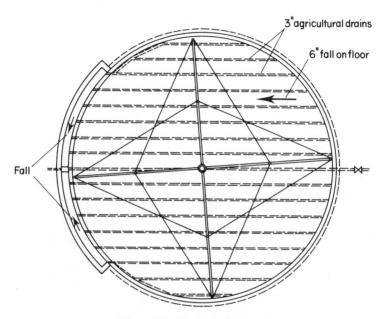

Figure 74   Typical trickling filter

seeing that filter tiles are expensive, adding an appreciable percentage to the cost of the filter, there would not appear to be much justification for using them as a rule. As a compromise, lines of square tiles may be laid, say 1 m centre to centre, either side of a centre channel, also covered by tiles, or in lines radiating from the central chamber or exterior outlet channel, as the case may be. Special junction tiles are available, and end-pieces to suit the radius of the bed.

Filter walls

At one time it was common practice to construct the walls of trickling filters of open honeycomb brickwork, dry-stone walling, or dry walling in boulder clinker so as to ventilate the sides of the bed. But this additional ventilation has been found unnecessary. Moreover, the exposure of the medium to the air in a place where it is not washed by the sparges encourages excessive development of flies. Further, boulder clinker is now not so easy to obtain, and dry wallers are scarce.

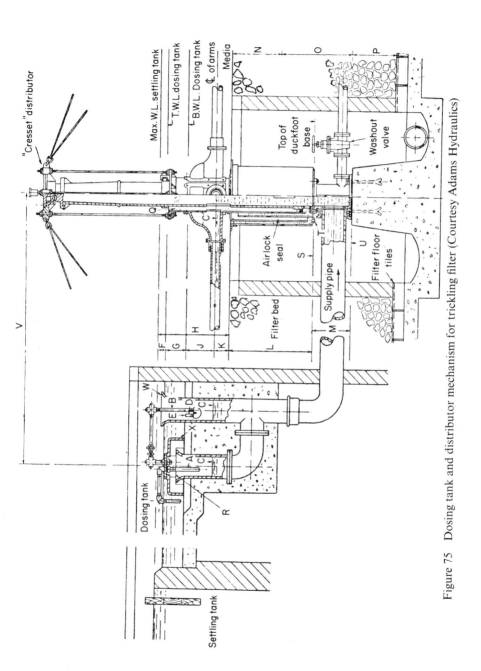

Figure 75   Dosing tank and distributor mechanism for trickling filter (Courtesy Adams Hydraulics)

For these reasons the commonest medium of construction for filter walls is brickwork, although in view of the many troubles experienced with it, due chiefly to frost effects, nothing less than a 'seconds' engineering brick should be used, and a plasticizer should be incorporated in the mortar, as this is where frost attack first takes effect. Low absorption is a valuable asset in the bricks themselves. Reinforced concrete is often used, particularly on larger works where practically all construction work is in this medium. Pre-cast concrete wall units with sides slightly splayed to suit the radius of the bed, and 'flanged' all round for convenience in bedding and bolting together, are readily available. Satisfactory bituminous-strip joints, and bolting techniques, are now available, making this type of construction very convenient. The particular units to be used should be detailed on the construction drawings, so that any holes for pipes passing through the wall units can be preformed, thus avoiding damage likely to arise in forming such holes *in-situ*. In Great Britain at least one machine-made hollow concrete block is available and makes an economical wall. To secure the necessary strength, in hoop-tension, reinforcing bars are laid through the hollow blocks every third course. Slight grooving on all contact faces make make the blocks easy and quick to assemble. It should not be forgotten, however, that no openings can be left in walls of this type, for bringing in the filter medium. The author has seen contractors thrown into some confusion by collapse of walls, where medium has been placed before the completion of the reinforcing rings. This point is, of course, valid with pre-cast units, or *in-situ* concrete, unless provided for in the design.

Coke and clinker are not heavy materials and are easily retained, as they weigh about 475 kg/m^3 (dry) or 25 % more (wet) and have an angle of repose of about 40° to the horizontal. Thus, where brickwork is used, the wall of a circular bed above ground level need be of no very great thickness, although every fourth course may be reinforced with a suitable brickwork reinforcement to prevent cracking. Broken stone can be taken as being three times as heavy as coke or clinker, but the angle of respose is somewhat similar.

Where beds are nested together in contact one with another, it is not considered necessary to build up the walls between them, but a dwarf wall is normally provided, so that should it be necessary to empty one bed of material, effluent from the adjacent beds will not drain on to the floor.

Dosing tanks and siphons and distribution pipes

The distributors of circular trickling filters, being rotated by the reaction of the jets of water, would stand still and dribble at low rates of flow unless something were done to prevent this. The usual arrangement is to provide a dosing tank having a low-draught flushing siphon arranged to discharge at the optimum rate of flow to the filters when the tank is full. When the flushing tank becomes partly full on discharge of the siphon the rate of flow will reduce but, provided that the siphon can take the optimum flow at the maximum head and discharge enough to rotate the distributors when the tank is nearly empty, this is satisfactory. Nevertheless, a weir must be provided to permit overflow should the siphon be overloaded at maximum rate of flow or fail to function.

When one siphon serves a number of filters, a distribution chamber may be constructed in order that the flow may be evenly distributed to all the filters or adjusted should it be found that some filters can accept more flow than others. The usual arrangement is to install adjustable weirs, each to feed one filter. Rules that have been used for determining the size of dosing tank are to make the tank capable of holding not less than 1 min's siphon discharge at the maximum rate of flow, or between 3 and 12 m³ 1000 m²) of the top-surface of the bed.

Where there are two filters the feed pipe to each and the central column of each distributor should be capable of taking the total maximum rate of flow to the works. Where there are three distributors each one should be able to take half the flow to the works to allow for one being laid off at a time. At large works it is best to have trickling filters in multiples of four, in which case every three filters should be capable of sharing the maximum flow that four would normally receive.

Figure 75 illustrates a typical arrangement of dosing tank, siphon, and distributor mechanism in which the central column of the distributor is fixed in a circular chamber in the centre of the bed.

Some distributors for small filters incorporate a dosing siphon in the central column of the distributor itself. These should never be used where a dosing tank has been provided from which several filters are served without control of distribution; not only are they unnecessary, but they would result in serious maldistribution of flow.

Where all flow to the works is pumped, or trickling filters are intended always to be used on recirculation, dosing siphons are unnecessary. The flow from the recirculating pumps can be distributed either from a chamber with adjustable weirs or by the process of progressive bifurcation of pipework in which the diameters and lengths of all the feed pipes are alike throughout, ensuring reasonably equal distribution to all filters.

The simplest method of arranging for recirculation at any desired rate is to deliver the settled sewage from the sedimentation tanks to a chamber in which floats or electrodes are placed so as to stop the recirculation pumps when the level rises to that which will cause the flow through the filters to be at the maximum desired rate. The pumps would deliver treated effluent to this chamber where it would mingle with the settled sewage. This arrangement will ensure that the rate of flow to the filters is maintained within narrow limits at the maximum desired rate of flow regardless of the rate of flow to the works. Another method of control commonly used, is to use the flow-recorder to control the recirculation pumps, so that they cut in and out as the inflow rises and falls, but maintain a steadily combined flow to the filters at all times. This implies, of course, that the bulk of the recirculated flow is pumped at night when inflow is low. Where 'peak' charges are in operation for electricity, the use of 'off-peak' current can achieve considerable economies at the medium or small works, where no other source of power is available. It should not be forgotten, however, that power should be available for recirculation at all times, so that the off-peak tariff will not apply throughout.

The feed pipes lead from the dosing or distribution chambers under or through the medium, and connect to the central columns of the distributors. Normally a wash-out is provided at the base of the central column, from which a length of pipe is laid to discharge to the sludge or humus-pumping station. Wash-outs should never discharge to filter outlet channels or manholes. Alternatively, wash-outs can be provided on the lowest part of each feed system, and connected to the nearest sludge-drain.

The total loss of head through the dosing tank, distribution piping and on to the surface of the filter medium varies, being greater in the case of large distributors and long pipelines. In the case of a small installation the loss between the floor level of the dosing tank to the holes in the filter arms can be as little as the friction in the pipework plus 75 mm, and the distance between the holes in the filter arms and the surface of the medium can be as little as 100 mm. Small dosing tanks can be constructed as shallow as 150 mm. Thus it is possible to arrange for top of medium to be no more than 375 mm below top-water level of dosing tank or, say, 600 mm below weir level of sedimentation tanks. But generally, and for larger filters, greater losses of head are involved. Excessive head does no harm.

The central column of the distributor consists of a cast-iron base plate which is bolted to a concrete or brick pier in the centre of the bed, a vertical pipe which supports the distributor and carries the flow, and a rotating casting to which the arms are fixed. Leakage between the fixed and moving portions of the central column is prevented in many different ways according to the design, and includes mercury seals, glands, and in one instance a design involving the principle of the venturi throat, the restriction of passage having the effect of reducing pressure and preventing leakage outwards. Some designs of distributors have two arms fed directly by gravity and two coming into operation only when the flow is so great as to back up over weirs incorporated in the mechanism.

Distributors of small size have only two arms: the majority of distributors of medium and large size have four arms. These arms consist of steel tubes. Removable caps are arranged at the ends to permit cleansing. The arms are supported by guy wires which run from the top of the central column to intermediate positions on the arms, which are also guyed one to another to prevent independent lateral movement. The holes in the arms are arranged at varying spaces, being closer together towards the outer ends so as to give even distribution over the surface of the bed. Their location on each arm is different, so that no two holes discharge at the same distance from the centre of the bed. In all good-quality distributors the holes are bushed with gunmetal, and often incorporate special deflectors so as to spread the discharge. The holes discharge horizontally, on one side of the arms only, for it is usually the reaction of the emitted water that causes the rotation of the distributor, although motor-driven units are sometimes employed.

With jet-propelled arms wind effect can be quite severe, but is avoided to a great extent by building up the walls of the beds say 50–75 mm above the tops of the arms.

There is much variety in the design of travelling distributors for rectangular beds, for these, not being supported on almost frictionless bearings, need a much more complicated mechanism to drive them. Some are cable-driven, powered by water-wheel or electric motor, and so arranged that distributors operate in pairs, moving in opposite directions, thus balancing out the effect of the wind.

All travelling distributors for rectangular beds are fed through siphons, which move with the sparges and suck from feed channels which are constructed down the centres or along the sides of the beds, according to the design.

Distributors for rectangular beds are comparatively costly, but where there is no recirculation they make the beds more efficient, because of their relatively slow motion. They involve losses of head in the region of 600–750 mm from top-water level in distribution trough to surface of medium when power-operated, and somewhat more if a water-wheel device is involved.

There are other methods of distribution apart from revolving and travelling distributors. Fixed-spray jets have been found useful for colder climates or where considerable head is available; for example, where tank effluent is pumped to the filters. They are not used for high-rate filters.

Humus removal

Humus is the name given to the organic matter which is discharged from trickling filters. It consists of sludge which the sedimentation tanks have failed to settle, changed and rendered settleable, the remains of decayed algae, moulds, bacteria, insects, larvae, worms, etc., and living organisms. It is inoffensive and most of it is easy to settle but, when settled, it is liable to gas and rise to form scum. The average amount of humus is about 3.5 litres per head per day. Humus has an average solids content of 0.028 kg dry matter per head per day and the average moisture content is 99.23 %. About 64.2 % of the dry matter is volatile. In designing pipework and pumping equipment allowance should be made for great seasonal variations in quantity, as a general rise or fall in temperature, such as occurs in spring and autumn, causes internal changes in the beds.

Common methods of removal of humus are:

(1) settlement in humus tanks;
(2) sand filtration;
(3) micro-straining;
(4) land irrigation.

The first of these is by far the most usual: the fourth is particularly applicable for very small works that are not liable to receive frequent attention. The second and third (which will be described under the head of tertiary treatment) are unusual and should always be preceded by settlement.

About fifty years ago British humus tanks were given capacities equal to 2 h DWF. Later this was increased to 3 h, while more recent practice has been to allow a capacity of 4 h. However, as in the case of sedimentation tanks, detail design is of more importance than capacity.

Two types of humus tanks are in general use: flat-bottomed mechanically sludged tanks and pyramidal-bottomed tanks. Mechanically raked humus tanks are generally similar to primary sedimentation tanks. The raking mechanisms can be of similar type and equipped to remove scum. Scum plates are necessary. The floors may be flat or sloping according to the design of raking mechanism. A central hopper should be provided to store humus settled between sludging periods unless continuous humus removal is the intention. In view of the high moisture content of humus, and consequent large quantity, the size of the central sump has a considerable influence on the design and even the total capacity of the tank.

Pyramidal-bottomed humus tanks are almost identical in design with primary sedimentation tanks of this type. Again, the storage capacity for humus and position of inlet must be studied if the humus is not to be prevented from settling by the turbulence of the inflow. The slopes of the sides of the hopper bottoms should be similar to those of primary sedimentation tanks.

Humus can be drawn off by manual control and discharged on to sludge-drying beds reserved for the purpose or, as humus is usually of good manurial value, it can be irrigated on to ploughed land which is afterwards harrowed and used for suitable crops. These methods are particularly applicable to small works.

At large works humus can be drawn off continuously at such rates as may be found satisfactory, and pumped to the feed of the primary sedimentation tanks. This has one disadvantage: the nitrified content of the humus, on being mingled with crude sewage, denitrifies and the free nitrogen given off may militate against settlement.

Improved type of tank

In the United Kingdom, particularly at smaller works, there has been considerable 'loss of faith' in humus tanks—probably because operators often fail to realize the importance of keeping them clean! There have been occasions when, in the spring when the trickling filters are 'sloughing', humus tanks have been allowed to become so full of sludge that the normal removal system has

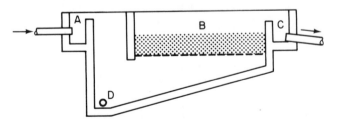

Figure 76   Banks clarifier (*Advances in Sewage Treatment, Instn. Civil Engnrs.*)

become inoperable, with disastrous effects! The upward-flow pebble-bed clarifier has been found to give much better results, and often it is possible to construct the necessary suspended bed in existing rectangular, or even pyramidal, humus tanks, though with the latter it may be difficult to obtain sufficient surface area. Such clarifiers, of course, often receive the flow from conventional humus tanks and thus become a tertiary stage of treatment. They will receive further mention in Chapter 16.

References

1. Ames Crosta Babcock 'Biodisc' Process, Ames Crosta Babcock Ltd., Heywood, Lancs. (World-wide design and contracting service.)
2. Imhoff, K. and Fair, G. M., *Sewage Treatment*, John Wiley & Sons, N.Y. 2nd edition (1956).
3. Babbitt, H. E. *Sewerage and Sewage Treatment*, John Wiley & Sons, N.Y. 7th edition (1953).
4. Metcalf, L. and Eddy, H. P., *American Sewerage Practice*, McGraw-Hill, N.Y. 3 vols (1928).
5. Pretorius, W. A. The rotating disc unit—a waste treatment system for small communities. *Wat. Pollution Control*, **72**, 6 (1973).
6. Ellis, K. V. and Banage, S. E. I. A study of rotating disc treatment units operating at different temperatures, *Wat. Pollution Control*, **75**, pt 1 (1976).
7. Porter, K. E. and Smith, E., Plastic media biological filters, *Wat. Pollution Control*, pt 3 (1979).
8. Mills, E. V. Treatment of settled sewage in percolating filters in series with periodic changes in order of the filters: results of operating the experimental plant at Minworth, Birmingham, 1940–44. *J. Inst. Sewage Purification*, **2** (1945).
9. Mills, E. V., The treatment of settled sewage by continuous filtration with recirculation of part of the effluent, *J. Inst. Sewage Purification*, **2** (1955).
10. Hunter, A. and Cockburn, T. Operation of an enclosed aeration filter at Dalmarnock sewage works, *J. Inst. Sewage Purification* (1944).
11. Hemming, M. L. and Wheatley, D. Low rate biofiltration systems using random plastic media. *Wat. Pollution Control*, **78**, pt 1 (1979).
12. Wheatley, A. D. and Williams, I. L., Pilot scale investigations into the use of random-pack plastics filter media in the complete treatment of sewage. *Wat. Pollution Control* **75**, pt 4 (1976).
13. Bruce, A. M. and Merkens, J. C., Further studies of partial treatment of sewage by high rate biological filtration. *Wat. Pollution Control*, **72**, pt 5 (1973).
14. Bruce, A. M., Merkens, J. C. and Keynes, B. A. D., Pilot-scale studies on the treatment of domestic sewage by two-stage biological filtration, with special reference to nitrification. *Wat. Pollution Control*, **74**, pt 1 (1975).
15. Kornegay, B. H. The modelling and simulation of fixed bed biological reactors for carbonaceous waste treatment. *Mathematical Modelling for Water Pollution Control processes*. Ann Arbor Sci. Publishers (1975).
16. Banks, P. A., Hitchcock, K. W. and Wright, D. G. Studies of high-rate biological treatment of Ipswich sewage on pilot filters using plastics media. *Wat. Pollution Control*, **75**, pt 1 (1976).
17. Pullen, K. G. Trials on the operation of biological filters. *Wat. Pollution Control*, **76**, pt 1 (1976).
18. Bruce, A. M. and Boon, A. G. Aspects of high rate biological treatment of domestic and industrial waste waters, *Wat. Pollution Control*, **70**, pt 5 (1971).
19. *Process Design Manual for Upgrading Existing Wastewater Treatment Plants*. Environmental Control Ag. Technol. Transfer (Oct. 1974).

342

20. Jt. Comm. Am. Soc. Civ. Engnrs. & Fed. Sewage Ind. Wastes. *Sewage Treatment Plant Design* (1959).
21. *Units of Expression for Wastes and Waste Treatment.* Man. of Practice No. 6, Fed. Sewage Ind. Wastes, **30** (5) (May 1958).
22. Velz, C. J. A basic law for the performance of biological beds. *Sewage Works J.*, **20**, pt 4 (July 1948).
23. Gurnham, C. F. *Principles of Industrial Waste Treatment.* John Wiley & Sons, N.Y. (1955).
24. Bruce, A. M. and Merkens, J. C. Pilot-scale studies on the treatment of domestic sewage by two-stage biological filtration—with special reference to nitrification. *Wat. Pollution Control*, **74**, pt 1 (1975).
25. Nealy, A. B. Chemical biological treatment with biological filters. *Wat. Pollution Control*, **74**, pt 2 (1975).
26. Levine, M., Luebbers, R., Galligan, W. E. and Vaughan, R. *Sewage Works J.* **8**, 699 (1936).
27. Truesdale, G. A. and Wilkinson, R. A comparison of the behaviour of various media in percolating filters. *J. Instn. Public Hlth. Engnrs.* **60**, pt 4 (Oct. 1961).

CHAPTER 15
Activated-sludge Processes

The term 'activated-sludge treatment' was originally applied to what is now known as the diffused-air system, but it has come to mean all methods of sewage treatment in which aeration is effected in tanks, the aerobic organisms being suspended in an organic floc. The activated-sludge principle is to develop bacteria and other organisms, to keep this floc in suspension, and to aerate the sewage so as to replace the oxygen as rapidly as it is taken up by oxidation of the organic content of the sewage.

The methods of activated-sludge treatment are classed under two heads:

(1) diffused-air system, in which air is passed through the sewage as bubbles,
(2) mechanical agitation, or surface-aeration system, in which the surface of the sewage in the aeration tank is agitated so as to encourage solution of oxygen from the air.

All the activated-sludge methods are preceded by the normal preliminary treatment processes, screening, grit removal, and sedimentation, but the standard of sedimentation of primary sludge is usually a little lower than that favoured for a trickling-filter scheme. Aeration by the activated-sludge system is always followed by final sedimentation to remove the activated sludge which would otherwise pass out with the effluent, rendering it unfit for discharge. This activated sludge is necessary to the process and, after being settled out from the effluent, it is returned in measured quantity to the aeration tanks. Surplus activated sludge is usually passed to the primary sedimentation tanks, settled out, and digested with raw sludge, or it can be dewatered and mixed with primary sludge.

There is considerable elasticity in the design of all the activated-sludge processes, because aeration can be effected in either a comparatively small tank, using a heavy concentration of activated sludge and a high rate of aeration, or, alternatively, in a larger tank with a lower concentration of activated sludge and aeration not so intense.

As in trickling-filter treatment, the amount of oxidation of the organic content of the sewage is in direct proportion to the quantity of healthy aerobic organisms. This was proved by Danson and Jenkins[1] who found by laboratory experiments that oxygen uptake was directly related to the weight of activated-sludge solids and the time during which they worked. It is therefore necessary to preserve as high a concentration of activated sludge as can be contained without spilling solids over the weirs of the final sedimentation tanks.

The capacities of the tanks of an activated-sludge plant are, of course, limited, but the aerobic organisms can be either widely dispersed or more congested and in a tank where the density of organisms as high, more aeration can be effected than where it is low. The sewage-works operator can build up the density of the activated sludge until the tanks will not hold it, or when such a point is reached that the replenishment of oxygen is not sufficient for the needs of the organisms. When the latter condition comes about, bulking occurs (of which more will be said) and the process is upset. If, however, it is possible to increase the rate of aeration of the mixed liquor (that is, the mixture of settled sewage and activated sludge) bulking will be prevented and the degree of treatment can be increased.

In the early days of the development of the activated-sludge process[2,3] it was considered that longer detention periods were necessary for surface-aeration methods than for the diffused-air system; for example, 16 h dry-weather flow (DWF) were allowed for the Sheffield system or for the original Simplex system while the horsepowers used were lower than for air diffusion. Since those days the detention periods of the surface-aeration methods have been reduced and the horsepowers increased, so that now they are more or less in line with those of the diffused-air process.

Practice in America and Britain is not alike. American engineers use shorter detention periods but greater horsepowers than are favoured in Britain. Perhaps the most usual figures for the United States are 7 h detention in the aeration tanks and 2 h average daily flow in the final sedimentation tanks, making a total of 9 h. In Great Britain the total detention period in aeration and final sedimentation tanks is about 12 h, or $1\frac{1}{3}$ times the United States figure. On the other hand, the input of free air to diffused-air plants in Great Britain averages $50 \text{ m}^3/\text{kg BOD}_5$ removed from the settled sewage, whereas in the United States 66.6 m^3 of free air per kg BOD_5 are allowed, or $1\frac{1}{3}$ times the British figure. Broadly this means that the electricity demand of an American scheme is $1\frac{1}{3}$ times that of a British scheme (see formula 34).

The above figures show that the product of the average detention period and the average electricity demand is generally the same on both sides of the Atlantic. The reason for this is explained in what follows.

The Manchester Rivers Department[2,3] made extensive large-scale experiments in an attempt to compare several methods of aeration. In this work the rule of scientific research that more than one parameter should never be altered at the same time was ignored and, as it happened, this error was fortunate because it brought to light a fact that previously had not been known. The author, on request, was provided with full particulars of the data and completed examining them in 1958. The most informative series of tests had been carried out (with the aid of a slight reasonable logarithmic-rule adjustment) to find what detention periods and horsepowers were required to produce a final effluent with a biochemical oxygen demand of approximately 15 mg/l.

When the author plotted biochemical oxygen demand load on the works against the product of the aeration-tank detention period and the horse-power used, he found that there were large discrepancies, as shown by the broken line in Figure 77. On investigation of this it came to light that, contrary to scientific research procedure, two parameters had been altered at the same time. When the detention period in the aeration tanks had been reduced the detention period in the final sedimentation tanks had been increased, but not in accordance with any rule. For example, the longest aeration-tank detention period was 8.65 h at which time the detention period in the final sedimentation tanks was 4.63 h, making the total detention in the works of 13.28 h (point B on the figure). When, however, shortest detention period in the aeration tanks was as little as 2.566 h, the detention period in the final sedimentation tanks had been increased to the very high figure of 8.924 h, making a total of 11.49 h (point E in Figure 77). The author then plotted the biological load against the product of the total detention period in aeration plus final sedimentation tanks and the aeration horsepower. This produced the almost perfectly straight line drawn in full on Figure 77.

While it must be admitted that additional sedimentation-tank capacity must improve settlement and thereby reduce the biochemical oxygen demand (BOD_5) of the final effluent, it cannot do so to such an extent as was recorded during the experiments in question. It therefore had to be concluded that the detention period of the final sedimentation tanks must be considered as additional aeration detention during which the organisms continue to work for as long as they have something to aerate and enough oxygen wherewith to effect aeration.

We are therefore faced with the fact that, in designing activated-sludge works, the total capacity of aeration tanks, final sedimentation tanks, and all connecting channels and pipework containing activated sludge must be included as effective aeration-tank capacity in the calculations.

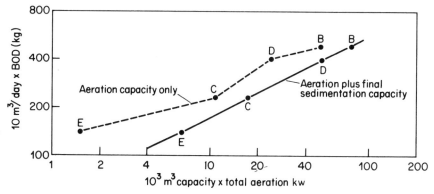

Figure 77   Graph showing load on activated-sludge works and power input

On the basis of this work the author devised a provisional formula for use between the limits of a final BOD of 8 to 30 parts per million. When, however, it was examined using data from several other works it was discovered that it could not be used except at works where nitrification was negligible and treatment was not carried to the degree of BOD removal beyond the well-known kink in the curve. (When BOD is plotted against detention period a regular curve is produced for so far, then there is a kink followed by another regular curve.) It was also found in some instances that the value of K changed after large changes of biochemical oxygen demand of the settled sewage. Accordingly, a modified formula was evolved as follows:

$$K = \frac{Q \times \sqrt{BOD_1} \times \log_{10}\left(\frac{BOD_1}{BOD_2}\right)}{78.1 \times \sqrt{C} \times \sqrt{KW}} \tag{33}$$

where: Q = cubic metres per day;
BOD_1 = biochemical oxygen demand of primary sedimentation tank effluent in milligrams per litre;
BOD_2 = biochemical oxygen demand of final treated effluent in milligrams per litre (the formula is applicable to values between 5 and 35 only);
K = a performance index dependent on efficiency of power utilization, nature of sewage, etc;
C = capacity of aeration plus final sedimentation tanks in cubic metres;
kW = aeration kilowatts (excluding electric power for sludge recirculation, unless this is by airlift).

This formula is still applicable only where nitrification is negligible and BOD not greatly reduced below Royal Commission standard but, with these limitations, when applied to average results of carefully collected data, it will give a good indication of the comparative performance of treatment plants. The higher the value of K the better the performance.[4]

To design an activated-sludge plant with the aid of formula (33), first the BOD_5 of the settled sewage should be obtained from records or, in the absence of these, estimated by the use of Table 41a, see p. 297 and then the future flow in m^3/min per day should be estimated. The value of K for any reasonably good diffused-air or surface-aeration plant should not be lower than 3, and this figure could be used for small installations in the absence of more accurate information. For plant with porous diffusers, see Table 48. British practice is to design municipal works with the intention of producing a final effluent having a BOD_5 averaging 15 mg/l at the estimated future DWF so as to be within the Royal Commission standard most of the time. If this should be achieved, fluctuation above or below it could be expected to fall to half the figure and comparatively rarely exceed twice this figure.

For diffused-air plants using porous aerators the following formula gives reasonable representation of both American and British practice

$$T = \frac{564}{A} \tag{34}$$

where: T = aggregate detention period in the aeration and final sedimentation tanks in hours, DWF (in US hours average flow);

A = air supply in m^3 of free air per kg of BOD$_5$ in the settled sewage as delivered to the aeration tanks.

The air supply indicated by A above is the average working quantity: in design of all equipment, allowance should be made for the possibility of 1.67 times this quantity being required.

Diffused-air systems

There are several varieties of the diffused-air system[5, 6] which can be classified under the heads of fine-bubble methods in which the diffusers are porous ceramic tiles, and coarse-bubble methods in which air is injected with the aid of either jets of water or perforated grids. Fine-bubble methods appear to be the more popular, the porous diffusers for which were originally in the form of flat tiles housed in cast-iron boxes and fixed at the bottoms of aeration tanks. Aluminium alloys have been tried, but corrode very rapidly on contact with activated sludge. Now porous tubes have become more usual in America, while in Britain, Activated Sludge Ltd use nothing other than dome diffusers. The diffusers are supplied with compressed air via cast-iron or polyvinyl chloride pipes as may be convenient according to the type of diffuser used.

Diffuser plates are usually 300 mm square and 25 mm thick. Tubes are available in sizes from 20 to 175 mm. A usual type of fixture is illustrated in Figure 78. Plate diffusers, when new, can pass about 6 m^3 of air per min per m^2. The performance of tubes is similar. Dome diffusers were once installed in two sizes: 100 mm and 175 mm diameter, but now only 175 mm diameter diffusers

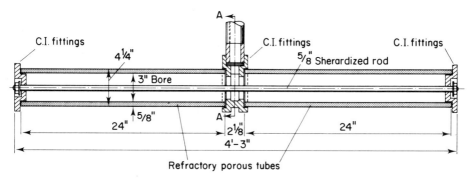

Figure 78   Arrangement of tube diffusers (Norton International Inc.)

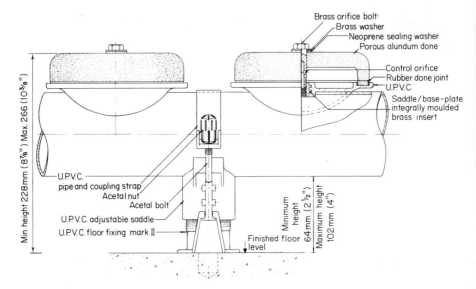

Figure 79   Arrangement of dome diffusers

are installed in new works. These can pass $0.85–4.25\,\text{m}^3$ of air per hour per dome, $1–3\,\text{m}^3$ being the general working range. The present design is for dome diffusers to be fixed to special polyvinyl chloride air mains (see Figure 79). One dome diffuser is allowed for every $0.2–0.25\,\text{m}^2$ of floor or, at the very most, $0.3\,\text{m}^2$.

In course of time porous diffusers may tend to become clogged by dust from the air that is delivered to them, and which coats the insides, and by a deposit of ferric hydroxide, lime, and organic matter on the outside. Diffusers may remain in service for several years provided that the air is thoroughly filtered. If they do become clogged they can be renovated by scrubbing with dilute hydrochloric acid or sand-blasting of the top surface. Another method is soaking them for 3–4 h in 33% hydrochloric acid, washing and then firing at about 1000°C.

In the diffused-air system in which porous diffusers are used the tanks or channels are usually long but not necessarily narrow, and of a depth of not more than 3.7 m or less than 1.5 m. Originally there were several ways in which the diffusers were arranged, including individual pockets, transverse ridge-and-furrow floors with the diffusers in the furrows, longitudinal ridge-and-furrow floors, and spiral flow. The first three arrangements were expensive to construct and are now outdated. In the spiral-flow method (still used in special circumstances) the diffusers are placed to one side of the tank or channel so as to rotate the flow laterally rapidly enough to prevent settlement of sludge. The preferred present-day method is to have a flat floor with the diffusers evenly spaced throughout and to rely on the turbulence due to the high rate of air flow to prevent any settlement.

Air is supplied in quantities of $10–15\,\text{m}^3/\text{m}^3$ of sewage treated, the blowers or compressors being designed on the larger figure where sewage is mainly

Table 46 Recommended size of air mains for diffused-air works

Internal diameter (nominal mm)	Free air discharge (m³/min)
100	3.75
150	11.05
225	32.2
300	69.7
375	126
450	204
525	311
600	437
675	601
750	802
825	1030
900	1300
975	1580
1050	1960

domestic, or still more where considerable proportions of trade waste have to be dealt with. Another rule is to allow from 45 to 50 m³ of free air per kg BOD_5 to be removed in aeration tanks 3 m deep, but to increase the air flow for shallow tanks or reduce it for deep tanks.

When calculating the air pressures, the losses that have to be taken into account are:

(1) friction losses through the distribution system;
(2) a pressure of 0.1 kg/cm² per m of depth of tank to overcome water pressure;
(3) loss of head through the diffuser (when new, 225 mm water gauge at 0.018 m³/min).

Table 47 Approximate efficiencies of low-pressure air blowers

Capacity of compressor (m³ of free air/min)	Overall efficiency (%)
28.4	53*
56.7	58*
85	62*
113	64*
140	66*
170	71†
198	72†
226	73†
284 upwards	74†

* Roots-type blowers.
† Turbo blowers.

Table 48 Performance of diffused air activated sludge plant

		1960-1	1961-2	1962-3	1963-4	1964-5	1965-6	1966-7	1967-8	1968-9	1969-70
Average flow	10^3 m³/d	278	282	300	314	323	300	309	314	373	328
	mgd	61	62	66	69	71	66	68	69	82	72
Aeration period	hr	9.0	8.8	8.3	7.9	7.7	8.3	8.0	7.9	6.7	7.6
Air supplied	m³/sec	38.0	38.2	38.0	34.2	34.7	33.2	34.5	32.5	33.6	30.6
	cu ft/min	80,600	81,000	80,500	72,400	73,500	70,400	73,100	68,900	71,200	64,800
	m³/m³	11.8	11.7	10.9	9.4	9.3	9.6	9.7	9.0	7.8	8.0
	cu ft/gall	1.90	1.88	1.75	1.51	1.49	1.54	1.55	1.44	1.25	1.29
Sludge volume index ($\frac{1}{2}$ hr)		141	158	139	120	167	173	110	105	114	143
Influent:											
Suspended solids	mg/l	91	97	99	90	86	87	91	90	97	108
BOD	mg/l	134	162	176	178	170	163	158	159	162	170
Ammoniacal nitrogen	mg/l	28.7	28.4	29.5	31.2	31.2	29.5	28.0	26.4	24.9	24.7
Effluent:											
Suspended solids	mg/l	13	11	12	10	10	11	11	11	11	10
BOD	mg/l	22	20	22	25	23	16	19	18	21	20
BOD (ATU)	mg/l	—	—	*7	*7	*6	5	5	6	7	7
Ammoniacal nitrogen	mg/l	11.5	9.8	10.5	11.0	15.0	5.1	6.0	6.9	10.9	11.2
Total oxidized nitrogen	mg/l	14.8	13.2	11.1	8.5	7.5	13.5	12.1	10.4	8.1	8.1
Effective oxygen load	tonnes/day	—	—	20	21	27	12	13	15	25	23
	ton/day	—	—	20	21	27	12	13	15	25	23
Surplus activated sludge:											
Volume	10^3 m³/d	3.89	4.21	5.04	4.98	4.86	4.67	4.19	5.16	6.18	6.43
	mgd	0.856	0.927	1.108	1.095	1.070	1.028	0.921	1.134	1.359	1.415
Total solids	per cent			0.62	0.62	0.65	0.60				0.54
Loading:											
kg BOD/m³ aeration capacity/d		0.35	0.45	0.51	0.55	0.53	0.47	0.47	0.48	0.58	0.55
lb BOD/10³ cu ft aeration capacity/d		22	28	32	34	33	29	29	30	36	34
Air:											
m³/kg BOD removed		108.9	87.9	65.4	56.4	58.7	64.9	65.7	61.1	52.4	49.4
cu ft/lb BOD removed		1745	1408	1048	903	941	1039	1052	978	839	792

* Sterilized.

Reproduced from Journal IPHE.

It is advisable to make allowance for the possibility of pressure building up above the last figure in case it should be found necessary to continue to use dirty diffusers at a time when labour for replacement is not available.

The air mains that supply the diffusers are of larger diameters than are usual in other air-supply problems because the pressure throughout the system must not vary much. Suggested figures are given in Table 46. Sizes in accordance with this table could give a loss of head of 0.063 m of water-gauge per 100 m length of pipe when the pressure at the compressor end is in the region of 6–7 m, or 0.058 m/100 m length of pipe when the pressure at the compressor end is about 10 m head of water.

It is better for air to be compressed by rotary blowers than by reciprocating compressors because the latter type requires lubrication which tends to choke the diffusers. There should be a sufficient number of compressors to make possible variation of rate of air flow and also to allow stand-by for maintenance purposes; where practicable, blowers of large size should be used for the sake of efficiency (see Table 47).

At small works the air can be filtered by mats of glass wool, hair, or metallic fibre covered with viscous oil, but electrostatic filters, although expensive, are much more effective and for this reason are used at large works.

Air mains should be provided with blow-out valves at the ends of all sections of pipework so that condensation water or oil may be removed or any large quantities of water should the mains have been accidentally flooded.

Table 48 gives the performance figures for the diffused-air plant at London's northern outfall works at Beckton over a 10-year period from 1960 to 1970, as discussed in a paper by Pearson and others.[5]

Coarse-bubble method

A method of diffused-air activated-sludge treatment not involving porous diffusers is the INKA system. This method, which was developed in Sweden, used in various places on the continent, and taken up in Britain by Dorr-Oliver,[7] differs from the process involving diffusers in the following respects. The aerators consist of either stainless steel or plastic grids which are submerged to a depth of 0.75 m to one side of a spiral-flow tank and which have 2.5 mm diameter holes on the underside. The tank is provided also with a longitudinal baffle of corrugated fibre-glass to assist spiral flow. The tanks vary in depth from 3 to 4.5 m and in widths from 3 to 9 m. The aeration grid is 1.85 m wide for tanks up to 6 m wide and 2.25 m wide for large tanks, and the depth of the fibreglass baffle varies from 1.5 to 3 m for tanks of 3 to 4.5 m depth respectively. Air is blown at the rate of 33 times the volume of sewage treated and at a working pressure of 0.9 m to 1 m water-gauge.

Returned activated sludge

For design purposes, allowance should be made for the return of activated sludge from the final sedimentation tanks to the aeration channels up to a maximum quantity of about 100% of the DWF of sewage. The normal working percentage will, however, probably be in the region of 67% (British practice) and allowance should be made for the possibility of dropping it to about 34%. In North America recirculation rates of 25–50% of average flow are usual.

The amount of sludge solids increases by accumulation of solids which the primary sedimentation tanks have failed to settle and by organic growth; therefore surplus sludge has to be withdrawn, otherwise it would fill the final sedimentation tanks. The quantity of this varies considerably according to the works. Some recent American figures suggest that the amount of dry solids in the surplus sludge averages in the region of 0.4 of the weight of BOD_5 in the crude sewage. Imhoff and Fair[8] said that the surplus sludge is normally about 1.5% of the sewage flow. English research figures give a daily weight of dry solids in the surplus activated sludge as averaging 0.026 kg per head of population which, at an average moisture content of 99.33%, is 3.9 litres per head per day. Of the dry matter, about 77.1% is volatile.[8]

The moisture content of the surplus sludge drawn off from an activated sludge tank is, of course, the same as that of the activated sludge in circuit. So variable is this figure that pipework, for both circulated activated sludge and surplus sludge, should be designed generously. The latter should also be capable of taking a flow equal to about one-quarter of the maximum rate of recirculation should it be necessary to rapidly remove the activated sludge on occasions of serious bulking or other breakdown of the treatment. The instruments measuring rate of flow of surplus sludge should be capable of accurately reading down to one-third of the average rate of flow.

Surplus activated sludge may be returned either to the incoming flow of crude sewage for resettlement in the primary sedimentation tanks, or passed to sludge-thickening plant such as the dissolved air thickening process (which will be described in Chapter 17 on Treatment and Disposal of Sewage Sludge).

Sludge can be recirculated with the aid of either air lifts or suitable pumps. Air lifts are often used where the diffused-air system has been adopted because the air supply for aeration is usually at the right pressure to operate the air lifts. (The calculation for design of air lifts of this kind is according to formula 20, p. 112). In surface-aeration schemes axial- or mixed-flow pumps are used. These should be so installed as not to impose a negative pressure on the activated sludge and large pumps of slow rate of rotation are to be preferred.

Recirculated sludge is drawn from the sludge-outlet pipes of the final sedimentation tanks, each of which should be capable of individual control. One satisfactory method is the use of telescopic weirs which can be raised or lowered by screws so that the quantity of sludge is drawn from all tanks equally.

There is a body of opinion that the best tanks for final sedimentation are conical-bottomed, having a slope set at 30° to the horizontal and a raking gear that trails or sweeps along the floor, actually touching, so that nothing can remain settled: a trailing chain will serve for this purpose (see Figure 69), p. 269. Diameters of much more than 18 m are not favoured. A 'theoretical' upward velocity of 2 m/h is usual at the maximum rate of flow, although some authorities are satisfied with 2.3 m/h or do not take returned activated sludge into consideration when allowing for 2 m/h.

In America the final sedimentation tanks have detention periods of $1\frac{1}{2}$–$2\frac{1}{2}$ h average daily flow, plus about half the rate of returned sludge (see Chapter 13). In Britain the detention periods are generally between 4 and 6 h DWF. A short detention period may be adequate to settle a healthy activated sludge but a longer detention period has the advantage of giving some elasticity to the plant at times of varying flows.

The capacity of the final tanks, if kept within reasonable limits, is a matter of convenience. When the total capacity of aeration plus final sedimentation tanks has been determined by either formula 33 or 34, pp. 346–7 the capacity of the aeration tanks can be arranged according to the floor area and depth required by the aerating equipment; some of the total capacity will be taken up by any channels between the units; what is left will then be available for the final sedimentation tanks.

The sludge withdrawn from the final sedimentation tanks can be delivered straight to the inlet ends of the aeration tanks but, in English practice, certain of the channels, amounting to from 10% to 20% of the whole, are often capable of being reserved for re-aeration or re-activation of the sludge before it is mixed with settled sludge to form mixed liquor.

Incremental loading and tapered aeration

Incremental loading, incremental feeding, or step-aeration is a modification of the process in which returned sludge is brought in at the inlet end of the aeration tanks where it is either aerated without any admixture of sewage or mixed with part only of the flow of sewage, further quantities of sewage being added farther along the tank in such stages as may be considered desirable. This method has been used in both the diffused-air and surface-aeration systems.

Tapered aeration is the diffusion of air at a higher rate near the inlet end of the tank and at decreasing rates towards the outlet end. Allowance is made in the arrangement of diffusers and pipes for about 45% of the air to be applied to the first third of the detention period, 30% to the next third, and 25% to the last third.

It is worth noting that the EPA, in their recent manual on upgrading existing works,[10] already quoted in the previous chapter, lists no fewer than six methods of improving the performance of a conventional installation. Application of step-aeration—whereby inflow keeps pace with aeration and thus evens out the oxygen demand, thereby making better use of the activated biomass—has been found capable of raising efficiency by up to 50%. Step-aeration also requires a change in feed location, involving the separation of biological sludge in a preliminary clarifier, and its separate aeration in a separate re-aeration tank. This involves less total aeration volume and could handle greater shock, or toxic, loads because of the biological buffering afforded by the re-aeration tank. Detention times in the re-aeration and contact tanks are interdependent—

Table 49 Operating data from conventional activated-sludge plants

Plant location	Influent flow (10³ m³/day)	Sludge recycle (%)	BOD Secondary influent (mg/l)	BOD Secondary effluent (mg/l)	Aeration tank MLSS (mg/l)	Organic loading $\frac{\text{kg BOD/day}}{\text{kg MLSS}}$	Volumetric loading $\frac{\text{kg BOD}}{\text{m}^3/\text{day}}$	Aeration detention time† (h)	Air supplied per kg of BOD removed (m³)	Secondary BOD removal efficiency (%)
Michigan	22.7	32	182	19	1844	0.34	0.62	7.0	48	90
Illinois	1309.0	48	129	11	1930	0.18	0.34	8.7	55	92
Ohio	395.0	25	92	12	2180	0.13	0.27	7.7	100	87
Indiana	67.7	30	161	14	2420	0.16	0.38	10.0	46	91
Maryland	17.7	32	254	33	1808	0.39	0.70	8.8	31	87
Michigan	36.4	16	118	6	2801	0.15	0.42	6.7	43	95
Wisconsin	34.5	52	157	36	1094	0.39	0.42	9.1	43	77
Indiana	17.7	31	134	14	2625	0.22	0.56	5.7	55	90
	25.0	29	113	6	1680	0.20	0.34	8.2	27	95
Maryland	36.4	26	155	10	2040	0.23	0.46	7.7	79	94
	35.0	25	148	15	2240	0.20	0.40	8.2	119	90
California	214.0	45	157	6	2449	0.19	0.45	8.4	99	96
	218.0	34	181	8	2111	0.23	0.46	9.2	84	96
Pennsylvania	15.6	20	175	20	1180	0.60	0.72	5.9	89	89
	12.3	26	161	14	1160	0.45	0.51	7.5	103	91
Illinois	910.0	38	119	13	2775	0.17	0.48	5.8	42	89

* MLSS = mixed liquor suspended solids concentration.
† Excluding sludge recycle.

Table 50 Operating data from step aeration activated-sludge plants

Plant location	Influent flow (10³ m³/day)	Sludge recycle (%)	BOD Secondary influent (mg/l)	BOD Secondary effluent (mg/l)	Aeration tank MLSS (mg/l)	Organic loading kg BOD/day / kg MLSS	Volumetric loading kg BOD / m³/day	Air supplied (m³/m³)	Air supplied per kg of BOD removed (m³)	Aeration detention time* (h)	Secondary BOD removal efficiency (%)
New York	500	24	74	12	1170	0.49	0.57	—	56.9	3.1	84
New York	94	49	137	3	3520	0.10	0.37	—	56.9	8.4	94
New York	418	35	100	8	1110	0.42	0.48	—	58.3	4.9	92
New York	227	28	120	6	3300	0.31	1.14	2.68	—	2.5	94
New York	432	28	115	16	3300	0.28	0.93	3.36	—	2.9	86
New York	141	28	100	12	4400	0.13	0.59	3.67	—	4.2	90
Maryland	77	24	140	11	1220	0.54	0.93	—	—	3.8	92
Indiana	58	92	124	15	2900	0.19	0.53	—	77.5	5.3	89
Indiana	88	50	139	17	2750	0.22	0.66	—	67.5	5.0	88
Indiana	156	52	131	18	3360	0.22	0.72	—	56.9	4.3	86
Pennsylvania	810	28	87	12	2780	0.23	0.64	3.61	57.9	3.2	86
Connecticut	171	34	121	17	2540	0.27	0.69	12.7	14.7	4.3	86
Ontario, Canada	832	16	115	11	1500	0.40	0.61	9.9	98.7	4.5	90

* Excluding sludge recycle.

Table 51 Operating data from complete mix activated-sludge plants

Plant location	Influent flow (10³ m³/day)	Sludge Recycle (%)	BOD Secondary influent (mg/l)	BOD Secondary effluent (mg/l)	Aeration tank MLSS* (mg/l)	Organic loading $\frac{\text{kg BOD/day}}{\text{kg MLSS}}$	Volumetric loading $\frac{\text{kg BOD}}{\text{m}^3/\text{day}}$	Aeration detention time† (h)	Air supplied per kg of BOD removed (m³)	Secondary BOD removal efficiency (%)
Illinois	7.2	21	102	8	6500	0.17	1.18	2.2	104	18
	8.8	21	80	13	6000	0.20	1.17	1.8	112	84
	8.7	25	80	19	6500	0.18	1.15	1.8	86	76
	7.0	25	108	18	6300	0.20	1.26	2.2	80	83
Minnesota	44.5	158	177	15	3750	0.31	1.17	3.6	−0.12	92
Nebraska	18.6	87	260	26	4400	0.27	1.28	4.5	33.7	90
	19.5	66	270	16	4460	0.32	1.55	4.3	28.1	94
	20.9	62	290	17	3920	0.43	1.92	4.0	31.2	94
	19.1	65	300	22	4020	0.41	1.76	4.4	29.9	93
	21.8	64	350	34	4280	0.51	2.11	3.8	34.9	90
	26.4	37	240	37	4040	0.49	1.82	3.2	35.6	85
	26.4	49	105	14	3400	0.24	0.81	3.2	56.2	87
Nebraska	15.5	50	250	15	4500	0.27	1.28	5.0	31.2	94
	18.6	100	270	13.5	4500	0.32	1.55	4.4	31.2	95
	22.7	200	280	6	4500	0.38	1.86	3.8	34.9	98
Nebraska	1.7	26	225	25	4230	0.48	2.02	2.6	—	89
	19.5	40	227	32	5460	0.42	2.27	2.5	—	86
Texas	1.32	82	115	9	3820	0.21	0.81	3.7	—	92
	1.32	100	141	25	5000	0.20	0.99	3.7	—	82
	1.36	145	123	19	5540	0.16	0.86	2.2	—	85
	1.68	100	180	17	5620	1.11	1.65	3.0	—	91

* Excluding sludge recycle.
† Mechanical aerators.

Table 52 Operating data from step aeration activated-sludge plants

Plant location	Influent flow (10³ m³/day)	Sludge recycle (%)	BOD Secondary influent (mg/l)	BOD Secondary effluent (mg/l)	Aeration tank MLSS (mg/l)	Organic loading kg BOD/day / kg MLSS	Volumetric loading kg BOD / m³/day	Air supplied (m³/m³)	Air supplied per kg of BOD removed (m³)	Aeration detention time* (h)	Secondary BOD removal efficiency (%)
New York	691	20	202	24	2000	0.9	1.70	—	—	2.6	88
Florida	181	8	205	62	360	5.1	2.21	50	6.0	2.3	68
	191	7	145	59	305	4.6	1.49	73	5.2	2.2	59
	254	6	125	38	275	6.2	2.00	62	4.5	1.7	69
	221	6	165	66	345	4.2	2.03	66	5.5	2.5	60
	243	8	175	62	310	4.9	1.71	66	6.2	2.3	64
	269	10	185	62	430	4.3	2.02	54	5.5	2.0	66
Washington, D.C.	1223	12	154	49	606	2.8	1.68	42	3.3	2.1	68
	1241	12	163	44	704	2.7	1.94	46	4.2	2.0	73
	1255	13	151	51	775	2.3	1.81	54	4.4	2.3	66
	1259	11	146	39	714	3.3	2.14	40	3.2	1.5	73

* Excluding sludge recycle.

the greater the fraction of soluble BOD_5, the greater the required contact time. Under American conditions, with domestic sewage it would require up to 1 h and 2–4 h re-aeration, as against conventional aeration times of 6–8 h. It has been found, however, that the air requirements of both processes are similar. Another method to employ is complete mixing, whereby influent and recycled sludge are introduced uniformly throughout the aeration tank, thereby creating a steady oxygen demand all through. This method has been found to add stability under varying load conditions. Next, modified aeration wherein, often without previous settling, the incoming wastes are brought into contact with the activated organisms for a brief period of sorption and synthesis. This process can achieve 60–75 % BOD_5 removal at twice at four times the rate in conventional applications. A significant saving in oxygen requirements has been achieved by two-stage aeration, where the second stage is used for nitrification, not achieved in conventional applications. In the process the addition of alum helps to maintain first-stage BOD_5 within the desired range of 30–60 mg/l.

Table 49 gives performance figures obtained at conventional plants whilst Tables 50–52 give similar figures for step-aeration, complete mix, and modified aeration, respectively.

In more general terms it can be said that whereas in the United States there are protagonists for several different activated-sludge systems, in the United Kingdom, probably due to the greater strength of sewage, some of these variations have not been found advantageous. Very careful study is being made, so that conclusions as to new processes, when made, will be thoroughly reliable, and success for a new generation of treatment plants will be guaranteed.

Surface aeration or mechanical agitation

An early method of surface aeration was the Sheffield system or bio-aeration, originated at Sheffield, England. The aeration tank was a long channel in the region of 1.25 m wide and 1.25 m deep and having a detention period of about 16 h DWF when treating sewage which, before settlement, had a BOD_5 of about 350 mg/l. The channel was folded against itself time and time again to occupy convenient space and to simplify the arrangement for agitation which was by means of a series of paddle wheels fixed onto a common axle laid across the battery of channels at the centre. The paddle wheels propelled the sewage along the channels at a velocity of 0.5 m/s and, at the same time, agitated the surface. Effluent drawn off from the end of the channel was passed to the final sedimentation tanks, the remainder being returned in measured quantity to the inlet end and recirculated. There are several Sheffield system works still in use but the method has gone out of favour.

The original Simplex surface aeration system which was developed by Ames Crosta Babcock in Britain,[11] but which is now used world-wide, incorporated a number of fixed, vertical uptake tubes having at their upper ends mixed-flow impellers known as aerating cones. The rotation of these impellers drew mixed liquor up the tubes and discharged it over the surface of the tank contents.

With the development of aerators of higher power, ensuring greater turnover of liquid, flat-bottomed tanks were found to be satisfactory and are now always employed. Similarly, in all but the most unusual cases the agitation is now sufficient for uptake tubes to be omitted.

Each aerator is driven by an electric motor through a speed reduction gearbox and rotates at a speed generally between 30 and 60 rpm.

The number of aerators needed for a particular application is divided into groups, usually of up to six, each group being located along the line of a separate channel. Alternatively, the aerators may be grouped in rectangular tanks, rather than channels, when a 'completely mixed' system is required.

Appreciable variation of horsepower absorbed—and therefore of oxygen input—can be effected by means of adjustable outlet weirs whose level can be varied by up to about 200 mm so as to increase or reduce the submergence of the aerating cones (see Figure 80). Adjustment of the outlet weirs may be effected either manually or automatically. In the latter case a dissolved oxygen electrode, located at a suitable point in the aeration channel, is used to send a signal to a control mechanism which actuates a motorized weir.

A further advantage of this surface aeration system is that at periods of low flow, or low loading, selected aerators may be switched off, thereby saving power. In such circumstances activated sludge will settle on the base(s) of the channel(s) but when the aerators are switched on again the sludge will be rapidly re-suspended in the mixed liquor. This is a useful facility at certain times of the day or when industrial effluent is received only on certain days.

It used to be said that surface aeration was limited to small works capable of taking flows up to about 3700 m^3/day, but the processes have advanced greatly since the days when such opinions could be held. The Crossness sewage-treatment works, London, England, were designed by the author to incorporate

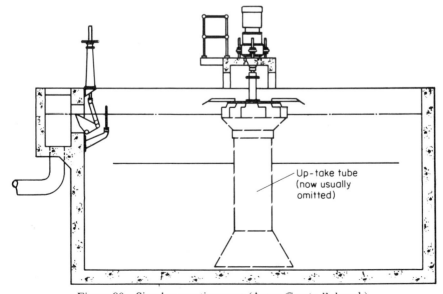

Figure 80   Simplex aerating cone (Amos Crosta Babcock)

the Simplex surface-aeration system and to treat a future DWF of 490,000 m^3/day and a maximum rate of flow of 980,000 m^3/day. These works have, in fact, been treating more than expected. For example, in the period 1 April to 30 September 1968 the average daily flow was 553,656 m^3/day of sewage which, after primary settlement, had a BOD_5 of 148 mg/l. An average final effluent of 13.5 mg/l was produced. The power required for aeration, pumping sewage and storm water, and other purposes on the works is derived from sludge gas used in dual-fuel engines. These works are believed to be the second-largest sewage-treatment works outside America and the largest surface-aeration activated-sludge works in the world.

The Kessener[12] process is similar in principle to the Simplex system in that it relies on a surface agitator to agitate the sewage and keep it in motion. The differences are in the form of the agitator and the shape of the tank. The tank is long, the flow being from end to end; and the agitator consists of a series of rotating stainless-steel combs which extends for the full length of the tank and is rotated at one side at the surface. These combs spray mixed liquor over the surface and have the effect of rotating the mixed liquor in the tank as it flows from end to end. To assist this rotation the bottom of the tank is rounded and a baffle board is arranged near the surface at the side of the tank opposite the comb, so as to round off the corner.

The popularity of surface aeration is increasing in North America.

The Passveer ditch

This process was first developed by Dr Passveer in Holland in 1953. The basic form of the plant is a continuous circuit, or ditch of trapezoidal cross-section, having a water depth of 1.0–1.8 m, and sidewall slope of 45°. One or more horizontal aeration rotors are mounted in the ditch to provide for oxygenation, and for circulation and mixing of the ditch contents. The ditch volume is such that the total amount of sludge in the plant is approximately five times that present in 'conventional' activated-sludge plants, and allows for retention time in excess of 1 day with domestic sewage. With continuous oxidation ditches, the mixed liquor passes, after aeration, to a standard settlement tank. The settled sludge is then returned to the ditch. Ditches are-sized to give a volumetric loading not exceeding 210 mg BOD_5/l of ditch volume/day. At a mixed liquor suspended solids concentration (MLSS) of 3500 mg/l, the sludge loading rate is 0.06 kg BOD_5/kg of sludge solids/day. The aeration rotors are required to dissolve at least twice the amount of oxygen required for BOD_5 removal alone, thus allowing for nitrification.

Denton and Spencer consider oxidation ditches to operate almost as completely mixed systems, capable of handling shock loads of high BOD_5, and also providing ideal conditions for the development of relatively slow-growing protozoa, considered to be largely responsible for the clarity of the high-quality effluent. They illustrate two types of rotor (the TNO and Mammoth, manufactured by Whitehead and Poole Ltd., Radcliffe, Manchester). The former is suited to liquid depths between 1 and 2 m, the latter, with blades radiating from the central shaft, is suitable for depths up to 4 m. This future enables economies in space to be made. Similar advantages can be gained by adding a second rotor if the plant has to treat loads considerably in excess of the design figure. Rotors are compared on the basis of kg of oxygen per kWh—usually 20–25, dependent on tank configuration and nature of waste being treated. An allowance must also be made for circulation of ditch contents, the minimum velocity being 0.15 m/s. Increased water velocity causes a decrease in rotor efficiency, and where several rotors are used it may be necessary to install baffles to reduce circulation rate. The rotors create turbulent flow, thus preventing sludge settlement in the aeration channel, which should not occur. Oxygen input can be varied by adjusting blade immersion, by means of dissolved oxygen control equipment which raises or lowers the outlet weir. Timers may also be fitted to operate the rotors intermittently, giving useful power savings and possibly preventing problems associated with over-aeration.

Probably the most important aspect of the process is the low sludge production. Usually sedimentation tanks are unnecessary, thus avoiding the production of primary sludge. Furthermore, due to the long sludge age, surplus activated-sludge production is considerably lower than with either a conventional plant or a packed tower system. Excess sludge, being well stabilized, can be readily concentrated in a holding tank, and may be further dewatered by filtration drying beds, or land disposal, without odour or fly nuisance.

Examples of the flexibility of oxidation ditches are to be found at holiday resorts and in extreme climates. At holiday resorts both quantity and strength of sewage become greatly increased (1000 mg/l BOD_5 and 100 mg/l NH^3). These conditions are dealt with by increasing rotor immersion, or by additional rotor capacity, even though at times they may become covered with grease and mixed liquor pH falls to 4.5–5.0, due to nitrification. Ditches operating at very low temperature have their inlets located close to the rotor and protection is provided for the rotor outlet weir and returned sludge pump. For operation in hot climates, or at high altitude, an occasional load factor greater than two is required to overcome the reduction in solubility of oxygen attainable with air, and the increased bacterial respiration, etc.

There are over 300 oxidation ditches operating in the UK. Besides domestic sewage they treat wastes from farming, food processing, tanneries, etc., and chemical wastes, achieving upwards of 97% BOD_5 reduction on influent strengths up to 2000 p.p.m. BOD_5. Some typical results with mainly domestic sewage are given below:

Oxidation ditches treating mainly domestic sewage

Site	Period	BOD_5	SS	NH_3-N	NO_3-N
Thornwood (TWA)	1970/74	4.5	7.3	1.0	21.5
Pytchley (AWA)	1974/75	5.8	12.0	1.1	19.6
Roydon (AWA)	1974/75	5.1	16.7	1.1	19.6
Southminster (AWA)	1974/75	6.4	10.0	7.5	8.0
Gt. Totham (AWA)	1974/75	6.8	14.0	2.9	17.0

(Results published with the permission of Anglian and Thames W. As.)

One of the latest types of oxidation ditch has been named the 'Carrousel', and employs vertical-type rotors with channel depth of 2.5 m or more. This system is being applied to plants varying in size between 6000 and 400,000 population equivalent.

Choice of conventional activated-sludge methods

The comparative costs of the activated-sludge methods can be calculated in terms of annual repayments of loan and annual running costs. But this is not altogether straightforward for, although the capital costs are not difficult to estimate, the actual running costs can be misleading in several instances. If electricity has to be purchased, costs of power, labour, and materials are not too difficult to ascertain; but where sludge digestion has been, or is to be, installed and the sludge gas used, there is a distinct possibility of much more gas being available than required for all power purposes at the works. On occasion it has been possible to sell power to the electricity authorities but, when this is not practicable, surplus gas may have to be burnt to waste. In this latter circumstance it may be better to use an aeration method which, while more extravagant in power consumption, is otherwise simpler to operate or particularly suited to the local conditions. The capital and running costs of the digestion plant must, of course, be taken into account.

Bulking

Bulking is the condition in which the suspended solids of the sludge require more volume for their accommodation and the sludge may become as much as twice the normal quantity, fail to settle readily, and a bad effluent is produced. Microscopic examination has shown that when this occurs, filamentous growths such as *Cladothrix* and also types of protozoa that are not normally present in the sludge have become evident. Bulking can be caused by excessive quantities of trade wastes, but the first things to suspect are either overloading or faulty operation. It has been said that works intended to produce a good effluent such as a Royal Commission effluent should not be loaded to more than 0.03 kg BOD_5/day per kg of volatile solids present in the mixed liquor (this opinion being based on aeration tanks of average detention period; the figure could be revised proportionately if the weight of volatile solids in the aeration tanks, final sedimentation tanks, etc., is to be considered).

If the load is too high in proportion to the volatile solids the density of the mixed liquor must be increased. If, however, the density of the mixed liquor cannot be increased because of bulking, more aeration must be given to make this possible. If it is not possible to give more aeration the load should be reduced.

Sludge-volume index and sludge age

The condition of activated sludge can be ascertained by testing it for sludge-volume index; that is, the volume in mm of sludge which, after 30 min settling, contains 1 g of dry solids. There have been several methods but Mohlmann's is perhaps the most extensively used. From a sample of mixed liquor taken from the aeration tank 1 l is poured into a graduated cylinder and allowed to settle for 30 min, then the percentage volume occupied by the sludge is recorded. Next, the suspended solids content of the original sample is found and the sludge-volume index determined by the formula

$$\text{Sludge-volume index} = \frac{\text{percentage settling by volume in 30 min}}{\text{percentage suspended solids}} \qquad (35)$$

A good sludge should have an index of not more than 100. The test is not too reliable, for the slightest disturbance of the sample may cause the index to change considerably.

The condition of the activated sludge can also be determined by finding the theoretical sludge age in days. There is more than one way of doing this but the following formula is, perhaps, the best:

$$N = TQ_1/24Q_2 \qquad (36)$$

where: N = sludge age in days;
$\quad Q_1$ = daily quantity of returned activated sludge;
$\quad T$ = detention period in aeration, final sedimentation tanks, etc., in hours;
$\quad Q_2$ = daily quantity of surplus activated sludge.

A difficulty when comparing the results of different works is that this and other formulae usually have been applied without allowance being made for the time the activated sludge has been retained in the final sedimentation tanks, which must contribute to the actual age.

Foaming

The increased use of synthetic detergents at one time caused serious foaming at activated-sludge works until the detergent manufacturers were persuaded to investigate the problem and change the composition of their products. Foaming is now no longer serious but it is well to mention the most effective method by which it can be reduced. This is the application of about 1.5–3.5 mg/l of special oil obtained from most of the large oil companies and sold under proprietary names. This can be applied to the various aeration channels or other points needing control by pumping with the aid of mechanical lubricators through small-diameter PVC or nylon tubing (3 mm to about 4 mm internal diameters have been used). The oil is purchased in bulk, stored in large tanks of sufficient size to hold twice the capacity of the appropriate tank-vehicle, and gravity-fed to ball-valve-controlled cisterns, locally supplying each mechanical lubricator. At large works the lubricators can be remote-controlled by push button.

Pure oxygen-activated sludge

The use of pure oxygen has, at any rate in the US, recently become competitive with conventional methods[13] due to the use of more efficient oxygen-dissolution systems. In addition to covered tanks in which oxygen is circulated until it becomes too impure for longer use, oxygenation may be performed in open aerators, wherein extremely fine diffusers emit bubbles so small as to be completely dissolved before breaking the surface. The second alternative is still in the early stages of development. Use of oxygen is found to result in a reduction in reactor volume and higher dissolved oxygen content in the effluent. Although its use does not allow activated sludge to go beyond its inherent capabilities, the process also guarantees effective odour control and possibly produces less and more highly concentrated sludge. Disadvantages lie in the increased complexity of the process, excessive pH depression with low alkalinity wastes, where alum is added for phosphorus removal, or where nitrification occurs.

An example of the results being obtained by the use of the oxygen process is given by the plant at Batavia, New York, where performance has been evaluated in parallel with a conventional plant. The details are shown in Table 53, which compares results from conventional plant with those from a Unox[14] pure oxygen one

Table 53 Batavia, NY pure oxygen activated-sludge study

	Phase I		Phase II	Phase III	
	Air	Pure O_2	Pure O_2	Air	Pure O_2
Sewage flow (10^3 m^3/day)	8.95	8.58	11.5	5.26	6.36
Aeration time (hrs)	3.5	3.4	1.2	2.6	2.0
BOD, in	159	159	220	262	262
BOD, out	16	11	23	30	14
MLSS	2440	3060	6980	3640	6190
SVI	76	64	36	63	49
Volumetric loading (kg BOD/m^3)	1.0	0.93	3.4	2.06	2.3
F/M (kg BOD/kg MLSS)	0.57	0.41	0.79	0.84	0.55
Excess sludge (kg VSS/kg BOD)	0.87	0.48	0.41	0.99	0.13
O_2 Utilization (%)	—	95.5	92.7	—	91.4

In Phase I half the activated sludge tank capacity was run with pure oxygen, the other half with air. In Phase II all flow was routed through the half with pure oxygen feed. In Phase III a quarter of the plant was on pure oxygen, a quarter on air, and the other half was out of service. Clearly, under much heavier volumetric loading, the oxygen produced a better effluent than the air system. The key is the development of a sludge which can settle quite well even at high concentrations. In effect, the pure oxygen system could operate at a greater sludge age than the conventional system. According to Gehm[15] total treatment costs in new plants, including aeration, will be lower with oxygen aeration than air aeration for all capacities above 50,000 m^3/day. Performance of pure

Table 54   Performance of pure oxygen plants

Plant	Loading (kg BOD/ kg MLVSS)*	Sludge (kg VSS†) kg BOD)	BOD removal (percent)	SS removal (percent)	Loading (kg BOD/ m³)
Newton Creek	0.65–2.44	0.6–0.94	87–94	84–92	1.78–6.13
La Virgenes	0.07–0.46	0.19–0.27	89–97	86–93	0.21–1.00
Denver	0.18	0.33	92	91	0.89

* MLVSS = Mixed liquor volatile suspended solids.
† VSS = Volatile suspended solids. Courtesy Reinhold.

oxygen plants at Newton Creek, Las Virgenes, and Denver, all show BOD_5 and suspended solids removals above 90% and future plants using submerged turbines are, or will be, installed at Detroit, New York, and Wyandotte (Michigan), and others, with surface aerators at Danville (Virginia), New Orleans and Speedway (Indiana).

At Las Virgenes a system primarily aimed at upgrading existing diffused air aeration plants is being demonstrated. Called the Simplex[16] system, it uses conventional air blowers and coarse bubble diffusers, with an inflated dome cover to contain the oxygen-rich atmosphere. Loadings in this type of system are lower than in the type described above, but extensive nitrification can be obtained.

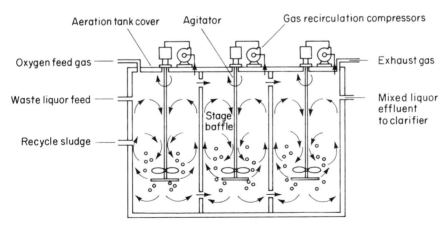

Figure 81   Schematic diagram of multi-stage oxygenation system (Van Nostrand Reinhold)

Position in the UK

Although keen interest is being shown in the UK in the use of oxygen in sewage treatment, economic restrictions have prevented much beyond experimental work. Boon[18] reviews this work and concludes that it has shown little sign of the benefits, including increased rate of treatment, increased density and rate of settlement of sludge, or reduced production of sludge reported in American literature. However, the first full-scale oxygen aeration plant was commissioned in August 1976 and has been fully operational since June 1977. The plant was designed for a DWF of 4500 m^3/day and to produce a 15:10:10 standard in suspended solids, BOD$_5$ and ammoniacal nitrogen, and has continually produced an effluent of this quality, by operation in two separate stages. Firstly, oxygen assisted to produce a 30:20 effluent and secondly a conventional aeration stage achieves full nitrification. Oxygen for the plant is produced by the pressure swing adsorption process.

The sidestream injection technique has also been developed by BOC[17] and installed at a treatment works to uprate the existing aeration capacity by about 25% during the summer months, when population increases, thus enabling the plant to produce a consistent nitrified effluent to the required standards.

In another experiment with a pilot-scale oxygen plant, the settling characteristics of sludge were found to be good when treating sewage which comprised about 50% of waste water from distilleries and from yeast production. In another, the settleability of sludge from an oxygen plant was little affected by changes in sludge loading, whereas settleability from an air plant with the same sewage deteriorated when the daily loading increased above 0.38 BOD$_5$/g activated sludge solids.

Boon[18] concludes that oxygenation has a significant economic advantage over aeration for a large plant, producing a non-nitrified effluent and situated in an area where sludge disposal costs are high. Waste waters of high BOD$_5$ (greater than 1000 mg/l) can be treated economically in an oxygen-activated sludge plant.

The deep shaft treatment process

Another version of activated sludge treatment arousing interest in Britain is the deep shaft oxygen transfer device consisting of a shaft typically 50–150 m deep.[19, 20] The shaft is vertically divided into a down-flow section (downcomer) and an up flow section (riser) as shown diagrammatically in Figure 82. It is surmounted by a gas-disengagement tank. The shaft may be 0.5–10 m diameter. Influent is fed in at the top, and for start-up air is injected into the riser to induce circulation. After circulation is established, air injection is progressively transferred to the air-sparge in the downcomer (process air). This is introduced at a depth sufficient to cause the contents of the shaft to circulate at a velocity of 1–2 m/s. At this velocity the rate of rise of air bubbles is overcome and they are therefore carried to the bottom of the shaft. The driving force for circulation arises from the difference in total bubble volume in the downcomer and the riser. In the latter, gas bubbles are released and grow as the pressure drops to escape to atmosphere at ground level. Some of the circulating liquid is displaced by the influent and the mixture of influent and circulating liquid passes back down the shaft. The injection of air into the downcomer reduces the rate of

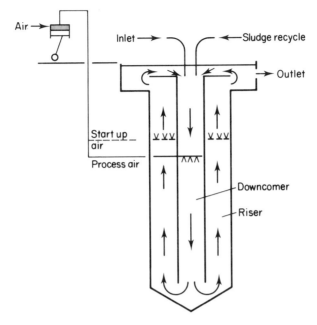

Figure 82   The deep shaft (*J. Instn. Public Hlth. Engnrs.*)

circulation, and the point of injection may be adjusted, thus minimizing power consumption by reducing the pressure required; this results in the maximum possible contact time between air and mixed liquor. After passage and re-circulation through the shaft and header tank, solids are separated from the shaft effluent either by settlement or flotation.

The deep shaft is probably principally effective in harnessing the oxygen transfer effects associated with enhanced pressure. The effect of this is firstly to raise oxygen saturation level in water at 20°C to 55 mg/l at 50 m depth, and to 100 mg/l at 100 m depth, compared with 8 mg/l at normal atmospheric pressure. Nitrogen solubility is of a similar order, whilst the carbon dioxide produced in the oxidation of organic compounds is readily retained in solution even on return to atmospheric pressure.

Complications arise from the pressure solubility fluctuations of the process, largely associated with a decompression stage and the emission of fine gas bubbles. This results in flotation separation or thickening of solids. More usually, however, clarification of the mixed liquor is effected by settlement of the activated sludge, and this necessitates the elimination of flotation effects by vacuum, or other means of accelerated degasification.

The process, based as it is on activated sludge principles, is suitable for the treatment of all biodegradable wastes, whether municipal or industrial. However, its unconventional nature brings out certain important advantages and disadvantages which can influence its suitability. Unless means are found, as they may be, for constructing a deep shaft plant above, rather than below, ground level, ground conditions are always likely to have an important influence. Whilst vacuum degassing and final settlement are considered essential to secure a 30 : 20 effluent, there is evidence that where suspended solids do not exceed 50 mg/l these processes may be dispensed with, or it may be that existing aeration and settlement units could be utilized to follow deep shaft treatment. These processes would also guarantee nitrification and denitrification, but there is some evidence that significant phosphorus removal can be achieved in a conventional shaft operation.

Much of the development work on the deep shaft process has been carried out by ICI[20] at Billingham Sewage Works, of the Northumbrian Water Authority. The plant was first commissioned in 1974 as a large-scale pilot plant, but since 1976 a full-scale plant treating domestic sewage has operated at Paris, Ontario. Plants in Britain now include that at Marsh Farm Sewage Works, Thurrock, Essex.

Electrolytic methods of treating sewage

What was probably the first attempt at electrolysis of sewage was made by Webster in 1888 at the London County Council Southern Outfall Works, Crossness, London, England (the site of the present Crossness Works referred to earlier). The effect of the electrolysis was to decompose the sodium, magnesium, and other chlorides and release free chlorine and oxygen which combined with the organic impurities.

It was found that 1 kWh was required to treat 10 m^3 of sewage and reduced the organic pollution by 61 %. The cost of the current was not the only expense, for the cast-iron electrodes were dissolved at the rate of 130 kg/50,000 m^3 of sewage.

Among other early methods was the Hermite process, in which sea water was electrolysed and then added to sewage. This was originated at Lorient in France. A similar method was invented by Albert E. Woolf and adopted by the Health Department of New York City, to treat sewage from a few houses in 1893. In the Landreth process, which was tried on a small scale in the United States in 1918, electrolysis was applied to sewage which had beed dosed with lime. The electrodes were closely spaced mild-steel plates with paddles revolving between them.

There has recently been a further revival of electrolysis; this time in Norway. The inventor is Dr Ernst Foyn and the method is being commercially exploited by Elektrokemisk A/S, Oslo. The Foyn method was developed chiefly to cope with pollution problems arising in Oslofjord, into which the sewage from a population of 500,000 was discharged. Not only was there concern with the danger to health, but also over-fertilization of the water by phosphorus and nitrogen compounds had been causing excessive algal growth, interfering with photosynthesis with resultant oxygen deficiency in the lower strata of the water.

It was considered that, whereas sewage treatment by biological oxidation would render the effluent stable and inoffensive and reduce the bacteriological count, it would not remove a sufficient proportion of phosphorus and nitrogen compounds. On the other hand, phosphates can be precipitated by magnesium and ammonium compounds in alkaline solution, and magnesium and alkali can be added to sewage by adding sea water and applying electrolysis. Laboratory experiments proved that the process would remove phosphates sufficiently to eliminate the fertilizer value of the effluent.

A pilot plant was constructed and, it is reported, tests with this proved that it was possible to treat 1000 l/kWh and that a detention period of 1 h was required.

The sludge from the electrolytic plant is carried to the surface by hydrogen bubbles and the scum removed mechanically. It is stated that it can be economically handled, being low in moisture content, and because of its high phosphate content is of greater manurial value than ordinary sewage sludge.

The Oslo authorities built in 1958 what is described as a semi-commercial plant which comprised one cell with vertical diaphragms and one non-diaphragm cell, and incorporated automatic control of the flows of sewage and sea water.

Activated carbon adsorption[10, 14, 15]

Some organisms, often referred to as refractory, cannot be removed by coagulation and sedimentation, or by biological oxidation. The ability of activated carbon to remove soluble organics from waste water is a consequence of the similarity in surface chemistry between activated carbon and organic molecules.[15] The characteristic of organic carbon that makes it unique is that is has a much higher adsorption capacity than other materials, due to the extensive internal microporous structure formed during the activation process. The usual treatment process involves a contact tank where carbon and waste water are retained sufficiently long for adsorption to take place. They are then separated and eventually the carbon becomes exhausted and is removed for regeneration. During regeneration some carbon is lost and make-up must be added. Commercial grades of activated carbon are granular (sizes 8 × 30 mesh or 12 × 40 mesh) or powdered (N 300 mesh). Both forms have been found equivalent in their ability to remove organics, but require a different method of containing and regeneration.

Granular carbon is contacted in columns, through which the sewage flows either under pressure or by gravity, the former method providing operational flexibility but the latter being more economical. Flow may be upward through an expanded bed or a packed bed, or downward through a packed bed. Packed bed operation provides filtration as well as adsorption but requires frequent backwash. Up-flow systems allow for a periodic removal of a portion of the carbon at the base, thereby providing counter-current contact in a single vessel—good flow distribution is more difficult in up-flow than down-flow. Columns can be connected in series or in parallel. Fundamental is contact time—a function of effluent quality desired and waste water characteristics—

as contact time increases organic removal decreases, reaching zero between 30 and 60 min. Contact time is rated on an empty bed basis.

Regeneration of granular carbon is conducted in a multi-hearth furnace, during reactivation temperature is 815–925°C. Regenerated carbon is quenched and returned to a contactor. Furnace ratings are 400–550 kg carbon/m² per day. An after-burner and wet scrubber, or cyclone, is usually required to purify the exhaust. Regeneration loss is 5–10% per cycle, due to physical attrition.

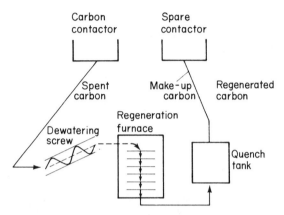

Figure 83   Activated carbon regeneration system (Van Nostrand Reinhold)

With powdered carbon contact is brought about by mixing carbon and waste water in a reactor-clarifier contractor. Carbon and waste water are flash-mixed and flocculated in the centre well, the mixture then flows into the outer section where gravity separation of carbon and waste water takes place. A pool of carbon in a thick slurry form is maintained at the bottom of the contactor. The carbon comes into equilibrium with the organics in the effluent waste water, whereas in a granular contactor it comes into equilibrium with the organic concentration in the incoming liquid. The organic adsorption capacity of activated carbon is proportional to the concentration of organics with which it comes into equilibrium, making it more necessary to achieve counter-current contact in powdered carbon systems than in granular systems. In a two-stage counter flow contacting system the carbon is thickened between stages to reduce pumping, and the effluent is filtered to remove residual carbon fines. In the most advanced system for powdered carbon regeneration, the slurry containing spent carbon is thickened, dewatered in a centrifuge or vacuum filter, and injected into a fluidized-bed furnace. The regenerated carbon is reslurried and pumped back into the system. In the furnace temperature is maintained at the same level as in the multiple hearth furnace. Contact time of a carbon particle averages several seconds—shorter time results in only partial regeneration—longer, in particle incineration. Thus, tight control must be exercised—losses average 15%.

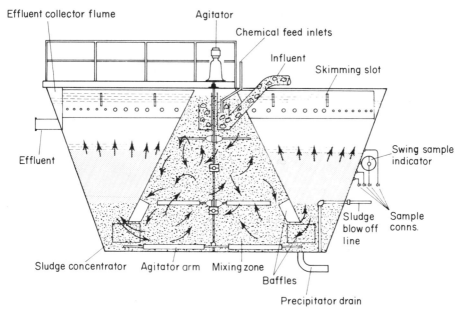

Figure 84   Solids contact clarifier with sludge blanket filtration (Van Nostrand Reinhold)

Advantages of powdered carbon units are inexpensiveness, requiring a much smaller carbon inventory, and that they can be dosed to meet demand. Granular carbon, however, exhibits higher practical adsorption capacity, does not require complex dewatering procedures prior to regeneration, and has been successfully regenerated on full-scale sewage applications.

Physiochemical treatment

This process consists of chemically aided clarification followed by carbon adsorption, and was evaluated as a technique in the United States in the 1940s, but abandoned on the score of cost. However, greater stringency in treatment standards, and the development of carbon regeneration techniques, have made it more viable. Moreover, impurity removal following chemical clarification has been greater than anticipated, thus aiding the economics of the process. As usually applied, the process included preliminary treatment, followed by chemical dosing to achieve the desired level of suspended solids and/or organics, and/or phosphorus removal. Due to coagulation, and possible chemical adsorption of colloidal and soluble organics, organic removal is higher than would be anticipated from suspended solids removal—up to 50% of 'soluble' organics can be removed by chemical coagulation. Following clarification the waste passes to carbon adsorption. Here again, biological action improves carbon adsorption, although the load in such plants is heavier than in those concerned only with tertiary treatment, and extra care is required in design to avoid development of the unsatisfactory features of this processes.

Table 55   Activated carbon performance in physiochemical treatment plants

Plant	Effluent organics (mg/l)	Carbon capacity	
		kg TOC / kg AC	kg COD / kg AC
Blue Plains (G)	TOC = 6	0.33	0.90
Lebanon	TOC = 6	0.48	1.10
Ewing-Lawrence (G)	TOC = 3–5	0.66	—
New Rochelle G	COD = 8	—	1.32
Owosso (G)	COD = 25	—	1.32
Salt Lake City (P)	COD = 22	—	0.79
Pomona (G)	COD = 20	—	7.80

Note: TOC = Total organic carbon.
COD = Chemical oxygen demand.
Ac = Actuated carbon.
G = Granular carbon
P = Powdered carbon

At the plant in Pomona, using granular carbon, the carbon capacity was found to be significantly higher than elsewhere. No regeneration was required over a 20-month period, and the effluent remained at 20 mg/l BOD (90% overall removal, 75% removal in carbon). The major reason for this was stimulation of biological activity on the carbon. Sodium nitrate was added (5 mg/l) to prevent sulphide formation. The nitrate was reduced to nitrogen gas. When the carbon was eventually regenerated, with only a 2.5% loss, effluent quality rose to only COD 8 mg/l, or considerably below the level before regeneration, thus proving that biological activity cannot produce complete regeneration.

Filtration is by way of being optional following adsorption, but it acts as a safety factor, bringing a high degree of reliability to the process. Expanded bed and powdered carbon systems usually require filtration after carbon contact. Packed bed systems will benefit from filtration before carbon contact.

Advantages of physiochemical treatment over conventional methods are quoted as: (1) less area required ($\frac{1}{2}$ to $\frac{1}{4}$); (2) lower sensitivity to demand variations; (3) plants are unaffected by toxic substances, and have a potential for heavy metal removal and superior removal of phosphorus compounds; (4) greater flexibility in design and operation and superior organic removal. On the strength of these advantages plants are either planned or under construction at Rocky River and Cleveland, Ohio; Fitchbury, Mass.; Cortland and Niagara Falls, N.Y.; Garland, Texas; and Owosso, Michigan.

Reverse osmosis

The process requires pressures upwards of 20 kg/cm^2 to achieve fluxes of the order of 0.5 m^3/m^2 per day. The degree of demineralisation increases with the pressure required and can approach 99% +. At present the best membranes are cast from a mixture of cellulose acetate, acetone, formamide, and magnesium perchlorate. A skin 1 μ thick provides the desalting surface and a 100 μ thick porous sublayer supplies structural backing. This membrane is compressible and hydrolyses at low and high pH; it is easily damaged. Even if stronger membrane materials are devised the engineering problems of a structural support system to absorb the high pressures across the membrane, and for control of fouling on the membrane surfaces will remain unchanged. The system should also provide high membrane area per unit volume. For dirty-water applications a system having the membrane position along the inner wall of a porous 13 mm diameter tube is the accepted method, and membrane area per unit volume is 66 m^2/m^3. Fouling of membranes is due to an accumulation of organic and inorganic materials at the membrane surface. These materials build up in the boundary layer next to the membrane surface and can cause precipitation, or significant biological activity. Control of fouling can be achieved by increased turbulence, at the expense of increased pressure drop. Pretreatment for removal of solids or organics can be effective, but the most economically successful method is *in situ* membrane cleaning with enzymatic detergent solutions, on a once-a-week basis. The system has to be shut down and depressurized for cleaning. Research for the development of a fouling-resistant membrane is continuing.

A number of studies indicate removals of 93–95% total dissolved solids, 95–99% phosphate, 80–90% of NH_3-N, 60–70% NO_3-N, 99–100% suspended solids, and 90–95% TOC. Development of higher flux membranes with lower salt rejection properties would reduce pressures and size of plant.

References

1. Danson, P. S. S., and Jenkins, S. S. *Sewage Works Journal*, **21** (4) (July, 1949).
2. Ardern, E. and Lockett, W. T., Experiments on the oxidation of sewage without the aid of filters. *J. Inst. Sewage Purification* **3** (1954).
3. Lockett, W. T. The development of the process using diffused air, with special reference to the Manchester pioneering work, *J. Inst. Wat. Pollution Control* **3** (1954).
4. Escritt, L. B. *Sewers and Sewage Works, with Metric Calculations and Formulae*, Chap. 16. George Allen & Unwin Ltd. (1971).
5. Pearson, R. F., Taylor, G., Wood, L. B., and King, R. P. The Beckton Sewage Treatment Works, Greater London Council: ten years operation, 1960–1970. *J. Instn. Public Hlth. Engnrs.* **70**, pt 4 (1971).
6. Coombs, E. P. Air diffusers—their history and use in the activated sludge process. *J. Inst. Sewage Purification* **4** (1955).
7. Dorr-Oliver Co. Ltd, Norfolk House, Wellesley Road, Croydon, Surrey.
8. Imhoff, K. and Fair, G. M., *Sewage Treatment*, John Wiley & Sons, N.Y. 2nd edition (1956).
9. Swanwick, J. D., Shurben, D. G. and Jackson, S., "Report on the Water Pollution Research Laboratory Survey of the Performance of Sewage Sludge Digestors throughout England, Scotland and Wales" (May 1968), *Wat. Pollution Control*, **68**, p. 369, 1969.
10. Environmental Protection Agency. *Process Design Manual for Up-grading Existing Wastewater Treatment Plants*. Environmental Prot. Agency Technological Transfer (Oct. 1974).

11. Bolton, J. The birth of the Simplex aeration process at Bury sewage works. *J. Inst. Sewage Purification* **3** (1954).
12. Kessener, H. Small bio-aeration plants. *J. Inst. Sewage Purification* (1930).
13. Bennett, F. W. and McWhirter, J. R. Performance of the Activated Sludge Process using high purity oxygen aeration. Int. Conf. on Ind. Waste Wat. (Nov. 1970).
14. Unox System, developed by Union Carbide.
15. Gehm, H. W. and Bregman, J. I. *Handbook on Water Resources and Pollution Control.* Chap. 14. Van Nostrand Reinhold Co., N.Y. and London (1976).
16. Simplex System, Ames Crosta, Babcock Ltd, Heywood, Lancs.
17. Sidestream injection technique—Brit. Oxygen Co., Hammersmith House, London.
18. Boon, A. G. Technical review of the use of oxygen in wastewater treatmt. Symp., 'The use of oxygen in public hlth. engnrng', Cranfield Inst. of Technol. (Apr. 1978), Instn. Public Hlth Engnrs.
19. Sidwick, J. M. and Staples, K. D. Deep shaft treatment process. Symposium, 'The use of oxygen in public hlth. engnrng', Cranfield Inst. of Technol., Bedford. Instn. Public Hlth. Engnrs. (Apr. 1978).
20. ICI—Imp. Chem. Inds. Ltd., Millbank, London, SW1.

CHAPTER 16
Treatment on Land and Tertiary Treatment

As methods of sewage treatment become more and more sophisticated it tends to be forgotten that early treatment works were invariably known as 'sewage farms', and did in fact use the manurial value inherent in sewage to produce crops. Three methods were available: land filtration, broad irrigation and intermittent sand filtration.

Land filtration

This is efficient, but applicable only on light porous soils. The sewage is irrigated into shallow trenches from which it soaks into the ground, eventually finding its way to underdrains which are best laid parallel with, and between, the irrigation ditches. Underdrains have been laid at right angles to the ditches, but this is not desirable as it makes possible short-circuiting of untreated or partially treated sewage at the crossing points. In passing through the soil the sewage is filtered and organic compounds in solution are oxidized by a biological film which develops below water level on the ground. Provided sufficient land is used and well maintained, the effluent should be adequately treated and may be passed to a watercourse.

Broad irrigation

This is less efficient, but has to be used if sewage is treated on clay or other impervious soil. The method consists in irrigation over the surface so that the sewage remains exposed to organic organisms or the air as long as possible before it drains away to collecting ditches. The usual procedure is to arrange a series of feeder channels with penstocks, or hand-stops, so that the sewage may be delivered onto different plots whilst other plots are being rested. To prevent rapid running over the ground, the land is ploughed parallel to the contours, so that sewage spills over from farrow to farrow. As a further precaution to ensure oxidation, the collected effluent should be distributed over additional land areas or re-collected and re-distributed as often as practicable.

The Royal Commission stated in their Fifth Report[1] that:

the total acreage of a farm must be relatively much greater when the sewage is purified by surface irrigation than when the method of filtration is employed, and a larger percentage of surplus area is also desirable in the former case. We are not able to lay down any rule as to what the ratio of surplus acreage to total acreage should be, but generally speaking, a large surplus area is advisable.

In a special Report to the Commissioners, McGowan, Houston and Kershaw wrote:

speaking in general terms, we doubt whether even the most suitable kind of soil worked as a filtration farm should be called upon to treat more than 30,000 to 60,000 gallons per acre [350–700 m^3/ha] per 24 hours (750 to 1,500 person per acre): or more than 10,000 to 20,000 gallons per acre [100 to 200 m^3/ha] per 24 hours, calculated on the total irrigable area (250 to 500 persons per acre [600–1200/ha]). Further, that soil not well suited for purification purposes, worked as a surface irrigation and filtration farm should not be called upon to treat more than 5,000 to 10,000 gallons per acre [50 to 100 m^3/ha] per 24 hours at a given time (125 to 250 persons per acre); or more than 1,000 to 2,000 gallons per acre [10–20 m^3/ha] per 24 hours, calculated on the total irrigable area (25 to 50 persons per acre) [60 to 120 per ha]. It is doubtful if the very worst kinds of soil are capable of dealing even with this relatively small volume of sewage. The population per acre is calculated on 40 gallons [0.2 m^3/ha] of sewage per head per day. It is here assumed that the sewage is of medium strength and is mechanically settled before going on to the land.

According to Imhoff, Müller, and Thistlethwayte,[2] agricultural areas growing crops and vegetables may treat waste water equivalent to ordinary settled municipal wastes from about 250 persons per hectare while grasslands may treat it, as the equivalent of settled wastes, from about 1000 persons per hectare of irrigation area, taking average unspoiled natural soil as a basis. It may be helpful to provide emergency basins for storage and partial treatment by other means before direct disposal into suitable natural waters.

Sewage farming as such fell into disrepute as a result of 'sickening' of the land, and increasing criticism of the possible effects of toxic and other undesirable constituents of sewage on the consumers of crops irrigated with sewage. Thus so far as the highly developed areas of Europe are concerned, sewage farming has rarely been practised of late years, although the use of grass plots, effluent ponds, or lagoons, is widely practised as a final treatment process, capable of producing effluents of very high standard.

American practice

In America land filtration, or intermittent sand filtration, was developed as a technique in the early years of this century and is still in use on suitable sandy soils. The technique is as already described, and consists of cyclical processes of flooding followed each time by a period long enough to permit complete drainage and natural aeration and recovery of the bed. Usually four to eight beds are used, not larger than an acre (0.4 ha) in extent. The surface is laid horizontally and the beds drained by 100 mm agricultural piping placed at some 10 m centres and 1 m deep. Influent and such effluent pipes are laid under the surrounding banks. Dosing tanks are often provided to ensure rapid flooding, using automatic siphons or valves to distribute flow to the various beds in turn.

Soil testing for percolation

Soil sampling for percolation purposes has become widely used in the US due to the developing interest in septic-tank systems for housing development beyond the reach of municipal sewerage systems. Undoubtedly, the techniques developed earlier for land filtration on a large scale will have proved of value for these more recent requirements. Particles above 0.1 mm are separated by sieving, and those of smaller size by sedimentation methods, from a sample of 100 g dry weight. Curves in which the particle sizes are plotted on a logarithmic scale against corresponding weight percentages of the sample are then prepared. Particle sizes corresponding to the 10% and 60% weight percentages are then read off. The rate of drainage of clean water through sand depends more or less on the particle characteristic of the finest 10%. The characteristic particle size is therefore taken as that size by which 10% dry weight of the sand is smaller. For suitable soils this figure lies between 0.2 and 0.5 mm—below this standard size soils are too tight, whilst those above cannot withstand uneven distribution during flooding, unless covered by finer material. The 'uniformity coefficient' is defined as the ratio of the particle sizes corresponding to the 60% and 10% weight percentages. Thus, where the two figures so obtained are respectively 2.1 mm and 0.3 mm the 'uniformity coefficient' is $(2.1/0.3) = 7$. Coefficients of 5 to 10 are common in soils, with lower figures for water graded sands. Co-efficients below 5 are recommended for municipal sewage. Following pretreatment by sedimentation, beds are operated by filling from 50 to 100 mm of water, or just between 500 and 550 m^3/ha or 0.05–0.10 m^3/m^2 allowing 5–15 min for filling. The waste floods the bed and should drain uniformly away, driving air downwards through the soil and drawing fresh air from above, the water disappearing within 1–2 h. The bed is normally charged once daily, but with strong wastes may require several days for recovery, whilst weak wastes may allow two to four fillings per day. Clogging may be overcome by aeration (resting) until a layer of dried sludge can be removed from the surface, which after light raking may be restored to the normal cycle. However, permeability diminishes gradually until replacement of the top layer eventually becomes necessary. After several decades the whole bed must be replaced. In colder climates 200–300 mm of flooding is necessary to maintain suitable conditions within the bed, operating under an ice covering through which vents must be maintained—loose sand-heaps reaching above water level may be used for this purpose.

Assuming a daily flooding, the air supply in a 1 m deep bed with 20% voids is replaced once daily, amounting to about 750 m^3 containing about 210 kg of oxygen per capita, giving a theoretical allowance of about 10,000 persons/ha. In practice, a loading of 5000 persons/ha, varying up or down according to strength of sewage, may be reckoned. Purification efficiency is high, BOD_5 reduction usually over 90% and bacterial reduction about 95%.

Oxidation ponds

In addition to the wide use of such ponds as a complete process, as described in Chapter 18, similar ponds are also used quite frequently on smaller treatment works for secondary or tertiary treatment. In these circumstances they are equivalent to the 'lagoons' employed in British practice.

Effluent polishing

In Great Britain concern at the condition of rivers polluted by either manufacturing or municipal waste water, not only brought to an end most attempts at land treatment of sewage, but soon began to bring about a demand for effluents of a higher standard than that set by the findings of the Royal Commission. Obviously an effluent of reasonably good standard, resulting from both primary sedimentation and a secondary biological process, is more amenable to any form of land treatment that may then be applied. Alternatively, some mechanical process or extended biological treatment can be applied, but in general such processes are likely to be expensive.

The commonest cause for the application of further treatment is lack of dilution in the watercourse to which the effluent is to be discharged. Next is the realization that certain constituents of the effluent may be harmful to the receiving waters. For instance, ammonia is toxic to fish particularly at high pH values, as in chalk streams, its demand for oxygen is much greater than that of organic matter as measured by BOD_5. Moreover, if the water is subsequently to be used for public supply, and will be subject to chlorination, serious problems can be created by high ammonia content. Thus a limit of 5–10 mg of nitrogen per litre would be prudent, and would normally require the breakdown of 75–90% of the ammonia during the biological stage of sewage treatment. Unfortunately the BOD_5 test gives no real indication of the degree of nitrification, and effluents meeting stringest BOD_5 standards may contain all the original ammonia. Only if this test is done with nitrification suppressed will the ammonia limit be revealed.[3]

Of the organic constituents of an effluent as measured by BOD_5 COD, permanganate value, or organic carbon, only 30% of the total weight has been classified into chemical groups. The remainder is largely of unknown composition and there must inevitably be reluctance in putting such water into public supply. Several of the inorganic constituents, such as nitrite, would be undesirable, whilst compounds of phosphorus and nitrogen accelerate eutrophication. Toxic heavy metals such as zinc, boron, and chloride may inhibit plant growth, and sodium ions may affect soil structures. Sewage effluents also

Table 56   Comparison of polishing methods when treating well-oxidized sewage effluents of at least 30:20 standard

Method	Rate of treatment, $(m^3/m^2$ per day)	Performance (% removal)		
		Suspended solids	BOD	Bacteria, coli-aerogenes
Grass plots	Up to 0.85*	60–75	55–60	90
Lagoons, short retention	0.25–0.45	35–55	35–50	70
Lagoons, long retention	0.28	80	65	99
Slow sand filters	Up to 3*	60–65	35–45	50
Rapid gravity sand filters	115–230	70–90	50–70	30
High rate upward flow sand filters	230–450	50–75	45–55	25
Microstrainers, 15,000 apertures/cm^2	350†	55–75	25–50	15
Upward flow gravel bed clarifier	14–23	50–60	25–40	25

* Based on area in use at any one time.
† Flow per unit area of fabric.

contain micro-organisms of many types, some of them pathogenic, so that even use of reclaimed water for industrial purposes may be undesirable. Table 56 indicates the efficacy of various polishing or tertiary processes that may be applied. It will be seen that where space is available the least sophisticated processes, such as simple grass plots or lagoons, can often produce the best results.

Grass plots

Reduction of both suspended solids content and BOD_5 of sewage that has already been treated by aeration and final settlement can be effected by irrigation over grass plots. The advantage of this method is that the BOD_5 reduction is greater than would be made by a mechanical filtration method only, because additional aeration occurs.

Humus tank effluent was treated in this way by the Birmingham Tame and Rea Drainage Board.[4] The irrigation area was divided into plots so that sections could be laid off, each plot being about 90 m long from inlet to outlet and having a fall of 1 in 60. After a period of use, each plot was dried out and the grass mown twice in the growing season to inhibit weeds. The area of grass land was about 1 m^2 per person served. The result was to reduce the suspended solids content of the humus tank effluent from 16.2 to 7.6 mg/l and the BOD_5 from 17.0 to 8.1 mg/l.

The author has used areas of grassland in lieu of humus tanks at small works and thereby obtained very good effluents. The area allowed for this purpose was in the region of 3.35 m^2 per head of population. Once prepared, the grass plots were not given any maintenance.

The method of distribution is by means of a ditch dug along a contour and provided with controllable outlets with their sills all at the same level. The effluent from the plots should be collected by a ditch which should discharge into a manhole through a pipe, the invert of which is above the invert of the ditch so as to provide a sampling point and in order that suspended solids washed down in wet weather should have an opportunity of settling before the sample is taken.

A common application of grass plots is for treatment of surplus flow, thus increased flow due to rainfall is often separated at the inlet to small works, and by-passed direct to land. Separation may be at 3 or 6 × DWF, according to the design of the treatment processes.

Lagoons

Shallow lagoons, or maturation ponds (about 1 m deep) with short retention periods (3–4 days, as distinct from the longer retention periods used in a similar form of more extensive treatment often employed in tropical countries, see Chapter 18) have been found moderately successful for effluent treatment. Reduction of suspended solids by about 50% may be anticipated, but rising sludge and algal growths may cause seasonal increases in solids content.[3] Windle Taylor describes a successful application of deeper lagoons (used in series) with up to 17 days retention time and removing about 80% of solids and a corresponding bacterial reduction of up to 99%. Again, there were seasonal variations.

Such growth may be turned to advantage by encouraging rapidly increasing algal population in shallow ponds, possibly by adding carbon dioxide to the effluent and then harvesting the growth in a live state so as to remove the solids before decay returns them to the effluent. This can be achieved by draining the ponds in turn and scraping away the algae from the sides and bottom. This method also reduces BOD_5 by virtue of the oxygen produced by the algae, and effects some reduction of bacterial content.

Sand filtration

When the land is of a suitable sandy nature, lagoons may be formed by removing the topsoil from a suitable area to form beds with level surfaces, and provided with underdrains. Alternatively the sand may be imported, and the procedures adopted for treating the effluent will be similar to those already described under the heading 'American practice' earlier in this chapter.

More generally, for effluent treatment in this context artificial beds are constructed, usually of concrete, and filled with a layer of sand resting on coarser material, and provided with an underdrainage system. The thickness of sand may be 300–750 mm and effluent may be applied at a rate up to 3 m^3/m^2 per day. Percolation continues until head-loss becomes excessive, say 300 mm or more, after which the bed is drained, the flow being diverted to another bed. The surface layer of sand and accumulated dirt is then removed, the sand topped-up, if necessary, and the bed returned to use. Beds used in this way are practically identical with the slow sand beds used in water treatment, and can be designed in the same way. The results obtained with them are due in part to the build-up of a 'smutsdec' on the surface, as well as to the straining out of suspended solids in passage through the bed. It is for this reason that they can achieve higher figures of BOD_5 removal than any purely mechanical process.

Hamlin, in South Africa, developed the Hamlin Filter,[6] which by dividing individual beds by a control channel—with side walls in the form of weirs at sand-level—incorporated a method of cleaning simply by opening a valve or penstock at the bottom end of the control channel and allowing the supernatant water to drain off, taking most of the dirt with it. This process was aided by supply of effluent from a common channel to a group of beds, and by travelling 'squeegees' moving across the surface. The arrangement is shown in Figure 85. Inevitably there is some loss of sand in the dosing process, but drying out is only necessary at long intervals.

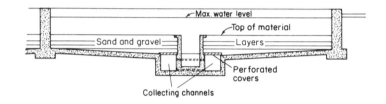

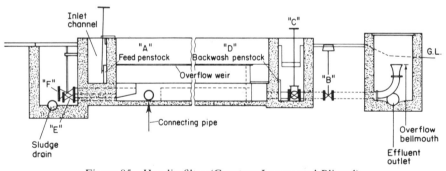

Figure 85   Hamlin filter (Courtesy Lemon and Blizard)

Rapid gravity filters

As applied to sewage works these filters consist of a layer of graded sand 1–1.5 m deep, supported on layers of gravel, resting in turn on special filter floors incorporating a series of nozzles for collecting the filtrate. The beds may be large metal cylinders as in waterworks practice, but are usually concrete beds with high side walls.

The drainage system is provided with means for washing and air scouring. The liquid flows down through the sand at a rate of 115–230 m^3/m^2 per day.

At a certain time each day, when the head loss has reached about 3 m, the filter is backwashed, using filtered effluent with air scour assisting separation of sludge and sand. Back-washings amounting, say, to $2\frac{1}{2}\%$ of the volume filtered are returned to works inlet. Unless pumping of the effluent is resorted to, the incorporation of rapid-type filters in a normal gravity-flow treatment works may well prove difficult as the head necessary just about doubles that required through the other processes, even including trickling filters! It may be possible, of course, to pipe the final effluent further down the stream course thus making extra 'head' available. This is an expedient worth examining at any works where improved standards are considered advisable.

Immedium upward flow filters

This type of filter, developed in Britain, bears a superficial resemblance to the rapid gravity filter, though the liquid flows upwards through the sand.[3] Grading of the sand occurs during back-washing, with coarser material at the bottom and finer at the top; thus during filtration the effluent meets progressively finer material. Solids removed by filtration are thus distributed through the sand rather than being concentrated near the surface. This provides for higher rates of treatment (up to $450 \text{ m}^3/\text{m}^2$ per day) and longer runs. To prevent expansion of the sand at these higher rates it is held down by a metal grid, consisting of parallel bars spaced 100–150 mm apart, just below the surface of the bed.

The filter consists of a casing containing a 1.5–2 m deep bed of sand, supported on 300 mm of gravel resting in turn on a special floor fitted with a liquid-distribution system. The bed is cleaned by increasing the upward flow of unfiltered effluent to about $800 \text{ m}^3/\text{m}^2$ per day to cause the bed to expand through the retaining grid. The back-wash water containing the solid matter is some 2–3% of volume filtered. The filter may be cleaned automatically when the pressure head against pumping reaches a certain figure.

The first full-scale application of this process, which is also in use at waterworks, was installed at Luton, East Hyde Works, as a means of complying with very stringent effluent quality requirements.

The upward-flow clarifier

Banks,[7] working in Battle Rural District, Sussex, developed the upward-flow clarifier, which has had quite wide application at small sewage works, at a time when engineers were tending to lose faith in the conventional humus tank. As Figure 76 (page 340) shows, the pebble bed could be quite easily incorporated in an existing rectangular humus tank, although a pyramidal tank would probably not provide sufficient surface area. Alternatively, clarifiers are sometimes employed as a tertiary process following humus tanks.

The filter, or humus tank effluent, is passed upwards through a 150 mm layer of pea gravel, supported on a perforated floor. This need be no more than two layers of welded steel fabric, with angle-iron supports at intervals, or can be a purpose-made floor of pre-cast concrete, with small perforation or made of wedge-wire units. The rate of percolation is 0.5–1.0 m^3/m^2 per hour. Solid material which accumulates in the bed is removed by back-washing. This is achieved by lowering the water level by running off effluent below the bed level, and is completed by washing the drained bed with a jet of water or effluent.

Experiments carried out by Truesdale and Birkbeck[7] showed a reduction in suspended solids of 57 %, and up to 40 % of BOD$_5$ (see Table 56), p. 379. The success of this comparatively simple process is probably explained by the work described in the following section.

Modular rapid-gravity filters

Recent work by Metcalfe and Stevenson[9] suggests that the biological solids removed in the tertiary treatment of sewage are very different from those found in potable water treatment. Particularly 'floc strength' appears to be very high, and 'breakthrough' in the sense used in the potable process does not seem to be a problem.

There is no more reason why a granular bed filter should not remove colloidal material in tertiary treatment of sewage, than in potable water treatment. The tertiary stage therefore depends on the satisfactory operation of the second stage.

Biological floc strength with conventional size media tends to give excessive head-losses and may prove difficult to clean by traditional methods. Logically therefore a coarser medium, which exploits the floc strength, should be used. This will also provide a greater shear stress on the particle surfaces during back-wash, to remove adherent material and prevent 'frog-spawn' coatings. Combined air scour/back-wash further improves cleaning efficiency.

The final result is a specific design tailored to meet the requirements of the sewage, which would not be particularly applicable in the case of water treatment. The experimental data confirm the validity of this approach.

One of these installations, at Eden Vale, East Grinstead — now under control of the Southern Water Authority — consists of 3 No. cylindrical units each of 20.7 m^3 capacity. Two of these are of pebbles and sand — upper zone 150 mm 23/56 mm size pebbles, then 1500 mm of sand 2.3/4.75 mm with a lower zone of 100 mm of pebbles, 6.7/13.2 mm. The third is filled with irregular granite medium, total depth of bed in all filters being 1750 mm. The works is so situated that these filters are gravity-fed from the humus tanks. They will deal with flows up to 8190 m^3/day and are fully automatic, back-washing at a head-loss of 1.8 m, or once every 24 h. The back-wash pumps are capable of 71 l/s, and operate in conjunction with positive displacement compressors, producing 140 l/s of air, at 0.5 bar. Average effluent obtained was 6 mg/l BOD$_5$ and 8 mg/l suspended solids. Even this standard was refined to about half the above figures after passing over grass plots, which had previously dealt directly with the humus tank effluent.

In the US the stringent standards lately demanded by some states led to what are known as the Hanover Park Tertiary Studies, in which three types of deep-bed filter and an ion-exchange unit were found capable of producing results within the 4–7 mg/l range.

Mechanical processes

The further reduction by filtration of the BOD_5 of a humus tank or final sedimentation tank effluent in mg/l is usually in the region of 30 % of the reduction of suspended solids content in the same units. Thus, if a humus tank effluent has a suspended solids value of 30 mg/l and a BOD_5 of 20 mg/l and by a filtration method of tertiary treatment suspended solids are further reduced to 20, the BOD_5 may perhaps be reduced to 17. If, however, the filtration is so good as to reduce the suspended solids to as low as 10, the BOD_5 may still be as high as 14 mg/l. For this reason mechanical filtration alone cannot be expected to produce an extremely low BOD_5 and if the latter is required, a tertiary method which effects aeration should be used.

Micro-straining is a method originally developed by Glenfield and Kennedy Ltd in connection with waterworks, and it has been applied to humus tank or final sedimentation tank effluents. The micro-strainer is a rotating drum covered with a very fine stainless-steel gauze which is back-washed by a sparge of effluent. The gauze has also to be cleaned more completely from time to time to remove organic growth that would otherwise develop and choke it. Micro-strainers of suitable grade can reduce the suspended solids content to about one-third of that of the influent; but, as there is no aeration, the BOD_5 is not likely to be reduced to less than 70 % of that of the influent; more or less according to the grade of the filter fabric.

One advantage of micro-strainers and screens is that they operate under very low head-loss. In fact head-losses above 150 mm are liable to distort the delicate mesh fabric, and overflow weirs come into operation to avoid continued operation under these conditions. Indicators are, of course, provided so that the operator knows when the washwater pump should be used. With the help of ultraviolet light equipment and careful speed control, the intervals between back-washings can be considerably increased.

Micro-screens manufactured by F. W. Brackett & Co. Ltd of Colchester, Essex, England, consist of horizontally mounted rotating drums, the surfaces of which are covered with stainless-steel fabric. These drums are of 1 m diameter and 1 m width for a packaged plant, and from 2 m diameter and 1 m width to 3 m diameter and 3 m width for erection in concrete chambers. The flow is brought into the open end of the drum and passes outwards through the mesh. As the drum rotates the mesh is brought under a high-pressure back-washing sparge at the top. The fabric is usually of 25, 35 or 60 μ rating. The capacity of the micro-screens depends on the solids content of the fluid to be treated, and the size and nature of the solids, but some indication of performance can be derived from the installation at Glemsford in Suffolk, England, where a 1 m diameter by 1 m wide drum dealt with 13.5 m^3/day of sewage-works effluent.

While back-washing by sparge adequately cleans the mesh usually, biological growth can build up and resist the flow. This can be combatted by spraying the fabric or rotating the drum in a 10% solution of hypochlorite.

References

1. Royal Commission on Sewage Disposal, 5th Report, HMSO (1915).
2. Imhoff, K., Müller, W. J. and Thistlethwayte, D. K. B. *Disposal of Sewage and Other Waterborne Wastes*. Butterworth, London (1956).
3. Truesdale, G. A. *Advanced Treatment*. Advances in Sewage Treatment Conference at Instn. Civ. Engnrs. Nov. 1972, Thomas Telford (1972).
4. Daviss, M. R. V. Treatment of humus tank effluent on grass plots. *Surv. Munic. County Engnr.* **116**, p. 613 (1957).
5. Windle Taylor, E. Improvement in quality of a sewage works effluent after passing through a series of lagoons. *J. Instn. Public Hlth. Engnrs* **65**, p. 86 (1966).
6. Hamlin, *Proc. Instn. Civ. Engnrs.* **1** (No. 5) part 1 (Sept. 1952).
7. Banks, D. H. *Surv. Munic. County Engnr.* **125** (2807) (1965).
8. Truesdale, G. A. and Barkbeck, A. E. Tertiary treatment processes for sewage works effluents. *J. Instn. Public Hlth. Engnrs.* **56** (Apr. 1967).
9. Metcalfe, S. M. and Stevenson, D. G. Tertiary filtration of sewage. *J. Instn. Public Hlth Engnrs.* **78**, July 1979.

CHAPTER 17

Treatment and Disposal of Sewage Sludge

The solids suspended in sewage, when settled and separated from the main body of the liquid, are known as sludge. Solids which float to the surface of either sewage or sludge are referred to as scum. There are several types of sludge, principally:

(1) primary sludge which is settled from crude sewage in the primary sedimentation tanks;
(2) humus which is settled from the effluent of trickling filters,
(3) activated sludge which is settled in the final sedimentation tanks of activated-sludge works.

When sludge is partially digested, prior to a further stage of digestion, it is referred to as semi-digested sludge. When digested to the maximum extent intended (which is never to the extent of complete digestion) it is known as digested sludge.

The disposal of sludge, either wet or dry, is one of the main problems at sewage-treatment works, involving an appreciable part of the total capital and annual running cost. Sewage-works managers attempt to dispose of sludge as manure but, while some sludge is not without manurial value, there is an increasing danger of toxic elements being present. There is also usually much more sludge than farmers are able or willing to accept.

The total amount of sludge to be expected at the works is made up of the quantities of solids coming down the sewer, plus the estimated solids produced by the aeration processes, minus the solids lost with the effluent and the solids removed prior to settlement in the preliminary treatment processes. In estimating quantities of sludge from analyses, the figures relating to average flow, and not dry-weather flow, are used. The weight of solids in the inflow and those lost in the effluent are easily calculated. Deciding how much new suspended matter is produced by the biological processes is not so easy and estimates vary considerably. Average of the figures for 5 years' operation at Crossness Works, gave 0.125 kg of suspended solids produced from dissolved solids per kg of BOD_5 removed by aeration. This is a low, but probably accurate, figure.

In trickling-filter schemes the loss and increase of organic solids during the process of aeration is more complex and uncertain than in activated-sludge schemes. In the trickling filter there is a process of building up of simple organisms which become the food of larger, more complex, animals and plants; and while there is aerobic growth, there is anaerobic decomposition. The resultant of these various conflicting processes is that the quantity of humus is far from constant throughout the year and may be very different at different works. Nevertheless the total amount of sludge of all kinds in the year at works with trickling filters appears to average about the same figure as found for activated-sludge works. Some authorities have said that there is much less humus on the average than surplus activated sludge, but as both humus and surplus activated sludge are in the main made up of the suspended solids which the primary sedimentation tanks have failed to settle, it would appear very unlikely that the quantity of humus could be much less than the quantity of activated-sludge solids at a similarly loaded activated-sludge works.

Quantities of sludge

A very extensive investigation was made by a team at the Water Pollution Research Laboratory, and the results were published in a Report of May 1968[1] on which the following figures are based:

The average quantity of sludge dry solids at all the works included in the investigation, regardless of type, was 0.0785 kg per head of population per day. The average moisture content was 95.52% giving about 1.76 l of sludge per head per day. Of the dry solids, 72.5% were volatile matter, or 0.06 kg per head per day. These figures include all the suspended solids that were in the crude sewage and those produced by the aeration process.

At works where no treatment was given other than preliminary processes and primary sedimentation, the sludge solids were 0.0545 kg per head per day. The average moisture content was 92.5% giving about 0.726 l of sludge per head per day. Of the dry solids, 72.2% were volatile matter, or about 0.04 kg per head per day.

The average quantity of humus settled in tanks following trickling filters was 0.0268 kg per head per day. The average moisture content was 99.23% giving about 3.48 l of sludge per head per day. Of the dry solids, 64.2% were volatile matter, or about 0.017 kg per head per day.

The average quantity of surplus activated sludge (diffused-air process) was 0.0253 kg per head per day. The average moisture content was 99.33% giving about 3.88 l of sludge per head per day. Of the dry solids 77.1% was volatile or about 0.02 kg per head per day.

It will be noted that the amount of humus solids per head of population per day was, according to the investigation, almost identical with the amount of surplus activated sludge. Assuming approximately average conditions as to load on the works, the amount of organic material produced by the activated-sludge works would be about 0.01 kg per head per day, which would mean that the solids which the primary sedimentation tanks failed to settle would be 0.017 kg per head per day. This would suggest sedimentation tank efficiency about average for British conditions.

In making calculations of sludge flow it is advisable to allow for the highest moisture content to be expected for the particular type of sludge because what may seem to be a small change of moisture content makes a great deal of difference in the amount of sludge to be accommodated. For example, an increase of moisture content from 90% to 95% doubles the quantity of sludge and a further increase of moisture content to 97.5% again doubles the quantity. (A moisture content of 97.5% is considered about usual for sludge from non-mechanized pyramidal-bottomed primary sedimentation tanks.)

Sludge drying

The method of sludge treatment or disposal to be adopted depends on the circumstances of the case. Broadly speaking, choice of method is decided by the size and location of the works. For example, sludge digestion is more common at large works than elsewhere.

Primarily, the method adopted should depend on the overall annual cost[1,2] to the community, provided that the method is practicable in the circumstances and not liable to cause nuisance. The least expensive methods applicable to small works only, and not requiring any construction, are disposal to permanent lagoons, drying on earth plots, or pumping in wet form to farmers. All other methods require constructional works and some need machinery and therefore must cost more. The method by far the most frequently used and, in nearly all circumstances, the most economical, is drying on properly constructed sludge-drying beds. However, increasing labour problems and poor climatic conditions have recently seen a decline in popularity of this method in the UK.

The approximate relative costs of the various means of sludge disposal are given in Table 57 in terms of percentage of the cost of drying on beds. Sludge digestion, used as an adjunct to drying on beds or other means of disposal,

Table 57   Comparative costs of sludge-drying methods

Method of disposal	Cost as a percentage of drying on beds
Disposal in permanent lagoons	25
Drying on earth plots	50
Pumping to farm land	50
Drying on sludge beds	100
Shipping to sea (where practicable)	100
Filter pressing	150
Heat-drying of press cake	300
Vacuum filtration and flash-drying	670

makes little difference to the economics where sludge is dried on beds or disposed to sea. At large works it may reduce the overall costs, and at small works slightly add to them. No attempt has been made to correct these figures to cover the increases in cost of fuel, particularly oil, which have taken place in the last few years.

Methods of dewatering

Sludge-drying beds

These are used at the majority of works. They consist of shallow beds or lagoons with or without concrete floors, according to the nature of the subsoil, the level of the subsoil water, and the disadvantages, if any, of the soakage of sludge liquor into the ground. The drying beds themselves consist of about 225 mm of gravel, broken stone or clinker graded from 25 to 38 mm, finished with 38 mm of fine-grade material such as will pass a 12.5 mm sieve but be held on a 6 mm sieve. This last surfacing material has to be replaced from time to time because portions of it are unavoidably removed with the dried sludge.

The floors of the beds are laid at a gradient of about 1 in 200 and on them are laid 75 mm agricultural drains, or 100 mm half-round perforated pipes which connect to manholes on the main sludge-liquor drain which, in most instances, discharges to a pumping station. Sludge liquor should be returned to the flow of sewage for full treatment because it is very foul.

The individual sludge beds should be so proportioned that a daily discharge of sludge completely fills one or more beds to a depth of about 225 mm. This means that a number of small beds should be constructed, not one or two large beds, and that the size of the beds should be calculated in accordance with the estimated sludge discharge. (The above, of course, applies only to works where there is no digestion tank, for where digestion tanks are installed the amount of digested sludge released from secondary or other open-topped tanks at any time can be adjusted to suit the capacity of drying bed.) What must be avoided is the construction of two or three beds only, for this would mean that if the sedimentation tanks of small works are sludged at frequent intervals, as they should be, it would be necessary to discharge wet sludge on top of sludge which had partly dried, which is objectionable because it militates against efficient drying. The number of drying beds at any works should never be less than six.

Sludge is usually fed to the sludge-drying beds by a pipeline with branches to each bed, controlled by sluice valves. The gradient of this pipeline should never be less than the figures given in Table 35, p. 263. The minimum size of sludge drain is taken as being 200 mm (US) and 150 mm (Britain): larger-diameter sludge pipes are not installed unless justified by the flow. Sometimes open channels are used in lieu of pipelines. This method requires some care in determining wall heights, so as to avoid overflow if the sludge being discharged is thicker than average.

The sludge is discharged on to aprons of concrete or stone which protect the surface of the beds from scour at the points of inlet. From these points the sludge floods over the surface of the clinker (which should be laid level and not to a fall) to a depth of about 225 mm. After a week or so in fine weather, the surface cracks and the sludge becomes spadable and can be barrowed to a dump, where it is left to thoroughly dry out. Various machines are available for removing sludge from the beds at large works.

During the process of drying supernatant water forms on the surface and this can be decanted to the manholes on the underdrainage system by means of special weir penstocks or by weirs formed of stacks of 50×25 mm boards let into chases in the sides of openings in the walls of the manholes and which can be adjusted for height by removing the boards as necessary.

Occasionally sludge-drying beds have been provided with moveable roofs which, running on rails, can be transported to those beds which most need protection from the rain. Some believe that moveable roofs do not justify their cost, and that money could be better spent on additional drying area.

The area of sludge-drying beds required depends on the climate. It can be comparatively small where there are long periods of hot sun and little rain. For temperate regions in the United States figures that have been recommended are 0.093 m^2 of bed per head of population for primary sedimentation tank sludge where there is no secondary treatment, 0.18 m^2 for works incorporating trickling filters, and 0.28 m^2 for activated-sludge works.[3] In Great Britain areas from 0.14 to 0.25 m^2 per head of population are allowed, the latter figure being for activated-sludge works. Present-day practice is to be lavish in the provision because, particularly in the wet regions of the country, air drying can be difficult at some times of the year.

Dewatering in filter presses

This was a method at one time largely used in Great Britain and which is returning to favour at present,[4] sometimes as an adjunct to other processes. A filter press is made up of a series of cast-iron plates with coarse filter cloths of jute or similar material between them. When the plates are screwed tightly together they form a series of cells. The sludge is forced by compressed air or pumping into every other cell at a pressure of about 414 kN/m^2 and the water passes through the filter cloths into the remaining cells and drains away. On release of the screw at the end of the press, the plates fall apart, the sludge cake falls out and is removed, to be further dried on a dump. The complete cycle of filling, compressing, and removal of cake occupies about $\frac{3}{4}$h. The cake has a moisture content of about 75%: in this respect the method is better than vacuum filtration.[5]

It is usual to dose sludge with $3\frac{1}{2}$–5% of lime[6] prior to filter pressing, because this assists the process, but some works consider the cost of lime treatment is greater than the annual repayment of loan on the cost of the additional presses that would be needed were the sludge to be pressed without lime. Polyelectrolytes have also been used for this purpose.[7]

Dewatering activated sludge

The dewatering of activated sludge has always been considered a problem, but this has been solved by the Komline-Sanderson Dissolved Air Flotation Thickening Process, (see Figure 86) marketed in Britain by Ames Crosta Babcock. It is claimed that this will reduce the moisture content of activated sludge from about 99.5 % to 96 % and that final moisture contents between 95 % and 93 % have not been unusual. Air is dissolved in water under pressure and this water (usually the recirculated effluent of the process) is added to the sludge. On the pressure being reduced, bubbles of air coming out of solution carry the suspended solids to the surface to be removed by a variable-speed mechanical skimmer. This scum has a minimum solids content of 4 % plus a large amount of air. After withdrawal it is retained for one or more days, in a hopper-bottomed tank or tank with rotating scrapers, for the air to be given off.[8-13]

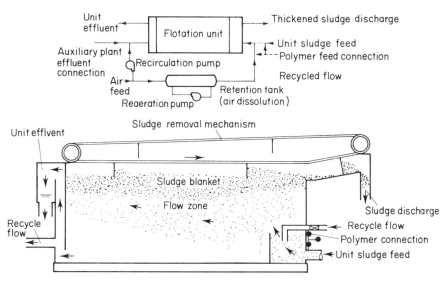

Figure 86   Komline-Sanderson Air Flotation unit (Courtesy Dorr Oliver Co.)

Sludge consolidation

Several of the processes just described, as well as the digestion process, which is dealt with in subsequent pages, operate better with a thicker sludge than with that normally available from sedimentation tanks.[14] For this purpose deep circular tanks, equipped with 'picket-fence' thickeners, are usually employed. Their design can be based on the information obtained from techniques such as that now described.

A primary technique for computing surface area requirements of thickening tanks for sewage and industrial sludges is that quoted by Lutkin[15] and Veselind,[16] which was developed in 1917 for ore slimes.

$$A = \frac{Q_0 C_0}{V} \left[\frac{1}{C} - \frac{1}{C_n} \right],$$
(37)

where: A = area of thickener (cm^2);
Q_0 = inflow ml slurry per minute;
C_0 = influent solids concentration (g/cm^3);
C = solids concentration of slurry (g/cm^3);
C_n = final solids concentration (g/cm^3);
V = settling velocity (cm/min).

The velocity V should be computed for at least 10 batch tests with solids concentrations varying from C_0 to C_n, solids interface level plotted against time readings for each test. Substituting the concentration (C) corresponding interface velocity (V) and the selected values of $C_0 Q_0$ and C_n into the equation, the required thickener surface area (A) can be computed for each test. The maximum value of A is considered the minimum design figure.

Sludge digestion

When sludge is stored and not aerated, it becomes septic; that is anaerobic bacteria that thrive in acid conditions digest the sludge and, in so doing, produce foul odours. Eventually bacteria that produce an alkaline reaction prevail, and once the alkaline reaction is established it normally remains. When the sludge has reached this condition it digests rapidly; its odour is unobjectionable, usually being described as tarry, and sludge gas is emitted.

At the beginning of a period in which sludge is seeded by being mixed with actively digesting alkaline sludge, gas is given off rapidly and consists largely of carbon dioxide, but as the digestion proceeds the proportion of carbon dioxide reduces and the proportion of methane increases. The mixture of methane and carbon dioxide, with small proportions of other gases, is known as sludge gas and, being combustible, this can be collected and used for power purposes.

Sludge gas consists of about 67% methane in average conditions and about 33% carbon dioxide (including small quantities of other gases). Because at the beginning of digestion the proportion of carbon dioxide is higher, if the digestion period in the primary (or gas-collection) tank is too short, the carbon dioxide content may be unduly high, militating against utilization. This can be rectified only by increasing the detention period.

Methane has a gross calorific value of 37,073 joules (J) and a net calorific value of 33,347 J/l. Thus, if there is 67% methane in the mixture the net calorific value of the sludge gas will be 5336 kcal or 22,343 kJ/m^3. Thus, 1 m^3 of sludge gas will produce, on average, 1.86 kWh, assuming dual-fuel engines and generators of normal efficiency.

In practice about $0.67 \, m^3$ of gas can produce at least 1 kW when used in a suitably designed gas engine. An advantage of power production in this manner at sewage works is that about 19% of the calorific value of the gas can be recovered from the cooling water of the engines and another 14% from the exhaust gases by calorifiers installed on the exhausts. Together these recoveries of heat are usually sufficient to establish and maintain a temperature in primary digestion tanks high enough for mesophilic digestion.

The gas, being produced by decomposition of organic solids—particularly animal and vegetable fats (mineral oils are not affected)—reduces the suspended-solids content of sludge and, as after this has occurred strata of comparatively clear liquor can be decanted, an important effect of sludge digestion is to reduce the quantity of sludge for final disposal. Other advantages of digestion are the elimination of the unpleasant odour of primary sludge, and converting the sludge to a substance which, when dewatered, is fibrous and more acceptable for agricultural purposes.

While the fact that sludge could be digested had been known for a long time, it was not until the process had been studied that the construction of sludge digestion units became popular; this was in the 1930s. At the present time the process seems to be becoming somewhat less popular: this may be due to too much having been expected of digestion, and tanks not being made large enough to give the desired results. The effect of detergents, and the increasing use of other chemicals, also inhibited the process at a number of works and gave rise to the investigations carried out by Sidwick and others.[25-27]

Whether or not sludge digestion is adopted is a matter of choice, in accordance with the advantages gained at the cost of construction and maintenance of works. It is considered that only at large works does the power afforded by sludge gas justify the capital and maintenance expenditures. Nevertheless, sludge digestion is usually provided at activated-sludge works because the power from sludge gas makes the works virtually independent of outside sources of power.

Sludge may be digested slowly at the day temperature of temperate countries, e.g. 15.5°C: this is known as cold digestion. If the sludge is heated to and maintained at a temperature between 26.7 and 35 °C digestion becomes more rapid. This process, which is known as mesophilic digestion, is the one most frequently adopted. If the temperature is raised to and maintained at between 43 and 49 °C the process becomes much more rapid but unstable and difficult to control. This process, known as thermophilic digestion, is not recommended.

Digestion is effected in two stages, for which separate batteries of tanks are used. In the primary stage the greater part of the gas is emitted and, where the gas is wanted for use, the tanks are provided with gas-tight roofs, sometimes floating roofs and sometimes in the form of gasholders. As emission of gas militates against separation of supernatant water, no provision is made for decanting water bands. In the secondary stage of digestion, little gas is given off and need not be collected, but supernatant water can be withdrawn at or near the surface. To facilitate this, valves are provided at various levels or arms that can be adjusted to draw off from any level where sludge liquor can be found. In mesophilic digestion it is usual to heat the primary tanks but not to heat the secondary tanks.

Digestion tanks are never made to give virtually complete digestion of organic solids, but only to digest the sludge to what is considered the economic proportion. It is here that the engineer has to decide on what detention period shall be adopted. Experiments in which seeded sludge is maintained at controlled temperatures give good indication of the proportion of digestion, reduction of suspended solids, and emission of sludge gas which may be expected at certain temperatures in certain periods of time with particular sludges. In actual sludge-digestion tanks in operation at sewage works, however, tanks are not so filled with sludge and left to digest, because this process would be inconvenient and difficult. The usual, and virtually unavoidable, procedure is to put a dose of raw sludge into every primary digestion tank once every day in order that all the tanks will be performing in a similar manner, giving off a reasonable steady quantity (and quality) of gas.

There must, of course, be some short-circuiting permitting sludge to leave the tanks before it is adequately digested, while some sludge is retained for an unduly long time. While attempts have been made to prevent this, they have never been altogether successful. A further difficulty with some designs of tank is that fine silt or detritus settles, robbing the tanks of their effective capacity. This has been known to happen to such an extent as to reduce the detention period until the proportion of carbon dioxide in the sludge is so high as to make the gas incombustible and useless for power-production purposes.

It is for the foregoing reasons that the most satisfactory way of determining the required capacities of primary and secondary sludge-digestion tanks is to use empiric figures determined by records of the performance of existing plant, and this is where the investigations of the Water Pollution Research Laboratory are particularly useful.[17-20]

The most practical method is to determine the capacities of the primary tanks on the basis of the required theoretical detention period. According to the research the average daily load on primary digestion tanks heated to about 32 °C was about 1.55 kg/day of volatile solids per m^3 of primary tank capacity. This was about 0.06 kg/day of total solids. At the average figure of 4.48% solids content of sludge before digestion, this gave a theoretical detention period of 21 days, or an average of 0.37 m^3 of tank capacity per head of population served.

This figure of 21 days could well be used for design purposes, for it would give a reasonable factor of safety in most cases. Its use would result in the digestion of about 40.8% of the volatile solids (say, 29.6% of the total solids) and about 0.96 m^3 of gas would, on the average, be produced per kg of solids destroyed. (Contrary to some opinions that have been expressed, the weight of sludge gas is usually equal to not greater than the weight of solids destroyed.) In the research the average worked out at 0.222 m^3 of gas per head of population per day.

The secondary tanks following heated primary digestion tanks should have $1\frac{3}{4}$ times the capacities of the primary tanks as calculated above if dewatering of 30% is to be effected, or as much as $2\frac{1}{2}$ times the capacity of the primary tanks if 50% or more dewatering is required.

Digestion at day temperature in temperate regions is employed at works serving populations of 10,000 to 50,000 persons. For this purpose the tanks should have a theoretical detention period of at least 2 months, and preferably 3 months.

If sludge is drawn off from the primary sedimentation tanks sufficiently dense for passing directly to the primary digestion tanks, no preliminary de-watering is necessary. But sludge from hopper-bottomed sedimentation tanks, or even mechanically swept tanks that are not sludged with due care, may be too thin and therefore may have to be de-watered. This can be done in a few hours by being passed to tanks from which supernatant water can be decanted. Dewatering reduces the quantity of sludge, and this means not only that smaller digestion tanks are required than would otherwise be needed, but also less heat is necessary for raising the temperature from sludge (or sewage) temperature to digestion tank temperature.

Reduction in volume of sludge by dewatering is calculated by Santo Crimp's formula

$$W_2 = \left(\frac{100 - P}{100 - Q}\right) \times W_1 \tag{38}$$

where: W_1 = original weight of wet sludge;
$\quad\quad W_2$ = weight of dewatered sludge;
$\quad\quad P$ = percentage moisture of wet sludge;
$\quad\quad Q$ = percentage moisture of dewatered sludge.

Primary mesophilic digestion tanks are usually circular on plan and constructed of reinforced concrete, because this is economical and suits the various types of stirring and heating mechanisms available. The shape is also convenient for tanks with floating roofs and spiral-guided gasholders. The floors of tanks that have scrapers to move the digested sludge to a central outlet usually slope at a very moderate gradient towards the centre. Formerly it was usual to have conical-bottomed tanks where no scrapers were provided so as to induce the sludge to gravitate towards the central outlet, but it was found that steeply sloping floors rendered the tanks difficult for maintenance because the men could not stand safely on them. The tendency now is not to have any floor sloping more steeply than about 1 vertical in 6 horizontal. In such flat-bottomed tanks silt that has failed to settle in the detritus-removal unit will eventually form a deposit, robbing the tank of effective capacity. The tank should therefore be increased in depth so as still to have adequate capacity after silt has deposited to a slope from the circumference to the centre at an angle of 40° to the horizontal.

If the rate of emission of sludge gas is more than about 9 m³/m² of surface area of tank there is a risk of foaming. If this occurs the tanks may spill over on to the ground and the mains taking the gas from the tank may become choked.

To overcome the risk of foaming it is usual to have a surface area great enough to ensure that the rate of gas emission per unit of surface area is well below the above-mentioned figure. This has the effect of limiting the depth of primary digestion tanks. It is also important that all primary tanks are given equal daily input.

Primary tanks are heated by either fixed or moving internal hot-water pipes, or by heating sludge drawn from the primary digestion tanks in a heat exchanger and returning the heated sludge to the tanks. The latter method is now favoured as internal pipework increases maintenance problems.

When tanks are heated it is usual to arrange for the sludge to be stirred or mixed by mechanisms as manufactured by firms such as Dorr-Oliver or Ames Crosta Babcock to prevent local overheating and to mix active with raw sludge. These stirring mechanisms must be so arranged as to break up and sink, or remove, any scum which collects on the surface. Ironwork in sludge digestion tanks generally requires no protection, as the sludge is alkaline and free from oxygen, although electrolytic troubles have been known.

Fresh sludge entering the primary tank should be mixed with at least 20% of its quantity of seeding sludge drawn off from about half-way down the primary digestion tank.

Dorr-Oliver have provided several types of digestion mechanisms, including those for flat-bottomed tanks with rotating scrapers to move the sludge towards a central outlet but without other stirring mechanisms, and with heating pipes fixed round the insides of the walls; circular tanks with scraping mechanism as described above, together with stirring and scum-breaking mechanism; also mechanisms as above described but with the heating pipes incorporated in the stirring arms instead of fixed to the walls. They also manufacture spiral-guided gasholders for use with tanks incorporating all types of their mechanisms. One Dorr-Oliver design is known as the 'B' type which, when used with a gasholder roof, is called the 'BGH' type (see Figure 87). This incorporates an axial-flow pump and heat exchanger. Sludge is drawn off from the tank and delivered into a manhole built on the side of the tank in which is installed a low-lift axial-flow pump of considerable delivery. The pump lifts the sludge a foot (300 mm) or so to impart sufficient head to force it through the annular spaces of a cylindrical heat exchanger which is heated by hot water. The sludge passes downwards through this heat exchanger and is discharged back into the tank tangentially at a fairly high velocity so as to cause the whole contents of the tank to rotate.

If heat exchangers are run at too high a temperature of hot water, boiler crust is deposited by the sludge (heating water is usually softened) and this has to be removed periodically. Where heating is by fixed pipes inside the tank, nothing in this respect can be done without emptying the tank, but the Dorr-Oliver mechanisms of the 'B' or 'BGH' types have the advantage that the heaters can be lifted out by a special crane and dipped into baths of acid, then neutralized with alkali. Suitable acid and alkali baths with lifting arrangement should be provided at the works when the plant is installed.

For removal of scum, a horn is provided with penstock control and water

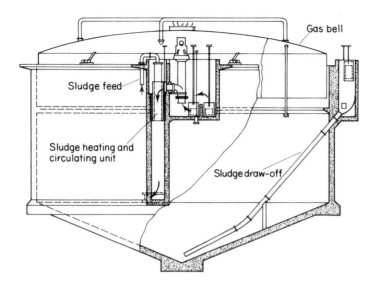

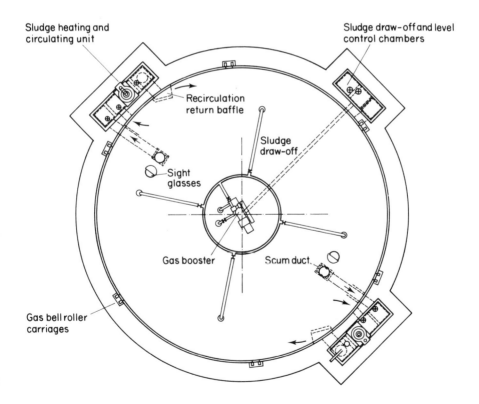

Figure 87   BGH Primary Digestion Tank (Courtesy Dorr Oliver Co.)

seal to prevent escape of gas. This discharges into the manhole containing the axial-flow pump, so that the scum can be broken up by the pump and returned to the tank at a low level or discharged to secondary digestion. One or more manholes containing this type of equipment are provided, according to the size of the tank. The pump motors run continuously but are arranged, by clock control, to reverse for a few minutes every hour so as to clear any fibrous matter from the inlet side of the heat exchangers. Raw sludge delivered from the primary sedimentation tanks is introduced to the flow of circulating sludge in the manhole which contains the pump.

The original Ames Crosta Babcock 'Simplex' digester consisted of a tank with either a fixed concrete roof or a floating steel roof rising and falling with the sludge level. On this was mounted a motor driving a pump fixed at the end of a long vertical tube. The motor was arranged to operate for a few minutes every hour, and could be run in either direction. Normally, it drew sludge from a low level in the tank and sprayed it over the surface of the scum, the purpose being not only to break up the scum but to circulate the contents of the tank.

Whilst this design is still available, the makers now favour the arrangement shown in Figure 88 with a separate gas-holder, and gas recirculated through diffusers on the floor of the tank. Digestion temperature is maintained by circulating the sludge through an external heat exchanger heated, in turn, by water from a conventional boiler with a dual-fuel option to allow for start-up and any other situation in which insufficient sludge gas is available.

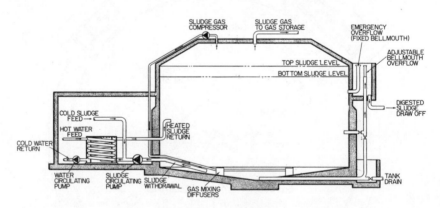

Figure 88   A digestion tank by Amos Crosta Babcock

With the Dorr-Oliver designs it is usual to provide some or all of the tanks with gasholders as roofs: those which do not have gasholder roofs have fixed roofs. With the Ames Crosta Babcock design it is usual to provide separate gasholders erected in gasholder tanks.

In tanks that do not have effective provision for scum removal, scum may accumulate to a depth of several feet. Ways of overcoming this are violent continual recirculation of sludge near the surface and recirculation of sludge gas. In the latter, gas is withdrawn from above the level of the scum and pumped back at a depth of 3 m or more below sludge level by a system of pipes and inlets, the inlets being about 50 mm diameter arranged in various positions throughout the tank, their lower ends being turned upwards to a J shape. Each inlet should be valve-controlled so that it can be used separately. Excess pressure valves should be provided on the delivery sides of the pumps, but these should be arranged with returns so that gas is not blown to waste but put back into the gas space below the roof or gasholder, otherwise there is a danger of creating a vacuum under the gasholder. Where there are several tanks it is generally best to provide a separate unit mounted on each gasholder, and not a central blowing station. It has been found that a delivery of $2 \, m^3$/min was satisfactory for a 23 m diameter digester. Mixing by gas-recirculation methods is common in North America.

The 'Heatamix' method is one in which sludge is recirculated inside the primary digestion tanks and heated at the same time with the aid of 'air-lifts' which are sludge-gas operated and are fitted with hot-water heaters. An installation at Mogden had four steel tubes each 9 m long and 300 mm diameter for a 23 m diameter tank having a capacity of $4000 \, m^3$. Each of these 'Heatamix' tubes delivered $6 \, m^3$ of sludge per minute for which $1.01 \, m^3$/min of gas were required per tube. The whole installation required $0.16 \, m^3$/min of hot water which was introduced at 60 °C and returned at 40 °C. The bottoms of the tubes had bellmouths and the top ends were turned over at right angles to give horizontal jets 1.2 m below surface level. The tubes were placed 3.6 m from the wall of the tank: the jets from the two opposite each other discharged at right angles to the radii, while one of the other two pointed at the tube ahead of it and the other pointed to half-way between the centre of the tank and the tube ahead of it.

Scum is liable to collect in chambers outside digestion tanks or anywhere where the sludge can become at all quiescent. It is advisable to provide penstocks or other means of drawing it off to the drainage system or other means of disposal.

The shape of secondary digestion tanks is not important. In many instances disused sedimentation tanks or other water-retaining structures of sufficient capacity can be used for this purpose; but when new tanks are constructed they are often circular on plan for the sake of economy. The tanks are open and the floors are sloped at a moderate angle to the outlet. Valves for decanting liquor should be provided near the surface at close intervals, and so arranged that the liquor can be inspected easily by daylight in order that sludge is not drawn off.

Gas collection

Methane will not burn in air unless it is in the proportion of about 5.8% to 13.3% of the total mixture. When mixed with carbon dioxide in the proportions of about 33% carbon dioxide to 67% methane, the combustion limits are still more reduced. If the quantity of carbon dioxide is greatly increased, as it may be when starting up a new plant, the gas will not burn at all. Nevertheless, sludge gas forms an explosive mixture with air, and care should be taken against the possibility of explosion. The use of floating roofs and gasholder roofs reduces the risk of air being drawn into the digestion tanks but does not altogether eliminate it. It is best to take the following precautions in addition:

(1) where gasholders are used the sludge in the primary tanks should be kept at a constant level, raw sludge being pumped into the tanks to displace semi-digested sludge which would then pass over weirs;

(2) to prevent a negative pressure being developed on gasholders, audible warning should be provided to indicate when the gasholders fall to near their rest level, also the holders should be provided with sealing plates to close the gas outlets before the holders come to rest;

(3) although this involves risk of introducing an explosive mixture, gasholders should be provided with relief valves to admit air should gas be sucked out to such an extent that there is danger of the structure being buckled by partial vacuum.

Where a gasholder is provided the gas is drawn off by means of a pipe carried under the gasholder to 1.2 m above top-water level, a dome being constructed, if necessary, in the gasholder over the entrance to the pipe. Where the tank has a fixed roof a fixed gas dome is constructed. In all gas leads connecting from tanks and to combustion points flame traps are provided to prevent the possibility of explosion should gas and air become mixed. There are various types of flame trap available. The type selected should have a low resistance to flow of gas yet effectively prevent the passage of flame. A thin gauze is not recommended, as this might blow out and permit an explosion: a more suitable type consists of a roll or pack of corrugated stainless steel sheet, the corrugations running in the direction of the flow of gas and the pack being of about 50 mm thickness and arranged for easy removal for cleaning. Isolating valves should be provided on each side of the flame traps.

Surplus gas is usually burnt in a burner (described later) some distance from the tanks. This can be designed as an attractive ornamental feature.

Gasholders for sludge gas are not altogether similar to gasholders for town gas, which have supporting trusses for the dome dipping into the water of the gasholder tank. Sludge digestion gasholders must have their trusses so designed as to come above the level of the sludge, even when they are resting in their lowest position, otherwise fibrous scum may cling to the trusses and resist the rising of the gasholders. The gasholders should be provided with inspection windows having wipers operated from outside; a manhole with skirt to prevent escape of gas; a grating to prevent accidents; and gas-tight, removable panels to facilitate maintenance, e.g. removal of any detritus after some years of operation.

There is a considerable difference of opinion as to what capacity should be allowed for gasholders. Imhoff and Fair have suggested 6 h average daily quantity of gas used for power purposes, whereas Babbitt suggests a capacity of between 26.4 and 27.6 h average daily yield where all the gas is used and the demand is constant, but that where the gas yield is greater than the demand, the gasholder capacity need not be more than 9.6 h average daily gas yield. In designing the gasholders for Crossness Sewage-treatment Works, where the yield was estimated to be in excess of demand, the author allowed 6.97 h storage of average production, which was convenient for the design: when the works had been put into operation this capacity was found to be more than adequate. The gasholder capacity at the Mogden works was about 5 h average production; during 10 years operation this was found adequate.

Gas can be stored under pressure of 30–45 lb/in^2 (2–3 kg/cm^2) in cylinders or spherical tanks, but floating gasholders are far more usual; in particular, circular spiral-guided gasholders which can serve as roofs for circular primary tanks. Gasholders give gas pressures of 75–250 mm of water gauge, according to design. Where the Dorr-Oliver 'BGH' plant is installed, all the tanks can be covered with such gasholders and then single-lift gasholders should provide all the capacity necessary. Where there are floating or fixed roofs and separate gasholders are necessary, these may be of multi-lift type.

Waste-gas burner

There is a variation of consumption of gas for power purposes throughout the day and also, unless digestion tanks are regularly fed with sludge, an hourly variation of gas production. In addition, the rate at which sludge is delivered to the sewage works varies from day to day, being at times, in the case of an old combined system of sewers, as much as 20% above and below the average. Thus, allowance should be made for at least a 20% variation of gas production on each side of the mean.

The hour-to-hour variation between production and consumption is normally accommodated in the gasholder capacity; but it is practically impossible to allow for variations of production over several days. For this reason, the practice is to make up deficiencies of gas supply by the use of other fuel or power, and dispose of surplus gas that cannot be used, by burning it in waste-gas burners.

Apart from rare exceptions, waste gas burners are simple Bunsen burners of large proportions. These are easily constructed from steel tubes erected like vent columns and fitted with nozzle inlets for the gas at the bottom and provided, as in an ordinary laboratory Bunsen burner, with a suitable air inlet.

The practical design of gas burners is empiric, the problem being to find the proportions of burner which would prevent the burner from lighting back, the flame blowing off, and excessive production of carbon monoxide. In the case of burners for methane, there is little chance of lighting back because of the low flame speed of the gas, and generally rule-of-thumb designs are adopted. One approach to the design is to find the diameter of orifice necessary to pass the quantity of gas to be disposed of at the working pressure. Table 58 gives approximate figures of discharge of orifices.

Table 58   Discharge of gas through orifice

Pressure of gas (mm of water)	m^3/min per 100 mm^2 of orifice
100	0.186
125	0.208
150	0.228
175	0.246
200	0.263
275	0.281

The steel column is made three to four times the diameter of the orifice. Provision is made for varying the diameter of the orifice and for varying the air inlet in order that efficient working conditions can be found by trial. On more than one occasion engineers, referring to gas-engineering textbooks, have pointed out that waste-gas burners for methane designed according to the above rule of thumb would not work because, owing to the low flame speed of sludge gas, the flame would blow off the top of the burner. However, the fact remains that this normally does not occur.

The burner should be made sufficiently high to keep the flame well above ground level: flames of 6 m length are not unusual, and in the case of one burner 12 m high some scorching of the grass was noticed. The burner should not be within 30 m of any building or trees, and a longer distance is recommended.

Waste-gas burners can be lighted by a sparking plug of the larger type, as used for lorries, but as they are in the open air this method may not be altogether reliable. Another method is to have halyards of non-ferrous stranded metal rope and to haul lighted oily cotton waste to the top, then retire and turn on the gas. There are more elaborate methods of arranging for pushbutton control, but these can easily increase the cost of the burner to as much as twenty times the amount that should be paid for a satisfactory installation.

The minimum amount of waste gas for which allowance should be made is 20% more than the maximum estimated production.

Heat balance

A rough rule for preliminary design of primary-tank heating arrangements is to allow that about one-third of the heat available from sludge gas is sufficient for heating the sludge in the digestion tanks to the mesophilic range of temperature and maintaining it there. Another rough rule is that if the whole of the gas is used in internal combustion engines (other than gas turbines), the amount of heat that can be economized from the cooling water and exhaust gases is again approximately one-third of the total. So, on the average, it can be assumed that all the gas can be used for power purposes and the digestion tanks heated by means of the cooling water and heat collected from the exhaust gases. A gas boiler needs to be installed, of course, to provide heat for the digestion tanks at times when the power plant is not in full operation.

The temperature of the sludge has to be raised from the coldest temperature of the sewage at winter time to the lowest workable desirable temperature of the primary digestion tanks in temperate parts of America and Great Britain, which means that the temperature has to be raised from 7 to 30 °C, requiring about 22,800 kcal/m³ of sludge put into the primary tanks.

The next heat demand is the amount required to make up radiation losses. This sounds equally straightforward and simple, and the first impression gained by an engineer used to dealing with heat exchange and insulation problems would be that this calculation should be capable of being made accurately. But this is not the case, for the thermal conductivities of some of the component parts of the external surfaces of digestion tanks have not been determined by experiment. Therefore guesses have to be made, with the possibility of a considerable degree of error.

For this reason, and the fact that usually ample heat is available, the heat losses from a large number of digestion tanks in Britain have been estimated by rule of thumb based on simple observations of existing tanks. The method of testing tanks has been to operate the tanks at normal working temperature for a period, then to stop admitting fresh sludge, stop heating, and measure the fall of temperature over a period of several days. The figures so obtained, plotted on a graph, give a fair indication of the initial rate of temperature drop, but do not fully allow for heat produced during rapid digestion of freshly introduced sludge.

Such experiments have shown that the initial loss of temperature per day during the winter in temperate areas varies from one site to another between the limits of 0.28 °C and 0.83 °C. The former figure was usual for cork-insulated tanks, the latter for a tank buried completely under the ground in waterlogged subsoil. No information is available as to how long the buried tank had been in operation before the test was made; therefore it is possible that the subsoil round the tank had not been completely heated at the time of the test, which would mean that an unduly high result had been obtained. Records of heat demand at works in operation are more valuable. Figures that have been recorded in this way for large tanks are 0.4 °C loss of heat per day for embanked tanks and 0.66 °C for tanks with gasholder roofs standing almost entirely out of the ground and having no insulation.

Heat exchange theory has been used in the design of heated sludge digestion tanks, but the estimates of heat losses have tended to be lower than the figures actually recorded after construction. The general formula is

$$Q = A\left[\frac{t_1 - t_0}{(1/f_1) + (b/K) + (1/f_0)}\right] \tag{39}$$

where: Q = kcal/h;
  A = area of wall in m²;
  t_1 = °C hot side;
  t_0 = °C cold side;
  f_1 = surface coefficient, inside (e.g. water to concrete);
  f_0 = surface coefficient, outside (e.g. concrete to air);
  b = wall or insulating material thickness in metres;
  K = thermal conductivity.

Where there are several layers of different materials, the additional value of b_1/K_1, b_2/K_2, etc. have to be added to the formula.

Loss of heat is in direct proportion to the area exposed, also with the difference of temperature from hot to cold side of the material traversed; and it is in inverse proportion to the sum of the heat-resisting properties of the component materials. The heat-resisting property of a solid wall of any single material varies in proportion with the thickness of the wall divided by its conductivity, or b/K.

Where, however, a space is occupied by a liquid or gas, or a liquid or gas in contact with the surface, the transfer of heat to or from the surface is complicated by the movement of the fluid. In immediate contact with the surface is a film of fluid which, moving slowly, has considerable heat resistance, while away from the surface convection currents rapidly transfer heat. Where liquid or gases are in contact with surfaces, surface coefficients (f_1, f_0, etc.), determined by experiment, have to be used. The heat resistance at the surfaces is generally so much greater than the heat resistance in the mass of liquid or gas in the remaining space that the latter can often be neglected.

Commencing with the loss of heat through the gasholder, the main resistances to heat loss are the surface resistance of the gas at point of contact with the sludge, the surface resistance of the gas in contact with the underside of the gasholder, and the surface resistance of the air on the outside of the gasholder. It is usual to neglect the surface resistance of the water, although if scum is present this may be considerable; the metal of the gasholder is so thin and its conductivity so high that it can be neglected.

As to the values of the surface resistance of gas and air, no figure has been given for methane, and those quoted for air are varied, but the following are suggested:

Surface of contact	Reciprocal of surface coefficient $(1/f)$
For sludge to methane	0.123
For methane to underside of gasholder	0.123
For gasholder to outside air, average weather conditions	0.05125
Sludge to inside of walls	negligible
Vertical wall to outside air	0.082

Table 69 gives a few values for the thermal conductivities of solid materials. The loss of heat through the floor or walls below ground level is a very uncertain quantity. Dry earth, once it has become heated, may act as a very thick heat-resisting mass, but ground which is waterlogged, and in which is a flow of subsoil water, may take heat away rapidly.

Heat exchange calculations have to be made with regard to the transfer of heat from the heating mechanism to the body of sludge in the primary digestion tank. When the sludge is heated by fixed coils there is little motion of the sludge

Table 59   Thermal conductivities

Material	K (approx.)
Brickwork (damp)	1.12
Expanded polystyrene, fibreglass	0.03
Concrete	0.83
Fibre board, balsa wood	0.05
Cork, slag wool	0.04
Asbestos, plasterboard, oak, deal	0.14
Clay (damp)	1.03
Loam	1.13

film in contact with the outer surface of the heating pipes, also these pipes develop an external crust. For these reasons a large heating surface is essential. A recommended figure is 1 superficial foot of external surface of pipe per 10 BTU per h per °F difference between hot-water and sludge temperature, or about 49 kcal/°C per m^3. For this purpose a maximum water temperature of 54–57 °C is usual.

When sludge is heated in an external heat exchanger the rate of exchange is much greater, for the sludge is kept in motion at a velocity of 1–1.2 m/s, and this greatly reduces the film resistance. Consequently, figures of five to six times the above are allowed. Heating-water temperatures in the region of 70 °C have been used but cause deposition of boiler crust from the sludge.

Vacuum filtration

The first vacuum filters consisted of cylinders divided internally into several sectors and covered with a filter fabric of metal gauze. These were rotated to pass through liquid sludge while a partial vacuum was applied to the insides of those sectors which were below the sludge and rising above it, so as to cause a coating of sludge to adhere to the gauze and to have water withdrawn from it. When the sectors had risen to a higher level the negative air pressure was altered to a slight positive pressure to loosen the sludge cake, which was then removed by a blade fixed near, but not touching, the gauze.

Prior to filtration the sludge from trickling-filter works was dosed with ferrous sulphate and lime[21] in the proportions of about 5% and 10% respectively of the suspended-solids content of the sludge. Vacuum filtration was sometimes preceded by elutriation, or washing the sludge with several times its own volume of clarified effluent in circular, upward-flow tanks so as to remove the finer particles. This, however, did not fully overcome the chief difficulty experienced with vacuum filters, which was choking of the filter fabric. In laboratory tests, with which the author was concerned, it was found that any filter fabric from fine gauze, cloth or filter paper to 2.5 mm gauze would become almost completely choked after six successive uses, but if it was turned over and back-washed it could be given six more uses before choking, and this procedure

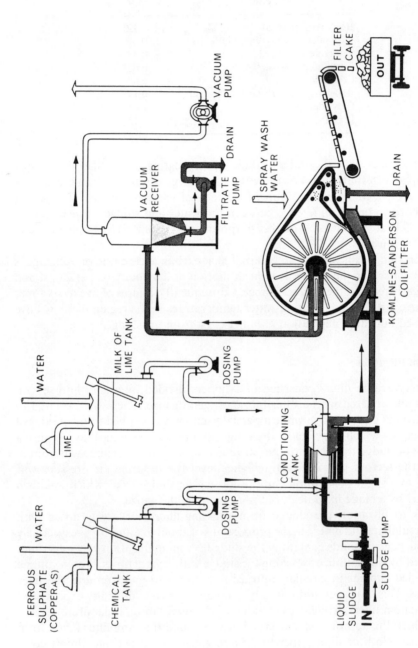

Figure 89 Komline-Sanderson coil filter (Courtesy Dorr Oliver Co.)

could be continued indefinitely. The difficulty was to find a practical way of providing this back-washing. However, the Komline-Sanderson Coilfilter, which is supplied by Dorr-Oliver of the United States and Great Britain, came on the market, and this has solved the problem.

The Coilfilter consists of two layers of small-diameter stainless steel helical coils. Each individual coil has its ends connected together to form a continuous loop that is carried round a filter drum and over guide rolls. The coils in the lower layer lie side by side but not touching, and those in the upper layer lie on the lower coils above the spaces between them. As the coils pass through the sludge and rise up from it they are round a drum to which a partial vacuum is applied. Then they leave the drum and part company just before the sludge cake is dropped off and again come together for the next cycle (see Figure 89). While the coils are apart they are washed by a sparge which cleans them so that their ability to filter the sludge is fully restored.[26] Coilfilters are manufactured in several sizes from 17 to 170 m² nominal, the nominal area being the effective width of the drum multiplied by the diameter multiplied by π. Diameters vary from 1.8 to 3.5 m, and the width from 1 to 5 m.

It is claimed that when handling undigested sludge from trickling-filter schemes the filters separate 15 kg of dry solids/h per m² of filter area. For deciding the size of a filter for works that are run on day shift only it is usual to allow one shift per day for 5 days per week and a net time of 7 h filtration per 8 h shift. The filter cake has a moisture content of about 80% or a little higher.

Centrifuging

While vacuum filtration has recently been the most popular method of sludge dewatering for use after either heat treatment or sludge digestion, there is an increasing interest in centrifuges for this purpose. Centrifuges do not require much space. The percentage of dewatering depends on the particular design.[22,23]

Rotary disc filters

This type of filter provides a much greater filtration area in a given space than does a drum filter. Recent trials indicate that it should be competitive in cost with drying beds on smaller works, besides being independent of the weather.[24,25] Conditioning with chlorohydrate, filter cakes of 22–29% dry solids were obtained at filtration rates up to 12 kg dry solids/m² per h with good solids recovery. In view of its compact structure portable units to service several small works can be adopted.

Sludge-liquor pumping station

The duties of the sludge-liquor pumping station vary. At some works a pump is installed to deliver humus back to the flow of sewage or to the sludge-drying beds, and this pump may also lift sludge liquor to the incoming flow of sewage through the same rising main as is used for humus. Where humus tanks are of the flat-bottomed type and are sludged by completely emptying, the suction well need not be very deep. Where humus tanks are of the pyramidal type, humus may be withdrawn under hydrostatic head to a comparatively shallow well; alternatively, the well may be made deep in order that the tanks may be completely emptied. Wherever practicable, primary sedimentation tanks should be arranged for complete emptying and the wash-out drains passed to the suction well of the humus pumping station. Thus, this pumping station becomes general wash-out, sludge-liquor and humus pumping station, taking all drainage that may be, or should be, returned to the incoming flow of crude sewage.

The capacity of the pump installed is arbitrary, between limits. The pump should not be so large that, at small works, it makes an undue increase of rate of flow to the sedimentation tanks. On the other hand the pump should not be small that it is unable to produce a self-cleansing velocity in a 100 mm rising main.

It is probable that the flow due to rainfall on the drying beds will need a higher pumping rate than that required for other purposes, though a larger unit installed for this purpose will often assist with tank-emptying operations. As an alternative to an enlarged pumping sump it may be possible to arrange for the larger pump to deliver through a separate rising main to a land treatment area and deal with the precipitation on the drying beds in this way. Sometimes pumps for this duty may be placed at higher levels than the others, thus creating emergency storage capacity to deal with the rainwater run-off.

If there is a sewage-pumping station at the works, a sludge-liquor pumping station may not be necessary.

The capacity may be considered adequate if it can deliver:

(1) 1 day's flow of sludge liquor in 6 h;
(2) the cubic contents of one sedimentation tank in 6 h;
(3) the cubic contents of one humus tank, or where humus tanks are sludged by complete emptying, the maximum flow produced on any one day by so sludging the tanks in 6 h;
(4) the rainfall run-off from the sludge-drying beds.

This last item is variable, dependent on the storage capacity of the suction well. It might not be practicable to allow for pumping the run-off due to the most intense short-period storm likely to occur. In this connection formula (15), (16) or (17), pp. 91–2 may be used as appropriate.

Heat processes

The Porteous process

Some years ago a good deal of interest was aroused by a process developed by W. K. Porteous.[27] This involved raising the temperature of liquid sludge and thereby breaking down its colloidal structure, thus reducing the water affinity of the solids and releasing much of the intercellular water. The effect is a time–

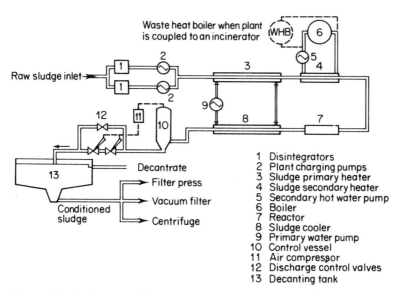

Waste heat boiler when plant (WHB) is coupled to an incinerator

Raw sludge inlet→

Decantrate
Filter press
Vacuum filter
Centrifuge

Conditioned sludge

1 Disintegrators
2 Plant charging pumps
3 Sludge primary heater
4 Sludge secondary heater
5 Secondary hot water pump
6 Boiler
7 Reactor
8 Sludge cooler
9 Primary water pump
10 Control vessel
11 Air compressor
12 Discharge control valves
13 Decanting tank

Figure 90   Continuous sludge heat treatment plant, Porteous type (W. E. Farrer)

temperature phenomenon, allowing higher temperatures to be attained under pressure and thereby reducing 'cooking' times. These are of the order of 20 min within the temperature range 180–200 °C, at a pressure in excess of 1 MN/m². All types of sludges may be conditioned by this method,[28] reducing the specific filtration resistance of the sludge by up to three times normal magnitude to about 10^{11} m/kg at 49 kN/m². The system is now operated as a continuous process, and heating may be achieved by direct steam injection or indirect systems using hot water or hot air jacketed heat exchangers (see Figure 90).

The process offers big economies in space, and about 30 plants[29] are operating in Great Britain. Because some of the solids content is rendered soluble, the process produces strong liquors (sometimes in excess of 10,000 mg/l COD and resistant to biodegradation). Efforts are being made to deal with these by means of multiple-effect evaporators, or activated carbon. Odour control by scrubbers or after-burners is essential. The resultant thickened sludge is usually dewatered in filter presses to a moisture content of about 40% and the final product, being sterile, may safely be used as a fertilizer, free from bacteria and weed seeds.

Heat-drying and incineration

In the multi-hearth incinerator comparatively dry sludge cake passes through a furnace which consists of a number of rotating hearths, one above another, alternating with a number of fixed hearths. The cake is raked so as to fall from one hearth to another until it is dried and eventually consumed. Gas or oil are used as fuel, but the heat produced by the sludge itself, in burning, greatly reduces the fuel bill.[30-32]

In the suspension-type drier sludge is mixed with dried sludge and passed through a flash drier, afterwards being collected as dust in a cyclone dust collector. Part of this dried sludge is used for mixing with wet sludge, and part is drawn off in sacks to be used as fertilizer or else to serve as fuel to aid the furnace. The exhaust air from the cyclone is passed through the furnace.

Fluidized-bed furnace

With the addition of heat the fluidized bed, already recognized as a filtration process (see Chapter 15) has application in the sludge treatment and disposal field. A refractory lined cylindrical combustion chamber has a 1 m deep layer of graded coarse sand (2–5 mm) supported on a heat-resistant grid. The bed is maintained in a state of turbulence by injecting combustion air at a pressure of 300–500 N/m^2 into the sand layer, along with the sludge. Auxiliary burners may also be directed into the bed where the resultant mixing in the hot sand evaporates the water and many organic compounds which burn at the surface of the expanded bed. The high thermal capacity ensures rapid heat transfer, and the abrading effect of the sand on the sludge particles favours complete combustion of organics with minimum clinker build-up. The method relies on the stripping of all fine ash from the bed by through-flow of gases, which after some heat recovery are cleaned by cyclones and scrubbing units.[33]

Dickens, Wallis and Arundel[34] give an interesting description of the fluidized-bed incinerator plant at Esher, Surrey, which deals with the sludge from a total of 114,000 population. Ram pumps raise the sludge into one of four thickening tanks, equipped with wedge-wire dewatering cells. Each cell is attached to the side wall of the tank and is wedge-shaped in section, tapering towards the bottom. The 2700 × 860 mm wedge wire screen is located on the inside slope of the wedge. When sludge is pumped final/effluent 'support water' is automatically discharged into the cell until settled liquid levels either side of the wedged wire screen are approximately equal. Once the tank is full the dewatering cell is switched from 'fill' to 'dewater', causing a pump to lower the support water level by 75 mm. Support water frees liquor from the sludge, to pass through the screen into the cell until the levels are almost equalized. The dewatering pump then restarts, further lowering the support water level. Continuation of the cycle results in about 15% reduction in sludge volume. A variable-speed vibrator helps to prevent blending and the tank contents is kept in motion by a vertical bar-type slow-speed stirrer. The retention period is 24 h.

The thickened sludge is pumped up and mixed with a polyelectrolyte conditioner in a baffled mixing drum prior to discharge to the vats of two coil filters. A slow-feed swing agitator prevents separation of solids and maintains a uniform sludge. The coil-filter drums are 3.5 m diameter and 4.26 m long, with a total surface area of 44 m^2. Vacuum is provided by a Mash-Hytor pump. The filtrate is separated in a vacuum receiver chamber and pumps to the sludge plant drainage system.

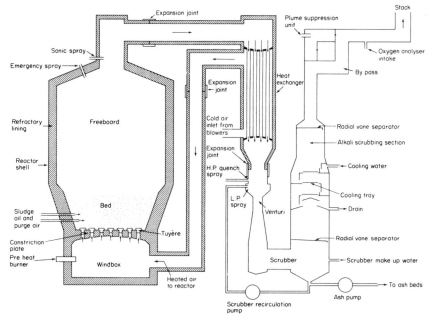

Figure 91 Diagrammatic section thro: incinerator, heat exchanger and gas scrubbing plant.

Sludge cake is discharged onto a reversible conveyor discharging either to the incinerator or to skips. The cake is monitored by a closed circuit TV system.

The BOD_5 load from the plant is about 12% of total daily average BOD_5. Ash residue represents about 20–23% of the sludge dry-solids content. It is almost completely inert, and although metal concentrations are high they are in completely stable form.

The Zimmerman process

This is a method of oxidation of organic matter with air while it is dissolved or suspended in water. The combustion produces heat in the same manner as if the dry matter were burnt in free air and this heat, which is also necessary for the purpose, can be utilized for the production of the power required to drive the mechanism. To obtain the necessary temperature, the liquid has to be under considerable pressure and consequently the plant involves pumps and also heat exchangers for economizing the heat.[35,36]

By co-operation between Stirling Drug Inc. and the Metropolitan Sanitary District of Greater Chicago, a Zimmerman process pilot plant was installed to treat a daily capacity of 1.8 tonnes of sewage sludge solids.

The sludge was heated in storage tanks to a temperature of 87 °C and pumped, first by a rotary pump and then by a high-pressure pump. Next, air was delivered into it by compressor at a pressure of 84 kg/cm², and the sludge and air passed to a preliminary heat exchanger in which steam was used to raise the temperature during the start-up only, then to a second heat exchanger which raised the

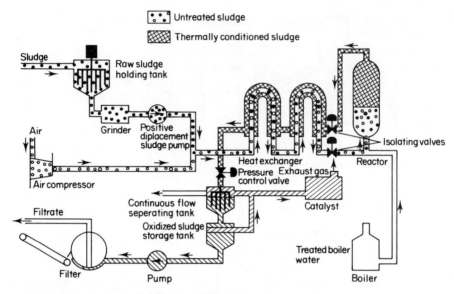

Figure 92   'Zimpro' wet air oxidation system (Courtesy Satec Ltd.)

temperature to 194 °C, and finally to a third heat exchanger where the temperature was raised to 238 °C. Heat exchangers 2 and 3 were economizers, which transferred heat from the liquor which had undergone the combustion process.

The hot sludge was passed from the third heat exchanger into the reactor, a pressure vessel. It entered this at the bottom and was oxidized as it passed upwards through a coarsely baffled path in a central body; then it passed out at the top and down through an outer shell. Next the liquid, burnt products, gases, etc., at a temperature of 268 °C passed through heat exchangers 3 and 2 in turn, dropping in temperature to 260 °C and 227 °C respectively, afterwards to be cooled by a water-cooled heat exchanger to 165 °C. In the final stage of treatment the gases were vented to atmosphere and the ash settled out and pumped to lagoons.

The method has proved itself and, slightly modified, is marketed as 'Zimpro' by Zimpro, Inc., Rothschild, Wisconsin 54474, USA.[35,36] There have been several installations in the United States and Canada. In Britain the equipment is manufactured under licence by Satec Ltd, PO Box No. 12, Weston Road, Crewe.[37] The first British plant has been constructed for the Guildford Rural District Council and another is proposed for Thurrock Urban District Council. The process has also been installed in Holland.

Recent practice in the United States

The methods of treatment described in the foregoing pages of this chapter refer, in particular, to British practice, although some of the processes have either originated, or been developed, in the US. However, the object of the chapter dealing with one of the most important aspects of the subject would not be achieved without a clear comparison with present practice in the US, which is admirably described in the recent volume edited by Gehm and Bregman,[38] from which much of the matter under this heading has been drawn.

Gravity thickeners

Gravity thickeners used in the US are simple types not involving washing or other processes, operating on either a 'fill and draw' or 'continuous' basis.[39–40] Hopper slope requires to be 60°. For continuous operation minimum detention should be 4 h; surface area is computed in the same way as for mechanical units. Mechanically operated thickeners are usually of the 'picket-fence' type (Figure 93). The bars suspended vertically to the collector rake assist the thickening process by applying gentle agitation to the slurry under compression and compaction, thus permitting the separation of more water from the sludge. It is important to provide good entry flow distribution and velocity dissipation arrangements. Adequate collector mechanism and collection sump, with properly sized underflow piping and overflow weirs are essential.[41] Tanks should be designed for a sludge volume ratio of two.[42]

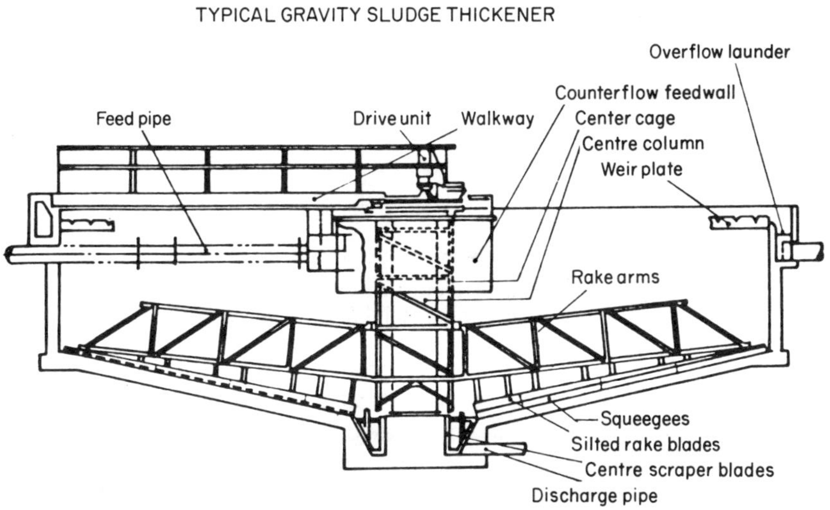

TYPICAL GRAVITY SLUDGE THICKENER

Figure 93   Typical gravity sludge thickener (Courtesy Dorr-Oliver Inc.,)

Fluctuating solids content in the underflow of mechanical thickeners, ascribed to cyclical effects within the compression and compaction zones, has a serious effect on performance.[43] This phenonemon is observed with mixtures of raw and modified activated sludge, other organic sludges, and inorganic sludges. A method employing the sludge volume index[44] has been devised as an operating control measure to minimize this effect. Experience shows that

414

deep circular units equipped with the 'picket-fence' mechanism are best for thickening very dilute and gelatinous sludges.[45]

Experiments with various chemicals in attempts to control gas evolution (which has an adverse effect on thickening) indicate them to be of little value for this purpose. Coagulants and weighting agents similarly have little effect.[46] Polyelectrolytes, although they have not been found to enhance the thickening of sewage sludge appreciably, frequently enhance overflow clarity[47-48] (see Figure 94).

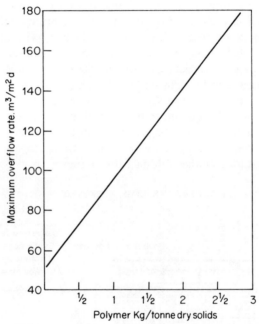

Figure 94   Effect of a polymer on thickening (Van Nostrand Reinhold)

Mechanical thickeners for organic sludges usually have detention times between 4 and 8 h, with solids loading 10–80 m^2 of surface area per tonne of dry solids per day. Hydraulic loadings may be from 160 to 200 l/m^2 of surface area, depending on the compression and compaction characteristics of the particular type of slurry. These loadings would be about doubled for mixed sludges except for the solids rate which could be considerably higher.

Underflow from thickeners is usually by positive displacement using a diaphragm or screw pump. Underflow pipe sizes should be conservatively selected, with straight-line flow if possible.

Air flotation

The dissolved-air flotation process[49-50] (one type of installation has already been described (see p. 391) has been applied for thin slimy sludges. Figure 95

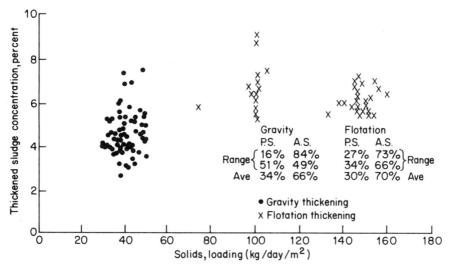

Figure 95 Comparison of gravity thickening and flotation thickening of mixtures (Van Nostrand Reinhold)

compares results of dissolved air and gravity thickening for mixtures of activated and primary sludges, Table 60 gives data for a number of different sludges. In Chicago such results were improved by the addition of polymer (5 kg polymer/tonne dry solids).[51]

Other flotation methods

In a gas flotation process[52] devised for thickening digested sludge prior to spray drying, sludge was pumped to a decantation tank along with an alum solution, which served both as a coagulant and acid reactant, liberating carbon dioxide from the bicarbonate present in the sludge. Final concentrations approaching 15% were obtained, and sludge consistency increased two to threefold.[53] However, the chemical cost was far too high.

Centrifuge thickening

Very hydrous sludges can be thickened by centrifuging.[54] Vertical disc-type centrifuges produce a clear concentrate: from less than 1% up to 6–12% solids content. The process allows solids to be further dewatered by other methods.

Sludge conditioning

Most techniques have been practised for many years, but some new ones are now being applied.[37,55] Conditioning commonly preceded dewatering on sand beds, pressure filters, centrifuges, also 'wet' machines and some special devices

Table 60   Flotation thickening of organic sludges

Type of sludge	Consistency (%)	
	Initial	Final
Activated sludge	0.5	4.0
	0.8	6.5
	0.5	4.6
	0.8	3.7
	0.8	4.9
Activated and primary	0.6	8.6
	2.3	7.1
	1.9	7.4
	1.3	6.1
Chemical coagulated	0.8	4.1
	0.5	4.1
Paper mill sludge	2.1	4.0
	1.9	3.9
Boardmill sludge	2.0	4.0
	2.1	3.8
	1.7	4.1
Boardmill activated sludge	0.9	4.0
	0.8	3.9

employing drainage and pressing. Conditioning also gained impetus as a result of the application of vacuum filtration.[56–60]

Chemical conditioning

Conditioners may be inorganic or organic, active or inert. Among chemically active inorganics trivalent cations are most effective. Ferric chloride and alum are most widely used. Aluminium chloride and chlorohydrate, ferric sulphate and copperas, chlorinated to make the iron trivalent, have all been used. Sulphuric acid and sulphur dioxide are effective for sludges high in fats and greases. Small plants use dry materials, dissolved for application either in mixing tanks or in the discharge of dry feeders. Large plants use solutions since their tank storage makes liquid handling and feeding easier.

The next most widely used conditioner is lime, used as a supplementary conditioner, in the ferric chloride process for raw and digested sewage sludges. For small plants calcium hydrate in bags is used for the preparation of slurry in agitated tanks. Large plants use calcium oxide in bulk from rail cars or trucks, converted into calcium hydroxide in slakers and passed to agitation tanks.[61,62]

Two new processes employing active inorganic chemicals are under development. One subjects sludges to chlorine, under pressure ($2–3 \ kg/cm^2$) for a few minutes, followed by neutralizing with lime and settling.[63,64] The second is

Table 61   Chemical conditioning

Type of sludge	Percentage dry solids		
	Al_2O_3	$FeCl_3$	CaO
Raw sludge	0.7–1.7	5.0–12.0	0–10
Digested sludge	2.1–5.1	8.0–16.0	0–15
Elutriated sludge	0.6–1.8	4.0– 8.0	—
Activated sludge	—	10.0–20.0	—

primarily for slimy sludges, either activated or from biological processes. The sludge is heated under pressure in the presence of a 0.5% concentration of sulphur dioxide, and partially hydrolysed. After treatment solids are readily filterable and the filtrate is a raffinate, which could be sold on the molasses market.

Active organic conditioning agents are for the most part synthesized long-chain polymers,[65–68] many acrylic. Complex carbohydrates and proteins, starches,[69] sugar gum,[70] and glue[71] are also used. Most of these developed as aids to clarification but some show strong coagulating activity, either alone or with iron salts or alum.[72] Synthetic polymers are anionic, cationic, or non-toxic according to the predominant charge they carry. Natural materials are generally non-ionic. Active organics are not as widely used as conditioners but their application is increasing. They offer the advantage of small volume handling, simple feed, and sometimes lower cost.[73] They usually improve cake retention on centrifuges and filters.[74] Observed dosages from seven works were 0.14–1.56% on dry solids,[75–78] with filtration rates from 12 to 40 kg/m² per hour.

Organic dewatering agents are usually furnished either in dry form or in strong solutions. In vacuum filtration metering pumps are generally used for feeding to the mixing tank. For centrifuge, feeding is direct to the bowl through a small diameter pipe. Performance cannot be accurately predicted and must be established by trial and error.

Of inert materials, fly ash has had most attention. It is mixed best by addition to slurries prior to dewatering.

The most effective use of inorganics has been as precoats to filters. Diatomites are most commonly employed, but the ratio of precoat to cake solids is about 5 to 1, making it a costly operation.

Elutriation is the commonest physical process, used only on anaerobically digested sludge, as an accessory process to chemical conditioning.[79–82] Digested sludge washed with effluent or other water is used to reduce the concentration of bicarbonates, and so prevent their reaction with the coagulant, which hinders its effectiveness.

Three different methods are employed in washing: (1) a single-stage system, (2) a two-stage system using two thickeners in separate washings, (3) a counter-current system, using two thickeners, in which the wash-water is introduced into the influent of the second thickener, and overflow from this unit is used as wash-water in the first one. The following equations determine the reduction in bicarbonate alkalinity.

$$\textit{Single-stage} \qquad \textit{Double-stage} \qquad \textit{Counter-current}$$

$$E = \frac{D + RW}{RH} \qquad E = \frac{D + W(R + 1)^n - 1}{(R + 1)^n} \qquad E = \frac{E + (R^2 - R)W}{R^2 + R + 1} \quad (40)$$

where: D = sludge liquor alkalinity (mg/l);
$\quad\quad\;\; E$ = elutriated sludge alkalinity (mg/l);
$\quad\quad\;\; W$ = elutriated water alkalinity (mg/l);
$\quad\quad\;\; R$ = ratio of elutriating water to liquor.

Elutriation eliminates the need for lime conditioning and alters the ferric salt and alum requirements.[83] Elutriated sludge has been successfully dewatered with polymers with as much as 200% increase in performance.[84,86]

Heat conditioning

The first full-scale operation was on a raw sludge heated to 143–188 °C with live steam,[87] held for $\frac{1}{2}$–$\frac{3}{4}$ h then evacuated from the reactor through heat exchangers, transferring heat to the incoming sludge. The heated sludge was dewatered on filter presses at 1400 kPa, with a sludge averaging 5% solids, the resultant cake averaged 52% solids, and was suitable for use as a fuel without supplementation. Performance could be enhanced by the use of chemicals, probably because of the high grease content.[85] Two other heat processes used are the Porteous and Zimpro (already described) and a third heats activated sludge in a heat exchanger.[88] The sludge is claimed to settle to one-quarter of its original volume and will centrifuge to cake of 25–40% solids content, without chemical additives. Heat treatment of some sludges is attractive as compared with conventional digestion and bed drying, and offers considerable saving in space.

Freezing and thawing

The first experiments actually took place in England in the 1950s.[89,92] Slow and complete freezing is necessary, but the rate of thawing is immaterial. Cakes have 23–30% solids content on vacuum filtration after thawing. The drainage rate can be raised by the use of chemicals, but the process is too expensive. More recently patents have been taken out in Poland[90] and in England.[91]

Filtration

Filters of the early vacuum drum type have generally been replaced by those of the bell type. These—whether continuous belt, wire, plastic filter media, or coil springs—afford superior cake removal and constant clean filtering surface. Vacuum filters may have 'cloudy port' compression roll, and precoat devices.

Compression rollers increase incrustation removal. Precoat devices deposit a thick layer of diatomite on a conventional drum, rotating under vacuum. Slurry is fed into the vat and a doctor blade removes a thin layer of filter aid with the deposited solids. The blade moves progressively nearer the drum until the layer of filter aid is no longer effective. The filter is then cleaned and the process repeated. The process is useful for oily and other difficult sludges.

Filters are available in a variety of materials depending on corrosion-resistance requirements. Vats may be in wood, mild steel, rubber, fibreglass, lined steel, and stainless steel. Filter drums and their decks are of similar materials. Drainage grids are commonly fabricated from plastics or rubber composition, in readily replaceable prefabricated sections. Piping is stainless steel or plastic. Vat agitators are generally designed for 13.5 strokes/min, but the necessity for their use, and the most desirable speeds, are determined by the nature of the slurry and the operating technique.[93] Whilst they tend to secure an even cake thickness they can cause deterioration in conditioning, particularly with ferric or aluminium chloride. Self-priming filtrate pumps are in common use, whilst vacuum is supplied by wet vacuum pumps which have largely replaced the old dry type, or by using a barometric leg. Vacuum breakers are almost universally used. For predicting filter performance, the filter leaf test and the Buchner funnel test 113 are usually employed. Figure 96 compares cycle

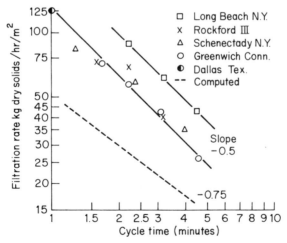

Figure 96 Filtration rate versus cycle time (Van Nostrand Reinhold)

time and filtration rates. Whilst the trend of the field data curves is parallel to the ideal, their slope is -0.5 rather than -0.75. The difference is ascribed to variation in individual sludge characteristics. Performance depended on cake porosity, and this varied widely. Cake moisture is a function of drying cycle time, although depending on the degree of vacuum and conditioning. Vacuuming should be maintained at a maximum during the cycle. Dual-vacuum receiver systems are recommended, to prevent large volumes of air being drawn through the cake. The method of coagulant addition and the entrained gas or air in the sludge may have a detrimental effect. It is advisable to pump sludges with the minimum of air entrainment. For sludges which contain gas, such as digested sludge, open head-boxes or agitation tanks should be provided in the feed system to aid removal. Conditioners (such as lime, ferric chloride or polymers) are necessary when activated sludge is involved, as this has a decidedly adverse effect on dewatering.

Pressure dewatering

So far filter processes have been applied only for dewatering heat-treated primary sludge, operating on a 2–4 h cycle at 1400 kPa, with a load concentration around 20% solids, to produce a cake about 50 mm thick with 45–55% solids. The process is not considered economical for dewatering digested or chemically conditioned sludges, owing to the long cycle time.

Compression rolls

These are not used directly on vacuum filters handling sewage sludges, except for the Roto-plug machine,[95] which has a mesh-covered drum and uses gravity filtration to pick up a cake which when doctored from the drum in wet form is dewatered on compression rolls. Wet cake leaving the drum at 8–10% solids content is pressed to 19–26% solids content in two stages.

Presses

Although used on sludges of a fibrous nature, presses are not used for handling sewage sludges.

Incineration

Incineration of sludges has been a common practice since 1935, and has become increasingly popular. Odour problems can be practically eliminated by suitable equipment.

Moisture and combustible content are major factors. Table 62 compares important parameters of sewage sludges.[96] The commonest types are flash-drying systems, multi-hearth furnaces, fluidized beds, and wet air oxidation units. All these systems involve drying and burning. Sludge is dried to a cake containing, on average, 3 kg of water/kg of dry solids. The latter then supply sufficient heat value to evaporate the water content. The drying process can be either separate, or combined with oxidation. The overall process raises the sludge temperature to the boiling point of water, which is evaporated off as vapour, at temperatures rising to that of the greases present. The combustible content of the sludge is also heated to ignition point. Excess air is needed to

Table 62   Important parameters of sewage sludges

	% Combustible	% Ash	kJ/kg Dry solids
Grease and scum	85–95	5–15	35,000–40,000
Fresh solids	75–85	15–25	21,000–25,500
Fine screenings	80–90	10–20	18,500–23,000
Digested sludge	55–60	40–45	11,500–14,000
Grit	30–40	60–70	7000– 9500

achieve complete oxidation, but its quantity must be carefully controlled as it tends to reduce burning temperature and increase heat losses. The type of equipment, the sludge characteristics, and the disposition of the off-gases all influence excess air requirements.

A proportion of the heat produced is lost in stack gases, but the remainder is often sufficient to sustain combustion once the incinerator has been brought up to temperature. The major end-products are carbon dioxide, water, sulphur dioxide, and ash.

Multiple-hearth units

These are probably the most commonly used,[97–101] and are an adaptation of the furnaces used for roasting ores. They consist of a refractory lined vertical steel shell with 4 to 11 refractory covered trays, or hearths, surrounding a central rotating shaft with air-cooled rabble arms raking the sludge across each hearth, breaking it up and exposing it to the air supply. Auxiliary fuel burners (oil or gas) are located in the shell surrounding the burning hearths. The flow of sludge cake is downward; heating and drying occurring in the upper hearths, incineration in the middle ones, and ash cooling at the bottom. Shaft cooling air is recycled to the bottom hearths. Temperatures are 540 °C at the top, 870–980 °C in the middle, and 310 °C at the bottom. The off-gases are sometimes heated to 675–815 °C to destroy odours; an expensive but effective process. The ash is removed mechanically, pneumatically, or hydraulically. The latter method is most satisfactory if the ash is disposed of in lagoons—probably the commonest method. When it is stored in basins and hauled away, dry conveyors are the most efficient form of transport. To prevent explosions separate feed openings are sometimes provided for grease and screenings. When used for grease a parallel flow of feed solids and hot gases should be provided.

Flash-drying and incineration

These processes have declined in popularity as there is no longer a market for sludge as a soil conditioner. The process is complex, and suffers from pollution potential and explosion hazards.

Fluidized-bed incineration

For sewage sludge a bed of hot sand is maintained in suspension above a distribution plate, by air from a plenum blower beneath it. The bed is heated to 649 °C with oil or gas, and the sludge introduced directly into it by screw or piston-type conveyors. Combustion is then self-supporting, the sludge being rapidly broken up, dried, and burned within the bed by the intense heat and violent mixing conditions.[102-107] The agitated bed retains organic particles long enough to produce complete oxidation at 650–815 °C. Fine ash is blown from the bed by the fluidizing air stream at a pressure of about 5 kg/m². The off-gases are scrubbed in wet units and the ash is disposed of on land (Figure 97). Combustion proceeds to a point where the gases are odourless and no

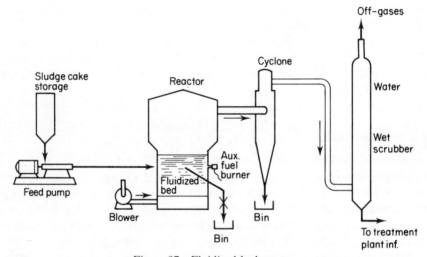

Figure 97   Fluidized-bed reactor

smoke is produced, when 10–15 % excess air is introduced. Raw primary sludge has been burned without difficulty in a residential area. Another advantage is the considerable period during which the sand retains its high temperature after shut-down—the drop is about 65 °C in an idle day, and restarting occupied only a very short period, thus the auxiliary fuel requirement is a function of shutdown time.

Sludge drawn from gravity thickeners is generally degritted and dewatered by centrifuge to a solids content of 20–30 % before being introduced to the bed. Operating power is estimated at about 250 kWh/tonne of dry solids handled. Capital and operating costs are considered to be less than those for sludge digestion systems.

Wet air oxidation

As already described the Zimmerman or Zimpro process[108-109] originated in the US. The degree of oxidation achieved is directly dependent on the degree of heat and pressure applied. In some cases a high degree (90 %) of oxidation is practised, the products being inorganic ash and a liquor consisting primarily of acetates. In others, a partial oxidation is achieved, destroying the putrescible compounds and leaving a carbonaceous easily dewatered residue, suitable for disposal on land.

Incineration of oily and greasy sludges

Such sludges are incinerated easily in fluidised bed units, but require special handling in multiple-hearth incinerators. Scum burning operations at Minneapolis–St Paul use a cascade-type incinerator which is equipped with cast-iron tuyers cooled by combustion air and a secondary chequered brick combustion chamber.[111] The unit starts at 705 °C after preheating by oil-burners, and operates at 871 °C burning some 500 kg dry solids/h.

Sludge disposal

Disposal on land

The simplest means of disposal of sludge at very small sewage works is to dig trenches into which the sludge is discharged. The trenches are then refilled and new trenches dug. At somewhat larger works less labour is involved if land is ploughed, sludge discharged on to the surface so as to fill the furrows and finally, when the sludge is dry, the ground harrowed or reploughed. About 0.4 ha of good land is sufficient to deal with the sludge from 5000 head of population.

The distribution of wet sludge to farm land by pipeline is a very simple means of getting rid of sludge, provided that there is sufficient land belonging to the sewage works for this purpose, or that nearby farmers co-operate in receiving the sludge as and when it is discharged. The method is little used in British, but disposal of sludge to fill low-lying land is not uncommon in America. There has, however been a tendency to return to method of disposal to farmers, usually by employing a tanker vehicle and spraying the sludge on the land.

Dumping at sea

This method of sludge disposal is, of course, limited to those localities where it is possible for wet sludge to be delivered into ships. For large works special ships are constructed for the purpose and it is necessary for the sludge to be transported sufficiently far out to sea for there to be no danger of its washing back to the estuary or foreshore.

Jetties are provided where ships can moor at high tides. Arrangements are made for sludge to be either pumped or gravitated from overhead storage tanks, so that ships can be filled in about an hour.

At the discharge points on the jetty are hinged arms controlled by penstocks and arranged so that they can be swung out over the loading hatches of the ships which, at commencement of receiving cargo, may be about 4.25 m above water level. From these loading hatches valves connect to the various storage tanks in the ship and are used during loading so as to trim the cargo. At smaller works tugs and lighters may be used.

So as to reduce the amount of sludge to be shipped, it is advisable to dewater the sludge by decanting liquor, and sludge digestion can be used to facilitate this process.

Sludge utilization

Sludge can be used as a manure[112-114] and the gas produced by digestion can be used for the production of power and in chemical industries. There has been a noticeable amount of lay writing on the need for full utilization of sewage sludge as a fertilizer, on the argument that if sludge is not returned to the land, the land will eventually become impoverished. Most of this writing ignores the facts that:

(1) sludge is sold or given to farmers, and that greater quantities would be used on the land if the farmers would accept it;
(2) the reasons that sludge is not too popular with farmers are that it is not a high-grade manure, it often contains grease in sufficient quantity to injure the land, and it frequently contains weed seeds and sometimes poisonous substances such as lead, cadmium, copper, zinc, and nickel which, even as traces, are harmful to plants, men, or livestock;
(3) sewage and sewage sludge used as manure can cause infection.

A good deal has been written on the composting of sewage sludge with household waste or other material having some manurial value. But tests have shown that the resulting product has no added manurial value to that of the individual materials before composting. There are also disadvantages of a practical nature: on the one hand the oil in the sewage sludge may spoil the compost to some extent; on the other hand, household wastes often contain broken glass which makes them unsuitable for spreading on fields.

The duty of the sewage-works operator is to dispose of sewage at the least cost and in the most sanitary manner. If he can do this and, at the same time, produce a fertilizer, he should do so; but he should not, for the sake of producing a fertilizer, either involve the ratepayer in excessive expenditure or introduce any danger to public health. It is from this point of view that sewage-sludge utilization should be studied.

The qualities of different kinds of sludge vary considerably, not only from town to town as a result of local trade wastes, but also according to the method of sludge treatment.

Raw sludge

Raw sludge from primary sedimentation tanks is most commonly available for disposal to farmers, but it is the least satisfactory as a fertilizer. The decomposition of raw sludge in the soil tends to cause acidity; industrial wastes and grease may poison the soil and crops; the sludge is not humified and therefore may interfere with the development of the crops. Raw sludge contains more pathogenic bacteria than any other form of sludge.

Digested sludge

Digested sludge has a lower nitrogen content than raw sludge. Nevertheless, it is more suitable for putting on land because it is humified and fibrous; the grease content is reduced and the sludge does not contain as many pathogenic bacteria. It has little odour.

Activated sludge

Activated sludge is difficult to dry separately on drying beds, but when so dried it is of greater value for manurial purposes than digested sludge, over which it has the advantages of higher nitrogen content and lower grease content.

Percolating-filter humus

Humus is a good manure, for which reason it should be dried separately where there is any demand for fertilizer.

Heat-dried activated sludge

Where there is a real need for the production of manure, the method of heat-drying activated sludge is the most effective, for the product of this process has a good nitrogen content, is believed to be free from harmful bacteria, contains comparatively little grease, and is free from weed seeds. It must, however, be kept dry until wanted for use, and generally the cost of production is prohibitive.

In considering the above methods of utilizing sludge, the danger of spreading disease should always be kept in mind.[115,116] It can be said that, generally, too little consideration has been given to this danger in this country and therefore the following rules, laid down many years ago in California, might be studied and applied by sewage works managers.

(1) Raw sewage containing human excrement shall not be used for irrigating growing crops.
(2) Raw or undigested sludge shall not be used for fertilising growing vegetables, garden truck (market-garden) or low-growing fruit or berries unless the sludge has been kiln dried, bed dried or aged in storage for at least 30 days, well digested (i.e. odourless, readily drainable and containing not over 50% of the total solid matter in volatile form).
(3) Settled or undisinfected sewage effluents are banned for growing vegetables, garden truck, berries or low-growing fruit, or for watering vineyards or orchards where windfalls or fruit lie on the ground. Such liquid may be used for watering nursery

426

stock, cotton, field crops such as hay, grain, rice, alfalfa, fodder corn, cow beets and fodder carrots, provided no milch cows are pastured on the land when moist with sewage, or have access to ditches carrying sewage. Use is also permitted for growing vegetables used exclusively for seed purposes.

References

1. Bolitho, V. Economic choices of sludge treatment and disposal. *Water Pollution Control*, **72**, pt. 2 (1973).
2. Burley, M. J. and Bayley, R. W. Sludge disposal strategy, processes and costs. *Water Pollution Control*, **76**, pt. 2 (1977).
3. Vosloo, P. B. B. Design of drying beds for surplus activated sludge. *Water Pollution Control*, **77**, pt. 1 (1978).
4. Simpson, J. R. and Gilbert, J. Dewatering of sewage sludge, design and operating experiences. *Water Pollution Control*, **72**, pt. 3 (1973).
5. Howorth, C. M. The thickening and centrifuging of sludge. *Water Pollution Control*, **77**, pt. 1 (1978).
6. Webb, L. J. A study of conditioning of sewage sludge with lime. *Water Pollution Control*, **73**, pt.2 (1974).
7. White, M. J. D. and Baskerville, R. C. Full-scale trials of polyelectrolytes for conditioning of sewage sludge for filter pressing. *Water Pollution Control*, **73**, pt. 5 (1974).
8. Pitman, A. R. Bio-flocculation as a means of improving the dewatering characteristics of activated sludges. *Water Pollution Control*, **74**, pt. 6 (1975).
9. White, M. J. D., Baskerville, R. C. and Lockyear, C. F. Continuous thickening of biological sludges and the influence of stability. *Water Pollution Control*, **76**, pt. 1 (1977).
10. *Winkelpress Sludge Dewatering*. Simon-Hartley Ltd., Stoke-on-Trent.
11. Maddock, J. E. L. and Tomlinson, E. J. The clarification of effluent from an activated sludge plant using dissolved air flotation. *Water Pollution Control*, **79**, pt. 1 (1980).
12. Langeneggar, O. and Viviers, J. M. P. Thickening of activated sludge with a dissolved air unit. *Water Pollution Control*, **77**, pt. 1 (1978).
13. Bratby, J. R. Aspects of sludge thickening by dissolved air flotation—*Water Pollution Control*, **77**, pt. 3 (1978).
14. Water Pollution Control Fed., Manual of Practice no. 20 (1969).
15. Lutkin, P. A. Proc. 1st N.E. Regional Anti-pollution Conference. University of Rhode Island, 1968.
16. Veselind, P. A. *Water and Sewage Works*, **115**, pt. 7, 302 (1968).
17. Swanwick, J. D., Shurben, D. G. and Jackson, A. S. Report on the Water Pollution Research Lab. survey of the performance of sewage sludge digestors throughout England, Scotland and Wales, May 1968. *Water Pollution Control*, **68**, p. 369 (1969).
18. Mosey, F. E. Assessment of the maximum concentration of heavy metals in crude sewage which will not inhibit digestion. *Water Pollution Control*, **75**, pt. 1 (1976).
19. Swanwick, J. D. and Faulks, M. Inhibition of anaerobic digestion of sewage sludge by chlorinated hydrocarbons. *J. Inst. Wat. Pollution Control*, **58**, pt. 1 (1971).
20. Swanwick, J. D. and Shurben, D. G. Effective chemical treatment for inhibition of anaerobic sewage sludge digestion due to anionic detergents. *J. Inst. Wat. Pollution Control*, **68**, pt. 2 (1969).
21. Mansfield, R. A. Vacuum filtration with particular reference to operational maintenance problems. *Water Pollution Control*, **77**, pt. 1 (1978).
22. Coackley, P. The theory and practice of sludge dewatering. *J. Inst. Public Hlth Engrs*, **64**, pt. 1, (1965).
23. Clark, E. I. and Fisk, B. Dewatering of mixed sludges. *Water Pollution Control*, **77**, pt. 5 (1978).

24. Walker, R. G. Operating experiences with the disc filter. *J. Inst. Wat. Pollution Control*, **68**, pt. 2 (1969).
25. Sidwick, J. M., Butler, B. C. and Ruscombe-King, N. S. Sludge dewatering trials at Banbury. *Water Pollution Control*, **74**, pt. 6 (1975).
26. Clifford, A. J. The dewatering of sludge by the vaccum coil filter. *Water Pollution Control*, **71**, pt. 1 (1972).
27. Macaulay, R. A. Chemical conditioning of sludge. *Water Pollution Control*, **77**, pt. 1 (1978).
28. Everitt, I. Heat treatment and pressing of digested sludge. *Water Pollution Control*, **77**, pt. 1 (1978).
29. Hurst, G. *et al.* The sludge heat treatment and pressing plant at Pudsey. *Water Pollution Control*, **71**, pt. 5 (1972).
30. Grove, A. Sludge incineration with particular reference to the Coleshill Plant. *Water Pollution Control*, **77**, pt. 3 (1978).
31. Baxter, R. J. and Bell, J. B. Commissioning and initial operation of the Coleshill Incineration Plant. *Water Pollution Control*, **77**, pt. 3 (1978).
32. Whitehead, C. R. and Smith, E. J. Sludge heat treatment operation and management. *Water Pollution Control*, **75**, pt. 1 (1976).
33. Gaillard, J. R. Fluidised bed incineration of sewage sludge. *Water Pollution Control*, **72**, pt. 2 (1973).
34. Dickens, R., Wallis, B. and Arundel, J. Fluidised bed incineration of sewage sludge at Esher. *Water Pollution Control*, **79**, pt. 4 (1980).
35. *Zimpro wet air oxidation unit*. Sterling Drug Inc. (1964).
36. Fisher, W. J. and Swanwick, J. D. High temperature treatment of sewage sludges. *Water Pollution Control*, **70**, pt. 1 (1971).
37. *Zimpro method*. Satec Ltd., PO Box 12, Weston Rd, Crewe.
38. Gehm, H. W. and Bregman, J. I. *Water Resources and Pollution Control*. Van Nostrand Reinhold Co., New York (1976).
39. Fitch, B. *Industrial and Engineering Chemistry*, **60**, pp. 7, 8 (1968).
40. Shannon, P. T. and Dehass, R. D. *Industrial and Engineering Chemistry Fundamentals*, vol. 3, p. 250 (1964).
41. Newton, D. Univ. of Michigan Continual Education publication No. 113, p. 4 (1965).
42. Brisbin, S. G. *Sewage and Industrial Wastes*, **28**, 158 (1956).
43. Torpe, W. N. *Proc. Am. Soc. Civil Engrs*, **433**, pt. 80, 1 (1954).
44. Dick, R. I. and Veselind, P. A. *J. Wat. Pollution Control Fed.*, **41**, 1285 (1969).
45. Sparr, A. E. and Grippe, V. *J. Wat. Pollution Control*, **41**, 1886 (1969).
46. Gourly, R. F. and Bennett, S. M. *Water Works and Sewerage*, **80**, 179 (1933).
47. Colin, F. *Terres Eaux (France)*, **21**, pt. 55, 27 (1968).
48. Burke, J. T. Proc. 21st Ind. Waste Conf., Purdue Univ. (1966).
49. Katz, W. J. Proc. 12th Ind. Waste Conf., Purdue Univ., p. 163 (1957).
50. Katz, W. J. and Gonopolis, A., Univ. of Michigan Continual Education Series, Bull. No. 133, p. 17 (1963).
51. Ettelt, G. A. and Kennedy, T. J. *J. Wat. Pollution Control Fed.*, **38**, 248 (1966).
52. Downs, J. R. *Wat. Wastes and Sewage*, **81**, 210 (1934).
53. Rudolfs, W. *Sewage Works Journal*, **15**, 642 (1943).
54. Woodruff, P. H. *et al.* Sanitary engineering conference, Vanderbilt Univ., Nashville (1967).
55. Burd, R. S. FWPCA, Pub. WP20-4-210 (1968).
56. Thompson, W. B. *Wastes Engineering*, **297** (1965).
57. Rudolfs, W. *et al. Sewage Works Journal*, **1**, 358 (1929).
58. Sparry, W. A. *Sewage Works Journal*, **13**, 855 (1941).
59. Keefer, C. E. *Sewage Treatment Works*. McGraw-Hill, New York (1940).
60. Conway, R. A. *Water and Sewage Works*, **109**, 342 (1962).
61. National Lime Association Bulletin no. 214, Washington D.C. (1964).

428

62. National Lime Association Bulletin no. 213, Washington D.C. (1945).
63. Purifax Inc., Descriptive Bulletin (1970).
64. Sparr, A. E. *J. Wat. Pollution Control Fed.*, **40**, 1438 (1968).
65. Sherbeck, J. M. *J. Wat. Pollution Control Fed.*, **37**, 1180 (1965).
66. Gehm, H. W. and Bregman, J. I. *Water Resources and Pollution Control*, Van Nostrand Reinhold Co., New York (1976).
67. Goodman, B. L. Univ. of Michigan Continued Educating Series, no. 113, p. 50 (1964).
68. Hopkins, G. T. and Jackson, R. L. Proc. 43rd Annual Conf., Wat. Pollution Control Fed.
69. Vogh, R. P. *et al.*, *J. Am. Wat. Ass.*, **G1**, 276 (1969).
70. Hahn, D. J. *et al. Wat. and Sewage Works*, **116**, 321 (1969).
71. Rudolfs, W. and Graham, H. W. U.S. Pat. 2,392,269 (1946).
72. Mogelnicki, S. and Gatze, E. H. U.S. Pat. 3,409,546 (1968).
73. Carr, R. L. *Wat. and Sewage Works*, **114**, R64 (1967).
74. Mogolnucki, S. 10th Sanit. Engng. Conf., Univ. of Illinois, p. 47 (1968).
75. Burd, R. S. Proc. 2nd Vanderbilt Univ. Sanitary Eng. Conf. (1968).
76. Morris, R. H. *Wat. Wks. and Wat. Engng.*, **2**, 68 (1965).
77. Sharman, L. *Wat. Wks. and Wat. Engng.*, **4**, 650 (1967).
78. Swanwick, J. D. and Davidson, M. F. *Wat. and Waste Treatment J., London.*, **8**, 386 (1961).
79. Genter, A. L. U.S. Patent 2,259,688 (1941).
80. Genter, A. L. *Sew. Wks. J.*, **13**, 1164 (1941).
81. Genter, A. L. *Sewage Wks. J.*, **18**, 580 (1946).
82. Genter, A. L. *Sewage Wks. J.*, **18**, 27 (1906).
83. Swanwick, J. D. Inst. of Sewage Purif. Water Purif. Lab. Paper, Stevenage (1962).
84. Kennedy, R. R. Univ. of Michigan Continual Education Series, No. 113, p. 37 (1964).
85. Lancond, F. *Bull. Tech. Suisse*, **91**, 24 (1965).
86. Goodman, B. L. and Wichter, C. P. *J. Wat. Pollution Control Fed.*, **37**, 1643 (1955).
87. Lumb, C. *J. Inst. Sewage Purif.*, Part 1, p. 5 (1951).
88. Evans, R. R. *Paper Trade Journal*, **154**, 31 (1970).
89. Clements, G. S. *J. Inst. Sewage Purif.*, pt. IV, 318 (1950).
90. Dozanska, W. *Gas Woda Tech. Sanit.* (Poland), **42**, 160 (1968).
91. Moss, G. Br. Pat. 1,097,900 (1968).
92. Benn, D. and Doe, P. W. *Filtration and Separation* (*British*) **6**, pt. 4, 383 (1965).
93. Eckenfelder, W. W. Jr. *Industrial Water Pollution Control*, McGraw-Hill, New York (1966).
94. Caron, A. L. NCASI Tech. Bull. No. 223 (1968).
95. Carpenter, W. L. NCASI Tech. Bull. No. 174 (1964).
96. Owens, M. B. Proc. Am. Soc. Civil Engnrs. Eng. Div. Paper No. 1182 (Feb. 1957).
97. Fair, G. M. and Moore, E. W. *Eng. News Record*, **114**, 681 (1961).
98. Fed. Wat. Pollution Control Admin. A study of sludge handling and disposal. *Water Pollution Control*, series FWPCA Pub. W.P. 20-4 (1968).
99. Gordon, C. W. Proc. 10th San. Eng. Conf., Univ. of Illinois (1868).
100. Bartlett-Snow-Pacific Inc., Bulletin No. 238 (1974).
101. Cardinal, P. J. *Wat. and Waste Treatment J., London*, **12**, 62 (1968).
102. Solir, W. H. *et al. Wat. Works and Waste Engng.*, **2**, 90 (1965).
103. Talalzke, G. H. 16th Annual San. Eng. Conf. Bull. No. 36, Univ. of Kansas (1965).
104. Laboon, J. F. *Civil Engng*, **24**, 45 (1954).
105. Miles, H. *Paper Trade Journal*, **26** (1970).
106. Gabaccia, A. J. *Compost Science*, 18 (Summer 1960).
107. Albertson, O. E. Proc. 5th Gt. Plains Sew. Works, Design Conf. (1965).
108. Copeland, G. G. *Wat. and Wastes Eng.*, **5**, 60 (1968).
109. Zimpro, 'Wet-Air Oxidation Unit'. Sterling Drug Inc. (1964).
110. Talalzke, G. H. Proc. 16th San. Eng. Conf., Univ. of Kansas (1966).

111. Sager, J. C. *J. Wat. Pollution Control Fed.*, **37**, 1243 (1965).
112. Rawcliff, E. W. and Sail, G. W. The agricultural use of Blackburn's sewage sludge. *J. Inst. Wat. Pollution Control*, **73**, pt. 2 (1974).
113. Williams, R. O. A survey of the heavy metal and inorganic content of sewage sludges. *J. Inst. Wat. Pollution Control*, **74**, 5 (1975).
114. Stead, P. A. The potential of composting admixtures of sewage sludge and domestic refuse. *J. Inst. Pollution Control*, **77**, pt. 4 (1978).
115. Barnard, J. A. Health aspects of the use of sewage sludge. *J. Inst. Wat. Pollution Control*, **77**, pt. 1 (1978).
116. Watson, D. C. The survival of Salmonella in sewage sludge applied to land. *J. Inst. Wat. Pollution Control*, **79**, pt. 1 (1980).

CHAPTER 18
Disposal of Sewage from Rural Areas

It so happens that, of late years, the interest in simple methods of sanitation has spread over a large part of the world. The reason for this is economics which, whilst it militates against sophisticated methods in the developing countries, is now having the same effect in highly developed ones. The US, having passed the Amended Water Pollution Control Act in 1972—committing themselves to higher standards of control and treatment of sewage to be largely financed from Federal Funds—find that present-day costs of sewerage are so high that it has become demonstrably uneconomic to extend piped sewerage to the spreading environs of some of the big centres of population. Moreover an increasing number of people have second homes, or are interested in living out of town either whole-time or part-time, thus putting pressure on development in beauty spots and other out-of-the-way places, many in geologically difficult locations, where conventional sewerage becomes ever more expensive to provide. Thus in 1977, despite the efforts of the Environmental Protection Agency (EPA) and various individual states, there remain some 60 million people in the US, to be provided with basic sanitary services, and for many of these conventional sewerage and sewage disposal had become uneconomic. One result has been a series of National Conferences[1-5] for 'Individual Onsite Waste-Water Systems', co-sponsored by the National Sanitation Foundation of Ann Arbor, Michigan, and the EPA Technology Transfer Programme instituted in 1974. Thus, largely the same problems which have occupied the attention of the World Health Organization, and led to the widely known 'Ten Commandments' are now receiving attention from American organizations previously geared to the most sophisticated forms of sewage treatment. In fact, no engineer working in the consulting field can afford to be without a thorough understanding of the more simple forms of sanitation, so although it may soon be very rare to find any habitation without water supply, it will be increasingly necessary to deal with the considerable range of problems posed by those properties enjoying the benefits of piped water, but without any prospect of connection to a piped sewerage system.

Conservancy sanitation in the UK

As an introduction to the various approaches to the problem just posed, it may be of interest to review the practices pursued in the UK. Here in its simplest form conservancy sanitation consists in the provision of closets or privies, the contents of which are periodically removed and disposed of by burial, or collected and carted some distance, commonly for disposal to a sewer, or at a sewage disposal works. The liquid wastes from kitchen sinks, etc., are drained to a cesspool for storage until they can be removed by a cesspool emptier, or to a septic tank with overflow to soakaway, or secondary treatment, before discharge to a ditch or stream. However, the provision of piped sewerage to over 90 % of habitations, has made such arrangements rare.

To comply with British public health legislation an earth closet must have a moveable receptacle for the reception of faecal matter and must have provision for 'deodorization' by means of earth, ashes, chemicals, or other methods. Thus it is clear that for all its advantages a chemical closet with a moveable receptacle falls under the definition of an earth closet, but one with a fixed receptable cannot do so. A water closet is a closet which has a fixed receptacle connected to a drainage system and a separate provision for flushing with clean water. It will be seen that a chemical closet with a fixed receptacle connected to a drainage system but flushed with any liquid other than clean water is neither a water closet nor an earth closet, and therefore its provision does not amount to provision of closet accommodation within the meaning of the Act.

Some years ago earth closets were available from several firms. They could also be improvised. The manufactured articles in their simplest form consisted of a casing with seat and (sometimes) lid and, underneath, a bucket which could be withdrawn through a door in the wall behind. Earth or ashes could be provided in a separate bucket and applied with a shovel to cover the contents of the bucket. Several improved models had a container for earth or ashes and a lever by which these could be caused to fall in convenient doses into the bucket. It is doubtful if this was, in fact, an improvement, for the ashes so deposited tended to fall in one place and not be spread where necessary, as they could be were a shovel used.

Chemical closets

To comply with the law chemical closets consist essentially of a container for the excreta and chemical, and a seat with lid. In addition, but not essential, is an outer case which supports the seat. This is provided in the better-quality or household models but not the industrial types. Many models also have a urine guard (a removable ring which, placed below the seat, helps to ensure that urine shall fall cleanly into the container). In order that closets may be used in caravans, other vehicles, and boats, models have been made so that the outer cases can be screwed down, and the containers are provided with lids to prevent spillage.

Containers have been made of galvanized steel, galvanized steel enamelled, vitreous porcelain-enamelled steel, and vitreous china. Perspex was used for a model no longer manufactured. All inner containers had handles for carrying. Effective capacities varied from 13.6 to 25 litres.

Outer cases consisted of stove-enamelled steel, galvanized-steel enamelled, and white vitreous china. Seats and lids were made of hardwood stained and polished, and black plastic, with non-corrodable fittings.

Containers should be made of non-absorbent material not liable to corrosion by either the sewage or the chemical used. They should be so shaped and finished that cleaning will not be difficult. Outer cases should also be non-absorbent and should have, both internally and externally, a corrosion-resisting finish. Suitable materials for plastic seats include phenolic synthetic resin, aminoplastic synthetic resin, and polymer or methyl methacrylate. Hinges may be of corrosion-resisting metal or plastic of adequate strength and reinforced if necessary. The outer case or, where there is no outer case, the container, should be of adequate strength to bear the weight of the heaviest person.

One of the drawbacks of the above-described type of closet is that no one has yet succeeded in inventing an anti-splash mechanism capable of use with it. Splashing, of course, can be prevented by placing a piece of paper on top of the liquid. In this connection it should be mentioned that ordinary toilet paper of the smooth or glazed type should not be used with chemical closets, for it tends to float and accumulate on the surface: special non-floating toilet tissue is available for use with chemical closets, and is generally to be recommended.

Types of closet are manufactured which do not comply with either the definition of water closets or earth closets in the Public Health Act, 1936. These are alike in one respect: they are arranged to overflow via a drain to a soakaway, and to comply with Section 49 of the Act must be located away from any living, sleeping, or work-room.

Although earth closets are not subject to the same restriction, the rooms in which they are placed should be entered from the external air only, or from a room which is entered from the external air only and is not used for habitation or for the manufacture or storage of food.

Occupiers of buildings are responsible for the proper maintenance of their closets and are liable to a fine for not keeping an earth closet or chemical closet supplied with suitable deodorizing material.

The chemicals used for the types of closet at present manufactured in Britain include:

(1) coal tar preparations with oil seal;
(2) coal tar preparations without oil seal;
(3) formaldehyde preparations.

The chemicals for use with oil seal are intended to effect a degree of disinfectant action, but the oil seal is included to prevent evaporation from the surface with the possibility of objectionable odour. The oil needs to be such that it will readily separate from the fluid to form a layer at the surface. The preparations without oil seal rely on their disinfectant action alone to effect deodorization.

The chemicals should not only deodorize excreta immediately, but should remain effective as they are diluted by further quantities or excreta and for a period of at least a week. They should not form, or permit the formation of, any obnoxious gases, and should be free of objectionable odour themselves. The colour of the fluid or the oil seal should be such as to obscure the contents of the closet. In view of the unavoidable possibility of splashing, the fluid should be non-injurious to the skin. The type of fluid should be such as not to make cleansing of the closet difficult. To avoid danger of fire should a lighted match be dropped into the closet, the oil should not have a low flash-point. The fluid should be capable of storage for at least a year without deterioration.

It should also be added that chemicals used in closets should not be of a type that would cause deterioration of the materials of which the closet is made. This point is generally covered by the use of chemicals marketed by the same firm that manufactures the closet.

The germicidal effects of disinfectants used in chemical closets should not be exaggerated. They are intended to deodorize, not sterilize. Complete sterilization of the closet contents would require the use of a strong disinfectant and the breaking up of solids so that the disinfectant could reach the organisms in the centres of them. In practice a small quantity of disinfectant is first diluted in water, after which it is further diluted as the closet is filled with excreta. It follows that the strength of the disinfectant is considerably reduced by the time the closet is nearly full, and it would be too optimistic to guarantee that unbroken faeces deposited at the end of the period would be sterilized. Finally, should the contents of the closet be deposited on land for disposal, dilution by rainwater would eventually destroy all disinfectant action and permit the continuance of the growth of micro-organisms: if this did not happen the sewage could not have the value as fertilizer which is claimed for it.

The best means for the disposal of the contents of earth or chemical closets is removal at intervals, not exceeding 1 week, by local authority's tank-vehicle. The most suitable type of vehicle is a cesspool-emptier with vacuum pump and hopper into which the closets can be tipped and the contents rapidly sucked away under vacuum on the operation of a quick-acting gate valve. The contents of such cesspool emptiers should be discharged to a sewer.

For estimating quantities, the following figures should be useful: the average amount of excreta is 1.63 litres per head per day. To this should be added about 4.55 litres per closet per week for chemicals and dilution water. These figures show that often more than one closet per household is required.

After emptying and cleansing the container, a quantity of water as specified by the maker should be poured into the closet and then the specified quantity of chemical added and stirred in. Chemicals with oil seal should be shaken before application. The closet is then ready for another week's use or until the container becomes filled.

Sullage drainage

The use of chemical closets still leaves the problem of the disposal of sullage from kitchen sinks, clothes, washing, and ablutions, and also of slops. These liquids, which should be properly conserved if dangers to water supply are to be prevented, are nevertheless disposed of far too frequently by soakage into the ground. It is often thought that sullage is not a seriously contaminating liquid, but in so far as it contains faecal matter from the washing of napkins and from slops, sullage is far from safe and is known to have caused several deaths.

Cesspool drainage

History shows that the development of waterborne sanitation led to the use of cesspools as means of storing all waste from buildings, including both night-soil and sullage, and, from this, a most undesirable practice has been developed and perpetuated. A cesspool is a tank for storage and nothing else. British law has laid down in the Public Health Act 1936, in former Model Bylaws under the Act, and, later, in the Building Regulations 1965, that cesspools shall be impervious to liquid from inside and to subsoil water or rain water from outside, and so sited as not to render liable to pollution any source of water likely to be used for drinking or domestic purposes. They must have no outlet for overflow or discharge other than for proper emptying or cleansing.

The Act states that if the contents of any cesspool should soak therefrom or overflow, the local authority may require the person responsible to execute such works or to take such steps as necessary by periodically emptying the cesspool or otherwise to prevent such soakage or overflow and, if he fails to comply with the notice requiring this action, he shall be liable to a fine for every day on which the offence continues.

A cesspool must be of such a depth that it can be completely emptied. (In practice this means a depth of not more than say 5 m below the nearest road on which a tank vehicle can stand.) It must be near enough to a road for the suction pipe of a tank-vehicle to be able to reach it. (In practice this means at most say 53 m—and a much shorter distance is preferable.) A cesspool should be covered, provided with adequate ventilation, and have a manhole cover for inspection and access for cleansing. The capacity of the cesspool must not be less than 18.2 m^3 as measured below the invert level of the inlet.

If the facts and figures of the case are not carefully examined and considered, the foregoing may seem very satisfactory. A house may be built, drains laid, and a cesspool constructed and then the premises may be occupied. However, the average amount of drainage from a cottage or small country house that has a proper water supply averages in the region of about 3 m^3/week or one full load for a cesspool-emptying vehicle. This would mean that, if a cesspool were constructed and maintained as required by the law, it would have to be emptied, on the average, by fifty-two vehicle loads per year. This would make cesspool drainage by far the most expensive of all forms of sewage disposal that are known, and this in itself should rule out the use of cesspools on the ground of cost alone. In addition cesspool-emptying services, either by local authority or private contractor, do not exist of sufficient scale to be capable of doing anything more than making a gesture towards this work. How then is it that cesspools are used?

The answer is that property developers or their agents show cesspools on plans, the cesspools are constructed and are tested for watertightness by the local authority, and then in many instances someone surreptitiously knocks a hole in the bottom of the cesspool to permit it to leak into the subsoil. The local authority, if it has the service and duty to empty cesspools, will make visits on the average once in 3 months to take away either one vehicle load or maybe something approaching the capacity of the cesspool: the rest of the sewage soaks into the ground for as long as the cesspool remains in use.

Despite the risks involved, the use of cesspools is still allowed in the UK, although in the US they are seldom permitted.

Sewage treatment works for small communities (UK practice)

Whilst the foregoing paragraphs trace the progress from the original forms of closet or privy, through to later types and the chemical closet, the almost universal acceptance of the water closet as by far the most sanitary of all the appliances invented for the purpose, could well have resulted in all other forms being declared illegal. However, water closets are not satisfactory in the absence of an adequate water supply, and consequently the Public Health Act 1936, and subsequent Acts, have permitted the substitution of other approved types.

Although in the UK there are few buildings without a water supply the use of the cesspool persists, often being accepted as a temporary measure on new estates where piped sewerage is likely to be made available in the near future. Some authorities provided a free emptying service, until such time as sewers are available. In many cases, however, developers are made to provide septic tanks with acceptable overflow arrangements, either to properly designed soakaway systems, or to some form of filter bed, before discharge to a ditch or stream course.

Sewage-treatment works for isolated habitations

The treatment of sewage to a good, often Royal Commission, standard for simple properties or small communities is not difficult or unduly expensive, although more costly per head of population than treatment at large municipal works. It is necessary only for the works to be designed with adequate first-hand experience of the design, construction, and maintenance of this type of plant and for the occupier of the property concerned to make some effort to ensure that the works have reasonable maintenance.

This is a section of sanitary practice which needs special study because it almost invariably suffers from neglect or abuse. In the first place the design of the works seldom becomes the responsibility of a qualified public health engineer and, even when it does, there is the possibility that the engineer concerned has knowledge of large works only, and knows nothing of the needs of very small works. More often the design is left to a builder, subcontractor, or at best a draughtsman who has learnt what little he knows of the subject from unsuitable publications. Speculative builders put in the cheapest arrangement that they can persuade the local authority to accept; and, here again, local authority officers have often in the past, accepted proposals for works that, even with careful maintenance, never could produce an acceptable effluent.

In the inspection of many existing small works serving country houses that had been converted or were scheduled to be converted to institutions, the writer found a few which, with minor attention, could be made serviceable. The rest varied from completely derelict to being capable of producing a fair effluent if largely reconstructed. In two instances the sewage was delivered to tanks, the foul effluent of which passed into streams after being carried for very long distances across land not in the ownership of the property-owner concerned. Another had a septic tank with completely corroded furniture and a trickling filter, the distributor of which had been reduced to scrap metal and, even if it had been in condition, could not have operated because two trees of 225 mm girth were growing through the medium! Other works had either rotating or tipper-tray distributors which were out of order and served filters far too small for their purpose. Some had septic tanks which discharged directly into streams, into the land drains of neighbours, or to sub-surface irrigation. Some of the septic tanks were so located that cesspool-emptying vehicles could not have access to them, and therefore they were never emptied. In one instance a septic tank discharged to a nearby field where the effluent was drunk by cows, whose milk was used by the occupiers of premises served by the septic tank. Even in an Inner London borough, premises were served by a sewage-treatment plant so designed that it could not possibly give effective treatment. The effluent from this was discharged into a series of public bathing pools and analyses of the water gave evidence of sewage contamination.

Modification of several of the works mentioned resulted in good effluents, sometimes very good, for as long as maintenance remained reasonable.

Apart from a minority of instances where untreated sewage can be discharged, e.g. into tidal waters, isolated buildings having good water supplies should have their own sewage-treatment works. The methods of sewage treatment are generally the same as those used at municipal works, i.e. trickling filter treatment, land treatment and, sometimes, adaptation of one of the activated sludge methods.

Isolated buildings and small communities should be drained on the separate system. The sewage will usually be strong and will not contain much grit; therefore neither screening nor separate grit removal need be incorporated, except in works of sufficient size to justify the part-time employment of an attendant.

Although the same processes are applied for the treatment of the small quantities of sewage that come from isolated buildings, institutions, and small communities as are used for treating large flows from towns and extensive built-up areas, the types of works differ considerably in detail. The reason for this is that, as contrasted with large installations where operational staffs are continually employed, small works receive little attention, at the best at intervals only, and therefore the engineer has to allow for this fact in his design.

The main difference between municipal and very small private works is that the septic tank, at one time used for the treatment of town sewage but now obsolete for that purpose, still continues to be incorporated in small sewage works under private ownership. These septic tanks are not installed in the hope of affecting appreciable digestion of sludge, as were early septic tanks; they are septic tanks in name only, being generally similar in design to early septic tanks. They are used in preference to modern sedimentation tanks because they are suitable for storing sludge over long periods of weeks or months. This makes them acceptable in circumstances in which frequent sludging is impracticable.

Septic tanks are either single- or double-stage. The usual water depth is in the region of 2 m. Shape is unimportant but rectangular tanks often have a total length from three to five times the width. The capacity is upwards of 16 h flow, never less than 2.73 m³, which is the minimum permitted by Building Regulations, and preferably not less than the capacity of the local cesspool-emptying vehicle.

An important difference between an ordinary sedimentation tank and a septic tank is that whereas a continuous-flow sedimentation tank is sludged by drawing off small quantities of sludge at frequent intervals, a septic tank *must* be sludged by *complete emptying* for the following reasons. After a period of about 3 months about one-third of the capacity of the tank is occupied by sludge at the bottom, one-third is occupied by scum at the top, and the middle third is taken up by the sewage which is flowing through, and from which the solids are being separated. If an attempt is made to draw off part of the capacity only, part only of the sludge is drawn off and the rest, unable to reach the sludge outlet, remains as a sloping bank. From then onwards it is sewage, not sludge, that is drawn off, and if the draw-off is not much more rapid than the rate at which the sewage enters the tank and not continued until the tank is empty, about one-half of the sludge and virtually the whole of the scum remains behind.

A small septic tank having the capacity of one tank-vehicle can be quickly emptied but, if from a large tank a second load has to be taken away, the tank could become partly filled by sewage before the vehicle has returned. Therefore there must be some practical means of isolating the sewage flow from the stored sludge while the vehicle is away. The author has installed works provided with two septic tanks in parallel (see Figure 98) with the intention that the one that

438

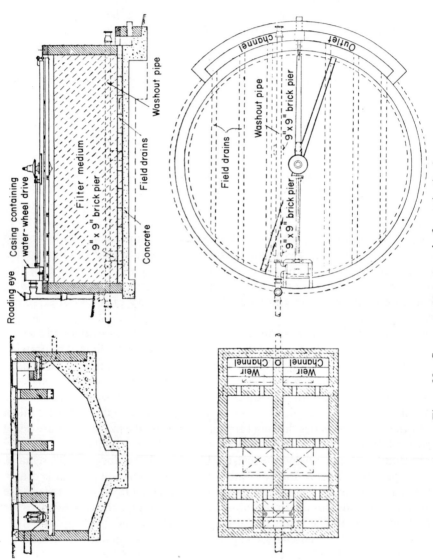

Figure 98   Sewage treatment works for an institution

is being sludged can be laid off by closing valves or penstocks; but the men who do the sludging will probably not operate the valves unless someone is there to see that they do it! Another method is to provide two septic tanks in series, the first equal in capacity to the tank-vehicle, the remainder of the capacity being in the second tank, and to make sure that the upstream tank is sludged first and after this, the downstream tank is completely emptied. (To facilitate complete emptying there should be a good slope on the floor.)

In designing a septic tank the first consideration is the determination of the capacity required. This is based on the sludge storage necessary to tide over the period between emptying dates. The amount of sludge to be expected is something less than 0.9 litre per head per day, and this figure can be taken as being reasonable for design purposes. Then, the period between emptying being known, the storage capacity for sludge can be calculated. This should be taken as being one-half the total capacity of the tank; thus the total capacity can be found. The figure so determined is then rounded off to the nearest multiple of the capacity of the local cesspool emptier (see Table 63).

Table 63   Proportions of septic tanks

Population served*	Total approx. capacity (m³)	Number of tanks	Dimensions of each tank (m)		
			Length	Breadth	Depth
20	3.4†	1	2.4	0.8	1.75
40	6.8	2	2.4	0.8	1.75
60	10.2	2	2.86	1.03	1.75
80	13.6	2	3.43	1.14	1.75

* On the basis of sludging at 3-monthly intervals.
† Minimum recommended capacity.

The inlet to a septic tank is usually a dip-pipe, a stock article sold by builders' merchants for the purpose. This can be improved upon by some form of low-velocity inlet, arranged to discharge into a small compartment of the tank where such inlet turbulence as remains is utilized in the breaking up of faeces to form sludge. The secondary and main portion of the tank should be practically quiescent.

Dip-pipes are used for septic tank outlets also, but a better arrangement is an outlet weir, protected by a scumboard set well away from the weir so as not to encourage upward velocity, and carried about 600 mm below top-water level. Where there are two tanks the outlet weir should be set to exactly the same level, otherwise one tank will do all the work. The outlet should *not* be lower than the inlet.

A baffle should be placed below the scumboard to prevent solids from rising and passing over outlet weirs (see Figure 98).

Septic tanks are covered for the purposes of preventing people from falling into them and of concealing their contents. The roof does not have much effect on odour, which is normally retained in the tank by the scum. On the other hand the odour of the effluent, as it passes from the septic tank to trickling filter or land irrigation, can be objectionable. A septic tank should be well ventilated to prevent the accumulation of gas, which could cause explosion. Access is by manhole cover, arranged above a sump and into which the suction pipe of the cesspool-emptying vehicle can be dropped. There should be access to inlets and outlet weirs also.

Occasionally septic tanks are sludged by emptying on to land, sludge being run into lagoons or, preferably, to trenches where it can be covered, or on to ploughed land where it can be ploughed in. The land on to which sludge is discharged must be below the lowest floor level of the tank in order that the sludge may gravitate to it, otherwise a pump is necessary. Sludge lagoons or trenches should have sufficient cubic capacity to receive at least the total contents of the tank or tanks.

Trickling-filter treatment

A primary consideration in the design of a trickling-filter scheme for a very small sewage works is the amount of fall available. Trickling filters should not be made much less than 1500 mm deep and this, together with loss of head through septic tank, dosing and distributing mechanism and outlet channels, etc., can amount to quite an appreciable head which may often not be available on a small site without the aid of pumping. If it is decided to pump, the type of pumping plant normally incorporated in small works is a very small automatic float-operated electric plant, including either reciprocating or centrifugal pumps capable of pumping settled effluent. The pumps suck from a tank which receives the settled effluent from the septic tank.

The capacities of small trickling filters need to be somewhat larger than those required for treatment of sewage at municipal works, for the efficiency of distribution at very small works with tipping-tray distribution is not infrequently poor (see Table 64).

Table 64   Recommended capacities for small trickling filters
(per person served)

Depth (m)	Rotating distributor (m³)	Tipping-tray distributor (m³)
1.83	0.4	0.6
1.68	0.44	0.65
1.63	0.47	0.7
1.37	0.53	0.8
1.22	0.6	0.9

In the UK the type of distributor used at very small works is a tipping-tray mechanism which doses tank effluent into a series of steel, or preferably cast-iron, distributor channels whereby the effluent is spread over the surface of the bed. These mechanisms are either single or double, i.e. the tipping-tray is placed at the end of the bed and doses the channels by tipping sideways when it is full; alternatively, it is placed centrally across a rectangular bed and tips alternatively from left to right, dosing each half of the bed in turn.

One advantage of tipping-tray mechanisms is that they seldom require more than 300 mm of head for their operation. A disadvantage is the noise that these distributors may make when tipping over during the night if the rubber buffers are worn away.

Rotating distributors for trickling filters usually cost less than tipping-tray mechanisms. They are obtainable for beds as small as 3 m diameter in several makes, and for 1500 mm diameter beds in at least one make.

Some designs have perforated sparge pipes, rotated by the reaction of the jets, and generally similar to those used at large works. But perforated sparges require regular cleaning if they are to remain in operation, and for this reason open-trough sparges with V-notch outlets that cannot choke, or other special devices, are to be preferred.

When perforated sparges are used the dosing siphons are either located (in the case of the larger beds) in dosing tanks outside the beds, or else (in the case of the smaller beds) incorporated in the central column of the distributor.

Open-trough sparges generally do not rotate under the reaction of the jets, but require some mechanism to drive them. In the Ames Crosta Babcock 'Simplette' the tank effluent passes over a water-wheel which is at the periphery of the bed and from which the power is transmitted to the centre by rotating shaft and gear wheels.

Rotating sparges of most designs for small-diameter beds are over-fed, i.e. the feeds to the distributors pass over the surface of the bed. Sparges for large beds of more than about 6 m diameter are usually under-fed, the feed pipes passing through the medium. The larger the circular trickling filter, the nearer it approaches in type the ordinary municipal installation.

Fixed perforated sparges dosed by siphon or tipping-tray have been used for distributing flow *below* the surface of rectangular beds. Such an arrangement is almost certain to give trouble unless very regular maintenance can be ensured because stoppages will happen unobserved.

Trickling filters for isolated buildings should have their side walls carried up above the top of the medium to protect the filter from the wind. They should never be covered over with boards or slabs, because this interferes with ventilation and encourages neglect. A wire mesh cover is, however, necessary as a protection from falling leaves where a filter is constructed near trees. All trickling filters should be adequately ventilated at the bottom as well as at the surface.

Humus removal is desirable at all works where a first-quality effluent is required, because the presence of humus in an effluent sample makes both the suspended solids and BOD_5 figures high. At very small sewage works humus is most easily removed by irrigation of the filter effluent through grassland or rough vegetation; an area of 20 m^2 per person served is ample for this purpose.

The performance of humus tanks varies considerably. At some sites they give good results; at others they do more harm than good. Generally humus tanks should be provided only where very regular maintenance can be given, or where land irrigation is not possible. The usual form of tank is one with 60° hopper bottom and arranged for sludging under hydrostatic head. A humus-drying lagoon should be provided, and this should be of sufficient capacity to accommodate at least the total contents of the tank.

Land treatment of settled sewage is not suitable in connection with isolated houses, for two reasons. In the first place, septic tank effluent can be offensive when exposed to the air and therefore an irrigation area in close proximity to a residence can be objectionable. Secondly, land treatment cannot be relied on to remain effective unless the irrigation area is properly rested and rotated, and watched to see that the sewage is not forming channels and finding its way to the outfall without treatment. Such proper attention to land irrigation is seldom given to the irrigation areas that serve private houses: it can, however, be ensured where the sewage works serving an institution is maintained by a competent handyman.

After humus removal, sewage works effluent is usually discharged to a watercourse, but where there are no watercourses it is necessary to soak the effluent into the ground. This can be done in several ways. In chalk and limestone districts, if the effluent is allowed to run over the surface of the land, it often finds a natural crevice in the structure and disappears underground without any special works having to be constructed.

In other permeable soils the effluent can be discharged into soakaway pits. In this case the pits should be made as deep as possible without excavating below the natural water table. Depth has far more effect on rate of discharge than diameter, which hardly increases initial discharge by a recordable amount.

Sub-surface irrigation is a permissible method for the disposal of fully treated effluent and such a system often lasts longer than a simple soakaway pit. It is possible to calculate rate of soakage after experiment (see pp. 377) but the theory is more elaborate than justified in the circumstances. Sub-surface irrigation should *not* be used as a substitute for aeration (see Chapter 16).

More often than not existing sub-irrigation systems are branching herring-bone systems of pipes; but, in fact, a single pipe laid in a straight line or carried round an area of land is as effective as a much more costly herring-bone system. The land irrigated should be level, or should rise to a crown inside the ring main, otherwise effluent may break surface when the water table is high.

BioDisc process

The BioDisc process, introduced by Ames Crosta Babcock is applicable to individual houses, groups of houses, or institutions, having populations from about 5 to 500 persons. As described in Chapter 14, the equipment consists of power-rotated discs dipping into the settlement compartment of a modified Imhoff tank and is capable of producing a Royal Commission effluent when under suitable biological loads. The plant is manufactured in sizes applicable to 5, 10, 30, 60, 100, 200, 300, and 500 head of population when serving normal residential buildings, or proportionately smaller or larger populations when used for hospitals, hotels, holiday camps, canteens, schools, etc. As the equipment is totally enclosed it is particularly applicable to the last type of premises.

Small activated-sludge plants

Whilst generally associated with large treatment works, the activated-sludge process has been quite widely applied in the form of 'package plants', for populations ranging from say 200 to 20,000. The initial objections of difficulty of maintenance and operation have been largely overcome. The concepts of 'extended aeration' and 'contact stabilization', briefly referred to in Chapter 15 are most commonly applied in small units. The former technique was explored by Downing, Truesdale and Birkbeck, in the UK, in 1964. They concluded that plants treating a typical rural sewage, having a BOD_5 of about 500 p.p.m., and a similar concentration of suspended solids, with a detention period of about 2 days could produce effluents containing 140–210 p.p.m., suspended solids. By increasing detention time to 3 days suspended solids would reduce a further 20–30%, but effluents from such plants would require some form of polishing to achieve acceptable limits. An effluent of 20 p.p.m. BOD_5 and 30 p.p.m. suspended solids could be expected if sludge concentration was maintained between 2000 and 3000 p.p.m. by removing the excess at intervals of 4–9 weeks for detention periods from 2 to 3 days. It was recommended that surface water be excluded from such plants. The Ministry of Health later laid down the following design criteria:

max flow rate to be 3 × DWF (dry weather flow);
aeration tank loading 7–10 kg BOD_5/30 m^3;
air supply 3.0 m^3/kg BOD_5 at 2 m aeration tank depth or 2.0 m^3/kg at 3 m depth;
final tank overflow rate not to exceed 35 m^3/m^2 per day.

Contact stabilization plants normally comprise four separate zones with total aeration times similar to those for conventional plants. The essential difference is that a considerable portion of total aeration time is used for reaeration and stabilization of return sludge, with only a minor part used for contact between the re-aerated sludge and the load before separation in the final tank. Surplus activated sludge is withdrawn at short intervals (15 min) and passed to a separate aerobic digestor for final oxidation of the surplus sludge. Loading criteria for contact stabilization plants would be:

Max flow rate 3–6 × DWF;
aeration tank loadings as high as 30 kg BOD_5/30 m^3;
air supply normally 60 m^3/kg BOD_5 per day.

Maintenance of small sewage-treatment works

Small sewage-treatment works should be inspected once a week or at least once a month to make sure that trickling filter distributors are in working condition. The septic tanks should be sludged by complete emptying with the aid of a tank-vehicle under supervision once every 3 months, more or less according to capacity, (see Table 63). If these not too difficult duties are done by the householder or his agent there should be little risk of trouble.

The twenty-eight works for which the author was at one time responsible were all visited by a sanitary officer according to a schedule which ensured that every one was inspected once a month. All septic tanks were sludged once every 3 months by the local authority or a contractor, and samples of treated effluent were taken about 6 weeks after the emptying date and tested for BOD_5 and suspended solids content. By this means it is thought that a higher general standard of performance was secured than is usual elsewhere.

Conservancy methods elsewhere

In the developing countries, waterborne sewerage must be regarded as a rarity and, in general, some form of earth closet or privy—often backed by a system of night soil collection on a village or small community basis—is run with widely varying degrees of hygienic protection. As early at 1956 Imhoff, Müller and, Thistlethwayte[6] mention a system used not only in Asia and Africa but also in Australia, where closet pans with tightly fitting lids are collected at least weekly, in exchange for empty pans which 'are cleansed and more or less disinfected ready for use again'. In Africa and India conditions tend to be even more primitive, although cleanliness is sometimes unintentionally achieved by keeping poultry which devour the flies. In the crowded cities of India communal latrines are usually the cheapest form of sanitation and these are readily acceptable in some areas. In small communities and villages there is considerable interest in communal sludge digestion, in simple tanks fed by night soil, and kitchen waste collection. These installations are community-run and have the simplest possible facilities for gas collection. The methane gas is used as a source of heat and light. The digested sludge is used as fertilizer, and is usually applied to the land in liquid form. In Kerala State, India, there are two full-scale municipal night soil digestors serving populations of 15,000–20,000.[7] In Lagos the night soil from 14,000 houses is collected in pails and taken to one of several collection points, whence it travels by tanker to the treatment works. The volume of night soil is 180 m^3/day and the BOD_5 load 8300 kg/day. At the treatment works the nightsoil is screened, diluted with make-up water and final effluent, macerated, and finally treated in two aerated lagoons (each 55 m square and 3 m deep and having four 56 kW aerators). The treated waste is discharged into Lagos Lagoon some 120 m below low water (see Fig. 99). More commonly night soil can be treated in facultative waste-stabilization ponds. Water lost by

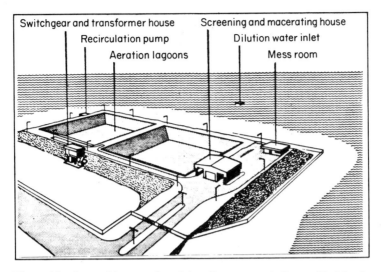

Switchgear and transformer house	Screening and macerating house
Recirculation pump	Dilution water inlet
Aeration lagoons	Mess room

Figure 99   Aerated lagoons for nightsoil treatment in Lagos (D. Mara)

evaporation must be regularly replaced and the pond must be properly constructed. A concrete ramp is provided which permits the night soil to be sluiced into the pond by a jet of water. The area required for such a pond may be calculated from the equation:

$$\lambda_5 = 20T - 120 \qquad (41)$$

where: λ_5 = max BOD_5 loading in kg per hectare per day;
    T = temperature in °C

The night soil BOD_5 contribution is assumed to be 27 g per head per day. The algae produced from a night soil pond may be harvested for use as livestock food or the pond effluent led to a fishpond. More information concerning design of ponds is given under a later heading (p. 449).

Not all highly developed countries are totally committed to water carriage systems. Some have preserved the composting type of toilet, as epitomized by the Swedish Multrum. Gutormsen[8] instances the dangers of eutrofication in lakes and rivers arising from discharge to them of human wastes, and describes the work of the Department of Microbiology, Agriculture University, As, Norway on a project entitled 'Alternatives to the flushing toilet in cabins' started in 1973, when he estimated the number of cabins as 200,000 increasing at a rate of 7000 per year. Bearing in mind the Ten Commandments of the World Health Organization a wide study of available means indicated that composting properly applied was the only method likely to meet the requirements. That recommended is aerobic; it starts with a layer of rich soil, good aeration,

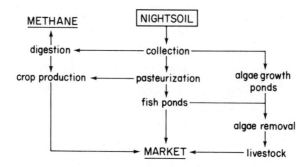

Figure 100   Nightsoil treatment alternatives (D. Mara)

moisture, and a minimum temperature of 10 °C. Wastes must be not too dry or too wet, excessive urine will stop decomposition which is restored by the sprinkling of peat or ground bark into the toilet; if it is too dry water must be added. The addition of kitchen waste (fish scraps, potato peels, cabbage leaves, etc.) increases the rate of decomposition.

The tests covered 21 different models of composting toilets. Three types are given special mention. The first is a large, box type, with a tank of 1200–1600 litres capacity divided into two compartments. One compartment is intended for 3 years' use, after which a flap is turned and wastes diverted to the second compartment. Wastes in the first compartment remain there for another 3 years before removal from the system. The model suffered from poor air flow, only the top layer being aerobic.

The second type, on the lines of the Swedish Multrum (see Figure 101) had sloping floors. The wastes collect initially under the toilet chair, then slowly move down towards the lower part where the emptying door is located. If the angle of slope is too great the wastes slide down too rapidly. Urine needs to be kept away from wastes near the emptying door or they will be become unhygienic.

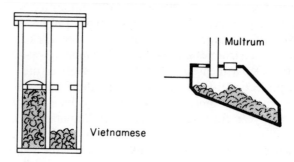

Figure 101   Compost latrines, Multrum type (*J. Instn. Public Hlth. Engnrs.*)

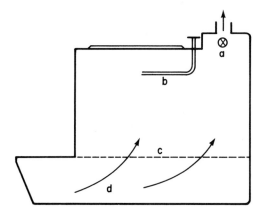

Figure 102   Small toilet with electric heating
elements (*Ann Arbor. Science Publs.*)

A third type is of small capacity, but is equipped with heating elements placed
either under the waste or in a space at the back of the toilet. Power requirements are between 130 and 300 W, with temperature controlled at 40–45 °C. A
fan moves air through the waste, using 15–25 W of power. There is a collecting
chamber at the bottom of the container, under a grate, and sometimes a tray
for removal of wastes, which is otherwise dug out. The process is rapid, and is
capable of producing waste reduction up to 90% as against only 20–25%
during the loading period with the large box-type, and 75% with the sloping-
floor type.

In general, experience in developing countries[9] has not favoured compost
latrines, which have often become malodorous. Double-vault latrines are used
in Victoria where ash and vegetable kitchen waste are added. In Japan and
Korea a waterproof vault under the latrine seat is emptied by vacuum truck
every 2–6 weeks.

The World Economic Development Council recommends aqua-privies for
developing countries (see Figure 103). Excreta falls directly into a septic tank,
and just sufficient water is added to make up evaporation losses—this may be
water which has been used for other purposes, such as ablution or laundry,

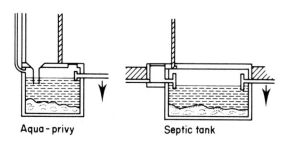

Aqua-privy   Septic tank

Figure 103   Aqua-privy (*J. Instn. Public Hlth. Engnrs.*)

and can be used where water is obtained from stand pipes. The capacity of an aqua-privy chamber 'S' in litres should be not less than

$$S = P\{rq + 40k[n + \tfrac{1}{4}(6 - n)^*]\} \qquad (42)$$

where: P = the number of persons served by the chamber;

$\quad\quad r$ = retention time for liquor in the chamber in days (may be one, if effluent is discharged to a drainage field but should be extended to allow for pathogens, where effluent discharges to open drains, streams, etc.)

$\quad\quad q$ = the waste water flow, in litres per person per day, in which allowance is made for the number and type of fittings discharging to a septic tank, and to the proximity of water supply to an aqua-privy;

$\quad\quad k$ = a factor not less than 1, to allow for local variations of diet and cleaning material causing increased sludge and scum accumulation;

$\quad\quad n$ = number of years between desludgings;

$\quad\quad$ * indicates that the term $(6 - n)$ should only be used when n is less than 6.

Waste-stabilization ponds

Waste-stabilization ponds have become the most important method of sewage treatment in hot climates, where sufficient land is normally available and temperature is most favourable for their operation. They are, of course, used at all latitudes, even as far north as Alaska, and in the United States account for nearly one-third of all municipal plants.

Of the three types,[7] facultative, maturation, and anaerobic, the first is the most common and although they normally receive sewage directly after preliminary treatment, they are becoming increasingly used to treat effluent from septic tanks and as anaerobic pretreatment ponds. The term 'facultative' refers to a mixture of aerobic and anaerobic conditions, the former being maintained in the upper layers whilst the latter exist towards the bottom. Although some of the oxygen comes from re-aeration through the surface, most of it is supplied by photosynthetic activity of algae, which grow naturally in the pond where considerable quantities of both nutrients and incident light energy are available. The algae growth is profuse, and the pond bacteria utilize this 'algal' oxygen to oxidize the organic waste matter. One of the major end-products is carbon dioxide, which is readily used by the algae during photosynthesis since their demand for it exceeds the supply from the atmosphere. Thus, there is an association of mutual benefit (symbiosis) between algae and bacteria. As photosynthesis is light-dependent, there is a diurnal variation in the amount of dissolved oxygen present in the pond and in the level of the 'oxypause', the point below the surface where the dissolved oxygen concentration becomes zero. The pH of the pond contents also follows a daily cycle, increasing with photosynthesis (maybe as high as 10). This happens because at peak demand algae remove carbon dioxide from solution more rapidly than it is replaced by bacterial respiration; as a result the bicarbonate ions present dissociate to provide not only more CO_2 but also the alkaline hydroxyl ions which increase the pH value:

$$HCO_3 \rightarrow CO_2 + OH.$$

Wind and heat are the two factors of major importance influencing the degree of mixing. Mixing minimizes hydraulic short-circuiting and the formation of stagnant regions, and it ensures reasonable uniform vertical distribution of BOD_5, algae, and oxygen. Mixing is the only means by which the large numbers of non-motile algae can be carried up into the zone of effective light penetration (the 'photic' zone), only the top 150–300 mm of the pond. Mixing is also responsible for transportation of oxygen produced in this zone, to the bottom layers of the pond. Good mixing thus increases the safe BOD_5 that can be applied to the pond.

The depth to which wind-induced mixing is felt is largely determined by the distance the wind is in contact with the water (the 'fetch'); an unobstructed contact length of about 100 m is required for maximum mixing. Proof of this is clear from a case in Zambia, where a 2 m high fence was erected around a pond, and within a few days the pond turned anaerobic; The fence was removed and aerobic conditions rapidly re-established. Diurnal mixing in a 1.5 m deep facultative pond in Lusaka, Zambia, has been thoroughly investigated and is probably typical of tropical and sub-tropical ponds.[13]

As sewage enters the pond most of the solids settle to the bottom to form a sludge layer. At temperatures $> 15\,°C$ intense anaerobic digestion of the sludge occurs; as a result the sludge layer is rarely more than 250 mm, and often much less. Desludging is rarely required, say once every 10–15 years. At temperatures greater than $22\,°C$ the evolution of methane gas is sufficiently rapid to buoy sludge particles up to the surface, where drifting sludge mats are formed. These must be removed (together with the floating debris) so that they do not prevent the penetration of light into the photic zone. The soluble products of fermentation diffuse into the bulk liquid of the pond, where they are oxidized further. The seasonal variation of the rate of fermentation (increasing approximately seven-fold with each $5\,°C$ rise in temperature) explains why the BOD_5 in the pond often remains sensibly constant throughout the year, despite changes in temperature.

A recent paper by Lumbers[10] reviews the various design methods advocated, but concludes that despite the large amount of literature on the subject, these methods can result in widely differing results. For facultative ponds Gloyna[11] gives a table of acceptable organic loadings based on the following assumptions: (1) the total evaporative and seepage losses are not greater than the influent volume; (2) the BOD_5 contribution is 50 g BOD_5/day; (3) waste water production is 100 litres per head per day. This is known as the global environmental design basis, as given in Table 65. McGarry and Pescod[12] analysed data from 143 primary facultative ponds and found a very definite relationship to exist between

Table 65 Global environmental design basis

Surface loading (kg BOD₅/ha per day)	Retention (days)	Environment
10	200	Frigid zones, icy low water temperature, variable cloud cover
10– 50	200–100	Cold seasons, icy temperate summer (short)
50–150	100– 33	Temperate to semi-tropical—no prolonged cloud cover; occasional ice cover
150–350	33– 17	Tropical, uniform distribution of sunshine and temperature, no seasonal cloud cover

aerial removal of BOD_5 and the influent BOD_5 value. The regression line took the form:

$$Ly = 10.37 + 0.725L_0 \tag{43}$$

where: Ly = aerial removal of BOD_5, kg BOD_5/ha per day;
L_0 = aerial loading of BOD_5 (kg BOD_5/ha per day).

A further equation describes the maximum feasible loading related to the mean monthly ambient temperature, taking the form

$$L_0 = 11.2(1.054)^{Tm} \tag{44}$$

where: Tm = mean monthly ambient temperature (°F).

Aerated lagoons are generally designed on the principle of a completely mixed steady-state reactor using the first-order kinetics theory[13]:

$$\text{Retention} = \frac{L_0/L_e - 1}{k_T} \tag{45}$$

where: L_0, L_e = influent and effluent BOD_5(mg/l′);
$k_T = k_{20}\theta(T - 20)$;
$k_{20} = 0.5$
$\theta = 1.053$

The oxygen input is related to the ultimate BOD_5 removed, and must be adjusted for temperature and atmospheric pressure (altitude) so that the aerator efficiency under standard conditions may be specified.[14] Although good performance of aerated lagoons has been observed in many situations there is need for additional information to determine:

(1) the validity of the first-order kinetics theory;
(2) the effects of basin geometry, aerator design, and power input on the BOD_5 removal coefficient;
(3) temperature effects.

Typical oxygen transfer efficiencies are the order of 0.8–1.0 kgO$_2$/kWh. Typical depths are 2–5 m, retention 3–6 days, and overall BOD$_5$ loadings around 0.016 kg BOD$_5$/m^3 per day.

Maturation ponds

The principal purpose of maturation ponds is the removal of enteric bacteria and viruses; other purposes include the reduction of BOD$_5$ and the removal of nutrients.

First-order kinetics theory is generally used to predict bacterial die-off, assuming complete mixing and steady-state conditions. For a number of ponds in series, the effluent faecal coliform concentration is predicted by the following equation:[15]

$$N_a = \frac{N_1}{(K_T R_1 + 1)(K_T R_2 + 1)}(K_T R_n + 1) \tag{46}$$

where: N_a = effluent faecal coliform count per 100 ml;
$\quad N_1$ = influent faecal coliform count per 100 ml;
$\quad K_T$ = faecal coliform removal rate constant at temp. $T\,°C$;
$\quad R_n$ = retention of nth pond;
$\quad K_T$ is calculated from $K_T = 2.6 \times 1.19^{T-20}$ for T between 5 and 21 °C.

Maturation ponds are generally designed to have a retention period of 3–7 days with a depth 1–1.5 m. An increased depth at outfall allows the effluent take-off level to be selected to achieve minimum algal content in the effluent; where the effluent is to be re-used for agricultural irrigation the effect of evaporation on the concentration of salts and toxics in the effluent should be checked.

In designing a pond or system of ponds in, any of these categories, knowledge of average rainfall figures throughout the year is necessary so that the effects of dilution and reduced retention period can be accounted for in the design.[16] For unlined ponds percolation into the subsoil should be checked, the initial rate will usually be very much higher than that finally achieved.[17,18] Such methods as rotation of pond usage and bottom scarification can be taken to maximize seepage. In hot climates evaporation losses can significantly reduce retention and cause water level to drop below that required for satisfactory operation. Concentration of pollutants may cause operational difficulties. A study of the combined effects of precipitation, percolation, and evaporation on overall flow should be undertaken to check level fluctuations and minimum retention. This analysis has particular reference to the early stages of operation, when flow rates much lower than the design rate may be encountered.

The directions of maximum wind-induced mixing and of maximum sunlight are principal factors in selection of the ideal orientation, although in many instances site area and topographical considerations will predominate. Rectangular ponds of 2–3:1 length to breadth minimize short-circuiting and simplify construction.[18] Generally it is preferable to divide a pond system into at least two parallel streams of ponds to facilitate draining down and allow flexibility of operation. On the question of wave protection, areas less than 12 ha do not usually develop wave erosion problems.[18] For environmental protection the distance of ponds from residential areas should be 0.5–1 km for faculative ponds and more for anaerobic ones. Pond inlets should be submerged, and placed at the centre of small ponds. Larger ponds should have multiple inlets positioned 25–60 m from the nearest bank.[10,18] An economic form would be a straight pipe sloping gently towards the base of the pond.

A bar screening facility for raw sewage should be provided. Drain-down chambers should be provided to allow emptying for maintenance and sludge removal. Two penstocks should be provided so that top-water only may be

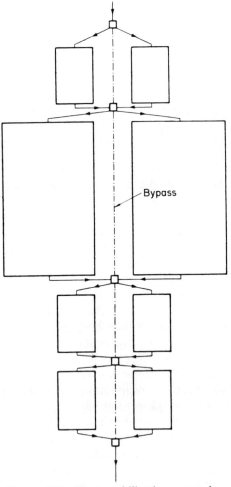

Figure 104 Waste-stabilization ponds—flexible operation pipework (*Instn. Public Hlth.Engnrs.*)

removed, if required. Interpond connections may be single pipes, but should always be valved. If water-level difference exceed 1 m an inlet chute should be used. A central feeder collector and by-pass (see Figure 104) provides maximum operational flexibility. Pond outlets usually consist of an overflow weir and a scum-board. Both should be adjustable so that water and extraction depths can be varied to suit seasonal changes of water quality at different depths. Embankment heights are related to maximum depth requirements for spare capacity, plus an allowance for sludge accumulation and expected wave height. Erosion is reduced by good compaction and a surface layer of stones on the inside slope. This slope should not be steeper than 1:3, and crest width between 3 and 5 m. Organic material should be excluded and size is often dictated by the balance of cut and fill (see Figures 105–108).

Wind-induced mixing is essential to satisfy operation as it increases the natural surface re-aeration and maintains an improved algal population and distribution of pollutants. The degree of mixing depends largely on the stretch or length of open water and surrounding obstructions, such as vegetation and buildings. A clear distance of 100 m around the pond should eliminate such

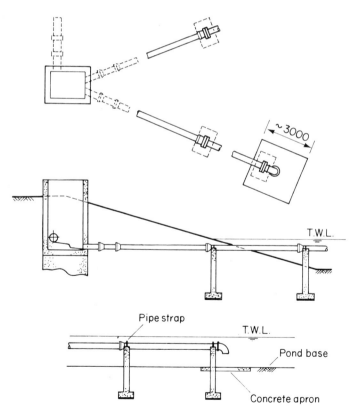

Figure 105 Waste-stabilization pond inlet (*Instn. Public Hlth. Engnrs.*)

effects. A fetch of 100 to 200 m is sufficient for adequate mixing provided wind speeds often exceed 20 km/h; below this artificial mixing should be considered. This can be by surface aerator (thus moving the design into the aerated field) or simple 'bubble-guns', rotary brushes, or horizontal screw pumps.

Sludge removal should be effected once the minimum desirable depths has been reached. After draining-down the typical moisture content will be between 94 and 80%. Removal may be by mechanical shovel or hand-barrow. Final disposal could be to agricultural land, land-fill or tip, depending on circumstances. Pescod has produced a graph relating total solids content to cumulative radiation.[12]

Where bottom sealing is necessary to avoid percolation, it may be achieved simply by a layer of clayey soil or bentonite, or by soil stabilization with cement, sprayed or rolled asphalt, and plastic lining materials, between sand layers.

Where land is not available for future extension, a system can be modified to include aerated lagoons, by deepening. The addition of small anaerobic lagoons for primary treatment may also achieve a useful increase in capacity, and re-circulation, feed, and withdrawal facilities may also assist in up-grading.

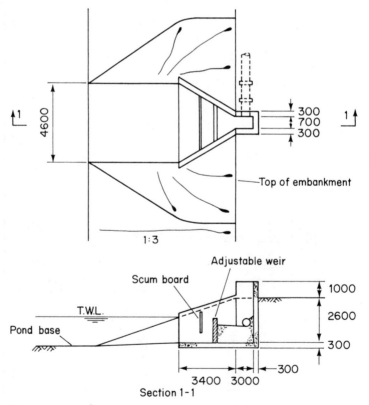

Figure 106   Waste-stabilization pond outlet (*Instn. Public Hlth. Engnrs.*)

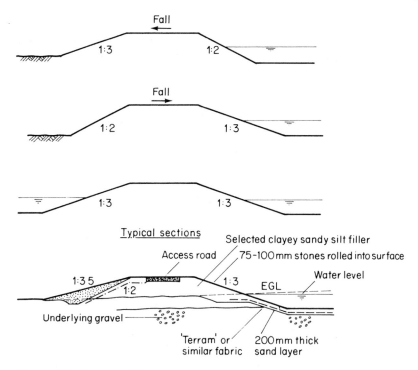

Figure 107 Waste-stabilization embankments (*Instn. Public Hlth. Engnrs.*)

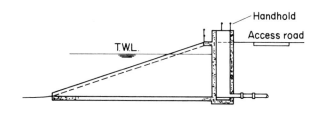

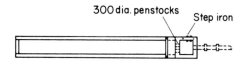

Figure 108 Waste-stabilization drain-down chamber
(*Instn. Public Hlth. Engnrs.*)

Anaerobic ponds

Anaerobic ponds are used mainly for short-period preliminary treatment of crude sewage. They are designed to take high organic loading and are completely devoid of dissolved oxygen. Solids settle and digest anaerobically: partially clarified supernatant liquor is discharged to a faculative pond for further treatment. Successful operation depends on a delicate balance between acid-forming bacteria and methanogenic bacteria. Temperature requires to be more than 15 °C and pond pH greater than 6 to ensure minimum sludge accumulation. Desludging is carried out when sludge volume reaches half the pond capacity—probably every 3–5 years. At temperatures below 15 °C such ponds are merely sludge storage basins. By their use, particularly in larger schemes, significant economies in land-use can be achieved. Their operation requires adequate maintenance facilities.

Experience in Israel

Watson,[19] describing the evolution of oxidation ponds in Israel, attributes the high degree of purification attained in that country to the work of algae, of which *Chlorella* is the most prolific. The photosynthetic action of light, and the heat that it generates, build new algal cell material. The oxygen thus released into the water enables aerobic purification to continue, whilst the products of sewage decomposition are, in turn, used by the algae. Depths have been standardized at between 900 and 1500 mm, with an additional 900 mm to top of embankment. Experience shows that it is difficult to control growth of aquatic weeds at depths less than 900 mm, whilst at depths over 1500 mm it is difficult to maintain dissolved oxygen. Any variation of depth leads to sludge accumulation and anaerobic conditions. Detention for 15 days in this type of aerobic pond would generally be preceded by 2 h pretreatment in an Imhoff tank, 16 h in a fully enclosed septic tank or 20 h in anaerobic ponds, 2.4 m in depth. The effluent produced after this process will be closely comparable with that from a conventional works employing activated sludge or trickling filter treatment. The minimum acceptable capacity for this type of pond is 5 days, at which a BOD_5 of 250 p.p.m. could be reduced to 50 p.p.m.

In the interests of accuracy he recommends that ponds should be watertight, so as to provide retention times corresponding to their design. Some ponds have been effectively protected by a bitumen felt lining on a clay blanket, others have embankments covered with bituminous felt, and soil cement is to be used for sealing pond bases. Soil samples should be examined before construction starts to determine soil type and degree of compaction required. All organic matter to a depth of 200 mm should be stripped and carted away. Compaction with a sheepsfoot roller, at optimum moisture to 90 % of maximum dry density, is normally satisfactory with a soil having a suitable clay content. The sewage itself tends to seal the wet area and planted grass protects the dykes, when no special precautions are necessary. In order to prevent mosquitoes breeding in weeds at the water's edge bank slopes should not be flatter than 1:2. The final disposal of effluent, in Israel, is rarely to a river, and a sufficiently large area of suitable land must be available for disposal by irrigation. 1 hectare/30 m^3 DWF per day is required by regulation. Such areas should be farmed by trained workers, and not be in the hands of smallholders, to avoid confusion of water supply connections. With regular maintenance aerobic oxidation ponds need only be desludged and cleaned at very long intervals.

High-rate ponds

These are designed for maximum algal growth and high protein yield, and are described by Oswald and Gotaas.[20] They are shallow ponds 0.2–0.6 m deep. The contents require daily mixing to re-suspend settleable solids. Removal of algae from the final effluent is also essential. The process is a very efficient one, capable of producing effluents of less than 30 mg/l BOD_5, from settled domestic sewage, and a weight of algae equivalent to that of BOD_5 (a pond in Thailand, depth 0.45 m, retention time 1 day, fed with domestic sewage at 450 kg BOD_5/ha per day, yielded 450 kg of algae from the same area). However, such ponds have the disadvantage of requiring highly skilled personnel to operate them. Considerable research and development work will be necessary before they can be recommended for general use.

Windle-Taylor[21] describes an interesting application of an effluent pond system in England to treat effluent from the Rye Meads Works, organized jointly by the Harlow Corporation and the Metropolitan Water Board, following experiments to ascertain the possibility of recharging the underground water resources of the Lee Valley, and to operate sand filters to handle surplus effluent that could not be dealt with by percolation. After 4 years the percolation rate had fallen from 0.45 to 0.15 million litres/ha per day, and to 0.1 after 8 years.

The ponds or lagoons were constructed in the gravel bed, being from 1.7 to 3.6 m deep. Their capacity was 135,000 m^3 and the flow rate 6 million litres/day, giving an average retention period of 17 days. They operated as maturation ponds and ensured an effluent of high dissolved oxygen content beneficial to the river and particularly to fish life. The buffering action was most valuable and although in summertime algal growth affected the effluent to some extent, this was regarded as preferable to an equivalent amount of humus, or other organic matter. A significant feature was the reduction in sewage-indicating bacteria— pathogenic organisms had virtually disappeared. Results showed that an effluent of excellent quality could be obtained after 8 days.

Latest developments in the United States

As already indicated, the ever-increasing cost of sewer construction is having repercussions in the developed countries; the US in particular where the widely accepted maxim that every property will be connected to a sewer some time in the future, is now acknowledged as economically unrealistic, and for 10 years interest in more economical alternatives has been growing. All this has developed alongside ever-increasing stringency in effluent treatment standards, and environmental controls. The highly organized research facilities, instead of concentrating entirely on advanced methods of sewage treatment, are giving attention to what they have termed 'individual on-site wastewater systems', and to specially economic forms of waterborne sewerage for small communities based on individual septic tanks, from which the overflow is either pumped or sucked by vacuum through a system of small-diameter pipes, following the run of the ground, for final disposal by community seepage fields or maturation ponds. Chlorination, now viewed with suspicion, could, according to Neilson[22] be continued with ozonization to secure the best effects from both. Ozone's utility and effectiveness could be greatly expanded if preceded by chlorination and then combined with fractionation. Not only would the bactericidal, viricidal, and parasiticidal characteristics be enhanced, but also the chemical and physiochemical treatment. The 'scrubbing' possible with just foam fractionation could be significantly improved by ozonating the air creating the foam. The process of foam fractionation is likely to be a very important treatment technique in the future. The use of gamma radiation as an advanced tertiary treatment, combined with ozone, would be unsurpassed at reasonable levels of energy consumption, in the production of potable water from waste water. The latter process has been added to a demonstration facility at Morgan Hill, California, treating the kitchen and rest-room waste from a large restaurant, and independent analysis showed SS, BOD_5 and ammonia removal as respectively less than 3.7, 5, and 0.1 p.p.m. In the interests of energy conservation experiments are being conducted into the use of gamma radiation to achieve top-level synergism microbe kill and molecular dissociation. It is pointed out that by the year 2000 the quantity of caesium in the US alone should be sufficient to treat all the sewage sludge being generated there.

The Farallones Institute of Berkeley, California[23] are committed to waste recycling and water conservation, and have developed a batch-type composting privy utilizing the principles of fast, hot compost. A 250 l drum is painted inside with epoxy or fibreglass resin, and fitted with a vent pipe; a scissors jack on a wheeled dolly is used to raise the drum up under the privy floor. A small amount of carbonaceous organic matter is added after each use. To mix and aerate the contents the drum is lowered, the lid clamped securely, and the drum set on its side and rolled round vigorously several times. When properly maintained some composting does occur during collection but further treatment by long-term storage and slow composting is necessary before the contents can be used for trenching around fruit trees or ornamentals. They have also developed a solar waste pasteurization oven in which the privy drums are placed. The drum heats up by direct solar gain during peak solar hours and can be sealed off by means of a hinged insulated lid to prevent heat loss. It is believed that this system shows real promise, capable of eliminating any pathogens or parasites in faecal material.

Papers presented at the 5th National Conference on Individual Waste Water Systems, 1978,[5] focused on information pertinent to the implementation of the latest Clean Water Act of 1977. This emphasizes the real need for recognizing that alternatives to centralized sewage-disposal systems have become essential

Table 66   Research programmes concerned with individual on-site wastewater systems

Subject	Researcher
Compost toilets and grey water systems	California Water Resources Control Board
Mound, evapotranspiration, electro-osmosis, drip irrigation, raised bed and fill systems	California Regional Water Quality Board, North Coast Region, and Somona County Health Dept. San Diego Region
Sealed evapotranspiration and evapo-transpiration infiltration systems	
Evapotranspiration	University of Colorado
Design improvements for conventional systems through detailed engineering assessment of all specific site parameters	Connecticut Department of Environ-mental Protection
Denitrification studies involving black and grey water systems	University of Connecticut
Evapotranspiration and mound systems	Indiana Department of Health in con-junction with 10 States committee of Great Lakes Upper Mississippi Basin Region
Electro-osmosis	Iowa State University
Evapotranspiration systems	University of Maryland
Electro-osmosis	University of Minnesota
Mound and evapotranspiration systems (irrigation of vegetation)	Cattanaugus County Health Department—New York
Alternating fields	Oklahoma (commercial firm)
Experimental permits for various types of alternatives with detailed departmental site evaluation	Oregon Department of Environment Quality
Sand mounds	Pennsylvania State University
Experimental alternate systems evaluation	Pennsylvania Department of Enviro-mental Resources
Evapotranspiration	City of San Antonio, Texas.
Evapotranspiration and innovative alternatives	Virginia Polytechnic Institute
Mound, aerobic, filter, and curtain drain systems	Washington Department of Ecology and University of Washington
Developments of design standards for mounds, elevated systems, composting toilets, recycling systems, home aerobic units, self-contained toilets, discharges from cluster development, grinder pumps and vacuum collection systems.	West Virginia Department of Health
Mound, aerobic, and subsurface sand filter systems	University of Wisconsin
Development of environmental impact statement regarding alternative designs	Wisconsin Department of Natural Resources

in the interests of real economies in capital cost, energy, and water savings. It emphasizes that small communities should get more help. EPA has established a national clearing house in Cincinnati to collect and distribute information on alternative treatment methods. There are to be special benefits for innovative and alternative technologies. If approved, they are eligible for a grant that funds 85%, rather than 75%, of costs. Privately owned treatment works serving one or more properties that were in existence in December 1977 subject to certain conditions will be eligible for grants. Types of research and evaluation being conducted include, but are not limited to, those listed in Table 66.

Clean Water Act 1977

The financial implications of this Act are described by Dearth,[24] who lists the following funding provisions

(1) establish the grant eligibility of individual systems;
(2) authorize 85% grants for alternative systems;
(3) establish a special fund to increase the grants from 75 to 85%;
(4) define the treatment works eligible for 85% grants;
(5) provided a back-up 100% insurance grant for failures;
(6) provide for a mandatory 4% set aside of the allocation for rural states to provide alternative systems for communities with up to 3500 persons.

Both publicly and privately owned individual systems are eligible for grant funds. Privately owned systems have several requirements that publicly owned systems do not have including:

(1) privately owned treatment systems are eligible only for existing principal residences constructed prior to, and occupied on or before, 27 December, 1977;
(2) small commercial establishments must pay back the federal state share of the cost of construction;
(3) non-profit and non-government institutional entities generally are treated the same as small commercial establishments.

Sand mound and evapotranspirational systems

The sand mound and the evapotranspirational systems are alternative approaches to on-site sewage disposal. Pennsylvania's Department of Environmental Resources[25] has permitted their application. The sand mound may be described as an 'above-ground' system, designed to take full advantage of the upper soil horizons to supplement the renovation processes which occur in the sand mound itself. It is dependent on lateral movement and dispersion of waste water, especially in soils of low permeability. Such movement promotes the uptake of nutrients and water by plants during the growing season. Wide dispersal is important to assist in dilution of pollutants (such as nitrates and chlorides).

The evapotranspiration system is a shallow system employing the concepts of evaporation of effluent to the atmosphere and transpiration of effluent by plants, as well as infiltration into the subsurface. Where tight soil, or high water table, precludes infiltration, an impermeable liner is used, and the system becomes 'total evapotranspiration of waste water'.

In Pennsylvania regulations require a 500 mm depth of suitable soil for 'on-site' sewage disposal. This excludes flood-plain soils and poorly drained soils on upland sites. Add to this slope limitations and only 50 % of the available land area remains. Sand mound and other alternate systems prevent further increase in this figure which could have reached 70 %.

In Erie County, New York state,[26] only 12 % of the area is suitable for conventional systems and the mound system was investigated. After extended comparisons there was found to be no statistically significant difference in performance between conventional absorption and sand mound systems during the first 25 years of system operation—though some soil associations appear to be better suited to mounds than others. The specification for mound systems is:

(1) depth of fill such that bottom of trenches will be at least 600 mm above ground water, rock, clay or other impervious stratum;
(2) fill material to be gravelly loam containing not less than 20 %, or more than 40 %, clay;
(3) trench area designed for a bottom dosing rate of 170 litres/m^2 per day and trench centres to be not less than three times trench width apart;
(4) shoulder of fill to be at least 3 m outside of centre line of outermost trench or end of pipe;
(5) mound surface to be graded to run off to side and ends of field;
(6) mound to be grass seeded or sodded.

This specification requires a mound with a base area of more than 670 m^2 for a three-bedroomed house with estimated flow of 1700 l/day. Certification, based on inspection and dye test, is required whenever ownership of property is to be transferred. Inspection ascertains that all and only sanitary sewage is discharged. Any appearance of dye on the ground surface is considered evidence of an unsatisfactory system.

The elevated sand mound

The sand mound was first introduced into Pennsylvania in the early 1970s, and in 1972 a technical advisory committee concluded that the system was environmentally sound and by 1976 some 2500 had been installed. 'Blending' of the mound within the building plot sometimes presents difficulties and has given rise to aesthetic objections, usually unfounded. The availability of sand has posed some problems—a dirty sand is required and the amount of fines is regulated so as to retard movement of effluent and give additional time for biological activity. Complexities of installation and maintenance are real, and must be carefully considered.

Evapotranspiration

This seemingly ideal system, which can be used to beautify gardens or increase food production, requires to meet three conditions to assure effectiveness:

(1) the waste water must contain dissolved oxygen (3.5 mg/l or more) because plant roots require oxygen-satisfied nutrient compounds;

(2) the evapotranspiration bed must be shallow (450–600 mm) so that plant roots can reach the nutrient containing liquid;

(3) the bed must be of sufficient area, depending on the climatically variable evapotranspiration rate (5 mm/day in the Lower Great Lakes Region).

Bernhart quotes actual installations to prove the method viable in North America, south of the 52nd or 55th parallel. Aerobic pretreatment increases the range of application. The same is true of a combination with soil infiltration. The method can withstand extreme cold—this is due to the energy produced by aerobic micro-organisms.

Alternatives to conventional sewerage

As an answer to unduly heavy expenditure on conventional sewers, particularly in difficult ground conditions, the use of small-diameter pipes forming either pressure or vacuum systems has had some success in the US[28] (see Figures 109 and 110). The pressure system originated from studies by the American Society of Civil Engineers (Ref. 6 in above paper) and have been taken up by the Environmental Protection Agency (EPA) who sponsor full-scale research and experimentation. This study has identified some 60 existing pressure sewer systems, and an equal number under consideration. Starting with a conventional septic tank, the system may employ a simple sump-pump, in a small chamber adjacent to the tank, to boost the overflow through a small-diameter pipe (say

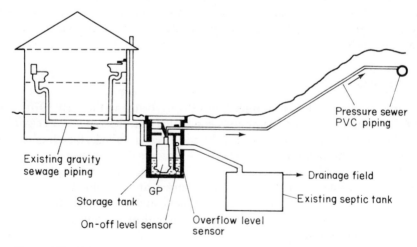

Figure 109 Pressure sewage—typical grinder pump connection (*Ann Arbor*)

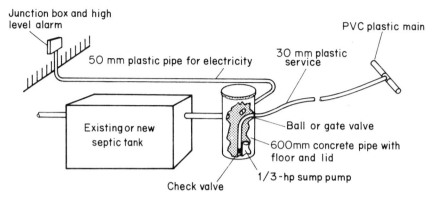

Figure 110   Pressure sewerage—typical STEP (septic tank, effluent pump) systems connection (*Ann Arbor*)

30 mm) to the pressure main. A refinement to this arrangement incorporates the introduction of a small storage tank *before* the septic tank, from which a grinder pump chops up the solids in the raw sewage and discharges direct to the pressure main. In this case flow to an existing septic tank would only take place in the event of failure of the pump, thus almost eliminating the need for septic tank emptying or maintenance. The previous arrangement only relieves the need for attention to the drainage field, or soakaway facilities originally provided.

The pressure mains conveying the flow to the central disposal point will be designed on the same basis as a rising main, and will generally be laid at constant depth below the ground. Although ostensibly simple, experience has already shown that such systems can only be expected to continue to operate satisfactorily if they are entirely under the control of the drainage authority. Whilst there is a trend towards central supervision of all individual systems, the introduction of a mechanical unit on every connection increases the general level of maintenance, and thereby the cost to the householder to some degree, although costs associated with faults on individual installations could well prove more expensive to remedy over short periods.

The vacuum system, like the pressure system, requires minimal depths of excavation and uses small diameter plastic pipes. The major difference is in the laying of the main, which, when going against the slope of the ground, must be arranged in a series of steps, consisting of downward sloping sections at 0.2% gradient, and short "lifts" at 45°, raising the pipe invert, say 300 mm, and keeping the depth of cover between about 900 and 1100 mm. This system would also have to apply on level ground, but on downward slopes, towards the pumping station, the pipes can follow at a regular gradient. The key item of equipment is a vacuum valve between the main and each connection. Sewage flows by gravity to a small sump, adjacent to the valve chamber. As the level rises in the sump it compresses the air in a censor pipe between the sump and the vacuum valve. At a predetermined level a diaphragm closes and signals the control unit, which then exhausts atmospheric air into the vacuum line, thus opening the vacuum valve. Sewage is then drawn into the system by differential pressure, normally two or three seconds, to empty the sump. The controller keeps the valve open for double this period. Fluid flow in the system is intermittent. With no valves open sewage fills the low points, whilst the downward slopes and risers are filled with air, leaving the bore of the pipe free for air to flow over the liquid, thus enabling high vacuum pressure to be maintained at the extremities. Isolating valves with vacuum gauges are provided on all main branches to allow for maintenance or repair. Vacuum valves require at least 125 mm of vacuum for operation, and this amount has to be deducted in determining the vacuum available for the conveyance of sewage in the system. A vacuum collector station is similar to a conventional sewage pumping station with collecting tank, vacuum reservoir, duplicate sewage discharge pumps, duplicate vacuum pumps and a standby generator. Other essential equipment is a re-set system of control so that the station is operational again after any electric supply failure. There must also be a complete shut-down control, as should both sewage pumps fail, it is vital that the vacuum pumps cease to operate so that sewage is not drawn up into the vacuum reservoir.

Treatment of sewage from pressure or vacuum systems presents no particular problems except that, if grinder pumps are used, the sewage will be stronger than average. If only the overflow from septic tanks has to be dealt with, average BOD_5 and suspended solids will be low, but the works will have to be designed to deal with the septage loads resulting from proper maintenance of the system.

In many cases, of course, it will be possible to deliver the flow to an existing gravity sewerage system. Flows per head should be comparatively low as, barring faulty work in the connections between properties and septic tanks, it should be possible to completely eradicate the effects of leakage of surface or ground water. Where treatment plants are required they may comprise either package units, or settlement, followed by seepage systems, sand beds (say $1.21/m^2$), or in lagoons, either aerated or evapotranspirational. Odours can produce problems, both at separate treatment works and at the discharge points to gravity sewers. A 30% solution of hydrogen peroxide has been found successful in counteracting odour. Recirculation of the flows to sand filters has also solved odour problems at treatment works.

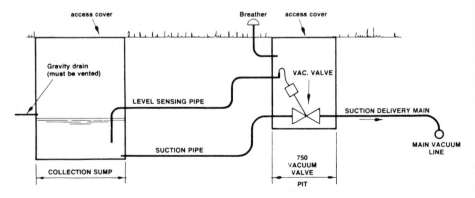

Figure 111   Vacuum valve installation (J. Instn. Public Health Engnrs.)

References

1. *Individual Onsite Wastewater Systems*. Proc. 1st National Conf., 1974, edited by Nina I. McClelland. Ann Arbor Science, Michigan (1975).
2. *Individual Onsite Wastewater Systems*. Proc. 2nd National Conf., 1975, Ann Arbor Science, Michigan (1976).
3. *Individual Onsite Wastewater Systems*. Proc. 3rd National Conf., 1976, Ann Arbor Science, Michigan (1977).
4. *Individual Onsite Wastewater Systems*. Proc. 4th National Conf., 1977, Ann Arbor Science, Michigan (1978).
5. *Individual Onsite Wastewater Systems*. Proc. 5th National Conf., 1978, Ann Arbor Science, Michigan (1979).
6. Imhoff, K. Müller, W. S. and Thistlethwayte, D. K. B. *Disposal of Sewage and other Waterborne Wastes*. Butterworth, London (1956).
7. Mara, D. *Sewage Treatment in Hot Climates*. John Wiley & Sons, Chichester (1978).
8. Gutormsen, D. Some aspects of composting toilets with specific reference to their function and practical applications in Norway.†
9. Pickford, J. Sanitation for buildings in hot climates. *J. Inst. Public Hlth Engrs*, **7** (4) (Oct. 1979).
10. Lumbers, J. F. Waste stabilisation ponds—design considerations and methods. *J. Inst. Public Hlth Engrs*, **7** (2) (Apr. 1979).
11. Gloyna, E. F. Waste stabilisation ponds. Monograph series No. 60. WHO, Geneva (1971).
12. McGarry, M. G. and Pescod, M. B. Stabilisation pond designs criteria for tropical Asia. 2nd Symp. for Waste Treatment Lagoons, Kansas City, June 1970. Fed. Wat. Quality Adm., Washington, D.C.
13. Bartsch, E. H. and Randall, C. W. Aerated lagoons—a report on the state of the art. *J. Wat. Pollution Control Fed.*, **43**, no. 4 (Apr. 1971).
14. Pearson, B. F. Aerated lagoons at high elevations. 2nd. Int. Symp. for Waste Treatment Lagoons, Kansas City, June 1970.
15. Marais, G. V. R. Faecal bacterial kinetics in stabilisation ponds. *J. Environmental Engnrs, Div. Am. Soc. Engnrs.*, **100**, EE1 (Feb. 1974).
16. Davis, C. V. *Handbook of Applied Hydraulics*. McGraw-Hill, New York (1965).
17. Arceivala, S. S. *et al.*, *Waste Stabilisation Ponds: design, construction and operation in India*. Central Public Hlth Engng Res. Unit, Nagpur, India (1970).
18. Sewage lagoons for developing countries—ideas and methods exchange. No. 62, Agency for Int. Dev., Washington, DC, 1966.
19. Watson, J. L. A. Oxidation ponds and use of effluent, in Israel. *Proc. Inst. Civil Engrs*, **22** (May 1962).
20. Oswald, W. J. and Gotaas, H. B. *Trans Am. Soc. Civ. Engnrs*, **81**, 686 (1955).
21. Wendle-Taylor, E., Improvement in the quality of a sewage works effluent after passing through a series of lagoons. *J. Inst. Public Hlth. Engnrs*, **LXV**, pt. 2 (Apr. 1966).
22. Neilson, N. E. Ozonation, irradiation, chlorination and combinations—a discussion on the practicabilities of their use in smaller water/wastewater management systems.†
23. Kroschel, M. L. Experiences with owner-built onsite waste management systems in California.†
24. Dearth, K. H. The impact of the Clean Water Act of 1977 on funding of individual wastewater systems.‡
25. Maurer, G. E. Field application: sand mound and evapotranspirational systems.*
26. Barry, D. E. and Donnelly, R. M., Experiences with mound systems in Erie County, N.Y.‡
27. Bernhart, A. P. Evapotranspiration—a viable method of re-use (or disposal) of wastewater in North America, south of the 52nd and 55th parallel.‡
28. Cooper, I. A. and Rezek, J. W. Treatability of pressure sewage.‡

* Paper presented at 3rd National Conference (see ref. 1–5).
† Paper presented at 4th National Conference (see ref. 1–5).
‡ Paper presented at 5th National Conference (see ref. 1–5).

CHAPTER 19

Control and Treatment of Industrial Wastes

Industrial wastes vary more in character than does domestic sewage; in fact, when a town sewage is in any way abnormal this is usually because of the trade wastes in it. Domestic sewage consists mostly of natural organic compounds, and it can be treated by sedimentation and oxidation with the aid of bacteria: trade wastes include inorganic substances in suspension or solution, some of which may be poisonous or liable to inhibit the action of bacteria; natural organic compounds and synthetic organic compounds. The last can involve problems both of sewerage and sewage treatment.

Owing to the great variety of trade wastes and methods of treating them, it is not possible, in this chapter, to give more than a general outline of trade-wastes treatment. The subject has been fully dealt with in specialist textbooks—of the nineteen listed,[1-19] British practice is probably best exemplified by Southgate,[2] Klein;[18] American, by Eckenfelder,[15] and Gehm.[19] In recent years the US Environmental Protection Agency has organized a large staff to report and advise on all aspects of sewage and trade-wastes treatment, to ensure compliance with the latest legislation. Manufacturing processes are dealt with in what are termed 'Development Documents for Proposed Effluent Limitation Guidelines and New Source Performance Standards'[90] of which a whole series are quoted in the references at the end of this chapter.

The wide use of natural gas, both in Great Britain and the US, has somewhat diminished the importance of the large amount of work done on the treatment of wastes arising from the manufacture of coal gas; however, large volumes of similar gases are still derived from the coal and oil distillation processes, so some references to such wastes have been retained.[20-25]

Although no specific allusion is made here to the methods of dealing with laundry waste, food-processing wastes, or those arising from plastics manufacture, information on these subjects will be found in the References.

The designer of works at which trade wastes—or sewages containing them in appreciable quantity—will be treated, should study the available literature, which includes monographs on specific wastes, and obtain the advice of a chemist with special knowledge of the subject.

Industrial wastes have been classified as follows:

Liquid wastes which are polluting by reason of the suspended matter which they carry with them, e.g.

coal-washing waste
waste waters from tin mines
waste waters from lead and zinc mines
waste waters from china clay pits
stone-quarrying waste
stone-polishing waste

Liquid wastes which are polluting mainly because of the suspended matter which they carry with them, but also to a considerable extent because of the dissolved impurities which they contain, e.g.

the cotton industry[90(c)]
the woollen industry
the manufacture of paper and cardboard[69,90(r,s)]

Liquid wastes which are polluting mainly because of the dissolved impurities which they contain, e.g.

waste liquor from wool scouring[46]
brewery waste[52]
steep water from maltings
waste liquor from the manufacture of sulphate cellulose[70]
the metal industries[9,40–42,44,90(a–d)]
fellmongers' waste[43,45]
tannery waste[90(u)]
dairy waste[30,31,58,60–2,90(p)]
margarine waste
waste liquor from shale-oil distillation
spent gas liquor[20–25]

This classification may still be applied, but many new and difficult industrial wastes have set up fresh problems for sewerage authorities.[35–39,90(c–m)]

The occupiers of trade premises may discharge industrial wastes to public sewers subject to the conditions which the sewerage authority may impose and the charges which they may make under the law in the country concerned. Charges for the treatment of industrial wastes at municipal treatment works are usually based on the estimated cost of the removal of suspended solids and the aeration of organic content in terms of annual repayment of loan on the capital costs of the extra works needed for this purpose and a fair proportion of the running costs of the settlement units, aeration units, and overheads.

Treatment at point of discharge

The works to be installed at factory premises so as to comply with whatever may be the sewerage authority's requirements may include balancing or storage tanks to regulate discharge, cooling tanks, sedimentation tanks, neutralization or other chemical-treatment tanks,[37,63,64] screens, grease traps, etc., the details of which depend on the nature of the trade waste. For example, a waste which contains a heavy but innocuous inorganic sediment might require no treatment other than sedimentation, but a waste from a metal works might involve the use of traps for oil removal, neutralization for acid content, and settlement to remove inorganic sediments.

The treatment of town sewage containing trade wastes is not necessarily difficult, for dilution with domestic sewage often renders oxidation of the trace wastes, with the aid of bacteria, practicable; the various trade wastes, some of which are acid and some alkaline, tend to neutralize each other. Nevertheless, in many instances an individual waste has been known to set up great difficulties at the treatment works. These difficulties are usually the result of wastes which interfere with the normal development of oxidizing bacteria.

When trade wastes are to be discharged, not to the public sewers but to natural watercourses, treatment is generally more difficult because there is no dilution with domestic sewage, therefore it may prove that the wastes are not amenable to biological oxidation.

Methods of industrial-wastes treatment

The first method of dealing with an industrial waste, which should always be considered where possible, is the reduction of the waste by the salving of materials in it. Very often it is found practicable to extract useful chemicals from trade-wastes water and thereby not only simplify the treatment problem, but at the same time effect an appreciable economy. Economy can also be effected by preventing salvable material from gaining access to liquid wastes, and problems of treatment can be reduced by altering methods of manufacture.

Once a trade waste has been produced it has to be treated in varying degrees according to whether it is to be discharged to a sewer or to a watercourse. Wastes discharged to sewers generally need to be treated only to such an extent as to ensure that they are not harmful to the structure of the sewer, that they are not liable to react with sewage causing deposition of sediment, that they will not cause organic growths in the sewer, and that they will not interfere with the processes of sewage treatment. Wastes discharged to watercourses must be purified so as to conform with the normal standards applicable to sewage effluents; they must also be non-poisonous.

The methods of treatment of trade wastes include bacteriological treatment[15,59] similar to the treatment of domestic sewage and special (e.g. chemical) treatment, dependent on the nature of the waste. These methods include land treatment, screening or straining, mechanical filtration, sedimentation, chemical precipitation[63] and/or neutralization, grease separation, anaerobic fermentation,[14,19,47] trickling-filter treatment,[52–57] activated-sludge treatment, and evaporation.

Land treatment

In some instances trade wastes are got rid of by soakage into the ground, and occasionally land filtration or broad irrigation can be used.[26-29,32] Generally, however, methods of land treatment are not to be recommended other than as temporary expedients.

Underground disposal

Discharge of a trade waste to a soakaway, borehole, shaft, or disused mine may seem superficially an attractive and cheap method of disposal. Unfortunately such wastes have a nasty habit of reappearing where they are not wanted, and in Great Britain it is an offence under the 1945 Wacter Act to discharge noxious effluents underground where there is a risk of polluting water supplies. Nevertheless this method of injecting into underground strata has been successfully used in the US for disposal of waste water from chemical industries. Suitable geological conditions are a deep porous layer underneath impervious rock, this separating the water from any shallow water-bearing zones. Depths as great as 4000 m have been used.[74-77]

Screening

The use of fine screens is advisable for removing solids from the effluents of many food and textile industries. The type of screen varies according to the size of particle to be removed and, in practice, bar screens of various grades and fine meshes, e.g. as small as 0.5 mm, are used. When fine-mesh screens are used, these are usually cleansed by back-washing with clarified effluent or water; the process of microstraining has distinct possibilities for trade-wastes treatment.

Mechanical filtration

Methods of filtration as used in the treatment of water, and including vacuum filtration, are applied to trade wastes for the removal or salvage of fine particles.[52,53]

Sedimentation

As in the treatment of sewage, sedimentation is one of the most commonly applied methods of removing suspended solids.[14] Continuous-flow sedimentation is adopted in many instances, but quiescent sedimentation is more common than it is at works for treating domestic sewage, being particularly applicable to the treatment of non-putrescible precipitants or to the operation of tanks which serve for holding-up purposes as well as for sedimentation.

Chemical precipitation and/or neutralization

Much of the content of trade wastes can frequently be removed by flocculation of suspended solids with the aid of precipitants, or precipitation of dissolved solids by reagents (including zeolites, etc.).[63] Chemicals used for the purpose of neutralizing acid or alkaline wastes, e.g. lime, frequently have to be settled out. Such chemical precipitation always involves the use of sedimentation tanks for the collection of sludge.

Adsorption

Adsorption, or the adherence of a substance (usually in solution) to the surface of another material, is sometimes of great value, and plays an important role in the widely used process of chemical flocculation. Important adsorbents are activated carbon, fullers earth, various natural clays, bauxite, and calcined magnesite. Activated carbon, used in small doses for removal of tastes and odour in water supplies, has also been found of use in the treatment of trade wastes, albeit in much larger quantities, rendering such treatment expensive. Activated-carbon filters have been used to complete the removal of phenols from gasworks effluent and for pre-treatment of strongly bactericidal wastes to render them suitable for treatment on trickling filters.[78–80]

Grease separation

Grease can be removed by chemical treatment, which causes it to be precipitated in sedimentation tanks. Generally, however, grease is separated in grease traps which are arranged for the skimming or decanting of solids or liquid grease from the surface of a quiescent trap. Such grease traps are preferably constructed, not as flat-bottomed chambers, but as inverted pyramids with the outlet at the apex, if it is desirable for them to pass on solids for removal in a further stage of treatment. Solid grease can be skimmed off the surface with the aid of skimmers consisting of wire gauze supported on wire frames and attached to long handles. Grease removal can be speeded up by aeration as in the diffused-air system.

Anaerobic fermentation

Anaerobic fermentation in septic tanks, sometimes aided by the addition of acid, gives partial treatment only, but is a useful adjunct to some treatment processes.[14,19,47–49]

Aeration

The methods of aeration can be applied to many trade wastes[50,51,71] if they are mixed with sewage, while some can be aerated alone. In a number of cases, such as citrus wastes, it is necessary to have a nutrient to support the organic life which otherwise cannot effect aeration. New cases need to be tested in the laboratory with small-scale model works before building the full-scale plant.

Flotation

Flotation provides a method for removing suspended solids and pseudo-colloidal suspension, or fatty, greasy, tarry, and oily matter, by causing them to rise to the surface of a liquid by the application of aeration, vacuum, or other suitable device. Komline, Sanderson, and Vokes Ltd.[84] supply a dissolved-air flotation unit available in various sizes, for use in treating tannery waste, packing-house and poultry wastes, and food processing. Eveson has patented a flotation method for coal-washing effluent, using an oxidizing or surface-active agent. Eliasson and Schulhoff[87] obtained good removal of BOD_5 and grease from laundry waste by a combination of chemical precipitation and flotation. Ames Crosta Babcock have developed an air-flotation process already applied to water and sewage treatment, which may also have its application in trade wastes treatment.

Trickling-filter treatment

Trickling-filter treatment can be applied to the oxidation of the organic content of many wastes of animal or vegetable origin, or of trade wastes mingled with domestic sewage.[52–58,62] The processes of alternating double-filtration, recirculation, etc. are considered particularly useful for treating some wastes.

Dialysis, electrodialysis, electrolysis

The development of the Graver acid-resistant vinyl film membrane (Graver Water Conditioning Co., New York) may well increase the number of applications of dialysis, especially where inorganic acids and salts have to be separated from organic materials.[82] Electrolysis, although seldom actually applied in trade wastes treatment (probably on account of expense), has a successful application in the recovery of copper from spent pickle-liquor[193,26] when copper is treated with sulphuric acid to remove scale. Silver has also been recovered at photo-processing premises; only a low voltage is required. Armour-Hess Chemicals Ltd. have a novel application—adding an electrolyte to trade wastes containing emulsified fatty materials and then passing a current of electricity and removing the fatty matter by air flotation. Laboratory experiments have shown electrodialysis has its chief use in desalination of brackish waters and it is used for this purpose also in gold mines.[238]

The activated-sludge process

The activated-sludge process has been applied with success to the treatment of organic trade wastes and sewages containing trade wastes.[64,65] Generally, however, it is not so elastic as trickling-filter treatment for dealing with sewages in which the trade-wastes content fluctuates.

Evaporation

When the moisture content of a waste is low and the waste has a salvage value, evaporation of water with the aid of heat or otherwise may be the cheapest means of disposal.[66-68,72]

Storage and/or disposal in sealed capsules in deep sea

These are methods applicable to radioactive wastes. Radioactive isotopes deteriorate at a reducing rate with time, and those which have a short half-life (the time in which they lose one-half of their activity), may be rendered harmless if stored for a sufficient time.[76,77] The method of disposal in deep sea is applicable to those wastes which have a long half-life. Radioactive effluents can be rendered less harmful by removal of radioactive sludge or by absorbing the dangerous substances with activated charcoal, which then has to be disposed of in a safe manner.

Chemical-resisting materials of construction

Lines of pipes, tanks, etc., accommodating flows of corrosive effluent need to be constructed in chemical-resisting materials, the qualities of which must be in accordance with the nature of the waste. Most often acid wastes are those which give trouble, but others can be harmful and, moreover, corrosive tendencies are not infrequently accompanied by abnormal temperatures. Chemical-resisting materials are very expensive; therefore they should not be used without adequate reason.

Pipework

For indoor purposes the materials available include chemical lead (the normal material for laboratory wastes), which should always be jointed with burned joints; also special high-grade ceramics. Out-of-doors, the normal substitute for ordinary vitrified clay pipes jointed in cement mortar are special pipes with chemical-resisting properties jointed with chemical-resisting cements. Pipework of suitable plastics is also used; such materials need protection against earth pressure.

Tank-work

Neutralizing, sedimentation, and other tanks, labyrinths, etc., for corrosive effluents have to be lined with chemical-resisting asphalts or high-quality brickwork jointed in chemical-resisting cement and asphalt-backed. The

Table 67

Silicate cements	Cement prodor	Grades for general and special acid purposes
Resin cements	Asplit 'CN' and 'O'	For mixed acids and alkalis
	Asplit 'CNS'	For mixed acids and alkalis
	Asplit 'CT'	For concentrated sulphuric acid and temperatures to 200 °C
	B100	For mixed acids and alkalis
	Furacin	For acid and strong alkalis and solvents
	CNSL standard	For mixed acids and alkalis
	CNSL, HF	For mixed acids
	Polyester P48	For mixed acids and alkalis
Special cements	Sulphur cement	For acids and mild alkali conditions
	Plasticized sulphur cement	For acids and mild alkali conditions
Mastics	Prodorphalts	For heating and spreading
	Prodorkitt	For heating and pouring
Rubber latex cement	Plassoleum	For waterproofing and underlay work for rubber, linoleum, etc., finishes

quality of brick used needs to be very high, for the bricks should not only be non-absorbent but should not break down under the influence of chemicals even at comparatively high temperatures. They should be heavy, vitrified, and smooth: in Great Britain Accrington 'Mori' bricks are considered excellent for this work.

Chemical-resisting cements and asphalts are manufactured by a number of firms; there is a variety of available materials, and a type of material needs to be chosen according to the reaction and concentration of the trade waste. Some examples available from Prodorite Ltd., Eagle Works, Wednesbury, Staffordshire, England are given in Table 67.

References

1. Eldridge, E. F., *Industrial Waste Treatment Practice*. McGraw-Hill, N.Y. and Lond. (1942).
2. Southgate, B. A. *Treatment and Disposal of Industrial Waste Waters*, HMSO, Lond. (1948).
3. Rudolfs, W., *Industrial Wastes, Their Disposal and Treatment*, HMSO, Lond. (1953).
4. Isaac, P. C. G. (ed.), *The Treatment of Trade Wastes Waters and the Prevention of River Pollution*, C.R. Books, Lond. (1957).
5. Isaac, P. C. G. (ed.), *Waste Treatment*. Pergamon Press, Oxford (1960).
6. Soc. of Chem. Ind., *Disposal of Industrial Waste Materials*, Lond. (1957).
7. Hodgson, H. J. N., *Report on Sewage and Trade Waste Treatment in America, Gt. Britain, Holland, Germany*. Govt. of South Australia, Trigg, Adelaide (1938).

8. Sierp, F., *Waste Waters from Trade and Industry*. Springer-Verlag, Berl.; Lange Maxwell and Springer Ltd, Lond. (1953).
9. Weiner, R., *Effluents in the Metal Industries*. 2nd edn. E. G. Leuze, Verlag, Saulgau, Germany (1958).
10. Besselievre, E. B. *Industrial Waste Treatment*, McGraw-Hill, N.Y. and Lond. (1952).
11. Gurnham, C. F., *Principles of Industrial Waste Treatment*. Wiley & Sons, N.Y. (1955).
12. Gurnham, C. F. (ed.), *Industrial Waste Water Control*, Academic Press, N.Y. and Lond. (1965).
13. Nemerow, N. L., *Theories and Practices of Industrial Waste Treatment*. Addison-Wesley, Reading, Mass. (1963).
14. McCaba, J. and Eckenfelder, W. W. (jr.), *Biological Treatment of Sewage and Industrial Wastes*, Vol. 1: *Aerobic Oxidation*; Vol. 2: *Aerobic Digestion and Solids–Liquids Separation*. Reinhold, N.Y. (1956; 2nd edn 1958).
15. Eckenfelder, W. W. and O'Connor, D. J. *Biological Waste Treatment*, Pergamon Press, Oxford (1961).
16. Rich, L. G., *Unit Operations of Sanitary Engineering*, Wiley & Sons, N.Y. (1961).
17. Rich, L. G., *Unit Processes of Sanitary Engineering*, Wiley & Sons, N.Y. (1963).
18. Klein, L., *River Pollution*, Vol. 3: *Control*, Butterworth, Lond.; Aust.: Sydney, Melbourne, Brisbane; Canada: Toronto; New Zealand: Wellington, Auckland; South Africa: Durban (1966).
19. Gehm, H. W. and Bregman, J. (eds). *Handbook of Water Resources and Pollution Control*. Van Nostrand Reinhold, New York, Toronto, London, Melbourne (1976).
20. Institution of Gas Engineers, 6th Report of Liquor/Effluents and Ammonia Committee, 1936. Comm. No. 142.
21. Webber, J. *Experiments on the use of Gas Liquor* (*as fertilizer*), Agriculture, London 571–4 (1957).
22. Law, W. W. The production and use of ammoniacal liquor as a nitrogeneous fertilizer.
23. Key, A. *Gas Works Effluents and Ammonia*, 2nd edn (revised by Gardiner, P. C.). Inst. Gas Engineers, London (1956).
24. Sherriff, F. C. Gas liquor as an agricultural fertilizer. *Gas J.*, **269** (1952).
25. Marsden, A. and Stansfield, I. L. Ammoniacal liquor as a fertilizer effect of various constituents on the activity of microbiological flora, *J. Sci. Food Agric.*, **10** (1959).
26. Scott, R. H. Disposal of high organic content wastes on land. *J. Wat. Poll. Control Fed.*, **34** (1962).
27. Sanborn, N. H. Spray irrigation as a means of cannery waste disposal. *Canning T.*, **74** (1952).
28. Nelson, L. E. Cannery Waste Disposal by Spray irrigation. *Wastes Eng.*, **23** (1952).
29. Dennis, J. M. Spray irrigation of food processing wastes. *Sewage Indust. Wastes*, **25** (1953).
30. McKee, F. J. Dairy waste disposal by Spray irrigation. *Sewage Indust. Wastes*, **29** (1957).
31. McDowell, F. H. and Thomas, R. H. *Disposal of Dairy Wastes by Spray Irrigation on Pasture Land*. New Zealand Pollution Advisory Council, Publ. No. 8 (1961).
32. Luley, H. G., Spray irrigation of vegetable and fruit processing wastes. *J. Wat. Poll. Control Fed*, **35** (1963).
33. Krieghoff, H. Microfiltration. *Gas—I. C. Wasserfach*, **104** (1963).
34. Spade, J. F. Treatment methods for laundry wastes. *Wat. Sewage Works*, **189** (1962).
35. Anon. Cone-Type Separator for difficult trade wastes. *Surveyor, London*, **121** (1963).
36. Anon. Compact solids separator for trade wastes, *Wat. Waste Treat. J.*, **9** (1963).
37. Gurnham, C. F. Chemical engineering aspects of industrial wastes control. *Sewage Indust. Wastes*, **23** (1951).
38. Jenkins, S. H. and Hewitt, C. H. Acid wastes, their source, composition, effect on sewage treatment, and their neutralization. (Symp. Trade Wastes), *Surveyor, Lond.*, **116** (1957).

39. Jacobs, H. L. Neutralization of acid wastes, *Sewage Ind. Wastes*, **23** (1951).
40. Eden, G. E. and Truesdale, G. A., Treatment of waste waters from the pickling of steel. *J. Iron Steel Inst.*, **164** (1950).
41. Hook, R. D. Acid iron wastes neutralization. *Sewage, Ind. Wastes*, **22** (1950).
42. Hook, R. D. Neutralization of spent pickle liquor. *Air and Water Pollution in the Iron and Steel Industry*. Special Rep. No. 61, Iron and Steel Inst., London (1958).
43. Lab, R. F. An acid iron wastes neutralization plant, *Sewage Ind. Wastes*, **22** (1950).
44. *Water Pollution Research, 1948*. HMSO, Lon. (1949).
45. Wheatland, A. B. and Borne, B. J., Neutralization of acidic waste waters in beds of calcined magensite. *Wat. Waste Treatment J.*, **8** (1962).
46. McCarthy, J. A. Method of treatment for wool scouring wastes. *Sewage Works J.*, **21** (1949).
47. Gehm, H. W. and Behn, V. C. High-rate anaerobic digestion of industrial wastes. *Sewage Ind. Wastes*, **25** (1953).
48. Dept. of Scientific and Industrial Research. Anaerobic digestion of industrial wastes. *Notes on Water Pollution*, No. 4. HMSO, London (1959).
49. Pettet, A. E. J., Tomlinson, T. G. and Hemens, J. The treatment of strong organic wastes by anaerobic digestion. *J. Public Hlth. Engnrs.* **58** (1959).
50. Waste and Pollution Control Fed. *Air Diffusion in Sewage Works*. Man. of Practice No. 5 (1952).
51. Green, R. V. and Mosec, D. V., Destructive catalytic oxidation of aqueous waste materials. *Sewage Ind. Wastes*, **24** (1952).
52. Tidswell, M. A. Sewage disposal research at Burton-on-Trent. Effluent and Water Treatment Convention, London (1962).
53. Porter, J. A. and Dutch, P. H., Phenol-cyanide removal in a plastic-packed trickling filter. *J. Wat. Pollution Control Fed.*, **32** (1960).
54. Egan, J. T. and Sandley, M. Evaluation of plastic trickling filter media. *Ind. Wastes*, **5** (1960).
55. Cawley, W. A. and Brouillette, R. W. Polyvinyl chloride for trickling filters. *Ind. Wastes*, **7** (1962).
56. Minch, V. A., Egan, J. T. and Sandlin, M., Design and operation of plastic filter media, *J. Wat. Poll. Control Fed.*, **34** (1962).
57. Minch, V. A., Egan, J. T. and Sandlin, M. Plastic trickling filters—design and operation. *Paper Trades J.* **146** (1962).
58. *Water Pollution Research 1952*, HMSO, Lond. (1953).
59. Pettet, A. E. J. and Mills, E. V., Biological treatment of cyanides with and without sewage. *J. Appl. Chem.*, **4** (1954).
60. *Water Pollution Research 1953*. HMSO, Lond. (1954).
61. Jenkins, S. H. Laboratory and larger-scale experiments on the purification of dairy wastes. *J. Inst. Sewage Purif.*, **1** (1937).
62. Jenkins, S. H., The treatment of waste waters from dairies. *J. Inst. Sewage Purif.*, **2** (1937).
63. Ayres, J. A., Treatment of radioactive waste by ion-exchange. *Ind. Engng. Chem.*, **43** (1951).
64. Murphy, R. S. and Nesbitt, J. B., Biological treatment of cyanide wastes. *Penn. St. Univ. Engng. Res. Bull.*, **B-88** (1964).
65. Waldmeyer, T., Treatment of formaldehyde wastes by activated sludge methods, *J. Inst. Sewage Purif.*, **1** (1952).
66. Kohlins, W. D. and Demarest, E. L. Waste utilization and disposal systems cost studies of evaporation and drying, *Chem. Engng. Progr.*, **46** (1950).
67. McCullough, G. E. Concentration of radioactive liquid waste by evaporation. *Indust. Engng. Chem.*, **43** (1951).
68. Horsch, C. E. How Hanford evaporates fission wastes. *Chem. Engng.*, **60** (1953).
69. Jolley, R. S. Burning of sulphite liquor, pulp and Paper. Ind., **26** (10) (1952).

70. Rootman, E. T. Viscose rayon manufacturing wastes and their treatment, Pts. I and II. *Waterworks and Sewerage*, **91** (1944).
71. Gauvin, W. H., Application of the atomised suspension technique, TAPPI, **40** (1957).
72. Austin, J. Submerged combustion evaporators. *Brit. Chem. Engng.*, **2** (1957).
73. Rody, J. J. Treatment and Disposal of Oilfield brine. *Sewage Works J.*, **13** (1940).
74. Lee, J. A., Throw your waste down the well. *Sewage Ind. Wastes*, **30** (1952).
75. Cecil, L. K. Underground disposal of process waste water. *Ind. Engng. Chem.*, **42** (1950).
76. Black, W. B. Underground waste disposal. *Sewage Ind. Wastes*, **30** (1958).
77. Anon. New process which discharges effluent underground. *Chem. Process*, **4** (1958).
78. Hyadshaw, A. Y. Treatment application points for activated carbon. *J. Am. Wat. Works Assoc.*, **54** (1962).
79. Snell, F. D. Treatment of trade waste with activated carbon, *Ind. Engng. Chem.*, **27** (1953).
80. Rudolfs, W. and Hanlon, W. D. Colour in industrial wastes. VI: Effect of biological treatment and activated carbon. *Sewage Ind. Wastes*, **25** (1953).
81. Sharp, D. H. and Lambden, A. E. Treatment of strongly bactericidal trade effluent by activated charcoal and biological means, *Chem. and Ind. (Rev.)*, **39** (1955).
82. Anon. Separation by dialysis. *Chem. Process*, **5** (1959).
83. Anon. Silver recovery from film processing wastes. *Wat. Waste Treatment J.*, **8** (1960).
84. Komline Sanderson Dissolved Air Flotation Units, Bull. No. KSM–8 2/62. Kom. Sand. Eng. Corp., New Jersey, USA.
85. Anon., Froth flotation at Frickley Colliery. *Colliery Engng.*, **27** (1950).
86. Rickert, E. E. and Bishop, W. T. Wash water treatment and fine coal recovery. *Ind. Engng. Chem.*, **42** (1950).
87. Eliasson, R. and Schulhoff, H. B., Laundry waste treatment by flotation. *Waterworks and Sewage*, **90** (1943).
88. *Standard Methods for the Examination of Water and Waste Water*, 13th edn. APHA, AWWA, and WPCF., New York (1971).
89. *Effluent Limitation Guidance for the Refuse Act Permit Program Steel Industry*, Environmental Prot. Ag. (Oct. 1972).
90. Environmental Prot. Ag., Development Documents for Proposed Effluent Limitations Guidelines and New Source Performance Standards, for:

    (a) The Iron and Steel Industry, EPA Contract No. 68-01 1507 (draft). June 1973.
    (b) Non Ferrous Metals Industry, EPA Contract No. 68-01-1518 (draft). June 1973.
    (c) Bauxite Refining Sub-category, EPA 440/1-73/019. Oct. 1973.
    (d) Copper Nickel Chromium and Zinc Segments, EPA 440/1 73-003. Aug. 1973.
    (e) Petroleum Refining Industry, EPA 440/1-73/014. June 1973.
    (f) Inorganic Chemicals Alkali and Chlorine Industry, EPA Contract No. 68-01-15-1513 (draft). June 1973.
    (g) Organic Chemical Industry, EPA Contract No. 68-01-1509 (draft). June 1973.
    (h) Fertilizer and Phosphates Manufacturing Industry, EPA Contract No. 68-01-1508 (draft). June 1973.
    (j) Plastic and Synthetics Industry, EPA Contract No. 68-01-1500 (draft). June 1973.
    (k) Major Inorganic Products Segment of the Inorganic Chemicals Manufacturing Point Source Category, EPA 440/1 73/007. Aug. 1973.
    (l) Phosphorus Derived Chemicals Segment of the Phosphate Manufacturing Point Source Category, EPA 440/1-73/006. Aug. 1973.
    (m) Synthetic Resins Segment of the Plastics and Synthetic Materials Manufacturing Point Source Category, EPA 440/1/73/010. Aug. 1973.
    (n) Canned and Preserved Fish and Seafoods Processing Industry, EPA Contract No. 68-01-1526. July 1973.
    (o) Meat Packing Industry, EPA Contract No. 68/01/0593 (draft). June 1973.

(p) Dairy Products Processing Point Source Category, EPA 440/1-73/021. Jan. 1974.

(q) Citrus Apple and Potato Segment of the Canned and Preserved Fruits and Vegetable Processing Industry, Federal Register 88-31076-31084. Nov. 1973.

(r) Building Paper and Roofing Felt Segment of the Builders Paper and Board Mills Point Source Category, EPA 440/1-74/-26. Jan. 1974.

(s) Umbleached Kraft and Semi-Chemical Pulp Segment of the Pulp Paper and Paperboard Mills Point Source Category, EPA 440/1-74/025. Jan. 1974.

(t) For the Textile Mills, EPA 440/1-74/022. Jan. 1974.

(u) Leather and Tanning and Finishing Industry, EPA Contract No. 68/01-0594 (draft). June 1973.

Appendix I

Metric Equivalents

Linear 1 inch = 25.4 millimetres
1 decimetre = 3.93701 inches.
1 foot = 3.048 decimetres.
1 metre = 3.28084 feet.

Square 1 square inch = 645.16 square millimetres.
1 square decimetre = 15.5 square inches.
1 square foot = 9.2903 square decimetres.
1 square metre = 10.7639 square ft.
1 hectare = 2.47105 acres.

Cubic 1 US gallon = 3.78541 cubic decimetres (litres) = 5/6th imperial gallon.
1 imperial gallon = 4.546 cubic decimetres (litres) = 1.2 US gallons.
1 cubic foot = 28.3168 cubic decimetres (litres).
1 cubic metre = 35.3147 cubic feet.

Pressure 1 kilogramme per square centimetre = 14.223 pounds per square inch = 10 metres head of water.
1 atmosphere = 1.03323 kilogrammes per square centimetre = 14.6959 pounds per square inch.
1 lb. per square inch = 6894.76 newtons per square metre

Mass 1 kilogramme = 2.20462 pounds

Temperature °Centigrade = (°Fahrenheit −32)/1.8 = °Kelvin −273.15.

Heat and energy 1 British thermal unit = 1.05506 kilojoules.
1 kilocalorie = 4.1868 kilojoules = 3.96832 British thermal units = 426.935 kilogramme metres = 3088.03 foot pounds.

1 foot pound = 1.35582 joules.

1 kilowatt hour = 3412 British thermal units = 859.845 kilocalories = 1.341 horsepower hours = 2,655,200 foot pounds = 367,100 kilogramme metres.

Power 1 kilowatt = 1 kilojoule per second = 1.34102 horsepower = 6.11832 cubic metres (of water) metres per minute = 101.972 kilogramme metres per second.

1 horsepower = 1.01387 metric horsepower.

1 metric horsepower = 75 kilogramme metres per second.

Standard gravity 9.80665 metres per second per second = 32.174 feet per second per second.

Flow Gal/min × 0.0758 = l/s (litres per second)

Gal/day × 0.004546 = m^3/day

Mil. gal/day × 4.546 = 10^3 m^3/day

BOD_5 *per head* lb. BOD_5/head/day × 0.4536 = kg BOD_5/head day

Screenings ft^3/mil. gal × 0.00623 = m^3/10^3 m^3

Tank surface loading gal/ft^2/day × 0.049 = m^3/m^2/day

Weir loading gal/ft./day × 0.0149 = m^3/m/day

Filter loading gal/yd^3/day × 0.005946 = m^3/m^3/day

Filter superficial gal/yd^2/day × 0.005437 = m^3/m^2/day

BOD_5 lb BOD_5/yd^3/day × 0.5932 = kg BOD_5/m^3/day

Activated sludge loading lb BOD_5/1000 ft^3/day × 0.016 = kg BOD_5/m^3/day

Air ft^3/gal × 6.23 = m^3/m^3

ft^3/lb BOD_5 removed × 0.0624 = m^3/kg BOD_5

ft^3/min/ft^2 area × 0.3048 = m^3/min./m^2

ft^3/min × 0.000472 = m^3/s.

Power hp/mil. gal/day × 0.164 = W/m^3/day

Tertiary filters gal/ft^2/hour × 1.176 = m^3/m^2/day

Grass plots mil. gal/acre/day × 1.123 × m^3/m^2/day

Sludge lb volatile matter or lb dry solids/head/dry × 0.4536 = kg/head/day

calorific value Btu/lb × 2.326 = kJ/kg

gas ft^3/lb volatile matter destroyed × 0.0624 = m^3/kg

loading lb volatile matter/ft^3/day × 16.03 = kg/m^3/day

dewatering lb dry solids/ft^2/hour × 4.881 = kg/m^2/hour

Appendix II

FLOW IN CONCRETE-PIPE SEWERS ACCORDING TO SCOBEY'S FORMULA

(Reproduce by courtesy of S. H. Dainty, B.Sc., C.Eng., F.I.C.E., M.I.W.P.C., Director of Public Health Engineering, Greater London Council)

Length/Fall	4 inch diameter		6 inch diameter		9 inch diameter		12 inch diameter		15 inch diameter		18 inch diameter	
	Velocity (ft./min.)	Discharge (cu. ft./min.)	Velocity (ft./min.)	Discharge (cu. ft./min.)	Velocity (ft./min.)	Discharge (cu. ft./min.)	Velocity (ft./min.)	Discharge (cu. ft./min.)	Velocity (ft./min.)	Discharge (cu. ft./min.)	Velocity (ft./min.)	Discharge (cu. ft./min.)
5	697	60.8	898	176								
6	636	55.5	820	161								
7	589	51.4	759	149	978	432						
8	551	48.1	710	139	915	404						
9	519	45.3	669	131	862	381						
10	493	43.0	635	124	818	361	979	769				
11	470	41.0	605	118	780	344	934	733				
12	450	39.2	579	113	747	330	894	702				
13	432	37.7	557	109	717	317	859	674	987	1212		
14	416	36.3	536	105	691	305	827	650	951	1168		
15	402	35.1	518	101	668	295	799	628	919	1128		
16	389	34.0	502	98.5	647	285	774	608	890	1092	997	1763
17	378	33.0	487	95.6	627	277	751	590	863	1059	968	1710
18	367	32.0	473	92.9	610	269	730	573	839	1030	940	1662
19	357	31.2	460	90.4	593	262	710	558	817	1002	915	1618
20	348	30.4	449	88.1	578	255	692	544	796	977	892	1577
21	340	29.7	438	86.0	564	249	676	530	777	953	870	1539
22	332	29.0	428	84.0	551	243	660	518	759	931	850	1503
23	325	28.4	418	82.2	539	238	645	507	742	911	832	1470
24	318	27.8	410	80.5	526	233	632	496	726	892	814	1439
25	311	27.2	401	78.8	517	228	619	486	712	874	798	1410
26	305	26.7	393	77.3	507	224	607	477	698	587	782	1383
27	300	26.2	386	75.9	498	220	596	468	685	841	768	1357
28	294	25.7	379	74.5	489	216	585	459	673	825	754	1332
29	289	25.3	373	73.2	480	212	575	451	661	811	741	1309
30	284	24.8	366	72.0	472	208	565	444	650	797	728	1287
31	280	24.4	360	70.8	464	205	556	436	639	784	716	1266

32	275	24.0	355	69.7	457	202	547	430	629	772	705	1246
33	271	23.7	349	68.6	450	199	539	423	619	760	694	1227
34	267	23.3	344	67.6	443	196	531	417	610	749	684	1209
35	263	23.0	339	66.6	437	193	523	411	601	738	674	1192
36	259	22.7	334	65.7	431	190	516	405	593	728	665	1175
37	256	22.4	330	64.8	425	187	509	399	585	718	656	1159
38	252	22.1	325	63.9	419	185	502	394	577	708	647	1144
39	249	21.8	321	63.1	414	183	496	389	570	699	639	1129
40	246	21.5	317	62.3	409	180	489	384	563	691	631	1115
41	243	21.2	314	61.6	404	178	483	379	556	682	623	1101
42	240	21.0	309	60.8	399	176	478	375	549	674	615	1088
43	237	20.7	306	60.1	394	174	472	371	543	666	608	1075
44	235	20.5	302	59.4	390	172	467	366	536	658	601	1063
45	232	20.3	299	58.7	385	170	461	362	530	651	594	1051
46	229	20.1	296	58.1	381	168	456	358	525	644	588	1039
47	227	19.8	293	57.5	377	166	451	354	519	637	582	1028
48	225	19.6	289	56.9	373	165	447	351	514	630	576	1017
49	222	19.4	286	56.3	369	163	442	347	508	624	570	1007
50	220	19.2	284	55.7	366	161	438	344	503	618	564	997
51	218	19.1	281	55.2	362	160	433	340	498	611	558	987
52	216	18.9	278	54.6	358	158	429	337	493	606	553	978
53	214	18.7	275	54.1	355	157	425	334	489	600	548	968
54	212	18.5	273	53.6	352	155	421	331	484	594	543	959
55	210	18.3	270	53.1	348	154	417	328	480	589	538	951
56	208	18.2	268	52.7	345	152	413	325	475	584	533	942
57	206	18.0	266	52.2	342	151	410	322	471	578	528	934
58	204	17.9	263	51.7	339	150	406	319	467	573	524	926
59	203	17.7	261	51.3	336	148	403	316	463	568	519	918
60	201	17.6	259	50.9	334	147	399	314	459	564	515	910
61	200	17.4	257	50.4	331	146	396	311	455	559	511	903
62	198	17.3	255	50.0	328	145	393	308	452	555	506	895
63	196	17.1	253	49.6	326	144	390	306	448	550	502	888
64	195	17.0	251	49.2	323	142	387	304	445	546	498	881
65	193	16.9	249	48.9	321	141	384	301	441	542	495	874
66	192	16.7	247	48.5	318	140	381	299	438	537	491	868
67	190	16.6	245	48.1	316	139	378	297	435	533	487	861

Length/Fall	4 inch diameter		6 inch diameter		9 inch diameter		12 inch diameter		15 inch diameter		18 inch diameter	
	Velocity (ft./min.)	Discharge (cu. ft./min.)	Velocity (ft./min.)	Discharge (cu. ft./min.)	Velocity (ft./min.)	Discharge (cu. ft./min.)	Velocity (ft./min.)	Discharge (cu. ft./min.)	Velocity (ft./min.)	Discharge (cu. ft./min.)	Velocity (ft./min.)	Discharge (cu. ft./min.)
68	189	16.5	243	47.8	313	138	375	295	431	529	484	855
69	188	16.4	241	47.4	311	137	372	292	428	526	480	849
70	186	16.3	240	47.1	309	136	370	290	425	522	477	842
71	185	16.1	238	46.8	307	135	370	288	422	518	473	837
72	184	16.0	236	46.4	305	134	365	286	419	515	470	831
73	182	15.9	235	46.1	302	133	362	284	416	511	467	825
74	181	15.8	233	45.8	300	132	360	282	414	508	463	819
75	180	15.7	231	45.5	298	132	357	280	411	504	460	814
76	179	15.6	230	45.2	296	131	355	279	408	501	457	809
77	178	15.5	228	44.9	294	130	353	277	405	498	454	803
78	177	15.4	227	44.6	293	129	350	275	403	494	451	798
79	175	15.3	225	44.3	291	128	348	273	400	491	449	793
80	174	15.2	224	44.0	289	127	346	272	398	488	446	788
81	173	15.1	223	43.8	287	127	344	270	395	485	443	783
82	172	15.0	221	43.5	285	126	342	268	393	482	440	778
83	171	14.9	220	43.2	284	125	340	267	390	479	438	774
84	170	14.8	219	43.0	282	124	338	265	388	476	435	769
85	169	14.8	217	42.7	280	124	336	263	386	474	432	764
86	168	14.7	216	42.5	279	123	334	262	384	471	430	760
87	167	14.6	215	42.2	277	122	332	260	381	468	427	756
88	166	14.5	214	42.0	275	121	330	259	379	465	425	751
89	165	14.4	212	41.8	274	121	328	257	377	463	423	747
90	164	14.3	211	41.5	272	120	326	256	375	460	420	743
91	163	14.3	210	41.3	271	119	324	255	373	458	418	739
92	163	14.2	209	41.1	269	119	322	253	371	455	416	735
93	162	14.1	208	40.8	268	118	321	252	369	453	413	731
94	161	14.0	207	40.6	266	117	319	250	367	450	411	727

95	160	14.0	206	40.4	265	117	317	249	365	448	409	723
96	159	13.9	205	40.2	264	116	316	248	363	446	407	719
97	158	13.8	203	40.0	262	116	314	247	361	443	405	716
98	157	13.7	202	39.8	261	115	312	245	359	441	403	712
99	157	13.7	201	39.6	260	114	311	244	357	439	401	708
100	156	13.6	200	39.4	258	114	309	243	356	437	399	705
105	152	13.3	196	38.4	252	111	302	237	347	426	389	688
110	149	13.0	191	37.6	246	109	295	231	339	416	380	672
115	145	12.7	187	36.7	241	106	288	226	332	407	372	657
120	142	12.4	183	36.0	236	104	282	222	325	398	364	643
125	139	12.2	179	35.2	231	102	277	217	318	390	356	630
130	137	11.9	176	34.5	226	100	271	213	312	383	350	618
135	134	11.7	172	33.9	222	98.4	266	209	306	376	343	607
140	132	11.5	169	33.3	218	96.6	261	205	300	369	337	596
145	129	11.3	166	32.7	214	94.9	257	202	295	362	331	585
150	127	11.1	164	32.2	211	93.3	252	198	290	356	325	575
155	125	10.9	161	31.6	207	91.8	248	195	286	351	320	566
160	123	10.8	158	31.1	204	90.3	244	192	281	345	315	557
165			156	30.7	201	89.0	241	189	277	340	310	549
170			154	30.2	198	87.6	237	186	273	335	306	540
176			151	29.8	195	86.4	234	183	269	330	301	533
180			149	29.3	192	85.2	230	181	265	325	297	525
185			147	28.9	190	84.0	227	178	261	321	293	518
190			145	28.6	187	82.9	224	176	258	317	289	511
195			143	28.2	185	81.8	221	174	255	312	285	505
200			142	27.8	183	80.8	219	172	251	309	282	498
205			140	27.5	180	79.8	216	169	248	305	278	492
210			138	27.2	178	78.8	213	167	245	301	275	486
215			136	26.8	176	77.9	211	165	242	298	272	481
220			135	26.5	174	77.0	208	164	240	294	269	475
225			133	26.2	172	76.2	206	162	237	291	266	470
230			132	26.0	170	75.3	204	160	234	288	263	465
235			131	25.7	168	74.5	202	158	232	285	260	460
240			129	25.4	167	73.8	199	157	229	282	257	455
245			128	25.1	165	73.0	197	155	227	279	254	450
250			127	24.9	163	72.3	195	153	225	276	252	446

Length / Fall	4 inch diameter Velocity (ft./min.)	Discharge (cu. ft./min.)	6 inch diameter Velocity (ft./min.)	Discharge (cu. ft./min.)	9 inch diameter Velocity (ft./min.)	Discharge (cu. ft./min.)	12 inch diameter Velocity (ft./min.)	Discharge (cu. ft./min.)	15 inch diameter Velocity (ft./min.)	Discharge (cu. ft./min.)	18 inch diameter Velocity (ft./min.)	Discharge (cu. ft./min.)
255			125	24.6	162	71.5	193	152	223	273	249	441
260			124	24.5	160	70.9	192	150	220	271	247	437
265			123	24.2	158	70.2	190	149	218	268	245	433
270			122	24.0	157	69.5	188	148	216	265	242	429
275			121	23.7	156	68.9	186	146	214	263	240	425
280			120	23.5	154	68.3	185	145	212	261	238	421
285					153	67.7	183	144	210	258	236	417
290					151	67.1	181	142	209	256	234	414
295					150	66.5	180	141	207	254	232	410
300					149	66.0	178	140	205	252	230	407
305					148	65.4	177	139	203	250	228	403
310					146	64.9	175	138	202	248	226	400
315					145	64.4	174	137	200	246	224	397
320					144	63.9	173	136	199	244	223	394
325					143	63.4	171	134	197	242	221	391
330					142	62.9	170	133	196	240	219	388
335					141	62.4	169	132	194	238	218	385
340					140	62.0	168	131	193	237	216	382
345					139	61.5	166	130	191	235	214	379
350					138	61.1	165	130	190	233	213	276
355					137	60.6	164	129	189	231	211	374
360					136	60.2	163	128	187	230	210	371
365					135	59.8	162	127	186	228	208	369
370					134	59.4	161	126	185	227	207	366
375					133	59.0	159	125	183	225	206	364
380					132	58.6	158	124	182	224	204	361
385					131	58.2	157	123	181	222	203	359

390	131	57.8	156	123	180	221	202	357
395	130	57.5	155	122	179	219	200	354
400	129	57.1	154	121	178	218	199	352
405	128	56.8	153	120	176	217	198	350
410	127	56.4	152	120	175	215	197	348
415	127	56.1	152	119	174	214	195	346
420	126	55.7	151	118	173	213	194	344
425	125	55.4	150	118	172	211	193	342
430	124	55.1	149	117	171	210	192	340
435	124	54.8	148	116	170	209	191	338
440	123	54.5	147	115	169	208	190	336
445	122	54.1	146	115	168	207	189	334
450	122	53.8	146	114	167	206	188	332
455	121	53.5	145	114	166	204	187	330
460	120	53.3	144	113	166	203	186	328
465	120	53.0	143	112	165	202	185	327
470			142	112	164	201	184	325
475			142	111	163	200	183	323
480			141	111	162	199	182	321
485			140	110	161	198	181	320
490			139	109	160	197	180	318
495			139	109	160	196	179	317
500			138	108	159	195	178	315
510			137	107	157	193	176	312
520			135	106	156	191	175	309
530			134	105	154	189	173	306
540			133	104	153	188	171	303
550			132	103	151	186	170	300
560			130	102	150	184	168	298
570			129	101	149	183	167	295
580			128	101	147	181	165	292
590			127	100	146	179	164	290
600			126	99.3	145	178	162	287
610			125	98.5	144	176	161	285
620			124	97.7	143	175	160	283
630			123	96.9	141	174	159	280

Length / Fall	4 inch diameter		6 inch diameter		9 inch diameter		12 inch diameter		15 inch diameter		18 inch diameter	
	Velocity (ft./min.)	Discharge (cu. ft./min.)	Velocity (ft./min.)	Discharge (cu. ft./min.)	Velocity (ft./min.)	Discharge (cu. ft./min.)	Velocity (ft./min.)	Discharge (cu. ft./min.)	Velocity (ft./min.)	Discharge (cu. ft./min.)	Velocity (ft./min.)	Discharge (cu. ft./min.)
640							122	96.1	140	172	157	278
650							121	95.4	139	171	156	276
660							120	94.7	138	170	155	274
670									137	168	154	272
680									136	167	153	270
690									135	166	151	268
700									134	165	150	266
710									133	164	149	264
720									132	162	148	262
730									131	161	147	261
740									130	160	146	259
750									130	159	145	257
760									129	158	144	255
770									128	157	143	254
780									127	156	142	252
790									126	155	142	250
800									125	154	141	249
810									125	153	140	247
820									124	152	139	246
830									123	151	138	244
840									122	150	137	243
850									122	149	136	241
860									121	149	135	240
870									120	148	134	239
880									120	147	133	237
890											133	236
900												235

Length / Fall	21 inch diameter		24 inch diameter		27 inch diameter		30 inch diameter		33 inch diameter		36 inch diameter	
	Velocity (ft./min.)	Discharge (cu. ft./min.)	Velocity (ft./min.)	Discharge (cu. ft./min.)	Velocity (ft./min.)	Discharge (cu. ft./min.)	Velocity (ft./min.)	Discharge (cu. ft./min.)	Velocity (ft./min.)	Discharge (cu. ft./min.)	Velocity (ft./min.)	Discharge (cu. ft./min.)
32	776	1868	844	2653	909	3614	970	4765				
33	765	1840	831	2612	895	3559	956	4693				
34	753	1812	819	2573	881	3506	941	4623	999	5937		
35	742	1786	807	2536	869	3455	928	4557	985	5852		
36	732	1761	796	2501	857	3407	915	4493	971	5770		
37	722	1737	785	2467	845	3361	902	4432	958	5692		
38	712	1714	775	2434	834	3316	891	4373	945	5616	998	7057
39	703	1692	765	2403	823	3273	879	4317	933	5544	985	6966
40	694	1671	755	2373	813	3232	868	4262	921	5474	973	6879
41	686	1650	746	2343	803	3193	857	4210	910	5407	961	6794
42	678	1631	737	2315	793	3154	847	4160	899	5342	949	6713
43	670	1612	728	2288	784	3117	837	4111	889	5280	938	6634
44	662	1593	720	2262	775	3082	828	4064	879	5219	927	6559
45	655	1575	712	2237	766	3047	818	4018	869	5161	917	6485
46	648	1558	704	2212	758	3014	809	3975	859	5105	907	6414
47	641	1541	696	2189	750	2982	801	3932	850	5050	897	6346
48	634	1525	689	2166	742	2951	792	3891	841	4997	888	6279
49	627	1510	682	2144	734	2920	784	3851	832	4946	879	6215
50	621	1494	675	2122	727	2891	776	3812	824	4896	870	6152
51	615	1480	668	2101	720	2862	769	3775	816	4848	861	6092
52	609	1465	662	2081	713	2835	761	3738	808	4801	853	6033
53	603	1451	656	2061	706	2808	754	3703	800	4755	845	5976
54	598	1438	650	2042	699	2782	747	3668	793	4711	837	5920
55	592	1425	644	2023	693	2756	740	3635	786	4668	830	5866
56	587	1412	638	2005	687	2732	733	3602	779	4626	822	5814
57	582	1400	632	1987	681	2708	727	3570	772	4586	815	5762
58	577	1387	627	1970	675	2684	721	3540	765	4546	808	5712

Length Fall	21 inch diameter		24 inch diameter		27 inch diameter		30 inch diameter		33 inch diameter		36 inch diameter	
	Velocity (ft./min.)	Discharge (cu. ft./min.)	Velocity (ft./min.)	Discharge (cu. ft./min.)	Velocity (ft./min.)	Discharge (cu. ft./min.)	Velocity (ft./min.)	Discharge (cu. ft./min.)	Velocity (ft./min.)	Discharge (cu. ft./min.)	Velocity (ft./min.)	Discharge (cu. ft./min.)
59	572	1376	621	1953	669	2661	715	3509	758	4507	801	5664
60	567	1364	616	1937	663	2639	709	3480	752	4469	794	5616
61	562	1353	611	1921	658	2617	703	3451	746	4433	788	5570
62	558	1342	606	1906	653	2596	697	3423	740	4397	781	5525
63	553	1331	601	1890	647	2575	691	3396	734	4362	775	5481
64	549	1321	597	1876	642	2555	686	3369	728	4327	769	5438
65	545	1311	592	1861	637	2536	681	3343	723	4294	763	5396
66	540	1301	588	1847	632	2516	676	3318	717	4261	757	5355
67	536	1291	583	1833	628	2497	671	3293	712	4229	752	5315
68	532	1281	579	1820	623	2479	666	3269	706	4198	746	5276
69	529	1272	575	1806	619	2461	661	3245	701	4168	741	5237
70	525	1263	571	1793	614	2443	656	3222	696	4138	735	5200
71	521	1254	566	1781	610	2426	651	3199	691	4009	730	5163
72	517	1245	563	1768	606	2409	647	3177	687	4080	725	5127
73	514	1237	559	1756	601	2393	642	3155	682	4052	720	5092
74	510	1228	555	1744	597	2376	638	3134	677	4024	715	5057
75	507	1220	551	1733	593	2360	634	3113	673	3998	710	5023
76	504	1212	548	1721	589	2345	630	3092	668	3971	706	4990
77	500	1204	544	1710	586	2330	625	3072	664	3945	701	4958
78	497	1196	540	1699	582	2315	621	3052	660	3920	696	4926
79	494	1189	537	1688	578	2300	617	3033	655	3895	692	4895
80	491	1181	534	1677	574	2285	614	3014	651	3871	688	4864
81	488	1174	530	1667	571	2271	610	2995	647	3847	683	4834
82	485	1167	527	1657	567	2257	606	2977	643	3823	679	4804
83	482	1160	524	1647	564	2244	602	2959	639	3800	675	4775
84	479	1153	521	1637	561	2230	599	2941	636	3777	671	4747
85	476	1146	518	1627	557	2217	595	2924	632	3755	667	4719

86	473	1139	515	1618	554	2204	592	2907	628	3733	663	4691
87	471	1133	512	1609	551	2192	588	2890	625	3712	659	4664
88	468	1126	509	1599	548	2179	585	2873	621	3690	656	4637
89	465	1120	506	1590	545	2167	582	2857	617	3670	652	4611
90	463	1114	503	1582	542	2155	578	2841	614	3649	648	4586
91	460	1108	500	1573	539	2143	575	2826	611	3629	645	4560
92	458	1102	498	1564	536	2131	572	2810	607	3609	641	4536
93	455	1096	495	1556	533	2120	569	2795	604	3590	638	4511
94	453	1090	492	1547	530	2108	566	2780	601	3571	634	4487
95	450	1084	490	1539	527	2097	563	2766	598	3552	631	4463
96	448	1078	487	1531	524	2086	560	2751	594	3533	628	4440
97	446	1073	485	1523	522	2075	557	2737	591	3515	624	4417
98	443	1067	482	1516	519	2065	554	2723	588	3497	621	4394
99	441	1062	480	1508	516	2054	552	2709	585	3479	618	4372
100	439	1057	477	1500	514	2044	549	2695	582	3462	615	4350
105	428	1031	466	1464	501	1995	536	2631	568	3378	600	4245
110	419	1007	455	1430	490	1949	523	2570	555	3301	586	4148
115	409	985	445	1399	479	1906	512	2514	543	3228	574	4057
120	401	964	436	1370	469	1866	501	2461	532	3160	561	3971
125	393	945	427	1342	459	1828	491	2411	521	3096	550	3891
130	385	927	419	1316	451	1793	481	2364	511	3036	539	3815
135	378	909	411	1291	442	1759	472	2320	501	2979	529	3744
140	371	893	403	1268	434	1727	464	2278	492	2926	520	3677
145	364	877	396	1246	427	1697	456	2238	484	2875	511	3613
150	358	863	390	1225	419	1669	448	2201	475	2827	502	3552
155	353	849	383	1205	413	1642	441	2165	468	2781	494	3494
160	347	835	377	1186	406	1616	434	2131	460	2737	486	3439
165	342	822	371	1168	400	1591	427	2098	453	2695	479	3387
170	337	810	366	1151	394	1568	421	2067	447	2655	472	3336
175	332	799	361	1134	388	1545	415	2037	440	2617	465	3288
180	327	787	356	1118	383	1523	409	2009	434	2580	458	3242
185	323	777	351	1103	378	1503	403	1982	428	2545	452	3198
190	318	766	346	1088	373	1483	398	1955	422	2511	446	3156
195	314	756	342	1074	368	1464	393	1930	417	2479	440	3115
200	310	747	337	1061	363	1445	388	1906	412	2448	435	3076
205	306	738	333	1048	359	1428	383	1882	407	2418	429	3038

Length / Fall	21 inch diameter Velocity (ft./min.)	Discharge (cu. ft./min.)	24 inch diameter Velocity (ft./min.)	Discharge (cu. ft./min.)	27 inch diameter Velocity (ft./min.)	Discharge (cu. ft./min.)	30 inch diameter Velocity (ft./min.)	Discharge (cu. ft./min.)	33 inch diameter Velocity (ft./min.)	Discharge (cu. ft./min.)	36 inch diameter Velocity (ft./min.)	Discharge (cu. ft./min.)
210	303	729	329	1035	354	1410	379	1860	402	2389	424	3002
215	299	720	325	1023	350	1394	374	1838	397	2361	419	2967
220	296	712	322	1011	346	1378	370	1817	393	2334	415	2933
225	293	704	318	1000	342	1363	366	1797	388	2308	410	2900
230	289	697	315	989	339	1348	362	1777	384	2283	405	2868
235	286	689	311	979	335	1333	358	1758	380	2258	401	2838
240	283	682	308	968	331	1319	354	1740	376	2234	397	2808
245	280	675	305	958	328	1306	350	1722	372	2212	393	2779
250	277	668	302	949	325	1293	347	1705	368	2189	389	2751
255	275	661	299	939	322	1280	343	1688	365	2168	385	2724
260	272	655	296	930	318	1268	340	1671	361	2147	381	2698
265	269	649	293	921	315	1255	337	1656	358	2126	378	2672
270	267	643	290	913	312	1244	334	1640	354	2107	374	2647
275	265	637	288	905	310	1232	331	1625	351	2087	371	2623
280	262	631	285	896	307	1221	328	1611	348	2069	367	2600
285	260	626	282	889	304	1211	325	1596	345	2050	364	2577
290	258	620	280	881	301	1200	322	1583	342	2033	361	2554
295	255	615	278	873	299	1190	319	1569	339	2015	358	2533
300	253	610	275	866	296	1180	317	1556	336	1999	355	2511
305	251	605	273	859	294	1170	314	1543	333	1982	352	2491
310	249	600	271	852	292	1161	311	1531	331	1966	349	2471
315	247	595	269	845	289	1151	309	1519	328	1950	346	2451
320	245	590	267	838	287	1142	307	1507	325	1935	344	2432
325	243	586	265	832	285	1134	304	1495	323	1920	341	2413
330	241	581	262	826	283	1125	302	1484	320	1905	338	2395
335	240	577	261	819	280	1117	300	1472	318	1891	336	2377
340	238	573	259	813	278	1108	297	1462	316	1877	333	2359

345	2342	331	1864	313	1451	295	1100	276	808	257	569	236
350	2325	329	1850	311	1441	293	1092	274	802	255	565	234
355	2309	326	1837	309	1430	291	1085	272	796	253	561	233
360	2293	324	1824	307	1420	289	1077	271	791	251	557	231
365	2277	322	1812	305	1411	287	1070	269	785	250	553	230
370	2261	320	1800	303	1401	285	1062	267	780	248	549	228
375	2246	317	1787	301	1392	283	1055	265	775	246	545	226
380	2231	315	1776	299	1383	281	1048	263	769	245	542	225
385	2217	313	1764	297	1374	279	1042	262	764	243	538	223
390	2203	311	1753	295	1365	278	1035	260	759	241	535	222
395	2189	309	1742	293	1356	276	1028	258	755	240	531	221
400	2175	307	1731	291	1347	274	1022	257	750	238	528	219
405	2161	305	1720	289	1339	272	1015	255	745	237	525	218
410	2148	303	1709	287	1331	271	1009	253	741	235	522	217
415	2135	302	1699	286	1323	269	1003	252	736	234	518	215
420	2122	300	1689	284	1315	268	997	250	732	233	515	214
425	2110	298	1679	282	1307	266	991	249	728	231	512	213
430	2098	296	1669	281	1300	264	985	247	723	230	509	211
435	2086	295	1660	279	1292	263	980	246	719	229	506	210
440	2074	293	1650	277	1285	261	974	245	715	227	503	209
445	2062	291	1641	276	1278	260	969	243	711	226	501	208
450	2050	290	1632	274	1270	258	963	242	707	225	498	207
455	2039	288	1623	273	1263	257	958	241	703	223	495	206
460	2028	287	1614	271	1257	256	953	239	699	222	492	204
465	2017	285	1605	270	1250	254	948	238	695	221	490	203
470	2006	283	1597	268	1243	253	943	237	692	220	487	202
475	1996	282	1588	267	1237	252	938	235	688	219	485	201
480	1985	280	1580	266	1230	250	933	234	685	218	482	200
485	1975	279	1572	264	1224	249	928	233	681	216	479	199
490	1965	278	1564	263	1217	248	923	232	278	215	477	198
495	1955	276	1556	262	1211	246	918	231	678	214	475	197
500	1945	275	1548	260	1205	245	914	229	671	213	472	196
510	1926	272	1533	258	1193	243	905	227	664	211	468	194
520	1907	269	1518	255	1182	240	896	225	658	209	463	192
530	1889	267	1503	253	1171	238	888	223	651	207	459	190
540	1872	264	1489	250	1160	236	879	221	645	205	454	189

Length/Fall	21 inch diameter		24 inch diameter		27 inch diameter		30 inch diameter		33 inch diameter		36 inch diameter	
	Velocity (ft./min.)	Discharge (cu. ft./min.)	Velocity (ft./min.)	Discharge (cu. ft./min.)	Velocity (ft./min.)	Discharge (cu. ft./min.)	Velocity (ft./min.)	Discharge (cu. ft./min.)	Velocity (ft./min.)	Discharge (cu. ft./min.)	Velocity (ft./min.)	Discharge (cu. ft./min.)
550	187	450	203	639	219	871	234	1149	248	1476	262	1855
560	185	446	201	634	217	863	232	1139	246	1463	260	1838
570	184	442	200	628	215	856	230	1129	244	1450	257	1822
580	182	438	198	623	213	848	228	1119	242	1437	255	1806
590	180	435	196	617	211	841	226	1109	240	1425	253	1791
600	179	431	195	612	209	834	224	1100	237	1413	251	1776
610	177	427	193	607	208	827	222	1091	236	1401	249	1761
620	176	424	191	602	206	821	220	1082	234	1390	247	1747
630	175	421	190	597	204	814	218	1074	232	1379	245	1733
640	173	417	188	593	203	808	217	1065	230	1368	243	1719
650	172	414	187	588	201	801	215	1057	228	1358	241	1706
660	171	411	185	584	200	795	213	1049	226	1347	239	1693
670	169	408	184	579	198	789	212	1041	225	1337	237	1680
680	168	405	183	575	197	784	210	1033	223	1327	236	1668
690	167	402	181	571	195	778	209	1026	221	1318	234	1656
700	166	399	180	567	194	772	207	1018	220	1308	232	1644
710	164	396	179	563	192	767	206	1011	218	1299	231	1632
720	163	393	178	559	191	761	204	1004	217	1290	229	1621
730	162	391	176	555	190	756	203	997	215	1281	227	1610
740	161	388	175	551	189	751	201	991	214	1272	226	1599
750	160	385	174	548	187	746	200	984	212	1264	224	1588
760	159	383	173	544	186	741	199	977	211	1255	223	1578
770	158	380	172	540	185	736	197	971	210	1247	221	1567
780	157	378	171	537	184	732	196	965	208	1239	220	1557
790	156	376	169	533	182	727	195	959	207	1231	219	1547
800	155	373	168	530	181	722	194	953	206	1224	217	1538
810	154	371	167	527	180	718	192	947	204	1216	216	1528

820	153	369	166	524	179	714	191	941	203	1209	214	1519
830	152	366	165	520	178	709	190	935	202	1201	213	1510
840	151	364	164	517	177	705	189	930	201	1194	212	1501
850	150	362	163	514	176	701	188	924	199	1187	211	1492
860	149	360	162	511	175	697	187	919	198	1180	209	1483
870	149	358	161	508	174	693	186	914	197	1173	208	1475
880	148	356	161	505	173	689	185	908	196	1167	207	1466
890	147	354	160	503	172	685	184	903	195	1160	206	1458
900	146	352	159	500	171	681	183	898	194	1154	205	1450
910	145	350	158	497	170	677	182	893	193	1147	204	1442
920	144	348	157	494	169	674	181	888	192	1141	202	1434
930	144	346	156	492	168	670	180	884	191	1135	201	1426
940	143	344	155	489	167	666	179	879	190	1129	200	1419
950	142	342	155	486	166	663	178	874	189	1123	199	1411
960	141	341	154	484	165	659	177	870	188	1117	198	1404
970	141	339	153	481	165	656	176	865	187	1111	197	1396
980	140	337	152	479	164	653	175	861	186	1106	196	1389
990	139	335	151	476	163	649	174	856	185	1100	195	1382
1000	138	334	151	474	162	646	173	852	184	1094	194	1375
1050	135	326	147	463	158	630	169	832	179	1068	189	1342
1100	132	318	144	452	155	616	165	812	175	1043	185	1311
1150	129	311	140	442	151	602	161	795	171	1021	181	1282
1200	126	305	137	433	148	590	158	778	168	999	177	1255
1250	124	298	135	424	145	578	155	762	164	979	174	1230
1300	121	293	132	416	142	567	152	747	161	960	170	1206
1350			130	408	139	556	149	733	158	942	167	1184
1400			127	401	137	546	146	720	155	925	164	1162
1450			125	394	135	536	144	708	153	909	161	1142
1500			123	387	132	527	141	696	150	893	158	1123
1550			121	381	130	519	139	684	148	879	156	1105
1600					128	511	137	673	145	865	153	1087
1650					126	503	135	663	143	852	151	1071
1700					124	495	133	653	141	839	149	1055
1750					122	488	131	644	139	827	147	1040
1800					121	481	129	635	137	816	145	1025
1850							127	626	135	804	143	1011

Length/Fall	39 inch diameter Velocity (ft./min.)	39 inch diameter Discharge (cu. ft./min.)	42 inch diameter Velocity (ft./min.)	42 inch diameter Discharge (cu. ft./min.)	45 inch diameter Velocity (ft./min.)	45 inch diameter Discharge (cu. ft./min.)	48 inch diameter Velocity (ft./min.)	48 inch diameter Discharge (cu. ft./min.)	51 inch diameter Velocity (ft./min.)	51 inch diameter Discharge (cu. ft./min.)	54 inch diameter Velocity (ft./min.)	54 inch diameter Discharge (cu. ft./min.)
59	842	6988	882	8489	921	10174	959	12053	996	14132		
60	835	6930	875	8418	913	10089	951	11952	987	14014		
61	828	6873	867	8349	906	10006	943	11854	979	13898		
62	821	6817	860	8281	898	9925	935	11758	971	13786		
63	815	6763	853	8215	891	9846	928	11664	964	13676	999	15890
64	808	6710	847	8151	884	9769	920	11572	956	13569	991	15765
65	802	6658	840	8088	877	9693	913	11483	949	13464	983	15644
66	796	6607	834	8026	871	9620	906	11396	941	13362	976	15525
67	790	6558	828	7966	864	9548	900	11310	934	13262	968	15408
68	784	6509	821	7907	858	9477	893	11227	928	13164	961	15295
69	779	6462	815	7850	851	9408	887	11145	921	13068	954	15183
70	773	6416	810	7793	845	9341	880	11065	914	12974	947	15075
71	767	6370	804	7738	839	9275	874	10987	908	12883	941	14968
72	762	6326	798	7684	833	9210	868	10911	901	12793	934	14864
73	757	6282	793	7632	828	9147	862	10836	895	12705	928	14762
74	752	6240	787	7580	822	9085	856	10762	889	12619	921	14662
75	747	6198	782	7529	817	9024	850	10690	883	12534	915	14563
76	742	6157	777	7479	811	8965	845	10620	877	12452	909	14467
77	737	6117	772	7431	806	8906	839	10550	872	12370	903	14373
78	732	6078	767	7383	801	8849	834	10483	866	12291	897	14281
79	728	6039	762	7336	796	8793	828	10416	860	12213	892	14190
80	723	6001	757	7290	791	8738	823	10351	855	12136	886	14101
81	719	5964	753	7245	786	8683	818	10287	850	12061	881	14014
82	714	5928	748	7201	781	8630	813	10224	845	11987	875	13928
83	710	5892	743	7157	776	8578	808	10162	839	11915	870	13844
84	706	5857	739	7114	772	8527	803	10101	834	11844	865	13761
85	701	5822	735	7072	767	8477	799	10042	830	11774	860	13680

86	13600	855	11705	825	9983	794	8427	763	7031	730	5788	697
87	13522	850	11638	820	9926	789	8379	758	6991	726	5755	693
88	13445	845	11571	815	9869	785	8331	754	6951	722	5722	689
89	13369	840	11506	811	9813	781	8284	750	6912	718	5690	685
90	13295	835	11442	806	9759	776	8238	745	6873	714	5658	682
91	13221	831	11379	802	9705	772	8192	741	6835	710	5627	678
92	13149	826	11317	797	9652	768	8148	737	6798	706	5596	674
93	13078	822	11256	793	9600	764	8104	733	6761	702	5566	671
94	13009	818	11196	789	9549	759	8061	729	6725	699	5536	667
95	12940	813	11137	785	9498	755	8018	726	6690	695	5507	663
96	12872	809	11079	781	9449	751	7976	722	6655	691	5478	660
97	12806	805	11022	776	9400	748	7935	718	6620	688	5450	657
98	12740	801	10965	773	9352	744	7894	714	6587	684	5422	653
99	12676	797	10910	769	9305	740	7854	711	6553	681	5395	650
100	12612	793	10855	765	9258	736	7815	707	6520	677	5368	647
105	12308	773	10593	746	9035	719	7627	690	6363	661	5238	631
110	12025	756	10350	729	8827	702	7451	674	6217	646	5118	617
115	11761	739	10122	713	8633	687	7288	659	6080	632	5005	603
120	11513	723	9909	698	8451	672	7134	646	5952	618	4900	590
125	11281	709	9709	684	8280	659	6990	632	5832	606	4801	578
130	11062	695	9520	671	8120	646	6854	620	5719	594	4708	567
135	10855	682	9342	658	7968	634	6726	609	5612	583	4620	556
140	10659	670	9174	646	7824	622	6605	598	5511	572	4536	546
145	10474	658	9014	635	7688	611	6490	587	5415	562	4457	537
150	10298	647	8863	624	7559	601	6381	577	5324	553	4383	528
155	10130	637	8719	614	7436	591	6277	568	5237	544	4311	519
160	9971	626	8581	604	7319	582	6178	559	5155	535	4243	511
165	9819	617	8450	595	7207	573	6084	550	5076	527	4179	503
170	9673	608	8325	586	7100	565	5994	542	5001	519	4117	496
175	9534	599	8205	578	6998	556	5907	534	4929	512	4057	489
180	9400	591	8091	570	6900	549	5825	527	4860	505	4001	482
185	9273	583	7981	562	6806	541	5746	520	4794	498	3946	475
190	9150	575	7875	555	6716	534	5669	513	4730	491	3894	469
195	9032	567	7773	548	6630	527	5596	506	4669	485	3844	463
200	8918	560	7675	541	6546	520	5526	500	4610	479	3795	457
205	8809	553	7581	534	6466	514	5458	494	4554	473	3749	451

Length / Fall	39 inch diameter		42 inch diameter		45 inch diameter		48 inch diameter		51 inch diameter		54 inch diameter	
	Velocity (ft./min.)	Discharge (cu. ft./min.)	Velocity (ft./min.)	Discharge (cu. ft./min.)	Velocity (ft./min.)	Discharge (cu. ft./min.)	Velocity (ft./min.)	Discharge (cu. ft./min.)	Velocity (ft./min.)	Discharge (cu. ft./min.)	Velocity (ft./min.)	Discharge (cu. ft./min.)
210	446	3704	467	4499	488	5393	508	6388	528	7490	547	8703
215	441	3661	462	4447	482	5330	502	6314	521	7403	540	8601
220	436	3619	456	4396	477	5269	496	6241	515	7318	534	8503
225	431	3578	451	4347	471	5210	491	6172	510	7236	528	8408
230	426	3539	446	4299	466	5153	485	6104	504	7157	522	8316
235	422	3501	442	4253	461	5098	480	6039	499	7081	517	8227
240	417	3465	437	4209	456	5044	475	5976	493	7007	511	8141
245	413	3429	433	4166	452	4993	470	5914	488	6935	506	8057
250	409	3395	428	4124	447	4942	465	5855	483	6865	501	7977
255	405	3361	424	4083	443	4894	461	5797	479	6797	496	7898
260	401	3329	420	4044	438	4846	456	5741	474	6732	491	7822
265	397	3297	416	4005	434	4801	452	5687	470	6668	487	7747
270	393	3266	412	3968	430	4756	448	5634	465	6606	482	7675
275	390	3237	408	3923	426	4712	444	5583	461	6546	478	7605
280	386	3208	405	3896	422	4670	440	5532	457	6487	473	7537
285	383	3179	401	3862	419	4629	436	5484	453	6430	469	7471
290	380	3152	398	3829	415	4589	432	5436	449	6374	465	7406
295	376	3125	394	3796	412	4550	428	5390	445	6320	461	7343
300	373	3099	391	3764	408	4512	425	5345	441	6267	457	7281
305	370	3073	388	3733	405	4475	421	5301	438	6215	454	7222
310	367	3048	384	3703	401	4438	418	5258	434	6165	450	7163
315	364	3024	381	3674	398	4403	415	5216	431	6116	446	7106
320	361	3000	378	3645	395	4369	411	5175	427	6068	443	7050
325	358	2977	375	3617	392	4335	408	5135	424	6021	439	6996
330	356	2955	373	3589	389	4302	405	5096	421	5975	436	6943
335	353	2932	370	3562	386	4270	402	5058	418	5930	433	6891
340	350	2911	367	3536	383	4238	399	5021	415	5887	430	6840

345	348	2890	364	3510	380	4207	396	4984	411	5844	426	6790
350	345	2869	362	3485	378	4177	393	4948	409	5802	423	6741
355	343	2849	359	3460	375	4148	391	4913	406	5761	420	6694
360	341	2829	357	3436	372	4119	388	4879	403	5721	417	6647
365	338	2809	354	3413	370	4090	385	4846	400	5682	415	6601
370	336	2790	352	3390	367	4063	383	4813	397	5643	412	6557
375	334	2772	350	3367	365	4035	380	4780	395	5605	409	6513
380	331	2753	347	3345	363	4009	377	4749	392	5568	406	6470
385	329	2735	345	3323	360	3983	375	4718	390	5532	404	6428
390	327	2718	343	3301	358	3957	373	4688	387	5496	401	6386
395	325	2700	341	3281	356	3932	370	4658	385	5461	399	6346
400	323	2684	338	3260	353	3907	368	4629	382	5427	396	6306
405	321	2667	336	3240	351	3883	366	4600	380	5394	394	6267
410	319	2651	334	3220	349	3859	363	4572	377	5361	391	6228
415	317	2635	332	3200	347	3836	361	4544	375	5328	389	6191
420	315	2619	330	3181	345	3813	359	4517	373	5296	386	6154
425	313	2603	328	3163	343	3791	357	4490	371	5265	384	6118
430	312	2588	326	3144	341	3768	355	4464	369	5234	382	6082
435	310	2573	324	3126	339	3747	353	4439	366	5204	380	6047
440	308	2559	323	3108	337	3725	351	4413	364	5175	378	6012
445	306	2544	321	3091	335	3704	349	4388	362	5145	375	5979
450	305	2530	319	3073	333	3684	347	4364	360	5117	373	5945
455	303	2516	317	3057	331	3663	345	4340	358	5089	371	5912
460	301	2502	316	3040	329	3644	343	4316	356	5061	369	5880
465	300	2489	314	3023	328	3624	341	4293	354	5034	367	5849
470	298	2476	312	3007	326	3605	339	4270	352	5007	365	5817
475	296	2463	310	2991	324	3586	338	4248	351	4980	363	5787
480	295	2450	309	2976	323	3567	336	4225	349	4954	361	5756
485	293	2437	307	2960	321	3548	334	4204	347	4929	360	5727
490	292	2425	306	2945	319	3530	332	4182	345	4903	358	5697
495	290	2412	304	2930	318	3512	331	4161	343	4879	356	5669
500	289	2400	303	2916	316	3495	329	4140	342	4854	354	5640
510	286	2377	300	2887	313	3460	326	4099	338	4806	351	5585
520	283	2354	297	2859	310	3427	323	4060	335	4760	347	5531
530	281	2331	294	2832	307	3394	320	4021	332	4715	344	5478
540	278	2310	291	2806	304	3363	317	3984	329	4671	331	5427

Length / Fall	39 inch diameter Velocity (ft./min.)	Discharge (cu. ft./min.)	42 inch diameter Velocity (ft./min.)	Discharge (cu. ft./min.)	45 inch diameter Velocity (ft./min.)	Discharge (cu. ft./min.)	48 inch diameter Velocity (ft./min.)	Discharge (cu. ft./min.)	51 inch diameter Velocity (ft./min.)	Discharge (cu. ft./min.)	54 inch diameter Velocity (ft./min.)	Discharge (cu. ft./min.)
550	275	2288	289	2780	301	3333	314	3947	326	4628	338	5378
560	273	2268	286	2755	299	3302	311	3912	323	4587	335	5329
570	271	2248	283	2731	296	3273	308	3877	320	4546	332	5282
580	268	2228	281	2707	293	3245	305	3844	317	4507	329	5237
590	266	2210	279	2684	291	3217	303	3811	315	4469	326	5192
600	264	2191	276	2662	288	3190	300	3779	312	4431	323	5149
610	262	2173	274	2640	286	3164	298	3748	309	4395	321	5106
620	259	2155	272	2618	284	3138	295	3718	307	4359	318	5065
630	257	2138	270	2597	281	3113	293	3688	304	4324	315	5025
640	255	2121	267	2577	279	3089	291	3659	302	4290	313	4985
650	253	2105	265	2557	277	3065	288	3631	300	4257	311	4947
660	251	2089	263	2538	275	3042	286	3603	297	4225	308	4909
670	250	2073	261	2519	273	3019	284	3576	295	4193	306	4872
680	248	2058	259	2500	271	2997	282	3550	293	4162	304	4836
690	246	2043	258	2482	269	2975	280	3524	291	4132	301	4801
700	244	2028	256	2464	267	2953	278	3499	289	4102	299	4767
710	242	2014	254	2447	265	2933	276	3474	287	4073	297	4733
720	241	2000	252	2430	263	2912	274	3450	285	4045	295	4700
730	239	1986	250	2413	261	2892	272	3426	283	4017	293	4668
740	237	1973	249	2397	260	2873	270	3403	281	3990	291	4636
750	236	1960	247	2381	258	2853	269	3380	279	3963	289	4605
760	234	1947	245	2365	256	2834	267	3358	277	3937	287	4575
770	233	1934	244	2349	255	2816	265	3336	275	3912	285	4545
780	231	1922	242	2334	253	2798	263	3315	274	3886	283	4516
790	230	1909	241	2320	251	2780	262	3293	272	3862	282	4487
800	228	1897	239	2305	250	2763	260	3273	270	3837	280	4459
810	227	1886	238	2291	248	2746	258	3253	268	3814	278	4431

820	4404	276	3790	267	3233	257	2729	247	2277	236	1874	225
830	4377	275	3767	265	3213	255	2712	245	2263	235	1863	224
840	4351	273	3745	264	3194	254	2696	244	2249	233	1852	223
850	4326	272	3723	262	3175	252	2680	242	2236	232	1841	221
860	4300	270	3701	260	3157	251	2665	241	2223	231	1830	220
870	4276	268	3680	259	3138	249	2649	239	2210	229	1819	219
880	4251	267	3659	257	3120	248	2634	238	2198	228	1809	218
890	4227	265	3638	256	3103	246	2619	237	2185	227	1799	216
900	4204	264	3618	255	3086	245	2605	235	2173	225	1789	215
910	4181	262	3598	253	3069	244	2590	234	2161	224	1779	214
920	4158	261	3578	252	3052	242	2576	233	2149	223	1769	213
930	4135	260	3559	250	3035	241	2562	232	2138	222	1760	212
940	4113	258	3540	249	3019	240	2549	230	2126	221	1750	211
950	4092	257	3521	248	3003	239	2535	229	2115	219	1741	209
960	4070	255	3503	246	2988	237	2522	228	2104	218	1732	208
970	4049	254	3485	245	2972	236	2509	227	2093	217	1723	207
980	4028	253	3467	244	2957	235	2496	226	2083	216	1714	206
990	4008	252	3450	243	2942	234	2483	224	2072	215	1706	205
1000	3988	250	3432	241	2927	232	2471	223	2062	214	1697	204
1050	3892	244	3350	236	2857	227	2411	218	2012	209	1656	199
1100	3802	239	3273	230	2791	222	2356	213	1966	204	1618	195
1150	3719	233	3201	225	2730	217	2304	208	1922	199	1582	190
1200	3640	228	3133	220	2672	212	2256	204	1882	195	1549	186
1250	3567	224	3070	216	2618	208	2210	200	1844	191	1518	183
1300	3498	219	3010	212	2567	204	2167	196	1808	187	1488	179
1350	3432	215	2954	208	2519	200	2127	192	1774	184	1461	176
1400	3370	211	2901	204	2474	196	2088	189	1742	181	1434	172
1450	3312	208	2850	200	2431	193	2052	185	1712	177	1409	169
1500	3256	204	2802	197	2390	190	2017	182	1683	175	1386	167
1550	3203	201	2757	194	2351	187	1985	179	1656	172	1363	164
1600	3153	198	2713	191	2314	184	1953	176	1630	169	1342	161
1650	3105	195	2672	188	2279	181	1924	174	1605	166	1321	159
1700	3059	192	2632	185	2245	178	1895	171	1581	164	1301	156
1750	3015	189	2594	182	2213	176	1868	169	1558	162	1283	154
1800	2972	186	2558	180	2182	173	1842	166	1536	159	1265	152
1850	2932	184	2523	177	2152	171	1817	164	1516	157	1248	150

Length / Fall	39 inch diameter Velocity (ft./min.)	Discharge (cu. ft./min.)	42 inch diameter Velocity (ft./min.)	Discharge (cu. ft./min.)	45 inch diameter Velocity (ft./min.)	Discharge (cu. ft./min.)	48 inch diameter Velocity (ft./min.)	Discharge (cu. ft./min.)	51 inch diameter Velocity (ft./min.)	Discharge (cu. ft./min.)	54 inch diameter Velocity (ft./min.)	Discharge (cu. ft./min.)
1900	148	1231	155	1495	162	1793	169	2124	175	2490	181	2893
1950	146	1215	153	1476	160	1769	166	2096	173	2458	179	2856
2000	144	1200	151	1458	158	1747	164	2070	171	2427	177	2820
2050	142	1185	149	1440	156	1726	162	2044	169	2397	175	2785
2100	141	1171	147	1422	154	1705	160	2020	166	2368	173	2752
2150	139	1157	146	1406	152	1685	158	1996	165	2341	171	2720
2200	137	1144	144	1390	150	1666	157	1973	163	2314	169	2689
2250	136	1131	142	1374	149	1647	155	1951	161	2288	167	2659
2300	134	1119	141	1359	147	1629	153	1930	159	2263	165	2629
2350	133	1107	139	1345	145	1612	151	1909	157	2239	163	2601
2400	132	1095	138	1331	144	1595	150	1889	156	2215	161	2574
2450	130	1084	136	1317	142	1578	148	1870	154	2193	160	2548
2500	129	1073	135	1304	141	1563	147	1851	153	2171	158	2522
2550	128	1063	134	1291	140	1547	145	1833	151	2149	157	2497
2600	126	1052	132	1278	138	1532	144	1815	150	2128	155	2473
2650	125	1042	131	1266	137	1518	143	1798	148	2108	154	2450
2700	124	1033	130	1254	136	1504	141	1781	147	2089	152	2427
2750	123	1023	129	1243	134	1490	140	1765	145	2070	151	2405
2800	122	1014	128	1232	133	1476	139	1749	144	2051	149	2383
2850	121	1005	126	1221	132	1463	138	1734	143	2033	148	2362
2900	120	996	125	1210	131	1451	136	1719	142	2015	147	2342
2950			124	1200	130	1438	135	1704	140	1998	146	2322
3000			123	1190	129	1426	134	1690	139	1981	144	2302
3050			122	1180	128	1415	133	1676	138	1965	143	2283
3100			121	1171	127	1403	132	1662	137	1949	142	2265
3150			120	1161	126	1392	131	1649	136	1934	141	2247
3200					125	1381	130	1636	135	1918	140	2229

Length / Fall	57 inch diameter Velocity (ft./min.)	Discharge (cu. ft./min.)	60 inch diameter Velocity (ft./min.)	Discharge (cu. ft./min.)	63 inch diameter Velocity (ft./min.)	Discharge (cu. ft./min.)	66 inch diameter Velocity (ft./min.)	Discharge (cu. ft./min.)	69 inch diameter Velocity (ft./min.)	Discharge (cu. ft./min.)	72 inch diameter Velocity (ft./min.)	Discharge (cu. ft./min.)
86	884	15674	913	17933	941	20384	969	23031	996	25882		
87	879	15584	908	17830	936	20266	963	22899	991	25733		
88	874	15495	902	17728	930	20151	958	22768	985	25586		
89	869	15408	897	17629	925	20037	952	22640	979	25442		
90	864	15322	892	17530	920	19926	947	22514	974	25300		
91	859	15237	887	17434	915	19816	942	22390	969	25161	995	28135
92	855	15154	883	17339	910	19708	937	22268	963	25024	989	27982
93	850	15073	878	17245	905	19602	932	22148	958	24889	984	27831
94	846	14992	873	17153	900	19497	927	22030	953	24756	979	27682
95	841	14913	869	17063	895	19394	922	21913	948	24625	973	27536
96	837	14835	864	16974	891	19293	917	21799	943	24497	968	27392
97	832	14759	860	16886	886	19193	912	21686	938	24370	963	27251
98	828	14683	855	16800	882	19095	908	21575	933	24246	958	27111
99	824	14609	851	16714	877	18998	903	21466	929	24123	954	26974
100	820	14536	847	16631	873	18903	899	21358	924	24002	949	26839
105	800	14185	826	16230	852	18448	877	20844	902	23423	926	26192
110	782	13859	807	15857	832	18023	857	20364	881	22885	905	25590
115	764	13554	789	15508	814	17627	838	19917	861	22382	885	25027
120	748	13269	773	15182	797	17256	820	19497	843	21911	866	24500
125	733	13001	757	14875	781	16907	804	19103	826	21468	849	24005
130	719	12748	742	14586	765	16579	788	18732	810	21051	832	23539
135	706	12510	729	14313	751	16269	773	18382	795	20657	817	23099
140	693	12285	715	14055	738	15976	759	18051	781	20285	802	22683
145	681	12071	703	13811	725	15698	746	17737	767	19932	788	22288
150	669	11868	691	13579	713	15434	734	17439	754	19597	775	21914
155	658	11675	680	13358	701	15183	722	17155	742	19279	762	21557
160	648	11491	669	13148	690	14944	710	16885	730	18975	750	21218

Length/Fall	57 inch diameter Velocity (ft./min.)	Discharge (cu. ft./min.)	60 inch diameter Velocity (ft./min.)	Discharge (cu. ft./min.)	63 inch diameter Velocity (ft./min.)	Discharge (cu. ft./min.)	66 inch diameter Velocity (ft./min.)	Discharge (cu. ft./min.)	69 inch diameter Velocity (ft./min.)	Discharge (cu. ft./min.)	72 inch diameter Velocity (ft./min.)	Discharge (cu. ft./min.)
165	638	11316	659	12947	679	14716	699	16627	719	18685	739	20894
170	629	11148	649	12755	669	14498	689	16381	708	18409	728	20584
175	620	10988	640	12571	660	14289	679	16145	698	18144	717	20288
180	611	10834	631	12396	650	14089	670	15919	688	17890	707	20004
185	603	10687	622	12227	642	13898	661	15703	679	17646	697	19732
190	595	10545	614	12065	633	13714	652	15495	670	17413	688	19471
195	587	10409	606	11909	625	13537	643	15295	661	17188	679	19220
200	580	10278	598	11760	617	13366	635	15103	653	16972	671	18978
205	572	10152	591	11615	609	13202	627	14917	645	16764	663	18745
210	566	10030	584	11476	602	13044	620	14739	637	16563	655	18520
215	559	9913	577	11342	595	12892	613	14566	630	16369	647	18304
220	553	9800	571	11212	588	12744	606	14400	623	16182	640	18095
225	546	9690	564	11087	582	12602	599	14239	616	16001	632	17892
230	540	9584	558	10966	575	12464	592	14083	609	15826	625	17697
235	535	9482	552	10848	569	12331	586	13933	603	15657	619	17508
240	529	9383	546	10735	563	12202	580	13787	596	15493	612	17324
245	524	9286	541	10625	557	12077	574	13645	590	15334	606	17147
250	518	9193	535	10518	552	11955	568	13508	584	15180	600	16974
255	513	9102	530	10414	546	11837	563	13375	578	15030	594	16807
260	508	9014	525	10314	541	11723	557	13246	573	14885	588	16645
265	503	8929	520	10216	536	11612	552	13120	567	14744	583	16487
270	499	8846	515	10121	531	11504	547	12998	562	14607	577	16333
275	494	8765	510	10028	526	11399	542	12879	557	14473	572	16184
280	490	8686	506	9939	521	11297	537	12764	552	14344	567	16039
285	485	8610	501	9851	517	11197	532	12651	547	14217	562	15898
290	481	8535	497	9766	512	11100	527	12542	542	14094	557	15760
295	477	8463	493	9683	508	11006	523	12435	538	13974	552	15626

300	473	8392	489	9602	504	10914	519	12331	533	13857	548	15495
305	469	8323	485	9522	500	10824	514	12230	529	13743	543	15368
310	465	8255	481	9445	495	10736	510	12131	525	13632	539	15243
315	462	8190	477	9370	492	10650	506	12034	520	13523	534	15122
320	458	8125	473	9297	488	10567	502	11939	516	13417	530	15003
325	455	8063	469	9225	484	10485	498	11847	512	13314	526	14887
330	451	8001	466	9155	480	10406	494	11757	508	13212	522	14774
335	448	7941	462	9086	477	10328	491	11669	505	13113	518	14663
340	444	7883	459	9019	473	10251	487	11583	501	13017	514	14555
345	441	7825	456	8953	470	10177	484	11499	497	12922	511	14449
350	438	7769	452	8889	466	10104	480	11416	494	12829	507	14346
355	435	7714	449	8826	463	10032	477	11336	490	12739	503	14244
360	432	7661	446	8765	460	9963	473	11257	487	12650	500	14145
365	429	7608	443	8705	457	9894	470	11179	483	12563	496	14048
370	426	7556	440	8646	454	9827	467	11103	480	12478	493	13953
375	423	7506	437	8588	450	9761	464	11029	477	12394	490	13859
380	420	7456	434	8531	447	9697	461	10956	474	12312	486	13768
385	418	7408	431	8476	445	9634	458	10885	471	12232	483	13678
390	415	7360	428	8421	442	9572	455	10815	468	12154	480	13590
395	412	7313	426	8368	439	9511	452	10746	465	12076	477	13504
400	410	7268	423	8315	436	9451	449	10679	462	12001	474	13419
405	407	7223	420	8264	433	9393	446	10613	459	11926	471	13336
410	405	7178	418	8213	431	9335	444	10548	456	11893	468	13255
415	402	7135	415	8163	428	9279	441	10484	453	11782	465	13174
420	400	7092	413	8115	426	9224	438	10422	451	11711	463	13096
425	397	7051	410	8067	423	9169	436	10360	448	11642	460	13019
430	395	7009	408	8020	421	9116	433	10300	445	11574	457	12943
435	393	6969	406	7974	418	9063	431	10240	443	11508	455	12868
440	391	6929	403	7928	416	9011	428	10182	440	11442	452	12795
445	388	6890	401	7883	413	8961	426	10125	438	11378	450	12723
450	386	6852	399	7840	411	8911	423	10068	435	11314	447	12652
455	384	6814	397	7796	409	8862	421	10013	433	11252	445	12582
460	382	6777	394	7754	407	8813	419	9958	430	11191	442	12513
465	380	6740	392	7712	404	8766	416	9904	428	11130	440	12446
470	378	6704	390	7671	402	8719	414	9852	426	11071	437	12380
475	376	6669	388	7630	400	8673	412	9800	424	11013	435	12314

Length/Fall	57 inch diameter Velocity (ft./min.)	Discharge (cu. ft./min.)	60 inch diameter Velocity (ft./min.)	Discharge (cu. ft./min.)	63 inch diameter Velocity (ft./min.)	Discharge (cu. ft./min.)	66 inch diameter Velocity (ft./min.)	Discharge (cu. ft./min.)	69 inch diameter Velocity (ft./min.)	Discharge (cu. ft./min.)	72 inch diameter Velocity (ft./min.)	Discharge (cu. ft./min.)
480	374	6634	386	7591	398	8628	410	9748	421	10955	433	12250
485	372	6600	384	7551	396	8583	408	9698	419	10898	431	12187
490	370	6566	382	7513	394	8539	406	9648	417	10843	428	12124
495	368	6533	380	7475	392	8496	404	9600	415	10788	426	12063
500	366	6500	378	7437	390	8453	402	9551	413	10734	424	12002
510	363	6436	375	7364	386	8370	398	9457	409	10628	420	11884
520	359	6374	371	7293	382	8289	394	9366	405	10525	416	11769
530	356	6314	367	7224	379	8211	390	9277	401	10425	412	11658
540	353	6255	364	7156	375	8134	386	9191	397	10328	408	11549
550	349	6198	361	7091	372	8060	383	9107	394	10234	404	11444
560	346	6142	357	7027	369	7988	379	9025	390	10142	401	11341
570	343	6088	354	6966	365	7917	376	8946	387	10053	397	11241
580	340	6035	351	6905	362	7849	373	8868	383	9966	394	11144
590	337	5984	348	6846	359	7782	370	8793	380	9881	390	11049
600	334	5934	345	6789	356	7717	367	8719	377	9798	387	10957
610	332	5885	342	6733	353	7653	364	8647	374	9718	384	10866
620	329	5837	340	6679	350	7591	361	8577	371	9639	381	10778
630	326	5791	337	6626	347	7531	358	8509	368	9562	378	10693
640	324	5745	334	6574	345	7472	355	8442	365	9487	375	10609
650	321	5701	332	6523	342	7414	352	8377	362	9414	372	10527
660	319	5658	329	6473	339	7358	349	8313	359	9342	369	10447
670	316	5615	327	6425	337	7303	347	8251	357	9272	366	10368
680	314	5574	324	6377	334	7249	344	8190	354	9204	364	10292
690	312	5533	322	6331	332	7196	342	8131	351	9137	361	10217
700	310	5494	320	6285	330	7144	339	8072	349	9072	358	10144
710	307	5455	317	6241	327	7094	337	8015	346	9007	356	10072
720	305	5417	315	6198	325	7044	335	7959	344	8945	353	10002

730	9933	351	8883	342	7905	332	6996	323	6155	313	5380	303
740	9866	348	8823	339	7851	330	6949	321	6113	311	5343	301
750	9800	346	8764	337	7799	328	6902	318	6072	309	5307	299
760	9735	344	8706	335	7747	326	6857	316	6032	307	5272	297
770	9672	342	8649	333	7697	323	6812	314	5993	305	5238	295
780	9610	339	8594	330	7647	321	6768	312	5954	303	5204	293
790	9549	337	8539	328	7599	319	6725	310	5917	301	5171	291
800	9489	335	8486	326	7551	317	6683	308	5880	299	5139	290
810	9430	333	8433	324	7504	315	6642	306	5843	297	5107	288
820	9372	331	8382	322	7458	313	6601	304	5807	295	5076	286
830	9316	329	8331	320	7413	312	6561	303	5772	294	5045	284
840	9260	327	8281	318	7369	310	6522	301	5738	292	5015	283
850	9205	325	8232	317	7326	308	6483	299	5704	290	4985	281
860	9152	323	8184	315	7283	306	6446	297	5671	288	4956	279
870	9099	321	8137	313	7241	304	6408	296	5638	287	4928	278
880	9047	320	8091	311	7200	303	6372	294	5606	285	4900	276
890	8996	318	8045	309	7159	301	6336	292	5574	283	4872	274
900	8946	316	8000	308	7119	299	6301	291	5543	282	4845	273
910	8897	314	7956	306	7080	298	6266	289	5513	280	4818	271
920	8848	312	7913	304	7041	296	6232	287	5483	279	4792	270
930	8801	311	7870	303	7003	294	6198	286	5453	277	4766	269
940	8754	309	7828	301	6966	293	6165	284	5424	276	4741	267
950	8707	307	7787	299	6929	291	6133	283	5395	274	4716	266
960	8662	306	7746	298	6893	290	6101	281	5367	273	4691	264
970	8617	304	7706	296	6857	288	6069	280	5339	271	4667	263
980	8573	303	7667	295	6822	287	6038	278	5312	270	4643	262
990	8530	301	7628	293	6788	285	6007	277	5285	269	4619	260
1000	8487	300	7590	292	6754	284	5977	276	5259	267	4596	259
1050	8282	292	7407	285	6591	277	5833	269	5132	261	4485	253
1100	8092	286	7237	278	6439	271	5699	263	5014	255	4382	247
1150	7914	279	7077	272	6298	265	5574	257	4904	249	4286	241
1200	7747	274	6928	266	6165	259	5457	252	4801	244	4196	236
1250	7591	268	6788	261	6041	254	5346	247	4704	239	4111	232
1300	7443	263	6657	256	5923	249	5242	242	4612	234	4031	227
1350	7304	258	6532	251	5813	244	5144	237	4526	230	3956	223
1400	7173	253	6414	247	5708	240	5052	233	4444	226	3884	219

Length / Fall	57 inch diameter Velocity (ft./min.)	Discharge (cu. ft./min.)	60 inch diameter Velocity (ft./min.)	Discharge (cu. ft./min.)	63 inch diameter Velocity (ft./min.)	Discharge (cu. ft./min.)	66 inch diameter Velocity (ft./min.)	Discharge (cu. ft./min.)	69 inch diameter Velocity (ft./min.)	Discharge (cu. ft./min.)	72 inch diameter Velocity (ft./min.)	Discharge (cu. ft./min.)
1450	215	3817	222	4367	229	4964	236	5609	242	6303	249	7048
1500	211	3753	218	4294	225	4880	232	5514	238	6197	245	6929
1550	208	3692	215	4224	221	4801	228	5425	234	6096	241	6817
1600	205	3634	211	4157	218	4725	224	5339	231	6000	237	6709
1650	201	3578	208	4094	214	4653	221	5258	227	5908	233	6607
1700	198	3525	205	4033	211	4584	218	5180	224	5821	230	6509
1750	196	3474	202	3975	208	4518	214	5105	220	5737	226	6415
1800	193	3426	199	3920	205	4455	211	5034	217	5657	223	6326
1850	190	3379	196	3866	203	4395	209	4965	214	5580	220	6240
1900	188	3334	194	3815	200	4336	206	4900	212	5506	217	6157
1950	185	3291	191	3766	197	4280	203	4836	209	5435	214	6077
2000	183	3250	189	3718	195	4226	201	4775	206	5367	212	6001
2050	181	3210	187	3673	192	4175	198	4717	204	5301	209	5927
2100	179	3172	184	3629	190	4125	196	4660	201	5237	207	5856
2150	176	3134	182	3586	188	4076	193	4606	199	5176	204	5788
2200	174	3099	180	3545	186	4030	191	4553	197	5117	202	5722
2250	172	3064	178	3506	184	3985	189	4502	194	5060	200	5658
2300	171	3030	176	3467	182	3941	187	4453	192	5004	197	5596
2350	169	2998	174	3430	180	3899	185	4406	190	4951	195	5536
2400	167	2967	172	3394	178	3858	183	4359	188	4899	193	5478
2450	165	2936	171	3360	176	3819	181	4315	186	4849	191	5422
2500	164	2907	169	3326	174	3780	179	4271	184	4800	189	5367
2550	162	2878	167	3293	172	3743	178	4229	183	4753	187	5315
2600	160	2850	166	3261	171	3707	176	4188	181	4707	186	5263
2650	159	2823	164	3230	169	3672	174	4149	179	4662	184	5213
2700	157	2797	163	3200	168	3638	173	4110	177	4619	182	5165
2750	156	2771	161	3171	166	3604	171	4072	176	4577	181	5118

2800	155	2747	160	3142	165	3572	169	4036	174	4536	179	5072
2850	153	2722	158	3115	163	3540	168	4000	173	4496	177	5027
2900	152	2699	157	3088	162	3510	166	3966	171	4457	176	4983
2950	151	2676	155	3062	160	3480	165	3932	170	4419	174	4941
3000	149	2653	154	3036	159	3451	164	3899	168	4382	173	4900
3050	148	2632	153	3011	158	3422	162	3867	167	4346	171	4859
3100	147	2610	152	2987	156	3395	161	3836	166	4310	170	4820
3150	146	2589	150	2963	155	3368	160	3805	164	4276	169	4782
3200	145	2569	149	2940	154	3341	158	3775	163	4243	167	4744
3250	143	2549	148	2917	153	3315	157	3746	162	4210	166	4707
3300	142	2530	147	2895	152	3290	156	3718	160	4178	165	4672
3350	141	2511	146	2873	150	3266	155	3690	159	4146	164	4637
3400	140	2492	145	2852	149	3241	154	3663	158	4116	162	4602
3450	139	2474	144	2831	148	3218	153	3636	157	4086	161	4569
3500	138	2457	143	2811	147	3195	151	3610	156	4057	160	4536
3550	137	2439	142	2791	146	3172	150	3584	155	4028	159	4504
3600	136	2422	141	2771	145	3150	149	3559	154	4000	158	4473
3650	135	2406	140	2752	144	3128	148	3535	153	3972	157	4442
3700	134	2389	139	2734	143	3107	147	3511	151	3945	156	4412
3750	133	2373	138	2715	142	3086	146	3487	150	3919	155	4382
3800	133	2358	137	2697	141	3066	145	3464	149	3893	153	4353
3850	132	2342	136	2680	140	3046	144	3442	148	3868	152	4325
3900	131	2327	135	2663	139	3027	143	3420	148	3843	152	4297
3950	130	2312	134	2646	138	3007	143	3398	147	3819	151	4270
4000	129	2298	133	2629	138	2988	142	3377	146	3795	150	4243
4050	128	2284	133	2613	137	2970	141	3356	145	3771	149	4217
4100	128	2270	132	2597	136	2952	140	3335	144	3748	148	4191
4150	127	2256	131	2581	135	2934	139	3315	143	3725	147	4166
4200	126	2242	130	2566	134	2916	138	3295	142	3703	146	4141
4250	125	2229	129	2551	133	2899	137	3276	141	3681	145	4116
4300	125	2216	129	2536	133	2882	137	3257	140	3660	144	4092
4350	124	2203	128	2521	132	2866	136	3238	140	3639	143	4069
4400	123	2191	127	2507	131	2849	135	3219	139	3618	143	4046
4450	122	2179	126	2493	130	2833	134	3201	138	3598	142	4023
4500	122	2166	126	2479	130	2817	134	3183	137	3578	141	4000
4550	121	2154	125	2465	129	2802	133	3166	137	3558	140	3978

Length/Fall	78 inch diameter Velocity (ft./min.)	78 inch diameter Discharge (cu. ft./min.)	84 inch diameter Velocity (ft./min.)	84 inch diameter Discharge (cu. ft./min.)	90 inch diameter Velocity (ft./min.)	90 inch diameter Discharge (cu. ft./min.)	96 inch diameter Velocity (ft./min.)	96 inch diameter Discharge (cu. ft./min.)	102 inch diameter Velocity (ft./min.)	102 inch diameter Discharge (cu. ft./min.)	108 inch diameter Velocity (ft./min.)	108 inch diameter Discharge (cu. ft./min.)
165	776	25779	813	31316	849	37533	884	44462	918	52132	952	60571
170	765	25398	801	30852	837	36977	871	43803	905	51360	938	59674
175	754	25032	790	30408	825	36445	858	43173	892	50621	924	58815
180	743	24682	779	29982	813	35935	846	42569	879	49913	911	57993
185	733	24346	768	29574	802	35446	835	41990	867	49234	899	57204
190	724	24024	758	29183	791	34977	824	41434	856	48581	887	56446
195	714	23714	748	28806	781	34525	813	40899	845	47955	875	55718
200	705	23415	739	28444	771	34091	803	40385	834	47351	864	55017
205	697	23128	730	28095	762	33673	793	39889	824	46770	854	54342
210	688	22851	721	27758	753	33270	784	39411	814	46210	844	53691
215	680	22584	712	27434	744	32880	774	38951	804	45670	834	53063
220	672	22326	704	27120	735	32505	766	38505	795	45148	824	52456
225	665	22076	696	26817	727	32141	757	38075	786	44643	815	51870
230	658	21835	689	26524	719	31790	749	37659	778	44155	806	51303
235	651	21601	681	26240	711	31450	741	37256	769	43683	797	50755
240	644	21375	674	25965	704	31121	733	36866	761	43226	789	50223
245	637	21156	667	25699	697	30802	725	36488	753	42782	781	49708
250	631	20943	661	25441	690	30492	718	36121	746	42352	773	49208
255	624	20737	654	25190	683	30192	711	35765	739	41935	765	48724
260	618	20537	648	24947	676	29900	704	35420	731	41530	758	48253
265	613	20342	642	24710	670	29616	698	35084	724	41136	751	47795
270	607	20153	636	24480	664	29341	691	34758	718	40753	744	47351
275	601	19969	630	24257	658	29073	685	34440	711	40381	737	46918
280	596	19789	624	24039	652	28812	679	34131	705	40019	730	46498
285	591	19615	619	23827	646	28558	673	33831	699	39666	724	46088
290	586	19445	613	23621	640	28311	667	33538	693	39323	718	45689
295	581	19280	608	23420	635	28070	661	33252	687	38988	712	45300

300	576	19118	603	23224	630	27835	656	32974	681	38662	706	44921
305	571	18961	598	23033	624	27606	650	32703	675	38344	700	44551
310	566	18808	593	22846	619	27383	645	32438	670	38033	694	44190
315	562	18658	588	22664	614	27164	640	32179	664	37730	689	43838
320	557	18511	584	22487	610	26951	635	31927	659	37434	683	43494
325	553	18368	579	22313	605	26743	630	31680	654	37145	678	43159
330	549	18229	575	22143	600	26540	625	31439	649	36863	673	42830
335	545	18092	571	21977	596	26341	620	31204	644	36587	668	42509
340	541	17959	566	21815	591	26147	616	30974	640	36317	663	42196
345	537	17828	562	21657	587	25956	611	30748	635	36053	658	41889
350	533	17700	558	21501	583	25770	607	30528	630	35794	653	41589
355	529	17575	554	21349	579	25588	603	30312	626	35541	649	41295
360	525	17453	550	21201	575	25410	598	30101	622	35293	644	41007
365	522	17333	547	21055	571	25235	594	29894	617	35051	640	40725
370	518	17215	543	20912	567	25064	590	29691	613	34813	635	40449
375	515	17100	539	20772	563	24896	586	29493	609	34580	631	40178
380	511	16987	536	20635	559	24732	582	29298	605	34352	627	39913
385	508	16876	532	20501	556	24571	579	29107	601	34128	623	39653
390	505	16768	529	20369	552	24413	575	28920	597	33909	619	39398
395	502	16661	525	20240	549	24258	571	28736	593	33694	615	39148
400	499	16557	522	20113	545	24106	568	28556	590	33482	611	38903
405	495	16454	519	19988	542	23957	564	28379	586	33275	607	38662
410	492	16354	516	19866	538	23810	561	28206	582	33071	604	38425
415	489	16255	513	19746	535	23666	557	28035	579	32872	600	38193
420	486	16158	510	19628	532	23525	554	27868	575	32675	596	37965
425	484	16063	507	19512	529	23386	551	27704	572	32483	593	37741
430	481	15969	504	19398	526	23250	547	27542	569	32293	589	37521
435	478	15877	501	19286	523	23116	544	27383	565	32107	586	37305
440	475	15786	498	19177	520	22984	541	27227	562	31924	583	37092
445	473	15698	495	19069	517	22855	538	27074	559	31744	579	36883
450	470	15610	492	18962	514	22727	535	26923	556	31567	576	36678
455	467	15524	490	18858	511	22602	532	26775	553	31393	573	36476
460	465	15439	487	18755	508	22479	529	26629	550	31222	570	36277
465	462	15356	484	18654	506	22358	526	26485	547	31054	567	36081
470	460	15274	482	18554	503	22238	524	26344	544	30888	564	35889
475	457	15194	479	18457	500	22121	521	26205	541	30725	561	35699

Length / Fall	78 inch diameter Velocity (ft./min.)	78 inch diameter Discharge (cu. ft./min.)	84 inch diameter Velocity (ft./min.)	84 inch diameter Discharge (cu. ft./min.)	90 inch diameter Velocity (ft./min.)	90 inch diameter Discharge (cu. ft./min.)	96 inch diameter Velocity (ft./min.)	96 inch diameter Discharge (cu. ft./min.)	102 inch diameter Velocity (ft./min.)	102 inch diameter Discharge (cu. ft./min.)	108 inch diameter Velocity (ft./min.)	108 inch diameter Discharge (cu. ft./min.)
480	455	15114	477	18360	498	22006	518	26068	538	30565	558	35513
485	453	15036	474	18265	495	21892	515	25933	535	30407	555	35329
490	450	14959	472	18172	493	21780	513	25801	533	30251	552	35149
495	448	14884	469	18080	490	21670	510	25670	530	30098	549	34971
500	446	14809	467	17989	488	21561	508	25541	527	29947	546	34795
510	441	14663	462	17812	483	21348	503	25290	522	29652	541	34453
520	437	14521	458	17640	478	21142	498	25045	517	29366	536	34120
530	433	14384	454	17473	474	20942	493	24808	512	29087	531	33796
540	429	14250	449	17310	469	20747	488	24577	507	28817	526	33482
550	425	14120	445	17152	465	20558	484	24353	503	28554	521	33176
560	421	13993	441	16998	461	20373	480	24134	498	28298	516	32879
570	418	13870	437	16848	457	20194	475	23922	494	28048	512	32589
580	414	13750	434	16703	453	20019	471	23715	490	27805	507	32307
590	410	13633	430	16560	449	19848	467	23513	485	27569	503	32032
600	407	13519	426	16422	445	19682	463	23316	481	27338	499	31764
610	404	13407	423	16287	441	19520	460	23124	477	27113	495	31502
620	400	13299	419	16155	438	19362	456	22937	473	26894	491	31247
630	397	13193	416	16026	434	19208	452	22754	470	26679	487	30998
640	394	13089	413	15900	431	19057	449	22576	466	26470	483	30755
650	391	12988	410	15778	428	18910	445	22401	462	26266	479	30518
660	388	12889	406	15658	424	18766	442	22231	459	26066	476	30285
670	385	12793	403	15540	421	18626	438	22064	455	25871	472	30059
680	382	12699	400	15426	418	18488	435	21901	452	25680	469	29837
690	379	12606	397	15313	415	18354	432	21742	449	25493	465	29620
700	377	12516	395	15204	412	18222	429	21586	446	25310	462	29407
710	374	12427	392	15096	409	18093	426	21434	442	25131	459	29200
720	371	12341	389	14991	406	17967	423	21284	439	24956	455	28996

730	28797	452	24785	436	21138	420	17844	403	14888	386	12256	369
740	28602	449	24617	433	20995	417	17723	401	14787	384	12173	366
750	28410	446	24452	430	20854	414	17604	398	14688	381	12091	364
760	28223	443	24290	428	20717	412	17488	395	14591	379	12012	362
770	28039	440	24132	425	20582	409	17374	393	14496	376	11933	359
780	27859	437	23977	422	20449	406	17262	390	14403	374	11857	357
790	27682	435	23825	419	20320	404	17153	388	14311	371	11781	355
800	27508	432	23675	417	20192	401	17045	385	14222	369	11707	352
810	27338	429	23529	414	20067	399	16940	383	14134	367	11635	350
820	27171	427	23385	412	19944	396	16836	381	14047	365	11564	348
830	27006	424	23244	409	19824	394	16734	378	13962	362	11494	346
840	26845	422	23105	407	19705	392	16635	376	13879	360	11425	344
850	26687	419	22968	404	19589	389	16536	374	13797	358	11358	342
860	26531	417	22835	402	19475	387	16440	372	13717	356	11292	340
870	26378	414	22703	400	19363	385	16345	370	13637	354	11227	338
880	26228	412	22574	397	19252	383	16252	367	13560	352	11163	336
890	26080	409	22446	395	19144	380	16160	365	13483	350	11100	334
900	25935	407	22321	393	19037	378	16070	363	13408	348	11038	332
910	25792	405	22198	391	18932	376	15982	361	13334	346	10977	330
920	25651	403	22077	389	18829	374	15895	359	13262	344	10917	329
930	25513	401	21958	386	18728	372	15809	357	13190	342	10858	327
940	25377	398	21841	384	18628	370	15725	355	13120	340	10800	325
950	25243	396	21726	382	18530	368	15642	354	13051	339	10743	323
960	25111	394	21613	380	18433	366	15560	352	12982	337	10687	322
970	24982	392	21501	378	18338	364	15480	350	12915	335	10632	320
980	24854	390	21391	376	18244	362	15401	348	12849	333	10578	318
990	24728	388	21283	375	18151	361	15323	346	12784	332	10524	317
1000	24604	386	21176	373	18060	359	15246	345	12720	330	10471	315
1050	24011	377	20666	364	17625	350	14878	336	12414	322	10219	307
1100	23459	368	20190	355	17220	342	14536	329	12128	315	9984	300
1150	22943	360	19747	348	16841	335	14217	321	11862	308	9765	294
1200	22460	353	19331	340	16487	328	13917	315	11612	301	9559	288
1250	22006	345	18940	333	16154	321	13636	308	11377	295	9366	282
1300	21579	339	18572	327	15840	315	13371	302	11156	289	9184	276
1350	21176	332	18225	321	15544	309	13121	297	10948	284	9012	271
1400	20794	326	17897	315	15264	303	12885	291	10750	279	8850	266

Length/Fall	78 inch diameter		84 inch diameter		90 inch diameter		96 inch diameter		102 inch diameter		108 inch diameter	
	Velocity (ft./min.)	Discharge (cu. ft./min.)	Velocity (ft./min.)	Discharge (cu. ft./min.)	Velocity (ft./min.)	Discharge (cu. ft./min.)	Velocity (ft./min.)	Discharge (cu. ft./min.)	Velocity (ft./min.)	Discharge (cu. ft./min.)	Velocity (ft./min.)	Discharge (cu. ft./min.)
1450	262	8696	274	10563	286	12661	298	14998	309	17586	321	20432
1500	257	8550	269	10386	281	12448	293	14746	304	17290	315	20089
1550	253	8411	265	10217	277	12246	288	14506	299	17009	310	19762
1600	249	8278	261	10056	272	12053	284	14278	295	16741	305	19451
1650	245	8152	257	9903	268	11869	279	14060	290	16485	301	19154
1700	242	8031	253	9756	264	11693	275	13852	286	16241	296	18870
1750	238	7915	249	9615	260	11525	271	13652	282	16007	292	18599
1800	235	7805	246	9481	257	11363	267	13461	278	15783	288	18339
1850	232	7699	243	9352	253	11209	264	13278	274	15569	284	18089
1900	228	7597	239	9228	250	11060	260	13102	270	15362	280	17849
1950	226	7499	236	9109	247	10918	257	12933	267	15164	276	17611
2000	223	7404	233	8994	244	10780	254	12770	263	14973	273	17397
2050	220	7313	230	8884	241	10648	250	12614	260	14790	270	17184
2100	217	7226	228	8778	238	10520	247	12463	257	14613	266	16978
2150	215	7141	225	8675	235	10397	245	12317	254	14442	263	16780
2200	212	7060	222	8576	232	10279	242	12176	251	14277	260	16588
2250	210	6981	220	8480	230	10164	239	12040	248	14117	257	16402
2300	208	6904	217	8387	227	10053	236	11908	246	13963	255	16223
2350	205	6831	215	8298	225	9945	234	11781	243	13813	252	16050
2400	203	6759	213	8211	222	9841	231	11658	240	13669	249	15882
2450	201	6690	211	8126	220	9740	229	11538	238	13529	247	15719
2500	199	6622	209	8045	218	9642	227	11422	236	13393	244	15561
2550	197	6557	207	7965	216	9547	225	11310	233	13261	242	15407
2600	195	6494	205	7889	214	9455	222	11200	231	13133	239	15259
2650	193	6432	203	7814	212	9365	220	11094	229	13008	237	15114
2700	192	6372	201	7741	210	9278	218	10991	227	12887	235	14973
2750	190	6314	199	7670	208	9193	216	10891	225	12769	233	14837

2800	188	6258	197	7602	206	9111	214	10793	223	12655	231	14703
2850	186	6203	195	7535	204	9031	212	10698	221	12543	229	14574
2900	185	6149	194	7469	202	8952	211	10605	219	12435	227	14448
2950	183	6096	192	7406	200	8876	209	10515	217	12329	225	14325
3000	182	6045	190	7344	199	8802	207	10427	215	12226	223	14205
3050	180	5996	189	7283	197	8729	205	10341	213	12125	221	14088
3100	179	5947	187	7224	196	8659	204	10257	211	12027	219	13974
3150	177	5900	186	7167	194	8590	202	10176	210	11931	217	13863
3200	176	5853	184	7111	192	8522	200	10096	208	11837	216	13754
3250	175	5808	183	7056	191	8457	199	10018	207	11746	214	13648
3300	173	5764	181	7002	189	8392	197	9942	205	11657	212	13544
3350	172	5721	180	6950	188	8329	196	9867	203	11569	211	13442
3400	171	5679	179	6898	187	8268	194	9794	202	11484	209	13343
3450	169	5637	177	6848	185	8208	193	9723	200	11400	208	13246
3500	168	5597	176	6799	184	8149	192	9653	199	11319	206	13151
3550	167	5557	175	6751	183	8091	190	9585	198	11239	205	13058
3600	166	5519	174	6704	181	8035	189	9518	196	11160	203	12967
3650	165	5481	173	6658	180	7980	188	9453	195	11084	202	12878
3700	164	5444	171	6613	179	7926	186	9389	194	11009	201	12791
3750	162	5407	170	6568	178	7873	185	9326	192	10935	199	12705
3800	161	5371	169	6525	177	7821	184	9265	191	10863	198	12621
3850	160	5336	168	6483	175	7770	183	9204	190	10792	197	12539
3900	159	5302	167	6441	174	7720	181	9145	188	10723	195	12458
3950	158	5268	166	6400	173	7671	180	9087	187	10654	194	12379
4000	157	5235	165	6360	172	7623	179	9030	186	10588	193	12302
4050	156	5203	164	6320	171	7575	178	8974	185	10522	192	12226
4100	155	5171	163	6282	170	7529	177	8919	184	10458	191	12151
4150	154	5140	162	6244	169	7484	176	8865	183	10395	189	12077
4200	153	5109	161	6207	168	7439	175	8812	182	10333	188	12005
4250	153	5079	160	6170	167	7395	174	8760	181	10272	187	11934
4300	152	5049	159	6134	166	7352	173	8709	179	10212	186	11865
4350	151	5020	158	6099	165	7310	172	8659	178	10153	185	11796
4400	150	4992	157	6064	164	7268	171	8610	177	10095	184	11729
4450	149	4964	156	6030	163	7227	170	8561	176	10038	183	11663
4500	148	4936	155	5996	162	7187	169	8513	175	9982	182	11598
4550	147	4909	154	5963	161	7147	168	8467	174	9927	181	11534

Length/Fall	78 inch diameter		84 inch diameter		90 inch diameter		96 inch diameter		102 inch diameter		108 inch diameter	
	Velocity (ft./min.)	Discharge (cu. ft./min.)	Velocity (ft./min.)	Discharge (cu. ft./min.)	Velocity (ft./min.)	Discharge (cu. ft./min.)	Velocity (ft./min.)	Discharge (cu. ft./min.)	Velocity (ft./min.)	Discharge (cu. ft./min.)	Velocity (ft./min.)	Discharge (cu. ft./min.)
4600	147	4882	154	5931	160	7108	167	8420	174	9873	180	11471
4650	146	4856	153	5899	160	7070	166	8375	173	9820	179	11410
4700	145	4830	152	5867	159	7032	165	8330	172	9767	178	11349
4750	144	4804	151	5836	158	6995	164	8286	171	9716	177	11289
4800	144	4779	150	5806	157	6958	164	8243	170	9665	176	11230
4850	143	4755	150	5776	156	6922	163	8201	169	9615	175	11172
4900	142	4730	149	5746	155	6887	162	8159	168	9566	174	11115
4950	141	4706	148	5717	155	6852	161	8117	167	9518	173	11058
5000	141	4683	147	5688	154	6818	160	8077	166	9470	172	11003
5100	139	4637	146	5632	152	6751	159	7997	165	9377	171	10895
5200	138	4592	144	5578	151	6685	157	7920	163	9286	169	10789
5300	137	4548	143	5525	149	6622	156	7845	162	9198	168	10687
5400	135	4506	142	5474	148	6560	154	7772	160	9112	166	10588
5500	134	4465	140	5424	147	6501	153	7701	159	9029	164	10491
5600	133	4425	139	5375	145	6442	151	7632	157	8948	163	10397
5700	132	4386	138	5328	144	6385	150	7564	156	8869	162	10305
5800	131	4348	137	5281	143	6330	149	7499	154	8793	160	10216
5900	129	4311	136	5237	142	6276	147	7435	153	8718	159	10129
6000	128	4275	134	5193	140	6224	146	7373	152	8645	157	10044
6100	127	4239	133	5150	139	6173	145	7312	151	8574	156	9962
6200	126	4205	132	5108	138	6123	144	7253	149	8504	155	9881
6300	125	4172	131	5068	137	6074	143	7195	148	8436	154	9802
6400	124	4139	130	5028	136	6026	142	7139	147	8370	152	9725
6500	123	4107	129	4989	135	5980	140	7084	146	8306	151	9650
6600	122	4076	128	4951	134	5934	139	7030	145	8242	150	9577
6700	121	4045	127	4914	133	5890	138	6977	144	8181	149	9505
6800	121	4015	126	4878	132	5846	137	6926	143	8120	148	9435

Supplement to Appendix II

EXTRA VALUES CALCULATED FOR AMERICAN SEWER SIZES, 8-INCH AND 10-INCH

Length Fall	8 inch diameter		10 inch diameter		Length Fall	8 inch diameter		10 inch diameter	
	Velocity (ft./min.)	Discharge (cu. ft./min.)	Velocity (ft./min.)	Discharge (cu. ft./min.)		Velocity (ft./min.)	Discharge (cu. ft./min.)	Velocity (ft./min.)	Discharge (cu. ft./min.)
16	601	210	691	377	88	256	89.5	295	161
17	583	204	670	366	89	255	89.0	293	160
18	567	198	652	355	90	253	88.5	291	159
19	552	193	634	346	91	252	88.0	290	158
20	538	188	618	337	92	251	87.5	288	157
21	525	183	603	329	93	249	87.0	287	156
22	513	179	589	321	94	248	86.6	285	156
23	501	175	576	314	95	247	86.1	284	155
24	491	171	564	308	96	245	85.7	282	154
25	481	168	553	302	97	244	85.2	281	153
26	472	165	542	296	98	243	84.8	279	152
27	463	162	532	290	99	242	84.4	278	152
28	454	159	522	285	100	240	83.9	276	151
29	446	156	513	280	105	235	81.9	270	147
30	439	153	505	275	110	229	80.0	264	144
31	432	151	496	271	115	224	78.3	258	141
32	425	148	489	267	120	219	76.6	252	138
33	419	146	481	262	125	215	75.1	247	135
34	412	144	474	259	130	211	73.6	242	132
35	406	142	467	255	135	207	72.2	238	130
36	401	140	461	251	140	203	70.9	234	127
37	395	138	454	248	145	200	69.7	230	125
38	390	136	448	245	150	196	68.5	226	123
39	385	134	443	241	155	193	67.4	222	121
40	380	133	437	238	160	190	66.4	219	119
41	376	131	432	235	165	187	65.3	215	117
42	371	130	427	233	170	184	64.4	212	116
43	367	128	422	230	175	182	63.5	209	114
44	362	127	417	227	180	179	62.6	206	112
45	358	125	412	225	185	177	61.7	203	111
46	355	124	408	222	190	174	60.9	201	109
47	351	122	403	220	195	172	60.1	198	108
48	347	121	399	218	200	170	59.4	195	107

49	343	120	395	215	205	168	58.6	193	105
50	340	119	391	213	210	166	57.9	191	104
51	337	118	387	211	215	164	57.2	189	103
52	333	116	383	209	220	162	56.6	186	102
53	330	115	380	207	225	160	56.0	184	101
54	327	114	376	205	230	159	55.3	182	99.4
55	324	113	373	203	235	157	54.8	180	98.4
56	321	112	369	201	240	155	54.2	178	97.3
57	318	111	366	200	245	154	53.6	177	96.3
58	316	110	363	198	250	152	53.1	175	95.4
59	313	109	360	196	255	151	52.6	173	94.4
60	310	108	357	195	260	149	52.1	171	93.5
61	308	107	354	193	265	148	51.6	170	92.6
62	305	107	351	191	270	146	51.1	168	91.8
63	303	106	348	190	275	145	50.6	167	90.9
64	301	105	346	188	280	144	50.2	165	90.1
65	298	104	343	187	285	142	49.7	164	89.3
66	296	103	340	186	290	141	49.3	162	88.5
67	294	103	338	184	295	140	48.9	161	87.8
68	292	102	335	183	300	139	48.5	160	87.0
69	289	101	333	181	305	138	48.1	158	86.3
70	287	100	330	180	310	137	47.7	157	85.6
71	285	99.6	328	179	315	135	47.3	156	85.0
72	283	98.9	326	178	320	134	46.9	155	84.3
73	281	98.2	324	176	325	133	46.6	153	83.6
74	280	97.6	321	175	330	132	46.2	152	83.0
75	278	96.9	319	174	335	131	45.9	151	82.4
76	276	96.3	317	173	340	130	45.5	150	81.8
77	274	95.7	315	172	345	129	45.2	149	81.2
78	272	95.0	313	171	350	129	44.9	148	80.6
79	271	94.4	311	170	355	128	44.6	147	80.0
80	269	93.8	309	169	360	127	44.2	146	79.5
81	267	93.3	307	168	365	126	43.9	145	78.9
82	266	92.7	305	166	370	125	43.6	144	78.4
83	264	92.1	303	165	375	124	43.3	143	77.9
84	262	91.6	302	164	380	123	43.1	142	77.3
85	261	91.0	300	164	385	123	42.8	141	76.8
86	259	90.5	298	163	390	122	42.5	140	76.3
87	258	90.0	296	162	395	121	42.2	139	75.9

Glossary

The following are definitions used in drainage and sewerage practice. Some of these are the same on both sides of the Atlantic but where there are differences these are indicated, and what would appear to be the most appropriate alternative has been selected for use throughout the foregoing text.

Backwater gate. The American term for tidal flap.

Backwater valve. The American term for a valve used to prevent flooding of a drainage system by back-flow from the sewers. The British term is anti-flooding ball-valve interceptor.

Building drain. A drain which receives soil waste and other drainage from inside buildings and/or surface water within the curtilage of a building.

Building sewer. A term used in America for the pipe which connects the building drains to the public sewer. This is known as a house connection in Great Britain.

Building storm drain. A building drain reserved for surface water.

Building storm sewer. A building sewer reserved for surface water.

Building trap. An American term for a running trap installed between the indoor sanitation and the drain. Such running traps are not permitted in Great Britain.

Combined drain. A drain which receives both surface water and soil or sanitary drainage.

Combined sewer. A sewer which receives both surface water and soil or sanitary drainage.

Crude sewage. Sewage which has not received any treatment.

Culvert. In British engineering usage this word always means a covered underground conduit constructed of brickwork, concrete, etc., as distinguished either from a pipeline or an open channel. This is contrary to some dictionary definitions. (See also the *Oxford English Dictionary.*)

Depressed sewer. An American term for an inverted siphon.

Domestic sewage. The drainage of private houses or the drainage from the lavatories, kitchens, etc., of industrial premises but to which no trade waste is discharged.

Drain. A drain which carries sanitary sewage, sullage, or surface water in the curtilage of a building. In England and Wales a drain which serves more than one curtilage is a sewer. (See Sewer.)

Fixture. An American term for an appliance which receives soil or waste drainage inside a building. In Great Britain the terms fitment and, more recently, appliance, have been used.

Fixture unit. An American unit used in the design of indoor sanitation. It has the value of about 1 ft^3/min. (28.32 1/min)

Grease trap. A receptacle to collect grease from kitchen waste to prevent it from choking the drains or sewer.

Gully. A British term for an inlet trapped or not trapped, according to circumstance, for receiving surface water from paxed areas; for receiving sullage from waste and vent pipes in the British two-pipe system of building sanitation; or for receiving drainage from floors in buildings. (See Road Gully and Street Inlet.)

Gutter. A British term for the channel that collects roof water and discharges to the leaders or rain water pipes. The American term is roof drain.

House drain. (See Building drain.)

House sewer. (See Building sewer.)

House trap. (See Building trap.)

Inspection chamber. A British term for a shallow manhole on a drain that permits access to the drain but is generally too small for a man to enter.

Intercepting sewer. A sewer which is constructed to collect flow so as to relieve overloaded sewers and to convey the sewage to a place where it can be received.

Interceptor. In America this term applies to a receptacle to intercept sand, grease, oil, gasoline, etc. In Great Britain the term interceptor means a trap on a drain particularly at the point at which the drainage system delivers to the sewer for the purpose of preventing flow of foul air from the sewer to the drainage system. Interceptors for this purpose are now in disfavour except in special circumstances. Interceptors in this sense are provided between surface-water drains and sanitary drains to prevent flow of foul air into the surface-water drains.

Inverted siphon. The British term for a sewer which is carried below the hydraulic gradient so as to pass under a river or valley. An American term is depressed sewer.

Leader. An American term for a vertical pipe to deliver roof water from the roof guttering to the drains. The British term is rain-water pipe.

Main sewer. The American term for public sewer.

Plumbing fixture. (See Fixture.)

Public sewer. British term for a sewer vested in the local authority.

Reflux valve. A valve which prevents back-flow: also a non-return valve.

Rising main. The delivery from a pumping station.

Road Gully. British name for a gully with a large receptacle of pre-cast concrete or vitrified clay-ware, having a water-sealed outlet and covered by a cast-iron grating to collect surface water from roads. These gullies are designed for easy cleansing, and special vehicles are available for sucking out settled solids and replenishing the water seals.

Rodding eye. A British term for an inlet to a drain that permits cleansing with drain rods where there is no manhole or inspection chamber.

Roof drain. American term for gutter.

Sanitary sewer. American term for a sewer which takes foul water but excludes surface water. The British term is soil sewer.

Sewage. The contents of sewers, or liquid discharged from sewers. Sewage may be foul water or surface water but being in, or having been in, a sewer it is (in Britain) legally sewage no matter how clean it may be.

Sewer. In British legal definition and accepted practice a sewer is a drain which receives flows from more than one curtilage. There may be private sewers that are the responsibility of individuals or groups of individuals and public sewers which are vested in the local authority and maintained at public expense.

Shoe. A clay-ware pipe with access door or grating, and with or without upward bend, to give access for cleansing to a rain-water drain, also a cast-iron drain pipe with upward bend and bolted access.

Site. The site of a building is the land that is covered by the building itself but does not include the curtilage.

Slop sink. An appliance for receiving sanitary sewage from bedpans, chamber-pots, etc.

Soil pipe. A pipe which conveys the discharge of upstairs water closets, urinals, or slop sinks.

Stack pipe. Any vertical soil, waste, or ventilating pipe secured to a building. In England and Wales all new stack pipes must be inside the building and not on the outside wall.

Storm sewage. Storm water that is or has been in a sewer.

Storm sewer. (See Surface-water Sewer.)

Street inlet. American name for a pit constructed of brickwork or concrete, and provided with a grating at the top, to collect surface water from the channels (gutters) of roads. (See Road gully.)

Subsoil drain. A drain laid below the subsoil to prevent waterlogging of the ground.

Sullage. A British term for the drainage of baths, bidets, lavatory basins, and sinks but excluding the drainage of water closets, urinals or slop sinks.

Surface-water sewer. A sewer reserved for surface water and from which sanitary sewage is excluded.

Tidal flap. A flap preventing back-flow and constructed for bolting to the face of a wall. (See Backwater gate.)

Trap. A construction to provide a waterseal that will prevent flow of foul air to or from a drain.

Vent pipe. A stack pipe which ventilates the drainage system and/or indoor plumbing. It may be combined with a soil pipe or a waste pipe.

Waste pipe. A stack pipe which receives sullage from baths, lavatory basins, bidets, and sinks but which excludes the drainage of water closets, urinals, or slop sinks.

Water closet. A sanitary closet which has a separate fixed receptacle connecting to a drainage system and separate provision for flushing from a supply of clean water.

Water table. The natural surface level of subsoil water.

Index

raw slodge, 425
Sludging
  continuous, 308
  intermittent, 307
  mechanical, 304
  under hydrostatic head, 307
SMART, P. L., 214
SMISSON, R. P., 96
SMITH, B., 123, 145
SMITH, E., 341
SNOOK, W. G. G., 211, 215, 223, 225
Soakaways, 93, 193, 196–197
Soil mechanics, references to, 263
Solar heat, for waste pasteurization, 458
Southern Water Authority, 383
SPANGLER, PROF. M. G., 161
Specification
  British Standard, 19
  definition of, 18, 20
  for pumping machinery, 140
  standard, 19
  writing, 19
STEVEN, W., 248
STEVENSON, D. G., 383
Standing wave flumes
  float chambers for, 253
  proportions of, 252–253
Standards of purification
  Australian, 231
  Canadian, 231
  of sewage effluents, 232
  Japanese, 231
  Royal Commission, 232
  secondary treatment, 232
  South African, 231
  Third World, 231
Storm(s)
  classification of, 74
  frequency of, 76, 92–93
  intensity of, 74, 77, 83
  intensity curves, 76
  overflows, 22, 92–93
  percentile, 94
Storm overflows, 92–93, 286
  in the Netherlands, 93
  Symposium on, 96
  Technical Committee on, 92, 94
Storm water
  American practice, 286
  discharge to grass plots, 293
  initial flush, 288
  Joint Committee Report on, 286
  separation, 250, 284–286
  storage formulae, 91, 92, 134

tanks, 285–286, 288–290
Streams, particulars of, 16
Structural protection of pipes, 161
Structures
  caisson-type, 134
  flotation of, 135
Subsoil surveys, 255
Suction wells, form and size, 130
Sullage, 434
  disposal of, 434
Sulphates, 149
Supernatant water, 393
  draw-off valves for, 393, 399
  heating, 396
  'Heat a mix' method, 399
  heat-exchangers, 396
  scum removal, 399
Surface water
  drains, 8
  disposal by soakaway, 93
  for coastal towns, 228
  sewerage, 70
Surge in rising mains, 123, 175
Suspended matter
  produced by aeration processes, 386
  in trickling filters, 387
Suspended solids, 223
SVENSSON, A., 200
SWANWICK, J. D., 426
Sweden, Chalmers University of
    Technology, 215

Talbot formula, 73
Tangent method, 88, 90
Tank sewers, 207, 225–226
Tanks
  hopper bottomed, 308
  semi-Dortmund, 308
Temperature checks of sludge, 250
Tenders
  advertisement for, 18, 20
  invitation of, 20
Testing
  air, 163
  lines of pipe, 163
  water, 164
Tests for acidity and alkalinity, 240
  albuminoid ammonia, 239
  chlorine, 239
  dissolved solids, 239
  free ammonia, 240
  nitrates and nitrites, 240
  oxygen absorbed, 240
  settleable solids, 239